T0291609

ESSENTIAL PRINCIPLES OF ORGANIC CHEMISTRY

ESSENTIAL PRINCIPLES OF ORGANIC CHEMISTRY

by

CHARLES S. GIBSON

O.B.E., M.A. (Oxon. & Cantab.), Sc.D. (Cantab.), F.R.S.

Professor of Chemistry in the University of London
at Guy's Hospital Medical School

CAMBRIDGE

AT THE UNIVERSITY PRESS

1936

CAMBRIDGE
UNIVERSITY PRESS

University Printing House, Cambridge CB2 8BS, United Kingdom

Cambridge University Press is part of the University of Cambridge.

It furthers the University's mission by disseminating knowledge in the pursuit of education, learning and research at the highest international levels of excellence.

www.cambridge.org
Information on this title: www.cambridge.org/9781316603864

First published 1936
First paperback edition 2016

A catalogue record for this publication is available from the British Library

ISBN 978-1-316-60386-4 Paperback

PREFACE

It is hoped that this book will facilitate the study of Organic Chemistry for those who will eventually specialize in the subject and those who will later engage in the study of those sciences for which a knowledge of the principles of organic chemistry is essential. The importance of a sound knowledge of organic chemistry for the study of biochemistry, physiology, pharmacology and therapeutics needs no emphasis, and the study of organic chemistry with reference to the future work of the students of these subjects is a feature of the method of presentation by the author.

It is assumed that the reader has some familiarity with the elementary principles of general chemistry and physics and, to avoid elaboration of detail, no attempt has been made to attain the comprehensiveness of a 'text-book'. Beilstein's *Handbuch der organischen Chemie* (of which the present incomplete edition has already reached its thirty-eighth volume)—the standard text-book on the subject—has necessarily been frequently consulted and the *Dictionary of Organic Compounds*, edited by Professor I. M. Heilbron, D.S.O., F.R.S., has proved most valuable as a book of reference. For the mode of presentation of certain parts of the subject, the author has made use of some methods adopted by other teachers, particularly Professor G. Barger, F.R.S., author of *Organic Chemistry for Medical Students.*

The study of Organic Chemistry can only be made efficient in so far as it is carried out along with a comprehensive course of work in the laboratory; and the practical side of the subject is specially emphasized. Types of apparatus commonly used in the preparation and investigation of organic compounds are described in the 'Appendix'.

The study of the subject beyond the scope of the present work may best be continued by means of the treatises now available on specialized portions of organic chemistry as well as by reference to original papers in the recognized journals of which the *Journal of the Chemical Society* may be regarded as typical. Some of the treatises are mentioned in the text and others are given in the References.

The author is very greatly indebted to Dr F. H. Brain, M.A., for his valuable assistance in making some of the drawings, in the reading of the proofs, in making the 'Index' and for his many useful suggestions. He also expresses his thanks to Mr S. Holt, M.Sc., for his assistance in making the 'Index' and to Mr James Nield for his personal help without which this book would not have appeared.

C. S. G

The Chemistry Department,
Guy's Hospital Medical School,
University of London, S.E. 1.

September 1936

CONTENTS

INTRODUCTORY

ORGANIC CHEMISTRY is the Chemistry of Carbon Compounds. It includes such compounds as carbon dioxide—the ultimate product of oxidation of the carbon in all organic compounds—carbon monoxide, carbon disulphide, cyanogen, hydrocyanic acid and its salts which may, for convenience, be studied along with typical inorganic substances. The great variety of organic compounds range from substances having relatively simple molecular constitutions, as those above mentioned, to the highly complex substances such as the polysaccharides (e.g. starch and cellulose) and rubber and the very great number of plant and animal products.

The great number of organic compounds is due to the capacity of the quadrivalent carbon atom in uniting with other carbon atoms and which no other element possesses in anything like the same degree. In this way, relatively simple organic compounds are built on such 'carbon skeletons' as

$$-\overset{|}{\underset{|}{C}}-,\quad -\overset{|}{\underset{|}{C}}-\overset{|}{\underset{|}{C}}-,\quad -\overset{|}{\underset{|}{C}}-\overset{|}{\underset{|}{C}}-\overset{|}{\underset{|}{C}}-,\quad -\overset{|}{\underset{|}{C}}-\overset{|}{\underset{|}{C}}-\overset{|}{\underset{|}{C}}-\overset{|}{\underset{|}{C}}-\ldots,\quad -C-\overset{-C-}{\underset{-C-}{C}}-\ldots,$$

$$\overset{|}{\underset{|}{C}}=\overset{|}{\underset{|}{C}},\quad \overset{|}{\underset{|}{C}}=C=\overset{|}{\underset{|}{C}},\quad -\overset{|}{\underset{|}{C}}-C\equiv C-\overset{|}{\underset{|}{C}}-\ldots,$$

$$-C\equiv C-,$$

open chain compounds

ring or cyclic compounds

in all of which the quadrivalency of the carbon atom is preserved.

These 'skeletons' can be varied in an almost infinite number of ways, of which a few simple examples are

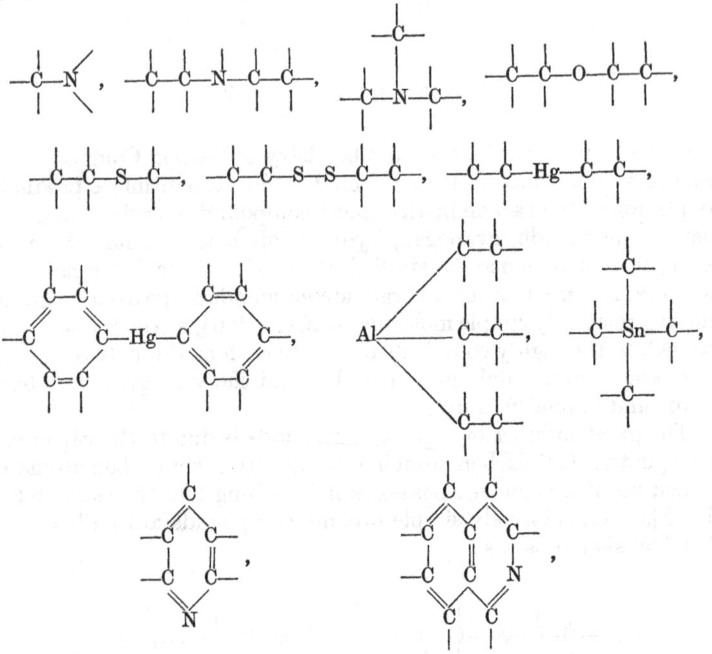

in which quadrivalent carbon is united to such elements as tervalent nitrogen, bivalent oxygen, bivalent sulphur and even to such metals as bivalent mercury, tervalent aluminium and quadrivalent tin.

Organic compounds obey the fundamental laws of chemistry, but in view of the capacity of the carbon atom to combine with itself, the Law of Multiple Proportions, in the case of most carbon compounds, cannot be stated so definitely as in the case of elements which form different series of compounds according to differences in the fundamental valency of the elements; thus, the above law in its simplest form is obeyed in the case of hydrocarbons (compounds containing carbon and hydrogen only) having the same number of carbon atoms in the molecule, such as those having the formulae:

C_2H_6	C_2H_4	C_2H_2
ethane	ethylene	acetylene

but, in its simple form, it does not apply in the case of such a series of hydrocarbons as

$$CH_4,\ C_2H_6,\ C_3H_8,\ C_4H_{10},\ C_5H_{12}\ \text{etc. (paraffins)}$$

the composition of which differ from member to member by $=CH_2$.

In so far as organic compounds include solids, liquids and gases, the general methods of their physical investigation are the same as in the investigation of chemical compounds in general. Solid organic compounds may be, for example, crystalloids and colloids, electrolytes and non-electrolytes (non-electrolytes are relatively far more common among organic than among inorganic compounds). Consequently, as far as their nature permits, such solid compounds may be submitted to the methods usually employed for investigating these types of substances. On the other hand, the methods for investigating what may be described as 'the architecture of the molecules' or, as it is generally termed, 'the chemical constitution' (i.e. how the various atoms in the molecule are related to each other) of organic compounds are peculiar to organic chemistry and necessitate the study of the chemical behaviour of the numerous types of individual substances.

Apart from metallic radicals in salts of organic acids, in organometallic compounds (compounds containing metallic radicals directly united to carbon), in haemin (the red blood pigment containing iron) and in chlorophyll (the green colouring material of plants containing magnesium), the different elements in typical organic compounds are comparatively few. All organic compounds must contain carbon, the great majority contain hydrogen and oxygen, a very large number contain nitrogen and halogens (particularly chlorine, bromine and iodine), fewer contain sulphur and much fewer phosphorus. In view of the rapid increase in our knowledge, almost any element may be found sooner or later in organic compounds and, for example, organic compounds containing fluorine and other non-metallic elements are rapidly increasing in number. The principles of determining qualitatively and quantitatively the more typical elements in organic compounds will illustrate the general methods for these purposes.

Qualitative detection of the more usual elements in organic compounds

By definition, all organic compounds contain *carbon*, and when heated, after being intimately mixed with dry cupric oxide, evolve carbon dioxide, which may be made manifest by its forming a precipitate of calcium carbonate when passed into an aqueous solution of calcium hydroxide. If the compound contains *hydrogen*, water will be produced at the same time; it is, of course, essential that initially the materials used for the test for combined hydrogen should be dry. Not a few organic compounds are inflammable and burn in the presence of air or oxygen; the products of this combustion always contain carbon dioxide and water, proving the initial presence of combined carbon and hydrogen in the compound.

There is no satisfactory test for *oxygen* in organic compounds. Compounds which on being heated in the absence of oxygen evolve oxides of carbon and water obviously contain carbon, hydrogen and oxygen. Theoretically, a compound having the general formula $C_xH_yO_z$ should, on complete oxidation, yield carbon dioxide and water according to the following:

$$C_xH_yO_z \rightarrow xCO_2 + \frac{y}{2}H_2O.$$

In the great majority of cases $\frac{1}{2}z$ molecules of oxygen are less than $x+\frac{1}{4}y$ molecules of oxygen and, therefore, for the complete combustion of such an organic compound oxygen has to be supplied from an external source. If the amount of oxygen to be supplied could be measured accurately, and this was found to be less than the amount of oxygen present in the carbon dioxide and water obtained in the quantitative oxidation, the presence of oxygen in the original compound would be proved.

The presence of *nitrogen* in an organic compound may be demonstrated by fusing the compound with sodium in a suitable hard glass tube, when sodium cyanide is produced. After getting rid of the excess of metallic sodium, the presence of sodium cyanide is demonstrated by extracting the product with warm water and filtering the solution. This strongly alkaline solution is treated with an aqueous solution of a ferrous salt, then with an aqueous solution of a ferric salt and, after addition of excess of hydrochloric acid to dissolve ferrous and ferric hydroxides, a blue precipitate of ferric ferrocyanide (insoluble Prussian Blue) is obtained if nitrogen is present in the original compound. Omitting the precipitation and solution of iron hydroxides, the reactions taking place are indicated:

Organic compound containing nitrogen + Na, and fused $\rightarrow$ NaCN

$$2NaCN + FeSO_4 \rightarrow Na_2SO_4 + Fe(CN)_2$$

ferrous cyanide, unstable—cannot be isolated

$$4NaCN + Fe(CN)_2 \rightarrow Na_4[Fe''(CN)_6]$$

sodium ferrocyanide

$$3Na_4[Fe''(CN)_6] + 2Fe_2(SO_4)_3 \rightarrow 6Na_2SO_4 + Fe_4'''[Fe''(CN)_6]_3$$

ferric ferrocyanide or Prussian Blue—insoluble in hydrochloric acid

A few organic compounds when heated with alkali hydroxide evolve ammonia which, in those cases, obviously indicates the presence of nitrogen in the original substance. All nitrogenous organic compounds when heated for some time with concentrated sulphuric acid undergo oxidation and the nitrogen is converted into ammonium sulphate from which ammonia can be obtained. Finally, from all nitrogenous organic compounds when heated after being mixed with a large excess of cupric oxide so as to oxidise them, the nitrogen may be evolved as such and, partly, as oxides of nitrogen. If the resulting gaseous products of the oxidation are passed over heated metallic copper so as to reduce the oxides of nitrogen and then through a concentrated aqueous solution of potassium hydroxide, the gas collected is nitrogen coming from the original compound. These last three methods are not so satisfactory as the 'sodium-fusion' method for the mere testing for the presence of nitrogen in an organic compound.

The *halogens* usually present in organic compounds are chlorine, bromine and iodine, and their presence may be detected in three ways. Whilst not conclusive, the first depends on the fugitive green colour imparted to the Bunsen flame by copper halides. A copper wire heated in the oxidising flame becomes oxidised and ceases to impart a colour to the flame; when such an oxidised copper wire is dipped into an organic compound containing halogen so that a little adheres and is then placed in the Bunsen flame, the latter shows the characteristic transient green colour. Another test depends on the fact that many halogen-containing organic compounds when heated with *pure* calcium oxide convert the latter into calcium halide. The solid product is extracted with water and the filtered solution tested for halogen ions in the usual manner—colourless (chlorine), pale yellow (bromine) or yellow (iodine) precipitates of silver halide, darkening on exposure to light and insoluble in nitric acid. The most satisfactory test for halogen in an organic compound is to fuse the latter with sodium as in the nitrogen test. When the vigorous reaction is over the product is treated carefully with water (excess of sodium may remain) and warmed. The warm strongly alkaline solution containing sodium halide is filtered, acidified with nitric acid and treated with an aqueous solution of silver nitrate, when the silver halide is precipitated. The precipitate identified as silver chloride, bromide or iodide indicates the presence of chlorine, bromine or iodine in the original compound. All organic compounds when completely oxidised by concentrated nitric acid in the presence of silver nitrate yield up their halogens quantitatively as silver halides.

The presence of *sulphur* in an organic compound may also be qualitatively determined by fusing the compound with sodium. The essential product in such cases is sodium sulphide in the presence of an excess of sodium hydroxide. The product is treated carefully with water (probable excess of sodium) and then the mixture is warmed. To the filtered alkaline solution is added a fresh aqueous solution of sodium nitroprusside or sodium nitrosoferricyanide,

$$Na_2[Fe'''(NO)(CN)_5].2H_2O,$$

when a characteristic intense purple colour is produced in the solution. The actual nature of the water-soluble intensely coloured compound is not definitely established; it may be a compound having the constitution

$$Na_3[Fe'''(O:N.S.Na)(CN)_5],$$

but its production is characteristic of alkaline sulphides. The sulphur in all organic compounds containing this element when completely oxidised with concentrated nitric acid is quantitatively converted into sulphuric acid, which may be identified by the production of barium sulphate, which is insoluble in water and mineral acids.

The presence of *phosphorus* in an organic compound may be detected by fusing the latter with a mixture of potassium nitrate and sodium carbonate when oxidation takes place, the phosphorus becoming converted into water-soluble phosphate. The filtered aqueous extract is tested for phosphate with an aqueous solution of ammonium molybdate or with magnesia-mixture (ammoniacal solution of magnesium chloride and ammonium chloride) in the usual way. The phosphorus in an organic compound may also be quantitatively converted into phosphoric acid by the complete oxidation of the compound with concentrated nitric acid.

The *metallic radicals* in organic compounds may be detected as metallic oxides or carbonates remaining after 'destroying' the compound by heating it in air. The 'ash' is, of course, submitted to the usual methods of analysis and the methods can be made quantitative.

As a general rule the *Quantitative Analysis of Organic Compounds* is carried out by determining the amount of products obtained by completely oxidising weighed quantities of the compounds. Micro-analytical methods are now extensively employed when possible and, although they involve special technique, they yield highly satisfactory results in the hands of trained operators. The principles involved are comparatively simple.

Determination of carbon and hydrogen. The weighed quantity of the substance is oxidised in a current of oxygen or air which is dry and free from carbon dioxide and also in the presence of a suitable oxidising agent such as cupric oxide or potassium dichromate. The products of oxidation or 'combustion', viz. water (formed from the hydrogen in the compound) and carbon dioxide (produced from the carbon in the compound), are collected in previously weighed tubes containing suitable absorbing material. The water is absorbed in calcium chloride or pumice saturated with concentrated sulphuric acid and then the carbon dioxide is absorbed either in solid potassium hydroxide or soda-lime. If nitrogen be present in the compound, the products of combustion, before being collected, are allowed to pass over heated copper which reduces any oxides of nitrogen and then the nitrogen passes through the absorption apparatus without change. If halogen be present in the original compound, lead chromate may be used as the additional oxidising agent, when the halogen is retained as lead halide; or, the products of combustion before passing to the absorption apparatus are allowed to pass over heated silver which retains the halogen as silver halide so that the halogen does not pass into the absorption apparatus. Generally, special precautions have to be taken in the quantitative oxidation or 'combustion' of substances which on oxidation may yield substances in addition to water and carbon dioxide and which, like these, may be retained in the absorption apparatus. The result of a typical analysis of a simple substance is the following:

7·200 mg. of an organic compound gave on combustion 4·320 mg. of water and 10·56 mg. of carbon dioxide.

Assuming the atomic weights of hydrogen, carbon and oxygen to be 1, 12 and 16 respectively,

7·200 mg. of the compound contains

$$\frac{2}{18} \text{ or } \frac{1}{9} \times 4{\cdot}320 \text{ mg. hydrogen and } \frac{12}{44} \text{ or } \frac{3}{11} \times 10{\cdot}56 \text{ mg. carbon,}$$

i.e. 100 parts of the compound contain

$$\frac{1 \times 4{\cdot}320 \times 100}{9 \times 7{\cdot}200} \text{ parts of hydrogen}$$

and

$$\frac{3 \times 10{\cdot}56 \times 100}{11 \times 7{\cdot}200} \text{ parts of carbon.}$$

That is, the compound contains 6·67 per cent. of hydrogen and 40·0 per cent. of carbon.

In the absence of further information, it must be assumed that the deficiency, viz. $100 - 6{\cdot}67 - 40{\cdot}0 = 53{\cdot}33$ per cent., is oxygen. The result of this analysis shows that the compound contains carbon $= 40{\cdot}0$ per cent., hydrogen $= 6{\cdot}67$ per cent. and oxygen $= 53{\cdot}33$ per cent.

The amount of *nitrogen* in a compound is always determined separately. For this purpose, the weighed quantity of the compound is intimately mixed with a large excess of a suitable oxidising agent (e.g. cupric oxide) and heated in a current of carbon dioxide; the resulting gaseous products are passed over heated copper and then through a concentrated aqueous solution of potassium hydroxide over which the nitrogen is collected in a suitable 'nitrometer'—an apparatus for measuring the volume of nitrogen. The potassium hydroxide absorbs all the carbon dioxide and, after making allowance for the vapour tension of the solution of potassium hydroxide, the weight of the nitrogen obtained and the percentage of nitrogen in the substance are calculated as illustrated in the following example:

6·720 mg. of a compound gave 1·91 ml. of nitrogen at 15° C. and 752 mm.

i.e. 6·720 mg. gave $\frac{1{\cdot}91 \times 273 \times 752}{288 \times 760} = 1{\cdot}791$ ml. of nitrogen at 0° C. and 760 mm. pressure.

Assuming that 1 litre of nitrogen at N.T.P. weighs 1·2507 grams, the compound contains

$$\frac{1{\cdot}2507 \times 1{\cdot}791 \times 100}{1000 \times 0{\cdot}00672} = 33{\cdot}3 \text{ per cent. of nitrogen.}$$

Another method of estimating nitrogen consists in oxidising a weighed quantity of the compound by digesting it with concentrated sulphuric acid, whereby, when the compound is completely oxidised, the whole of the nitrogen is retained as ammonium sulphate. The ammonia is liberated from this by making strongly alkaline and boiling and is absorbed in a known excess of a standard aqueous solution of sulphuric acid. This is Kjeldahl's method; it is rarely used for the analysis of pure compounds, but is the routine method for determining the nitrogen content of highly complex materials such as proteins, feeding stuffs, etc.

The amount of *halogen* may be determined by Carius' original method by sealing in a suitable glass tube a weighed quantity of the compound with an excess of concentrated nitric acid and silver nitrate and heating the whole in a closed or 'bomb' furnace at a temperature of about 250° when, after a time, the oxidation is complete. The halogen is converted into silver halide which, by careful manipulation, can be weighed directly. An example of the analysis of a compound containing bromine is the following:

0·2680 gram of the compound gave 0·2507 gram of silver bromide.

Assuming the atomic weights of bromine and silver to be 80 and 108 respectively,

the substance analysed contains

$$\frac{0{\cdot}2507 \times 80 \times 100}{188 \times 0{\cdot}2680} = 39{\cdot}8 \text{ per cent. of bromine.}$$

The amount of *sulphur* may also be determined by Carius' original method, in which the weighed quantity of the compound is heated in a suitably sealed tube with an excess of concentrated nitric acid. The sulphur is quantitatively converted into sulphuric acid, which is precipitated as barium sulphate and weighed as such. A typical example of the results of such a determination is

0·1440 gram of a compound gave 0·2801 gram of barium sulphate.

Assuming the atomic weights of oxygen, sulphur and barium to be 16, 32 and 137·4 respectively,

the substance contains

$$\frac{0{\cdot}2801 \times 32 \times 100}{233{\cdot}4 \times 0{\cdot}1440} = 26{\cdot}7 \text{ per cent. of sulphur.}$$

These examples indicate the principles involved in determining the percentage composition of organic compounds. From such results, the relative number of the different atoms in a molecule of a compound can be determined as indicated by the following simple example:

An organic compound contained 9·1 per cent. of hydrogen and 54·5 per cent. of carbon.

In the absence of further information it must be assumed that the deficiency, viz. $100-9{\cdot}1-54{\cdot}5=36{\cdot}4$ per cent., must be oxygen.

Therefore if the molecular weight of the compound were 100, there would be in such a molecule

$$\frac{9{\cdot}1}{1}=9{\cdot}1 \text{ atoms of hydrogen,}$$

$$\frac{54{\cdot}5}{12}=4{\cdot}54(2) \text{ atoms of carbon}$$

and

$$\frac{36{\cdot}4}{16}=2{\cdot}27(5) \text{ atoms of oxygen,}$$

the atomic weights of hydrogen, carbon and oxygen being assumed to be 1, 12 and 16 respectively. Such fractions of atoms cannot exist in the molecule of a compound, and in the molecule there must be at least one atom of the element present in the least atomic proportion, in this case, oxygen. These numbers also indicate that the molecular weight of the compound cannot be 100. By inspection, the above atomic proportions are almost completely divisible by 2·27, thus

$$\frac{9{\cdot}1}{2{\cdot}27}=4, \qquad \frac{4{\cdot}54}{2{\cdot}27}=2, \qquad \frac{2{\cdot}27}{2{\cdot}27}=1,$$

and therefore the relative numbers of atoms of carbon, hydrogen and oxygen in the compound are expressed by the formula C_2H_4O. At the same time, there is no apparent reason why the above atomic proportions should not be divided by 1·135, or by 0·756 or by 0·567 or by any submultiple of 2·27, in which cases the formulae arrived at would be $C_4H_8O_2$, $C_6H_{12}O_3$, $C_8H_{16}O_4$, etc. instead of the simplest formula C_2H_4O.

The simplest formula which can be deduced from the results of analysis of a compound is known concisely as the Empirical Formula, and, from the above, the empirical formula may be the molecular formula or a submultiple of it; and therefore to determine the molecular formula it is necessary to determine the molecular weight of the compound.

The methods of *Determination of the molecular weights of organic compounds* are those in general use for all types of compounds. If the particular compound can be vaporised without decomposition or dissociation, the vapour density can be determined by Victor Meyer's method, by Dumas' method or by Hofmann's method, whichever may be found the most convenient and suitable for the compound under investigation. Victor Meyer's method is frequently used and, although the value of the vapour density obtained by this method is only approximate and hence the molecular weight calculated from this (by multiplying by 2) according to Avogadro's Law is also approximate, it will be sufficiently accurate to determine whether it is the molecular weight corresponding to the empirical formula or to one of its multiples. In the case of the compound having the empirical formula C_2H_4O, it was found that the vapour density as determined by Victor Meyer's method was approximately 43, indicating an approximate molecular weight of 86. This indicates that the molecular formula must be twice the empirical formula, i.e. $C_4H_8O_2$, of which the true molecular weight is 88 ($H=1$, $C=12$, $O=16$).

In the case of compounds whose molecules dissolve without dissociation or association in liquids and which obey the gas laws as applied to dilute solutions, the molecular weight may be determined from (*a*) the depression of the freezing point of a suitable solvent, (*b*) the elevation of the boiling point of a suitable solvent, (*c*) the lowering of the vapour pressure of a suitable solvent and (*d*) the osmotic pressure of a suitable solution. In actual practice, the usual methods employed are determinations of the depression of the freezing point and elevation of the boiling point of suitable solvents (cf. p. 507).

Further information concerning the molecular weights of organic compounds can be obtained in special cases by what may be described as chemical methods. For example, the equivalents of compounds which are acids can generally be determined by titration with a suitable alkali. The equivalent of a monobasic acid of the general formula, AH, is the same as its molecular weight; generally, the molecular weight of an acid is $n \times$ its equivalent weight, where n represents the basicity of the acid. Again, the silver salts of all typical organic acids when completely decomposed by heat leave a quantitative residue of silver, and the equivalents of such organic acids can be determined from the relationship

$$A\mathrm{Ag} - \mathrm{Ag} + \mathrm{H} = A\mathrm{H},$$

where AAg represents the equivalent weight of silver salt or the molecular weight of the silver salt of a monobasic acid, AH. If the basicity, n, of the acid is known, the molecular weight of an organic acid of the formula, $A'\mathrm{H}_n$, is determined from the relationship

$$A'\mathrm{Ag}_n - n\mathrm{Ag} + n\mathrm{H} = A'\mathrm{H}_n,$$

where $A'\mathrm{Ag}_n$ represents the molecular weight of the normal silver salt of the n-basic acid, $A'\mathrm{H}_n$. It is only necessary to prepare the normal silver salt of the acid, determine the amount of silver in a weighed amount and calculate the amount of the silver salt which contains the atomic weight of silver or n times the atomic weight of silver and apply either of the above relationships according to the available information.

The chloroplatinate and chloroaurate derived from ammonia have the respective formulae:

$$[NH_3.H]_2PtCl_6 \quad \text{and} \quad [NH_3.H]AuCl_4,$$

and these compounds when decomposed completely by heating leave a quantitative residue of platinum and gold respectively. Organic bases (a molecule of which may be represented by *B*) form analogous compounds which may be represented by the respective formulae:

$$[B.H]_2PtCl_6$$

chloroplatinate of organic base, *B*

$$[B.H]AuCl_4$$

chloroaurate of organic base, *B*

These 'organic' chloroplatinates and chloroaurates also decompose on being heated leaving a quantitative residue of platinum and gold respectively. It is thus possible, having determined the amount of platinum or gold contained in a weighed amount of the chloroplatinate or chloroaurate, to calculate the molecular weight of the particular compound, i.e. the amount of the compound (in grams) which contains the atomic weight (in grams) of platinum or gold. The molecular weight of the base, *B*, can then be calculated from the following relationship:

in the case of the chloroplatinate

$[B.H]_2PtCl_6$	−	H_2PtCl_6	=	$2B$
molecular weight of the chloroplatinate		molecular weight of chloroplatinic acid		2 × molecular weight of base

in the case of the chloroaurate

$[B.H]AuCl_4$	−	$HAuCl_4$	=	B
molecular weight of the chloroaurate		molecular weight of chloroauric acid		molecular weight of base

Molecular weights of certain other types of organic compounds may also be determined by what may be described as chemical methods, and these will be referred to in, or will follow from, the chemical reactions of the particular types of compounds. Chemical methods of determining molecular weights are of considerable importance and have been applied in the case of starch and cellulose, colloidal substances having large molecules, and they may be useful in determining the molecular complexity of compounds of high molecular weights, such as many of those of animal and plant origin.

Such types of investigation outlined above, whilst important as giving information about the nature of the molecule of a compound as a whole, do not, of course, give (except where the compound has, e.g., acidic or basic properties) any information about what has been described above as 'the architecture of the molecule' or (as it is more usually described) the constitution of the molecule. The 'architecture of the molecule', i.e. the spatial relations of the atoms to each other in the molecule, determines the chemical and physical behaviour of the compound in question; and, by investigating the chemical and physical properties of a compound, the 'architecture of the molecule' of the compound can be deduced. This determination of the 'architecture' of the molecules of organic compounds or the determination of their constitution is the fundamental concern of the study of Organic Chemistry.

CHAPTER I

THE HYDROCARBONS

A. Aliphatic Hydrocarbons

FROM the standpoint of organic chemistry, the simplest compounds are those which contain a single atom of carbon in the molecule. The fundamental valency of carbon being four, it follows that the simplest compound is the substance, methane, having the formula CH_4, and whose *constitutional* formula can be simply represented as

```
    H
    |
H—C—H
    |
    H
```

This formula graphically represents a molecule of methane consisting of a carbon atom united to its equivalent of four atoms of hydrogen.

Methane, CH_4, occurs to the extent of 90 per cent. in the so-called 'natural gas' obtained from petroleum wells. It is also known as *marsh gas* and is the original 'carburetted hydrogen' of Dalton, being formed by the putrefaction of vegetable matter under water. The explosive fire-damp of coal mines is a mixture of methane and air. Methane, being produced by the destructive distillation of carbonaceous matter, wood and coal, forms 40 per cent. by volume of coal gas.

It is important to realise that this fundamental organic compound can be synthesised from its elements. By passing pure hydrogen over highly purified carbon at 1100–1200°, Bone and Coward in 1908 confirmed the previous work of Bone and Jerdan and obtained methane in a yield of 73 per cent. of the theoretical amount. When hydrogen and carbon monoxide are mixed in the proper proportions and passed over finely divided nickel at a temperature of 180–250°, almost pure methane is obtained. Similarly carbon dioxide is reduced by hydrogen at a temperature of 300–400° in the presence of nickel. These reactions may be represented by the following equations:

$$C + 2H_2 = CH_4$$
$$CO + 3H_2 = CH_4 + H_2O$$
$$CO_2 + 4H_2 = CH_4 + 2H_2O$$

These equations represent the simplest reactions which may take place, but the actual products vary considerably with the conditions employed. For example, it has been shown that when purified water gas is heated at 200–300° at atmospheric pressure in contact with finely divided nickel, methane is produced according to the reaction

$$2CO + 2H_2 \rightleftarrows CO_2 + CH_4$$

and this is by no means the only reaction which is possible between carbon monoxide and hydrogen.

Preparation. The gas can be conveniently prepared in the laboratory from aluminium carbide,* Al_4C_3, which is made by heating aluminium oxide and carbon in the electric furnace (Moissan, 1912). The aluninium carbide is placed in a suitable dry flask fitted up as shown (Fig. 1) and the gas collected in jars over water:

$$Al_4C_3 + 12H_2O = 3CH_4 + 4Al(OH)_3$$

To avoid difficulties due to the insolubility in water of aluminium hydroxide, it is convenient to replace the water by dilute hydrochloric acid, which is slowly added from the tap funnel.

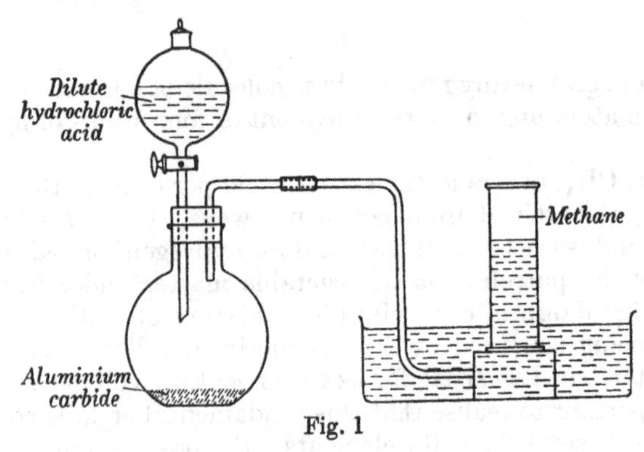

Fig. 1

As illustrating the production of methane by the putrefaction or fermentation of vegetable matter and also as a method by which the gas can be obtained in quantity and in a state of reasonable purity, cellulose offers a convenient source of the gas. Filter paper, almost pure cellulose, is thoroughly wetted with water so as to make a pulp. The pulp is mixed with horse and/or goat dung which ordinarily contains the anaerobic bacteria responsible for the degradation of the cellulose. If the mixture be kept at a little above the ordinary temperature, a mixture of methane and carbon dioxide is evolved. If the evolved gas is passed through an absorbent for the carbon dioxide—an aqueous solution of potassium hydroxide is suitable for laboratory purposes—the methane is obtained in a high state of purity. The temperature of the reaction or fermentation must be carefully controlled, since at a temperature of 60° it stops entirely.

* Aluminium carbide is more systematically described as 'aluminium methanide'.

The particular species of bacterium which causes the reaction to take place is not known definitely and it is probable that at least two species are concerned. Assuming a simple formula for cellulose and for sugar, the reaction may be assumed to take place in the following, stages:

$$(a)\ \underset{\text{(cellulose)}}{C_6H_{10}O_5} + H_2O = \underset{\text{(non-reducing sugar)}}{C_6H_{12}O_6}$$

$$(b)\ C_6H_{12}O_6 = 3CH_4 + 3CO_2$$

This method is actually used for the large-scale production of methane, and it is clear that its production by the above process is much more economical than by any ordinary chemical method.

Other methods of preparation of methane will be mentioned subsequently.

Properties. Methane is a colourless, odourless* gas of marked stability. Having a boiling point of $-160°$, it condenses when cooled to the temperature of liquid air; it is stable at a temperature of 1200°. It is unaffected when passed through fuming sulphuric acid, nitric acid and aqueous solutions of potassium permanganate. Bromine water is likewise without action on the gas. The gas can be dried by phosphorus pentoxide. Methane burns in air or oxygen with a pale blue flame, carbon dioxide and water being the products of combustion.

Methane forms an explosive mixture with air or oxygen, and its composition may be verified by determining the volume of oxygen required to burn a known volume of methane. The volume of carbon dioxide produced is measured; this is found to be the same as that of the methane taken, under the same conditions of temperature and pressure; therefore one carbon atom is present in the molecule of methane. It is also found that the volume of oxygen required to burn the methane is twice the volume of the latter. The equation representing the reaction, and the complete oxidation of methane, is

$$\underset{\text{1 vol.}}{CH_4} + \underset{\text{2 vols.}}{2O_2} = \underset{\text{1 vol.}}{CO_2} + \underset{\text{(condensed)}}{2H_2O}$$

The vapour density of methane having been found by experiment to be eight times that of hydrogen, it follows that the molecular weight of methane is 16, which confirms the formula, CH_4, determined by the combustion of the gas.

Since the molecule of methane consists of one carbon atom united with its full equivalent of four hydrogen atoms, it constitutes what is known as a *saturated* compound, which can only form derivatives by *substitution* of the hydrogen atoms by equivalent atoms or groups of atoms. This may be instanced here not only by the action of

* By some authorities it is said to possess a pleasant leek-like odour, but this is not generally recognised.

oxygen on methane according to the above equation, where the four hydrogen atoms are replaced by their equivalent of two oxygen atoms forming carbon dioxide—a substitution product of methane—but also by the action of chlorine on methane.

The reaction between chlorine* and methane is a photochemical one. Little or no action takes place between dry chlorine and dry methane in the dark; the constituents of the mixture react explosively in sunlight, whilst a gradual reaction between the two substances takes place in diffused daylight. Under these last conditions, no matter in what proportions the two gases are present, no less than four substances are produced. These substances are *monochloromethane* or *methyl*† *chloride*—a colourless gas, b.p. $-24{\cdot}1°$, whose composition is represented by the formula CH_3Cl; *dichloromethane* or *methylene chloride*—a colourless liquid, b.p. $41{\cdot}6°$, having the formula CH_2Cl_2; *trichloromethane* or *chloroform*—the well-known anaesthetic, a colourless liquid, b.p. $61{\cdot}2°$, having the formula $CHCl_3$; and *tetrachloromethane* or *carbon tetrachloride*—a colourless liquid, b.p. $76{\cdot}8°$, its formula, CCl_4, being analogous to that of methane. Hydrogen chloride is also formed at the same time. The reactions taking place are represented thus:

$$(a)\ CH_4 + Cl_2 = CH_3Cl + HCl$$

$$(b)\ CH_3Cl + Cl_2 = CH_2Cl_2 + HCl$$

$$(c)\ CH_2Cl_2 + Cl_2 = CHCl_3 + HCl$$

$$(d)\ CHCl_3 + Cl_2 = CCl_4 + HCl$$

While the hydrogen atoms are replaced in stages by their equivalent and monovalent chlorine atoms, it should be emphasised that the action never stops at any one of the first three stages. Whatever the relative volumes of the interacting gases may be at the beginning, all four chlorinated products are formed, and methylene chloride and chloroform actually form the greater proportion of the final mixture.

Applications of methane. Where methane in 'natural gas' is abundant (for example in Roumania), it is used for lighting and as a source of heat and power. The abundance of methane in Roumania enables it to be used as a means of obtaining nitrogen from the atmosphere, the nitrogen being then used for the preparation of calcium cyanamide—an artificial manure—from calcium carbide. The methane is burnt in a restricted supply of air, when the following reaction takes place:

$$CH_4 + \text{Air} \rightarrow CO_2 + 2H_2O + \text{Nitrogen, etc.}$$

* Similar reactions take place less readily with bromine vapour, but do not proceed with iodine.

† 'Methyl' is the simplest monovalent organic radical. It may be compared with 'ammonium'. It is ordinarily incapable of existing by itself (compare p. 26), but the same group of atoms exists in a large number of compounds, e.g. CH_3Br, CH_3I, $(CH_3)_2SO_4$, etc. 'Methylene' is similarly a bivalent organic radical.

and the resulting gas having been freed from carbon dioxide and water is then passed over heated calcium carbide, when a mixture of calcium cyanamide and carbon results:

$$CaC_2 + N_2 = \underset{\text{calcium cyanamide}}{CaN.CN} + C$$

By the partial oxidation of methane under carefully regulated conditions commercial processes for the manufacture of methyl alcohol and formaldehyde, two highly important organic compounds, have been developed. These compounds, substitution products of methane, will be described in detail later. Their production from methane may be expressed by the following reactions:

$$2CH_4 + O_2 = \underset{\text{methyl alcohol}}{2CH_4O}$$

$$CH_4 + O_2 = \underset{\text{formaldehyde}}{CH_2O} + H_2O$$

These reactions may be represented briefly and graphically:

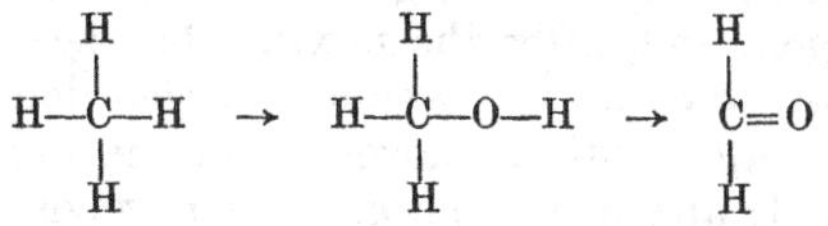

```
    H              H               H
    |              |               |
H—C—H   →   H—C—O—H   →   C=O
    |              |               |
    H              H               H
```

The industrial importance of the production of these two fundamental compounds from methane and hence from the precursors of methane cannot be too strongly emphasised.

The conversion of methane by pyrolysis into other hydrocarbons of industrial importance will be referred to later. The wastage of 'natural gas' which has gone on for many years will soon stop as the industrial applications of methane are carefully explored.

Ethane, C_2H_6. Associated with methane in the 'natural gas' from many petroleum wells is another gas with very similar properties. This gas can be obtained by the following method.

Corresponding to methyl chloride is a liquid compound, methyl iodide, CH_3I, which cannot, however, be prepared from methane directly. There is every reason to believe that the reaction described below would proceed with methyl chloride, but the latter compound, being a gas, is much more difficult to manipulate than the analogous liquid compound, methyl iodide. If methyl iodide be added gradually to powdered sodium suspended in dry ether (p. 30)—the ether does not take any part in the reaction—a vigorous reaction takes place, sodium iodide is precipitated, and the gas evolved may, like methane, be collected over water. The reaction taking place is illustrated as follows:

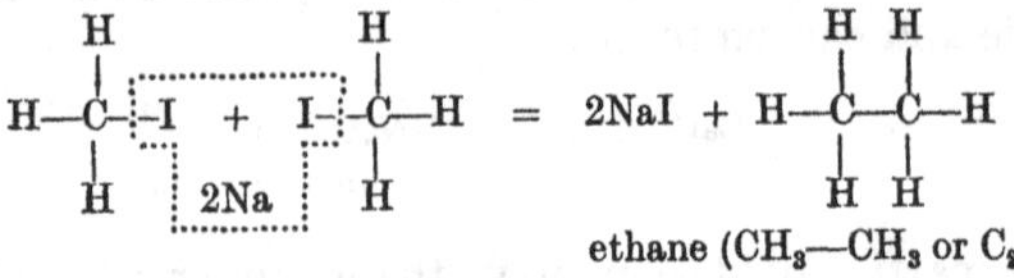

```
       H              H                          H   H
       |  ...........:|                          |   |
    H—C—:—I  +  I—:—C—H   =   2NaI  +  H—C—C—H
       |  :...    ....:|                          |   |
       H     : 2Na :   H                          H   H
             :.....:
```

ethane (CH_3—CH_3 or C_2H_6)

The gas so obtained has properties very similar to those of methane. It is a colourless gas having a higher boiling point (−89·5° at 735 mm.) than that of methane. It burns in air or oxygen with a slightly luminous flame, forming carbon dioxide and water, the final oxidation products. The reaction taking place is

$$2C_2H_6 + 7O_2 = 4CO_2 + 6H_2O$$

Vol. of CO_2 = twice that of the ethane, therefore 2 atoms of carbon in ethane mol.

When mixed with oxygen it forms an explosive mixture, and on the above reaction depends the method for determining the composition of the gas. A known volume of the gas is mixed with a known volume (excess) of oxygen, and, after the mixture has been exploded, the volumes of carbon dioxide and residual oxygen determined. From these figures, together with a knowledge of the vapour density of the gas (15, H = 1) and hence the molecular weight (30), the composition of the gas as C_2H_6 is verified.

Like methane, ethane is an extremely stable substance, being unaffected by such powerful chemical reagents as concentrated sulphuric acid, nitric acid, potassium permanganate, phosphorus pentoxide, etc. Like methane, it can only form derivatives by substitution. When the gas is mixed with chlorine,* progressive substitution of the hydrogen atoms in the molecule by chlorine atoms takes place and the reactions may be written similarly to the corresponding ones in the case of methane, thus:

$$C_2H_6 + Cl_2 = C_2H_5Cl + HCl$$
$$C_2H_5Cl + Cl_2 = C_2H_4Cl_2 + HCl$$
$$C_2H_4Cl_2 + Cl_2 = C_2H_3Cl_3 + HCl$$
$$C_2H_3Cl_3 + Cl_2 = C_2H_2Cl_4 + HCl$$
$$C_2H_2Cl_4 + Cl_2 = C_2HCl_5 + HCl$$
$$C_2HCl_5 + Cl_2 = C_2Cl_6 + HCl$$

If the formulae of these substitution products of ethane be

* The reaction goes similarly but less readily with bromine vapour.

written graphically in the manner already indicated, the first compound is

$$\begin{array}{ccccccc} & & H & & H & & \\ & & | & & | & & \\ H & — & C & — & C & — & Cl \\ & & | & & | & & \\ & & H & & H & & \end{array}$$

This represents the constitution of *monochloroethane* or *ethyl chloride* (a colourless liquid, b.p. 14°, an important general or local anaesthetic). The constitutional formula of ethyl chloride may be written more concisely as $CH_3.CH_2Cl$. This is the only formula possible for a compound of the composition C_2H_5Cl, since the six hydrogen atoms, in ethane, are obviously chemically equivalent to each other, and hence it does not matter which one of them is replaced by chlorine in the first instance. In ethyl chloride, however, the five hydrogen atoms are not equally placed with reference to each other in the molecule. There are the three hydrogen atoms attached to one carbon atom and the other two hydrogen atoms are attached to the carbon atom to which the chlorine atom is attached. When a second chlorine atom takes the place of a hydrogen atom, there is the possibility of two compounds being formed, and the constitution of these may be written

$$\begin{array}{ccccccc} & & H & & Cl & & \\ & & | & & | & & \\ H & — & C & — & C & — & Cl \\ & & | & & | & & \\ & & H & & H & & \end{array} \quad \text{and} \quad \begin{array}{ccccccc} & & H & & H & & \\ & & | & & | & & \\ Cl & — & C & — & C & — & Cl \\ & & | & & | & & \\ & & H & & H & & \end{array}$$

$(CH_3.CHCl_2)$ $(CH_2Cl.CH_2Cl)$

In actual fact, two different compounds, both of which have the molecular composition represented by $C_2H_4Cl_2$ and are *dichloroethanes*, are formed. These two compounds will be referred to later; the first one is generally known as *ethylidene chloride* (or unsymmetrical dichloroethane), a colourless liquid, b.p. 57·5° at 751 mm., and the second one as *ethylene chloride* (or symmetrical dichloroethane), a colourless liquid, b.p. 83·7°. The existence of different compounds each having distinct properties and possessing the same molecular formula but different constitutional formulae is a characteristic and common feature of organic chemistry. Ethylidene chloride and ethylene chloride are described as *isomeric compounds* or *isomers*. In isomeric compounds the atoms occupy different relative positions in the molecule and this phenomenon is described as *isomerism*.

Similarly, there are two isomeric *trichloroethanes*:

$$\begin{array}{ccccccc} & & H & & Cl & & \\ & & | & & | & & \\ H & — & C & — & C & — & Cl \\ & & | & & | & & \\ & & H & & Cl & & \end{array} \quad \text{and} \quad \begin{array}{ccccccc} & & H & & H & & \\ & & | & & | & & \\ Cl & — & C & — & C & — & Cl \\ & & | & & | & & \\ & & H & & Cl & & \end{array}$$

$(CH_3.CCl_3)$, b.p. 75° $(CH_2Cl.CHCl_2)$, b.p. 114°

and two isomeric *tetrachloroethanes*:

$$\begin{array}{ccccc} & H & & Cl & \\ & | & & | & \\ H- & C & - & C & -Cl \\ & | & & | & \\ & Cl & & Cl & \end{array} \qquad \text{and} \qquad \begin{array}{ccccc} & H & & H & \\ & | & & | & \\ Cl- & C & - & C & -Cl \\ & | & & | & \\ & Cl & & Cl & \end{array}$$

($CH_2Cl.CCl_3$), b.p. 130·5°

($CHCl_2.CHCl_2$), b.p. 147°
This colourless liquid is largely known as 'Westron'. It finds commercial application as an insecticide and more particularly as a non-inflammable solvent for varnishes and especially cellulose acetates

There is only one possible *pentachloroethane* and only one *hexachloroethane*, just as there is only one monochloroethane (ethyl chloride) and one ethane:

$$\begin{array}{ccccc} & H & & Cl & \\ & | & & | & \\ Cl- & C & - & C & -Cl \\ & | & & | & \\ & Cl & & Cl & \end{array} \qquad \text{and} \qquad \begin{array}{ccccc} & Cl & & Cl & \\ & | & & | & \\ Cl- & C & - & C & -Cl \\ & | & & | & \\ & Cl & & Cl & \end{array}$$

($CHCl_2.CCl_3$), b.p. 159°

($CCl_3.CCl_3$), m.p. (in a closed tube) 187°; b.p. 185·5° at 777 mm.

The two simple hydrocarbons, methane and ethane, illustrate another characteristic feature of organic chemistry. Carbon may combine with other atoms or groups of atoms and it has the power of combining with itself. Thus in methane and ethane:

$$\begin{array}{ccc} & H & \\ & | & \\ H- & C & -H \\ & | & \\ & H & \end{array} \qquad \begin{array}{ccccc} & H & & H & \\ & | & & | & \\ H- & C & - & C & -H \\ & | & & | & \\ & H & & H & \end{array}$$

it is seen that the valencies of a carbon atom may be satisfied by other atoms, in this case hydrogen atoms, or groups of atoms, or partially by other carbon atoms. In any case, the carbon atoms always remain quadrivalent.

Another characteristic feature of organic chemistry, viz. the similarity of reactions of analogous compounds, is also illustrated. It has been shown that the two hydrocarbons, methane and ethane, behave similarly towards chlorine and bromine, although the products of the reaction become more complicated and the number of isomerides increases with increasing complexity of the initial compound.

Ethane can be regarded as a substitution product of methane just as is monochloromethane or methyl chloride. Ethane might be described as methylmethane or dimethyl:

$$\begin{array}{ccc} & H & \\ & | & \\ H- & C & -Cl \\ & | & \\ & H & \end{array} \qquad \begin{array}{cccc} & H & H & \\ & | & | & \\ H- & C- & C & -H \\ & | & | & \\ & H & H & \end{array}$$

Since ethane yields halogen substituted derivatives like methane, the similar properties of methane and ethane would lead one to expect that derivatives of the latter can be formed just as ethane is formed from methane via methyl iodide. Actually a complete series of hydrocarbons can be built up by extending the reaction described on p. 17.

From ethyl iodide and methyl iodide is obtained the hydrocarbon *propane*, C_3H_8, b.p. $-45°$:

$$\begin{array}{cccc} & H & H & \\ & | & | & \\ H- & C- & C- & I \\ & | & | & \\ & H & H & \end{array} + \begin{array}{ccc} & H & \\ & | & \\ I- & C & -H \\ & | & \\ & H & \end{array} \;(2Na) = 2NaI + \begin{array}{ccccc} & H & H & H & \\ & | & | & | & \\ H- & C- & C- & C & -H \\ & | & | & | & \\ & H & H & H & \end{array}$$

$$(CH_3.CH_2.CH_3)$$

a hydrocarbon having properties analogous to those of methane and ethane and which also occurs in common petroleum.

On complete oxidation, propane reacts according to the equation:

$$C_3H_8 + 5O_2 \rightarrow 3CO_2 + 4H_2O.$$

Since the volume of carbon dioxide formed is three times that of the propane taken, there must be three atoms of carbon in the molecule of propane.

Propane can be considered to contain two types of hydrogen atom in its molecule, viz. that of the two end methyl or CH_3— groups and the other of the middle —CH_2—* group. Consequently propane will yield two isomeric monosubstitution derivatives, two monochloropropanes, two monobromopropanes, two monoiodopropanes, etc. The two isomeric monoiodopropanes have the following constitutional formulae and are known as *normal*†-*propyl iodide* and *iso*†-*propyl iodide* respectively:

$$\begin{array}{ccccc} & H & H & H & \\ & | & | & | & \\ H- & C- & C- & C & -I \\ & | & | & | & \\ & H & H & H & \end{array} \;(C_3H_7I) \qquad \begin{array}{ccccc} & H & H & H & \\ & | & | & | & \\ H- & C- & C- & C & -H \\ & | & | & | & \\ & H & I & H & \end{array} \;(C_3H_7I)$$

($CH_3.CH_2.CH_2I$), b.p. 102·5°, *n*-propyl iodide

($CH_3.CHI.CH_3$ or $[CH_3]_2CHI$), b.p. 89·5°, *i*-propyl iodide

* —CH_2—, methylene group.
† '*normal*' is usually abbreviated to '*n*' and '*iso*' to '*i*'.

These compounds have properties entirely analogous to those of methyl and ethyl iodides, and consequently the sodium or Wurtz reaction can be applied to these compounds also, thus:

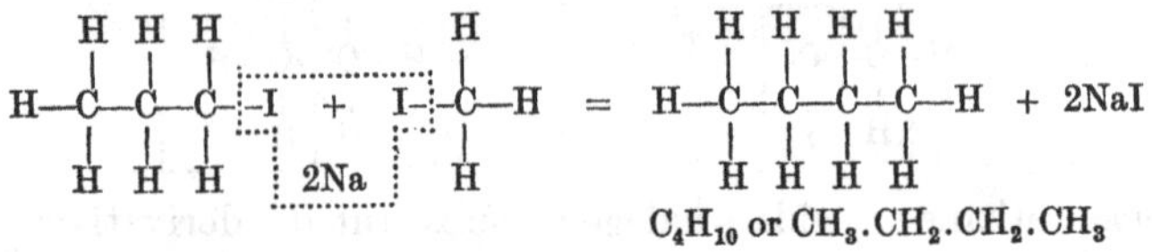

$$\begin{array}{ccccccccccccccccccc} & & H & & H & & H & & & & & & H & & & & H & & H & & H & & H & & \\ & & | & & | & & | & & & & & & | & & & & | & & | & & | & & | & & \\ H & - & C & - & C & - & C & - & I & + & I & - & C & - & H & = & H-C & - & C & - & C & - & C & -H & + 2NaI \\ & & | & & | & & | & & & & & & | & & & & | & & | & & | & & | & & \\ & & H & & H & & H & & & 2Na & & & H & & & & H & & H & & H & & H & & \end{array}$$

$$C_4H_{10} \text{ or } CH_3.CH_2.CH_2.CH_3$$

$$\begin{array}{ccccccccccccc} & H & & & & & & & & H & & & \\ & | & & & & & & & & | & & & \\ H- & C & -H & & & & H & & H- & C & -H & H & \\ & | & & & & & | & & & | & & | & \\ H- & C & -I & + & I- & & C-H & = & H- & C & ---- & C-H & + 2NaI \\ & | & & & & & | & & & | & & | & \\ H- & C & -H & 2Na & & & H & & H- & C & -H & H & \\ & | & & & & & & & & | & & & \\ & H & & & & & & & & H & & & \end{array}$$

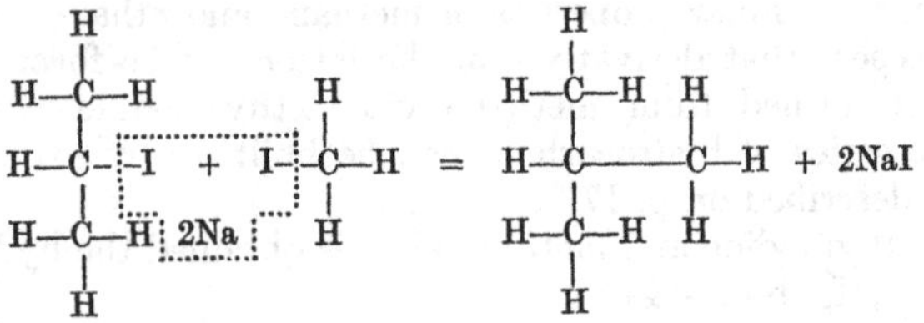

$$C_4H_{10} \text{ or } (CH_3)_3CH$$

From the two monoiodopropanes, two hydrocarbons are derived, both having the molecular formula, C_4H_{10}, but as shown above different constitutional formulae. They are therefore isomeric compounds. That having the constitutional formula, $CH_3.CH_2.CH_2.CH_3$, is known as *normal-* (or *n-*) *butane* (b.p. 1°), and the one having the constitutional formula, $(CH_3)_3CH$, is known as *iso-* (or *i-*) *butane* (b.p. −10·2°). Both hydrocarbons occur naturally in petroleum.

The above reaction can be continued almost indefinitely, and it will be realised that *n*-butane can not only be prepared from *n*-propyl iodide and methyl iodide but also from two molecules of ethyl iodide by the same reaction.

By extending this reaction one stage further three isomeric hydrocarbons, pentanes, having the molecular formula, C_5H_{12}, are known. The constitutional formulae of these pentanes may be written thus:

$CH_3.CH_2.CH_2.CH_2.CH_3$ or $CH_3.[CH_2]_3.CH_3$, *n*-pentane, which occurs in Pennsylvanian petroleum and has b.p. 36·3°

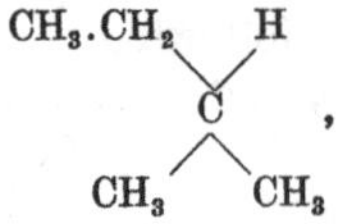

$$\begin{array}{ccc} CH_3.CH_2 & & H \\ & \diagdown\ \diagup & \\ & C & \\ & \diagup\ \diagdown & \\ CH_3 & & CH_3 \end{array}$$, dimethylethylmethane, β-methylbutane, b.p. 30·4°

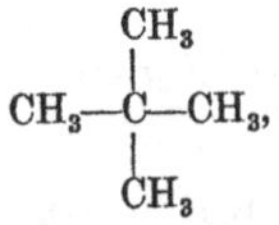

$$\begin{array}{ccc} & CH_3 & \\ & | & \\ CH_3- & C & -CH_3 \\ & | & \\ & CH_3 & \end{array}$$, or $C(CH_3)_4$, tetramethylmethane, $\beta\beta$-dimethylpropane, b.p. 9°

The generality of the reaction of the alkyl* iodides with sodium

* 'Alkyl' is the general name given to radicals such as methyl, ethyl, propyl, *i*-propyl, etc., or to radicals of the general formula C_nH_{2n+1} (*v.* below).

indicates that when two different alkyl iodides are used together for the reaction the product will not be homogeneous. Together with the propane obtained by the action of sodium on a mixture of equimolecular quantities of methyl and ethyl iodides will be ethane and *n*-butane due to the action of the sodium on some of the methyl iodide and some of the ethyl iodide respectively. Similarly, the *n*-butane obtained by the action of sodium on a mixture of equimolecular quantities of methyl iodide and *n*-propyl iodide will be accompanied not only by ethane but also by *n*-hexane, C_6H_{14}, $CH_3.[CH_2]_4.CH_3$, b.p. 69°, which occurs in Pennsylvanian petroleum; and the *i*-butane from methyl iodide and *i*-propyl iodide will be mixed with an isomeric hexane, b.p. 58°, having the following constitutional formula:

$$\begin{array}{ccccc} & CH_3 & & CH_3 & \\ & | & & | & \\ H- & C & - & C & -H \\ & | & & | & \\ & CH_3 & & CH_3 & \end{array}$$
, di*iso*propyl, *sym*-tetramethylethane, *βγ*-dimethylbutane

From the hydrocarbons mentioned it will be realised that as the number of carbon atoms in the molecule increases the number of isomeric hydrocarbons and of their monosubstitution derivatives (alkyl halides, for example) increases rapidly, thus:

Hydrocarbons	Isomers possible	Monosubstitution products	Isomers possible
CH_4	—	CH_3X	—
C_2H_6	—	C_2H_5X	—
C_3H_8	—	C_3H_7X	2
C_4H_{10}	2	C_4H_9X	4
C_5H_{12}	3	$C_5H_{11}X$	8
C_6H_{14}	5	$C_6H_{13}X$	17

The series of these hydrocarbons can be extended considerably. The higher members of the series, those having the carbon atoms in what may be described conveniently as a straight chain as in *n*-hexane, are described by the stem of the Greek numeral followed by the termination 'ane'. Omitting intermediate hydrocarbons we have, for example, *n*-triacontane, $C_{30}H_{62}$ or $CH_3.[CH_2]_{28}.CH_3$, a colourless crystalline substance, m.p. 66°, and hexacontane, $C_{60}H_{122}$, m.p. 101–102°.

Such a series of compounds, as methane, ethane, propane, the butanes, etc., the members of which differ in constitution from individual to individual by $—CH_2$, constitutes a *homologous series*, and ethane, propane, the butanes, etc. are described as *homologues* of methane. The chemical properties of all members of a homologous series are very similar and, generally, are described by those of the first three members of the series. The physical properties and constants vary from member to member fairly regularly with the change in complexity. A homologous series of compounds can be described by means of a *general formula*, which in the case of the series of compounds under discussion is C_nH_{2n+2}, where n = the number of carbon atoms in the molecule.

Owing to the general chemical stability of the compounds of this series, C_nH_{2n+2}, they have long been known as the paraffin hydrocarbons or paraffins (from *parum affinis*, indicating their chemical indifference). Compounds, such as the paraffins, which yield derivatives only by substitution, are described as *saturated* compounds, in contradistinction to those *unsaturated* compounds which yield saturated compounds by direct combination with other substances. The essential difference between a saturated and an unsaturated compound may be illustrated as follows:

Saturated compound + Halogen (chlorine) = Substituted compound + Hydrogen halide (chloride)

Unsaturated compound + Chlorine = Saturated compound

Petroleum

Although other types of hydrocarbons, which will be described later, are present along with compounds not hydrocarbons in natural petroleum, the paraffins occur to a very considerable extent in most of the products from all the well-known oil fields (United States of America, Borneo, Iran, Roumania, Iraq, Galicia, Mexico, etc.), and the collection and refining of petroleum is one of the most important chemical industries in the world. The natural gas consists largely of methane and to a much less extent of ethane, and the somewhat thick, fluorescent, dark-coloured liquid petroleum contains many of the higher members of the paraffin series. The usually unpleasant odour which characterises crude petroleum is due largely to the presence of organic compounds containing sulphur.

Several theories as to the origin of petroleum have been put forward. Mendeléeff suggested that it was due to the action of steam on subterranean carbides (compounds of metals and carbon). Other theories suggest a plant or animal origin for petroleum. Engler suggested marine animals as a possible source, and it is known that fish oils when heated under pressure in absence of air yield a petroleum-like product. Murray-Smith, again, suggested that petroleum has its origin in buried forests of coniferae; coniferae in ordinary circumstances give rise to more complicated types of hydrocarbons known as terpenes. How petroleum may originate from such sources is illustrated by the modern process of *cracking*. In this process, the higher boiling portions of the petroleum are heated to a considerable temperature in closed retorts, whereby a partial breaking down to lighter oils is effected.

Before coming into industrial use, petroleum is subjected to a process of refining, which is essentially one of very careful fractional distillation. The boiling point of particular fractions determines their industrial use, whether as solvents, motor spirit, illuminants, lubricants, etc. The highest boiling fractions are thick oils, and from these the commercial *vaseline* is obtained. Paraffin wax is a mixture of higher paraffins which are solid at the ordinary temperature. It is the residue from the refining and, after being freed from liquid material by pressure and cooling, is purified by treatment with chemical reagents, such as sulphuric acid, sodium hydroxide, etc., which do not react with the paraffin hydrocarbons themselves.

The constitutions of the various paraffin hydrocarbons have, in the preceding account, been represented by two-dimensional or planar formulae. A nearer approximation to the actual constitution of these and all compounds is obviously to consider them as existing in space, which necessitates the use of spatial formulae for a more adequate representation of their constitution. The molecule of methane, the simplest case, may be considered as consisting of a carbon atom occupying a mean position at the centre of a regular tetrahedron and the four hydrogen atoms occupying mean positions at the corners of the tetrahedron and therefore all equally placed relative to the carbon atom, as in Fig. 2.

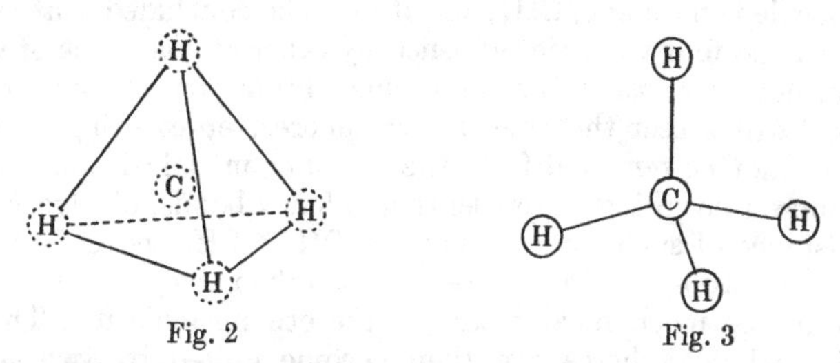

Fig. 2 Fig. 3

A better method of representation is to consider the carbon atom in a similar mean position and connected symmetrically in space in some way to the four hydrogen atoms, as in Fig. 3. Mechanical models using flexible springs for joining differently coloured spheres (representing valency bonds and atoms respectively) will be found convenient in studying the spatial formulae of compounds.

Using planar formulae it might be concluded that, for example, two dichloromethanes are capable of existence represented thus:

```
     H              H
     |              |
H—C—Cl       Cl—C—Cl
     |              |
     Cl             H
```

Spatial formulae, however, show that these two are identical, and, actually, disubstituted methanes never show isomerism.

The constitution of ethane can be represented either by joining two tetrahedra corner to corner, as in Fig. 4, or by replacing one sphere representing a hydrogen atom (in the spring model of methane) by a sphere representing a carbon atom to which are attached three spheres representing hydrogen atoms, as in Fig. 5.

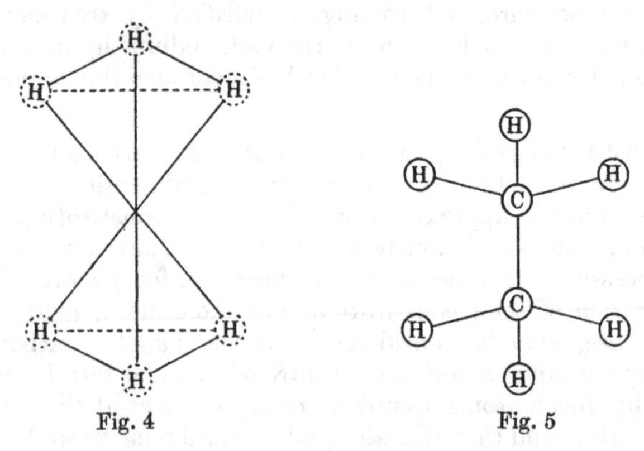

Fig. 4 Fig. 5

The only hydrocarbon known containing one atom of carbon in the molecule is methane, CH_4, and it must be concluded that carbon always has its four valencies satisfied by other atoms, some of which may be, however, other carbon atoms. From the above-described models, it will be seen that one or more spheres, representing hydrogen atoms, cannot be removed from the methane model without leaving one or more free springs, representing valency bonds, illustrating the non-existence of such hydrocarbons as CH_3,* CH_2 and CH. On the other hand, two hydrogen spheres can be taken away, one from each half, from the mechanical model of the ethane molecule. The two central 'carbon' spheres can then become united by two springs representing valency bonds. Fig. 6 represents the mechanical model of the molecule of a hydrocarbon, C_2H_4, which might be derived from ethane by taking away two atoms of hydrogen, one from each of the methyl groups:

$$CH_3.CH_3 - 2H \rightarrow CH_2:CH_2$$

The two carbon atoms in the hydrocarbon C_2H_4 are described as being united by a *double* bond. The mechanical representation of the double valency between two carbons unfortunately is not an adequate representation of the chemical nature of this linkage between carbon atoms (p. 39). Fig. 7 represents the mechanical model of a hydro-

* The methyl radical has, however, been shown to be capable of existence although its 'life period' is very short (p. 501).

carbon, C_2H_2, which theoretically would be derived from ethane by taking away two hydrogen atoms from each of the methyl groups, or from the hydrocarbon C_2H_4 by taking away two hydrogen atoms, one from each of the methylene groups:

$$CH_3.CH_3 - 4H \rightarrow CH \vdots CH$$
$$CH_2 : CH_2 - 2H \rightarrow CH \vdots CH$$

the carbon atoms then becoming united each to each by a *triple* bond.

In actual fact the hydrocarbons C_2H_4 and C_2H_2 do exist. They are known respectively as ethylene and acetylene and are the parent members of two homologous series of hydrocarbons the members of which have the general formulae C_nH_{2n} and C_nH_{2n-2} respectively.

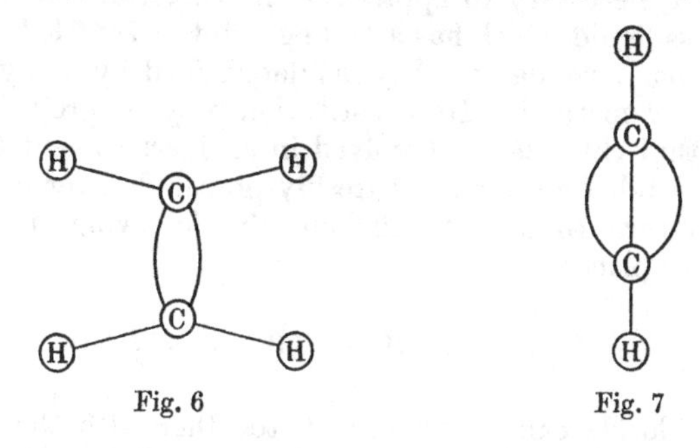

Fig. 6 Fig. 7

The Ethylene (or Olefine) Series of Hydrocarbons, C_nH_{2n}

*Ethylene**

The conversion of ethane into ethylene can be accomplished by indirect methods. Ethyl chloride, bromide or iodide, when heated with a concentrated solution of potassium hydroxide in alcohol, loses a molecule of hydrochloric, hydrobromic or hydriodic acid and ethylene is produced, the alkali neutralising the halogen acid liberated. Since ethyl chloride or ethyl bromide can be produced from ethane by direct substitution, ethylene is obtained indirectly from ethane, thus:

$$\begin{matrix} CH_3 \\ | \\ CH_3 \end{matrix} + Cl_2 \rightarrow \begin{matrix} CH_3 \\ | \\ CH_2Cl \end{matrix} + HCl$$

$$\begin{matrix} CH_3 \\ | \\ CH_2Cl \end{matrix} + KOH \rightarrow \begin{matrix} CH_2 \\ \| \\ CH_2 \end{matrix} + KCl + H_2O$$

* The 'bicarburetted hydrogen' of Dalton. Ethylene does not occur naturally. Coal gas may contain as much as 2·5 per cent. by volume of the gas. The gas obtained in the low temperature production of coke may contain as much as 6 per cent. The ethylene from these two sources has been used commercially.

If ethyl bromide, which is a liquid, be used ethylene can be obtained by this method in a high degree of purity. The alcohol is used in this preparation merely as a means of producing a homogeneous solution which promotes more rapid reaction. A concentrated aqueous solution of potassium hydroxide can be used, but the reaction would take place more slowly on account of the sparing solubility of the ethyl halide in water and aqueous solutions generally.

The most convenient source of ethylene, both from the laboratory and industrial points of view, is, however, *ethyl alcohol*. This fundamental organic compound is produced industrially by fermentation processes which will be described later. For the immediate purpose it is, however, necessary to appreciate the chemical constitution of this colourless liquid which has a boiling point of 78°. Ethyl alcohol has a molecular formula of C_2H_6O as determined by analytical and vapour density methods. Its constitution may be proved by synthesis. If ethyl chloride be dissolved in an inert solvent (ether) in which is suspended an excess of freshly prepared silver hydroxide* and the mixture thoroughly shaken, the following quantitative reaction takes place:

$$C_2H_5\boxed{Cl + Ag}OH \rightarrow C_2H_5.OH + AgCl$$

The silver chloride can be filtered off together with the excess of silver hydroxide from the reaction mixture, and the filtrate which can be tested and proved free from chlorine will be found to contain a colourless liquid having a boiling point of 78°. The ethyl alcohol has been formed from the ethyl chloride by taking away a chlorine atom and replacing it by the equivalent —O—H group (commonly described as the *hydroxyl* group). Hence ethyl alcohol† can be described as hydroxyethane just as ethyl chloride may be described as chloroethane, and the constitution of the former written:

```
      H
      |
  H—C—H
      |
  H—C—O—H
      |
      H
```

As a molecule of ethylene may be considered to be derived from a molecule of ethyl chloride by taking away a molecule of hydrogen chloride, it appears reasonable that a molecule of ethylene may be

* Sodium or potassium hydroxide in *dilute* aqueous solution can be used, but not so conveniently.

† It will be shown later that ethyl alcohol is the most convenient starting material for the preparation of ethyl chloride, ethyl bromide and iodide.

obtained from a molecule of ethyl alcohol by taking away a molecule of water:

```
      H                                    H
      |  .........                         |  ........
  H—C—:-H       :                      H—C—:-H    :
      |  :.......:      compare             |  :      :
  H—C—:-O—H     :                      H—C—:-Cl   :
      |  :.......:                          |  :......:
      H                                    H
```

The methods employed for the production of ethylene depend on the utilisation of ethyl alcohol. The dehydration (abstraction of water) can be accomplished in different ways: (*a*) by the use of concentrated sulphuric acid, (*b*) by the use of concentrated orthophosphoric acid, (*c*) by catalytic processes. By whichever of these methods ethylene is produced, the ***final result*** may be expressed:

$$C_2H_5.OH \rightarrow C_2H_4 + H_2O$$

(*a*) When a mixture of ethyl alcohol with about three times its volume of concentrated sulphuric acid is heated to 160–170° (the bulb of the thermometer should be immersed in the mixture) ethylene is given off. The reaction has been shown to take place in two stages, a substance known as ethyl hydrogen being the intermediate product:

$$\text{(i)}\quad C_2H_5.OH + H.SO_4H \rightleftarrows^{*} C_2H_5.O.SO_3H + H_2O$$

When ethyl hydrogen sulphate is heated, ethylene is produced and sulphuric acid regenerated:

```
          H
          |
(ii)  H—C—H                 H—C—H
          |          →          ||        +  H2SO4
      H—C—O.SO3H            H—C—H
          |
          H
```

$$[\text{(i)}+\text{(ii)}\; C_2H_5.OH \rightarrow C_2H_4 + H_2O]$$

The gas so produced can be collected over water, but requires preliminary purification from by-products of the reactions. Concentrated sulphuric acid is a powerful oxidising agent especially towards organic substances; in the present case, it oxidises part of the alcohol forming oxidation products of the latter and sulphur dioxide is produced by the simultaneous reduction of part of the sulphuric acid. Further, ethyl hydrogen sulphate itself can react with ethyl alcohol in the following manner, whereby a new and very important

* The reaction between acids and alcohols is not complete (see p. 92). The product of the reaction, in this case ethyl hydrogen sulphate, is known as an *ester*. Diethyl sulphate, $(C_2H_5)_2SO_4$, is also an ester; compare $NaHSO_4$ and Na_2SO_4.

substance, diethyl ether*—a colourless liquid, b.p. 34·6°—is produced:

$$
\begin{array}{ccccccc}
 & \mathrm{H} & & \mathrm{H} & & \mathrm{H} & \mathrm{H} \\
 & | & & | & & | & | \\
\mathrm{H-} & \mathrm{C} & \mathrm{-H} \quad + & \mathrm{H-C-H} & & \mathrm{H-C-H} & \mathrm{H-C-H} \\
 & | & & | & \rightarrow H_2SO_4 + & | & | \\
\mathrm{H-} & \mathrm{C} & \mathrm{-O-}\boxed{\mathrm{SO_3H} \quad \mathrm{H-O-}} & \mathrm{C-H} & & \mathrm{H-C} & \mathrm{-O-C-H} \\
 & | & & | & & | & | \\
 & \mathrm{H} & & \mathrm{H} & & \mathrm{H} & \mathrm{H}
\end{array}
$$

$$C_2H_5{-}O{-}C_2H_5$$
diethyl ether

If the reaction proceeded entirely according to equations (i) and (ii), a small amount of sulphuric acid would theoretically be capable of converting an indefinitely large amount of alcohol into ethylene (and diethyl ether). Apart from the production of by-products

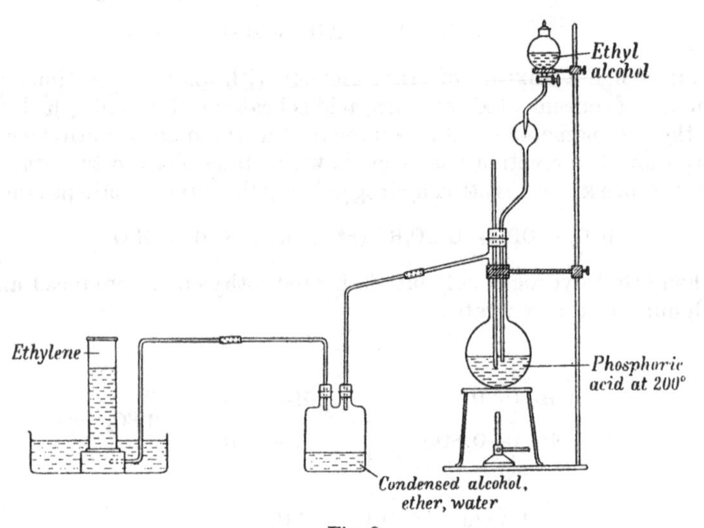

Fig. 8

mentioned above, it will be seen that water being produced in the reaction dilutes the sulphuric acid and the latter can only react in the ways described above when concentrated. Moreover, the necessary purification of the ethylene renders the preparation of the pure gas by this method somewhat cumbrous. Ethylene was, however, formerly prepared for laboratory purposes by this method, which was discovered by Weimann and other Dutch chemists in 1795.

* Generally known as 'ether'. This is the well-known anaesthetic and solvent and is the best known member of a large number of ethers. The constitutional formulae of the ethers are all analogous and consist of an oxygen atom united to two hydrocarbon radicals.

(*b*) The most convenient method for the preparation of ethylene in the laboratory is that devised by Newth (1901). It consists in the gradual addition of ethyl alcohol to syrupy orthophosphoric acid at 200°, the apparatus used being as shown (Fig. 8).

As in the case of the preparation of ethylene by the action of sulphuric acid on ethyl alcohol, it is probable that the reaction between phosphoric acid and ethyl alcohol takes place in two stages, viz. (i) the formation of an ester, and (ii) the decomposition of the latter by heat:

$$\text{(i)}\quad \begin{array}{c} H \\ | \\ H-C-H \\ | \\ H-C-OH \\ | \\ H \end{array} + H\,PO_4H_2 \rightleftarrows \begin{array}{c} H \\ | \\ H-C-H \\ | \\ H-C-O.PO_3H_2 \\ | \\ H \end{array} + H_2O$$

ethyl dihydrogen phosphate

$$\text{(ii)}\quad \begin{array}{c} H \\ | \\ H-C-H \\ | \\ H-C-O.PO_3H_2 \\ | \\ H \end{array} \rightarrow \begin{array}{c} H-C-H \\ \| \\ H-C-H \end{array} + H_3PO_4$$

In this method little or no oxidation of the ethyl alcohol takes place, but a certain amount of diethyl ether is formed as a by-product, due to a reaction between ethyl dihydrogen phosphate and ethyl alcohol comparable with that between ethyl hydrogen sulphate and ethyl alcohol. The ethylene can be freed from the ether and any alcohol carried over by suitable cooling before collecting the gas over water.

Newth's method for the preparation of ethylene is the basis of a method for the industrial manufacture of the gas. In this process, the flask containing the orthophosphoric acid is replaced by iron tubes containing pumice or other absorbent material which has been saturated with the phosphoric acid. The tubes are heated in a suitable furnace and ethyl alcohol vapour is led into the tubes at such a rate that the contents of the tubes can be kept at a known constant temperature. The evolved gas is passed through cooling towers to free it from diethyl ether and any unchanged ethyl alcohol, then through drying towers containing sulphuric acid and finally it is stored in suitable gasometers. For some industrial purposes the gas, after purification, is stored in cylinders under pressure.

(*c*) The dehydration of many alcohols can be effected by catalytic processes. On this has been built up an important process for the manufacture of ethylene from ethyl alcohol, chiefly by French chemists and particularly Sabatier and Mailhe. The catalysts which may be used are varied, but metallic oxides, phosphates and silicates are the chief. For example, thoria, kaolin and even silica itself have been employed. The ethyl alcohol vapour is passed upwards through tubes, usually heated electrically, containing the catalyst, and the gas issuing from the top after being freed from by-products, consisting of some oxidation products of the alcohol and diethyl ether, is collected in some such manner as is indicated above.

The mechanism of the reaction is probably explained by the formation of a compound between the catalyst and the alcohol, the compound subsequently undergoing decomposition producing ethylene and regenerating the catalyst.* In the particular case, using thoria, Sabatier and Mailhe suggest the following as the stages of the reaction:

$$ThO_2 + 2HOC_2H_5 \rightarrow ThO\begin{cases}OC_2H_5\\OC_2H_5\end{cases} + H_2O$$

ethyl thoriate (unstable)

$$ThO\begin{cases}OC_2H_5\\OC_2H_5\end{cases} \rightarrow ThO\begin{cases}OH\\OH\end{cases} + 2C_2H_4$$

$$ThO\begin{cases}OH\\OH\end{cases} \rightarrow ThO_2 + H_2O$$

Properties. Ethylene is a colourless gas, with a characteristic sweetish odour.† It boils at −105° and solid ethylene melts at −169°. It is sparingly soluble in, and can therefore be collected over, water. When ignited, the gas burns in air with a luminous flame, and, like other gaseous hydrocarbons, it forms an explosive mixture with air or oxygen. The products of combustion are carbon dioxide and water, the reaction expressing the complete combustion of the gas being

$$C_2H_4 + 3O_2 \rightarrow 2CO_2 + 2H_2O$$

The composition of the gas prepared in the laboratory by method (*b*) was determined as follows:

5·2 c.c. of the gas which had stood over water and was therefore saturated with aqueous vapour were placed in the eudiometer (Fig. 9). Oxygen was admitted and the total volume of the mixed gases was 29·0 c.c. After explosion,‡ the volume of the resulting gases was

* Catalytic processes for the industrial production of organic compounds are finding increasing application. A similar explanation of these processes to the above is generally accepted.

† Ethylene has definite and useful anaesthetic properties. Unfortunately, its odour or taste renders it obnoxious to patients, and the risk of explosion has caused its use for this purpose to be abandoned in spite of many favourable reports on ethylene anaesthesia. It is stated that green fruit can be ripened in an atmosphere of ethylene, the same changes taking place as occur in the plant.

‡ All measurements were taken at the same temperature and pressure. Since water is formed during the reaction, the gas examined was saturated with water vapour at the beginning of the experiment. For safety, the pressure of the reacting gases was reduced by lowering the levelling bulb as far as possible and covering the eudiometer with a wire cage prior to the explosion.

18·6 c.c. After introduction of concentrated aqueous solution of potassium hydroxide to absorb the carbon dioxide, the volume of the remaining gas (oxygen) was 8·2 c.c. From these results, it follows

a = eudiometer
b = wire gauze cage for covering eudiometer
c = cylinder containing oxygen
d = cylinder containing ethylene
e = concentrated aqueous solution of potassium hydroxide
f = induction coil
g = accumulators

Fig. 9

that the volume of carbon dioxide formed was (18·6 − 8·2 c.c.) 10·4 c.c. and that the volume of oxygen used for the reaction was (29·0 − 5·2 − 8·2 c.c.) 15·6 c.c. Assuming Avogadro's hypothesis, one molecule of the gas requires three molecules of oxygen for complete combustion and yields two molecules of carbon dioxide. Since two molecules of carbon dioxide each contain one atom of carbon, a molecule of the gas must contain two atoms of carbon. Further, two molecules of carbon dioxide require two molecules of oxygen for their formation, leaving one molecule of oxygen for the combination of the hydrogen in one molecule of the gas. Since one molecule of oxygen combines with two molecules or four atoms of hydrogen, there must be four atoms of hydrogen in one molecule of the gas, whose formula must therefore be C_2H_4, the formula already assigned to ethylene:

C_2H_4	+ $3O_2$	= $2CO_2$ + $2H_2O$
5·2 c.c.	15·6 c.c.	10·4 c.c.
1 mol.	3 mols.	2 mols.

The formula for ethylene is further verified by determination of its vapour density, which has been found to be 14 (H = 1). The molecular weight is therefore 28, corresponding with the formula C_2H_4. This illustrates the general method of determining the composition of gaseous hydrocarbons.

Ethylene differs markedly from ethane and other paraffins in its chemical activity. Ethylene was formerly known as *olefiant gas*, from the fact that it combines readily with chlorine yielding a colourless liquid (oil) which was known as Dutch liquid. This colourless liquid is now known as *ethylene chloride* (b.p. 84°). It has been shown that equal volumes of ethylene and chlorine combine directly forming one product, the equation expressing the reaction being

$$C_2H_4 + Cl_2 = \underset{\text{ethylene chloride}}{C_2H_4Cl_2}$$

This reaction under ordinary circumstances takes place with great rapidity. Adopting the constitutional formula for ethylene, the reaction may be graphically represented:

$$\begin{array}{c} H-C-H \\ \| \\ H-C-H \end{array} + \begin{array}{c} Cl \\ | \\ Cl \end{array} \rightarrow \begin{array}{c} H \\ | \\ H-C-Cl \\ | \\ H-C-Cl \\ | \\ H \end{array}$$

The addition of chlorine at the double bond (ethylenic or ethenoid linkage) is a characteristic reaction of all substances which possess this kind of carbon-carbon linkage, and the unsaturated hydrocarbon is thereby converted into a saturated compound. The ethylene chloride produced is actually symmetrical dichloroethane, one of the two isomeric dichloroethanes, $CH_2Cl.CH_2Cl$ and $CH_3.CHCl_2$, produced by the slow and substituting action of chlorine on ethane. When chlorine has added on to ethylene in equimolecular properties, the saturated compound, ethylene chloride, will react slowly with more chlorine just as paraffins, and saturated compounds generally, do, with the attendant formation of hydrogen chloride along with the chloro-substituted compound.

The reaction of the addition of chlorine to ethylene to form ethylene chloride is a particularly interesting one, and there is every reason to believe that the reaction between ethylene and bromine is quite analogous. Under ordinary circumstances, i.e. in a glass vessel and at the ordinary temperature, ethylene and bromine vapour react together almost instantaneously and the product, *ethylene bromide*, $CH_2Br.CH_2Br$ or $C_2H_4Br_2$, is a heavy colourless liquid, b.p. 131°:

$$\begin{matrix} CH_2 \\ \| \\ CH_2 \end{matrix} + \begin{matrix} Br \\ | \\ Br \end{matrix} \rightarrow \begin{matrix} CH_2Br \\ | \\ CH_2Br \end{matrix}$$

During recent years, the addition of halogens, especially bromine, to compounds containing the carbon-carbon double linkage has been closely studied. Norrish (1923) has shown that the addition of bromine to ethylene is almost completely suspended if the gaseous reactants are brought together in a glass vessel coated internally with paraffin wax. The reaction, on the other hand, is accelerated beyond the rate at which it occurs in ordinary glass vessels if the wax coating (a neutral substance) is replaced by other substances which have a definite basic or acidic character. Such surfaces, of which glass is one, are described as polar substances, and it is suggested that for any chemical reaction to take place the molecules interacting require activation. This activation in certain chemical reactions may be produced by the presence of moisture. In the present case, it appears that the reaction takes place at the surface of the containing vessel and that the reaction is accelerated or retarded according to the nature of the surface. The surface, if polar, may cause the appearance of polar charges on the reacting molecule, and the reaction taking place may be written

$$\underbrace{\begin{matrix} CH_2 \\ \| \\ CH_2 \end{matrix} + \begin{matrix} Br \\ | \\ Br \end{matrix}}_{\text{resting stage}} \rightarrow \underbrace{\begin{matrix} CH_2^+ \\ | \\ CH_2^- \end{matrix} + \begin{matrix} Br^- \\ \\ Br^+ \end{matrix}}_{\substack{\text{activated stage} \\ \text{polar forms}}} \rightarrow \underbrace{\begin{matrix} CH_2Br \\ | \\ CH_2Br \end{matrix}}_{\text{final product}}$$

Since ethylene combines with two atoms of bromine,* it can be considered as a bivalent radicle and compared with a bivalent metal such as (mercuric) mercury. The analogy becomes all the more complete since just as mercury can be displaced from its salts by more electro-positive metals (such as zinc) so ethylene bromide vapour

* The facility of the addition of halogens to ethylene is in the following order:

$$Cl_2 > Br_2 > I_2$$

when passed over heated zinc dust is converted into ethylene and zinc bromide is obtained:

$$\begin{array}{l} CH_2Br \\ | \\ CH_2Br \end{array} + Zn \rightarrow ZnBr_2 + \begin{array}{l} CH_2 \\ \| \\ CH_2 \end{array}$$

The remarkable chemical activity of ethylene due to the presence of the double linkage is shown in a large number of reactions. When ethylene and hydrogen are together passed over finely divided nickel at 150°, the nickel catalyses the reaction, which under ordinary conditions takes place excessively slowly (and in practice does not proceed), and ethane is produced:

$$\begin{array}{l} CH_2 \\ \| \\ CH_2 \end{array} + \begin{array}{c} H \\ | \\ H \end{array} \rightarrow \begin{array}{l} CH_3 \\ | \\ CH_3 \end{array}$$

The halogen acids (HCl, HBr, HI) combine directly with ethylene and ethyl halides are produced:

$$\begin{array}{l} CH_2 \\ \| \\ CH_2 \end{array} + \begin{array}{c} H \\ | \\ I \end{array} \rightarrow \begin{array}{l} CH_3 \\ | \\ CH_2I \end{array}$$

The order of facility of the combination of these acids with ethylene is

$$HI > HBr > HCl$$

When ethylene is shaken with a dilute solution of potassium permanganate rendered alkaline by means of sodium carbonate, the colour of the permanganate is discharged, a brown precipitate of a hydrated oxide of manganese is produced, and the colourless aqueous solution contains a colourless, sweet-tasting liquid, known as *ethylene glycol** (*symmetrical* dihydroxyethane), b.p. 195°:

$$\begin{array}{l} CH_2 \\ \| \\ CH_2 \end{array} + H_2O + O \rightarrow \begin{array}{c} H \\ | \\ H-C-O-H \\ | \\ H-C-O-H \\ | \\ H \end{array}$$

Just as chlorine and bromine can, as it were, saturate the double linkage in ethylene, so ozone can combine directly, forming an *ozonide*:

$$\begin{array}{l} CH_2 \\ \| \\ CH_2 \end{array} + O_3 \rightarrow \begin{array}{l} CH_2-O \\ | \\ CH_2-O \end{array}\!\!\!\!>O$$

This ozonide is unstable and is easily decomposed when warmed with

* Frequently described under the simple name 'glycol'.

dilute acid, yielding a gaseous compound, formaldehyde (p. 144) and hydrogen peroxide:

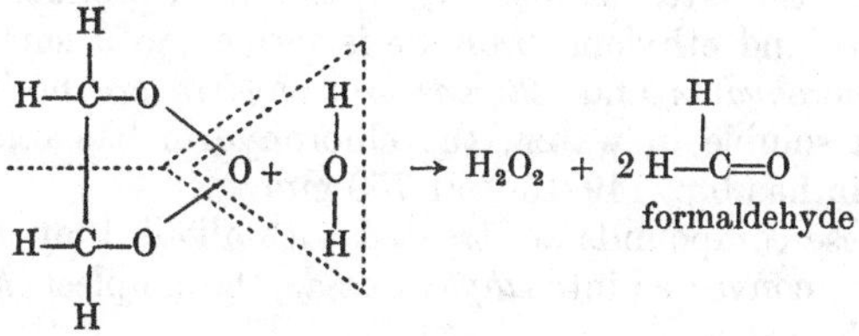

$$\begin{array}{c} H \\ | \\ H-C-O \\ | \quad\;\; \searrow \\ \cdots\cdots | \cdots\cdots \quad O \\ | \quad\;\; \nearrow \\ H-C-O \\ | \\ H \end{array} + \begin{array}{c} H \\ | \\ O \\ | \\ H \end{array} \rightarrow H_2O_2 + 2\,\begin{array}{c} H \\ | \\ H-C{=}O \end{array}$$

formaldehyde

Bromine, aqueous alkaline potassium permanganate solution and ozone are the three chief reagents for the detection of the double linkage between adjacent carbon atoms in organic compounds.

Sulphuric acid containing an excess of sulphuric anhydride (fuming sulphuric acid) rapidly absorbs ethylene, forming ethyl hydrogen sulphate:

$$\begin{array}{c} CH_2 \\ \| \\ CH_2 \end{array} + \begin{array}{c} H \\ | \\ SO_4H \end{array} \rightarrow \begin{array}{l} CH_3 \\ | \\ CH_2.O.SO_3H \end{array}$$

This reaction is frequently employed in the determination of ethylene in the presence of other gaseous organic substances which do not react with fuming sulphuric acid. Ethylene can be, and frequently is, dried by passing the gas through ordinary concentrated sulphuric acid. When ethylene is allowed to be in contact for some time with ordinary concentrated sulphuric acid, the gas is slowly absorbed owing to the above reaction taking place. Since ethyl hydrogen sulphate undergoes hydrolysis in the presence of water or, more rapidly, dilute alkalies:

$$\begin{array}{l} CH_3 \\ | \\ CH_2.O.SO_3H \end{array} + H_2O \rightarrow \begin{array}{l} CH_3 \\ | \\ CH_2.OH \end{array} + H_2SO_4$$

it has been suggested that commercial processes for the production of ethyl alcohol might be based on the above reactions, using as the source of ethylene, for example, coal gas which is known to contain ethylene in quantities which vary not only with the type of coal used but also with the conditions of production of the coal gas itself.

Hypochlorous and hypobromous acids combine directly with ethylene forming important addition compounds, known as ethylene chlorohydrin and ethylene bromohydrin respectively:

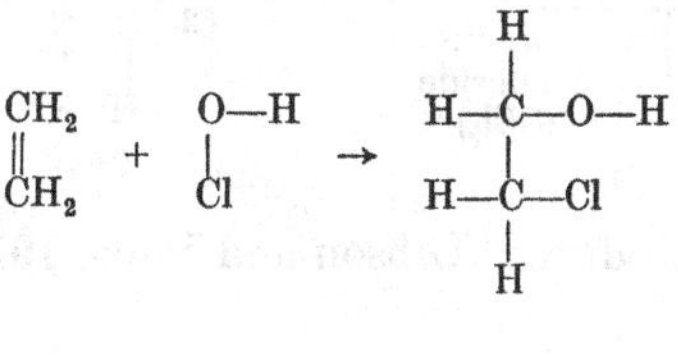

$$\begin{array}{c} CH_2 \\ \| \\ CH_2 \end{array} + \begin{array}{c} O-H \\ | \\ Cl \end{array} \rightarrow \begin{array}{c} H \\ | \\ H-C-O-H \\ | \\ H-C-Cl \\ | \\ H \end{array}$$

Read (1917) has shown that in the rapid reaction which takes place between ethylene and bromine water, preferably in the cold, whereby the colour of the latter is discharged, the main product is ethylene bromohydrin and ethylene bromide is formed to a smaller extent. *Ethylene chlorohydrin* and *ethylene bromohydrin* are both colourless liquids and soluble in water; the chlorohydrin has b.p. 132°, the bromohydrin has b.p. 149–150° at 750 mm.

When these compounds are treated with alkali, they lose halogen acids and are converted into *ethylene oxide*, the simplest '*hetero*cyclic' compound known:

```
     H                        H
     |                        |
  H—C—Cl                   H—C\
     |       + KOH  →         | >O + KCl + H2O
  H—C—OH                   H—C/
     |                        |
     H                        H
```

Ethylene oxide is an inflammable gas, b.p. 13·5° at 746 mm. It can be regarded as an 'inner'-ether (p. 30) derived from the dihydric alcohol glycol (p. 36).

When ethylene is brought into intimate contact with either of the two well-identified chlorides of sulphur, direct combination takes place, the reactions being represented diagrammatically:

```
   H H                                  H H
   | |                                  | |
 H—C=C—H                            Cl—C—C
                Cl                      | | \
          +        >S=S   →             H H  \
                Cl                            >S + S
 H—C=C—H                                H H  /
   | |                                  | | /
   H H        sulphur mono-         Cl—C—C
              chloride (S2Cl2)          | |
                                        H H
```

```
   H H                                  H H
   | |                                  | |
 H—C=C—H                            Cl—C—C
                Cl                      | | \
          +        >S     →             H H  \
                Cl                            >S
 H—C=C—H                                H H  /
   | |                                  | | /
   H H        sulphur               Cl—C—C
              dichloride                | |
              (SCl2)                    H H
```

The compound so produced (Gibson and Pope, 1918) was known as *mustard gas** during the war of 1914–1918. It is a colourless, highly toxic and vesicant liquid, b.p. 108° at 14 mm. This compound can be prepared, starting with ethylene chlorohydrin (V. Meyer, 1886), by the following series of reactions:

```
    H H                                   H H
    | |                                   | |
H—O—C—C—Cl                            H—O—C—C
    | |                                   | |  \
    H H        + Na2S → 2NaCl +           H H   \    thiodiglycol or
    H H                                   H H    S   ββ'-dihydroxy-
    | |                                   | |   /    diethyl sulphide
H—O—C—C—Cl                            H—O—C—C  /
    | |                                   | |
    H H                                   H H
```

```
           H H                          H H
           | |                          | |
Cl—H  H—O—C—C                       Cl—C—C
           | |  \                       | |  \
           H H   \                      H H   \
                  S →  2H2O +                  S
           H H   /                      H H   /
           | |  /                       | |  /
Cl—H  H—O—C—C                       Cl—C—C
           | |                          | |
           H H                          H H
```

The chemical properties of ethylene being so different from that of ethane, and the tendency of ethylene to yield substitution derivatives of ethane, indicate that whatever be the nature of the carbon-carbon linkage (the so-called double bond) in ethylene it is quite different from the carbon-carbon linkage (the so-called single bond)

* Its chemical name is ββ'-dichlorodiethyl sulphide, which is explained by analogy as follows:

```
  H H            H H            β  α            H H
  | |            | |                            | |
H—C—C          H—C—C           —C—C          Cl—C—C
  | | \          | | \              \           | | \
  H H  \         H H  \              \          H H  \
        O              S              S               S
  H H  /         H H  /              /          H H  /
  | | /          | | /              /           | | /
H—C—C          H—C—C           —C—C          Cl—C—C
  | |            | |            β' α'           | |
  H H            H H                            H H
```

diethyl ether or diethyl oxide

diethyl thioether or diethyl sulphide

carbon atoms distinguished according to their position relative to fundamental atom

ββ'-dichlorodiethyl sulphide

in ethane. The ease of production of substituted derivatives of ethane from ethylene also indicates that the single bond is actually stronger than the double bond. Although a too literal interpretation must not be applied to it, a mechanical conception of the double bond can be obtained from the 'spring' models already referred to.* The angle between any two of the four valencies of carbon in the model representing methane is easily calculated; it is 109° 28′, and in ethane we assume there is the same angle between any two adjacent valency directions since the model is without strain. When this model, represented diagrammatically by Fig. 10, is converted into that represented by Fig. 11 each of the two valency directions forming the double bond can be considered to be displaced through an angle of 54° 44′. An examination of the mechanical representation of the carbon-carbon linkage or double bond reveals considerable strain corresponding to the chemical properties of ethylene and the marked tendency of ethylene to yield substituted derivatives of ethane.

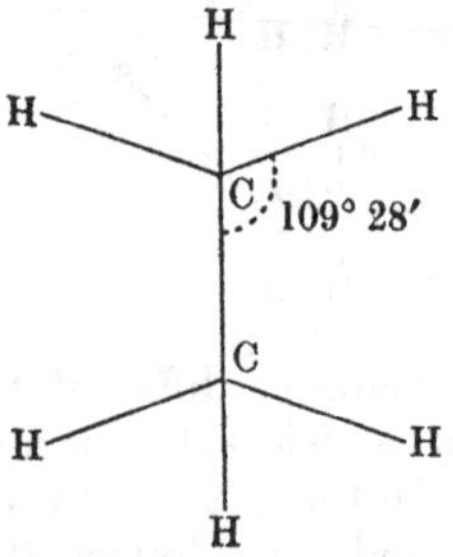

Fig. 10

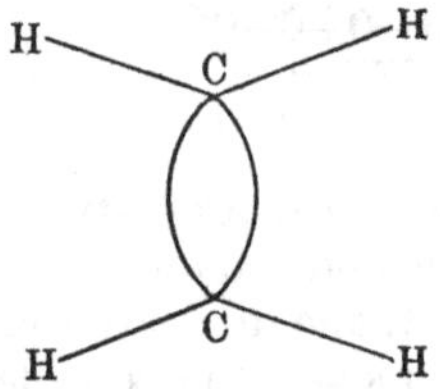

Fig. 11

If the model representing the ethylene molecule (Figs. 6 and 11) be carefully examined, it will be found to be rigid, and it is impossible to rotate the upper half of the model without rotating the lower half. In the case of the ethane model the upper half can be rotated independently of the lower half. This corresponds with chemical evidence of the possibility of free rotation about a single bond and no rotation about a double bond. It will also be found in the ethylene model that the centres of the spheres representing the atom in ethylene are all in the same plane, and that two models can be constructed to a compound, $\begin{matrix}CHA\\ \|\\ CHA\end{matrix}$, where two hydrogen atoms on the different carbon atoms in ethylene have been replaced by two monovalent atoms or mono-

* The views expressed here are based on Baeyer's Strain Theory (1885) and the tetrahedral configuration of the carbon atom (Kekulé, van't Hoff and le Bel).

valent groups or radicles. These models can be represented thus (Figs. 12 and 13):

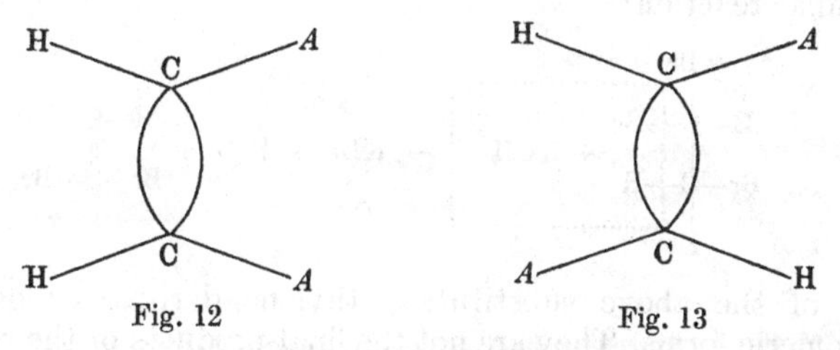

Fig. 12 Fig. 13

They are quite distinct, and one cannot be made identical with the other because of the impossibility of free rotation on account of the double linkage between the two carbon atoms. Compounds of the types $\begin{array}{l}\mathrm{CH}A\\ \|\\ \mathrm{CH}A\end{array}$ or, more generally, $\begin{array}{l}\mathrm{C}XY\\ \|\\ \mathrm{C}XY\end{array}$, $\begin{array}{l}\mathrm{C}XY\\ \|\\ \mathrm{C}XZ\end{array}$ and $\begin{array}{l}\mathrm{C}XY\\ \|\\ \mathrm{C}WZ\end{array}$ are capable of existing in two isomeric forms. When there are not more than three different atoms or groups on the carbon atom, those with the same atoms or groups on the same side are usually described as the *cis*-isomers and the others, the *trans*-isomers. Fig. 12 represents a *cis*-compound and Fig. 13 its *trans*-isomeride. This type of isomerism is usually described under the self-explanatory term *stereoisomerism* (spatial isomerism from στερεός = space). Stereoisomerism of ethylene derivatives—and it is by no means confined to them—was investigated by Wislicenus in 1887.

Halogen substituted derivatives of ethylene

Since the first action of halogens on ethylene is to produce symmetrical disubstituted derivatives of ethane, halogen substituted derivatives of ethylene cannot be produced by the direct method. They can be obtained by the partial removal of hydrogen halide from symmetrical di- and tetra-halogenated substitution products of ethane by means of alcoholic potassium hydroxide, as in the reaction (p. 27) for the preparation of ethylene from ethyl bromide. Thus, *monobromoethylene* or *vinyl bromide* ($CH_2 : CHBr$) is obtained:

$$CH_2 : CH_2 + Br_2 \rightarrow CH_2Br.CH_2Br$$

$$\begin{array}{c}\mathrm{H}\\ |\\ \mathrm{H{-}C{-}Br}\\ |\\ \mathrm{H{-}C{-}H}\\ |\\ \mathrm{Br}\end{array} + \mathrm{KOH} \rightarrow \mathrm{KBr} + \mathrm{H_2O} + \begin{array}{c}\mathrm{H{-}C{-}H}\\ \|\\ \mathrm{H{-}C{-}Br}\end{array}$$

vinyl bromide

Vinyl bromide is a colourless liquid, b.p. 16°. The corresponding vinyl chloride (monochloroethylene), $CH_2 : CHCl$, is a colourless gas.

Starting with symmetrical tetrabromoethane, *tribromoethylene*, $CHBr:CBr_2$, a colourless liquid, b.p. 163–164°, has been obtained by a similar reaction:

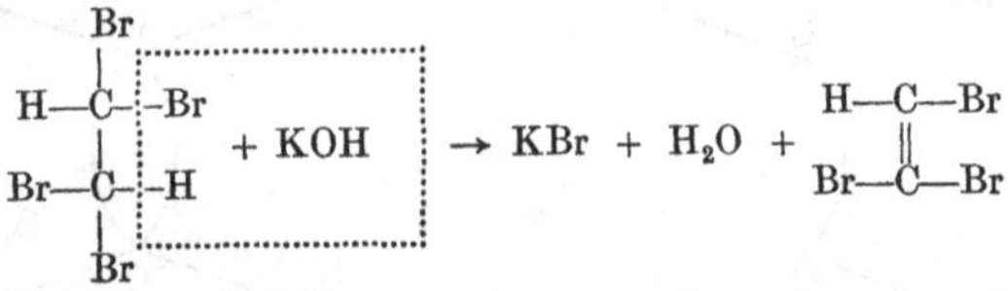

$$\begin{array}{c} Br \\ | \\ H—C—Br \\ | \\ Br—C—H \\ | \\ Br \end{array} + KOH \rightarrow KBr + H_2O + \begin{array}{c} H—C—Br \\ \| \\ Br—C—Br \end{array}$$

None of the above substituted ethylene derivatives occur in stereoisomeric forms. They are not the final products of the reaction by which they are obtained (p. 46). Apart from their containing halogen their chemical properties are similar to those of ethylene itself, because of the presence of the double bond or ethylene (or ethenoid) linkage.

Homologues of ethylene

The true homologues* of ethylene are the substituted ethylenes in

* Any compound of the general formula C_nH_{2n} and having more than two carbon atoms in the molecule may have a double bond between two neighbouring carbon atoms—true homologues of ethylene—*or* the carbon atoms may be joined up, as it were, in a ring. Thus, a hydrocarbon of the formula C_3H_6 may be monomethylethylene or propylene, having the constitution $CH_3.CH:CH_2$, or it may have a constitution which can be represented as

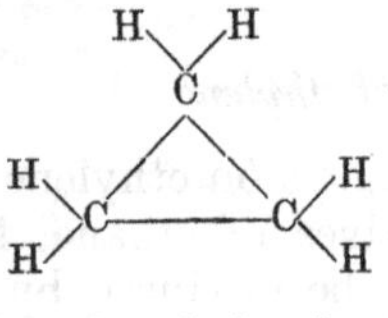

H H
C
H H
C——C
H H

or $(CH_2)_3$

Such a compound would be described under the name of trimethylene, because it contains three $=CH_2$, or methylene, groups. The possibility of the existence of such a compound follows from an extension of the theory outlined on p. 40. In order to build up a model representing the constitution of trimethylene, it is necessary to strain each of the springs (representing valency bonds) joining the carbon spheres very considerably. Actually, if the spheres representing the carbon atoms are to be at the corners of an equilateral triangle, the two adjacent springs normally at an angle of 109° 28′ to each other will have to be strained each through an angle of $\frac{109° 28' - 60°}{2}$ or 24° 44′. Similarly, a compound of the formula C_4H_8 might have a constitution represented by

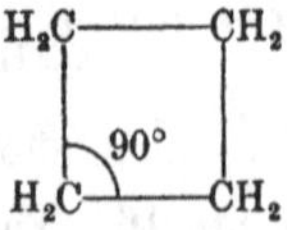

H_2C——CH_2
90°
H_2C——CH_2

or $(CH_2)_4$, tetramethylene

where the strain between the valency bonds uniting neighbouring carbon atoms

which hydrogen atoms have been replaced by alkyl (methyl, ethyl, propyl) radicals.

would be through an angle of $\frac{109° \, 28' - 90°}{2}$ or 9° 44′. For pentamethylene, C_5H_{10}, having a constitution represented by

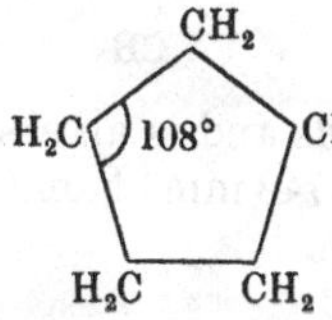

$$\begin{array}{c} CH_2 \\ H_2C \; (108°) \; CH_2 \\ H_2C \quad CH_2 \end{array} \quad \text{or} \quad (CH_2)_5$$

the strain between the valency bonds uniting neighbouring carbon atoms would be through an angle of $\frac{109° \, 28' - 108°}{2}$ or 0° 44 . For hexamethylene, C_6H_{12}, assuming the ring to be planar, the strain would be through an angle of

$$\begin{array}{c} H_2C \quad CH_2 \\ 120° \\ H_2C \qquad\quad CH_2 \\ H_2C \quad CH_2 \end{array}$$

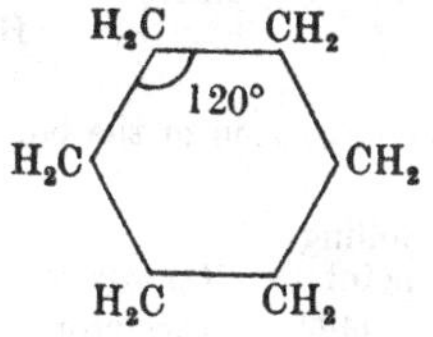

$\frac{120° - 109° \, 28'}{2}$ or 5° 16′ outwards or in the opposite direction from the strain in the previous cases.

These and higher polymethylene compounds and their numerous derivatives are known largely through the work of W. H. Perkin junr. The hydrocarbons themselves are usually known as *cyclo*paraffins. They are not unsaturated hydrocarbons in the strict sense of the term, since an unsaturated hydrocarbon adds on halogen to neighbouring carbon atoms. The ease with which these *cyclo*paraffins react with halogens (bromine and chlorine) is related to the strain mentioned above. Thus trimethylene or *cyclo*propane reacts with bromine more readily than tetramethylene or *cyclo*butane, and so on. Moreover, when *cyclo*propane reacts with bromine, as one would expect, trimethylene bromide or 1 : 3-dibromopropane is formed:

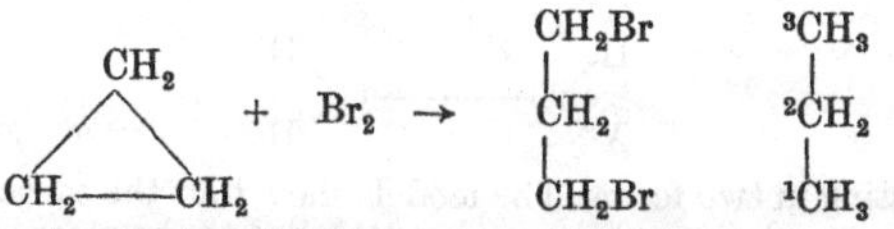

$$\begin{array}{c} CH_2 \\ CH_2 - CH_2 \end{array} + Br_2 \rightarrow \begin{array}{c} CH_2Br \\ | \\ CH_2 \\ | \\ CH_2Br \end{array} \qquad \begin{array}{c} {}^3CH_3 \\ | \\ {}^2CH_2 \\ | \\ {}^1CH_3 \end{array}$$

the carbon atoms in propane being numbered as shown. Similarly, *cyclo*butane forms tetramethylene bromide or 1 : 4-dibromobutane:

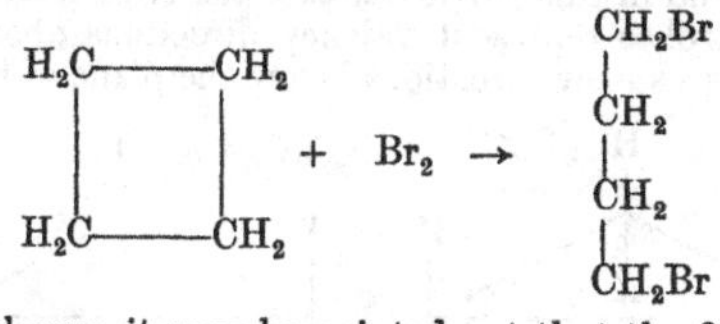

$$\begin{array}{c} H_2C - CH_2 \\ | \qquad | \\ H_2C - CH_2 \end{array} + Br_2 \rightarrow \begin{array}{c} CH_2Br \\ | \\ CH_2 \\ | \\ CH_2 \\ | \\ CH_2Br \end{array}$$

Conforming to this theory, it may be pointed out that the five- and six-membered cyclic compounds are generally the most stable.

Trimethylene, *cyclo*propane, was first obtained in 1882 by the action of sodium

Monomethyl ethylene, generally known as *propylene*, $CH_3.CH:CH_2$, is derived from propyl alcohol, $CH_3.CH_2.CH_2OH$, the next homologue to ethyl alcohol:

$$\begin{array}{l}CH_3\\ |\\ CH_2\\ |\\ CH_2OH\end{array} \quad - \quad H_2O \quad \rightarrow \quad \begin{array}{l}CH_3\\ |\\ CH\\ \|\\ CH_2\end{array}$$

It is obtained by analogous methods and is an easily liquefiable gas; liquid propylene has b.p. −48·2° at 749 mm. Homologues higher than propylene, C_3H_6, exhibit isomerism.

or zinc dust on trimethylene bromide (1 : 3-dibromopropane). (Compare the action of zinc dust on ethylene bromide whereby ethylene is produced.)

$$\begin{array}{l}CH_2Br\\ |\\ CH_2\\ |\\ CH_2Br\end{array} \quad + \quad Zn \quad \rightarrow \quad ZnBr_2 \quad + \quad \begin{array}{c}CH_2\\ \diagup \quad \diagdown\\ H_2C\text{———}CH_2\end{array}$$

The following table enables a comparison of the boiling points of the normal and *cyclo-* paraffins to be made.

Normal paraffin	Formula	Boiling point	*Cyclo*paraffin	Formula	Boiling point
Propane	C_3H_8	−44·5°	*cyclo*propane	C_3H_6	*c.* −35°
n-Butane	C_4H_{10}	+1°	*cyclo*butane	C_4H_8	11—12°
n-Pentane	C_5H_{12}	36·3°	*cyclo*pentane	C_5H_{10}	49°
n-Hexane	C_6H_{14}	69°	*cyclo*hexane	C_6H_{12}	81°
n-Heptane	C_7H_{16}	98°	*cyclo*heptane	C_7H_{14}	117·5°
n-Octane	C_8H_{18}	125·8°	*cyclo*octane	C_8H_{16}	145–148°
n-Nonane	C_9H_{20}	150°	*cyclo*nonane	C_9H_{18}	170–172°

Stereoisomerism is possible among *cyclo*paraffin derivatives, where at least two hydrogen atoms on different carbon atoms have been replaced by equivalent atoms or groups of atoms. For example, any derivative of *cyclo*propane of the type

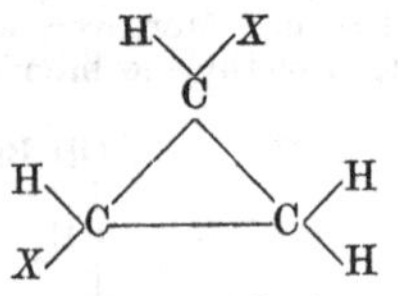

is capable of existing in two forms. The models show that the carbon atoms forming the ring are all in one plane, and the groups attached to them are in two places, one above and the other below the plane of the ring. Consequently *cis-* and *trans-*isomerism is possible, and in this particular case the two forms may be represented thus, where upright strokes represent valency directions above the plane of the ring and dotted strokes valency directions below the plane of the ring:

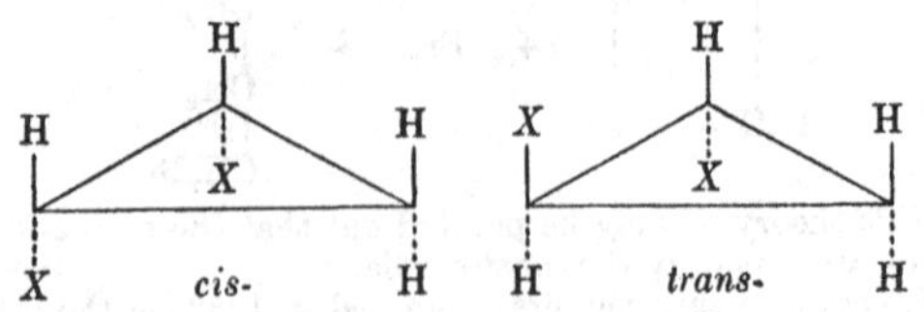

Of the hydrocarbons of the formula C_4H_8, the butylenes, ethylethylene, $C_2H_5.CH:CH_2$, has b.p. $-5°$. Isomeric with this is symmetrical dimethylethylene, $CH_3.CH:CH.CH_3$. Symmetrical dimethylethylene is capable of existing in stereoisomeric forms, the constitutions of which are represented:

$$\begin{array}{c} CH_3\text{—}C\text{—}H \\ \| \\ CH_3\text{—}C\text{—}H \end{array} \qquad \begin{array}{c} CH_3\text{—}C\text{—}H \\ \| \\ H\text{—}C\text{—}CH_3 \end{array}$$

These have been isolated and the former, or *cis*-modification, has b.p. $+1°$, and the latter, or *trans*-modification, has b.p. $+2{\cdot}5°$. The remaining isomeric butylene, unsymmetrical-dimethylethylene, $(CH_3)_2C:CH_2$, b.p. $-6°$, is incapable of existing in stereoisomeric forms. All these and higher homologues of ethylene possess the characteristic chemical properties of ethylene on account of the presence of the double bond in each of them.

Isomerism among the homologues of ethylene is then very pronounced and increases considerably with the number of carbon atoms in the compounds. While ethylene hydrocarbons do not occur in nature in anything like the same abundance as the saturated hydrocarbons, being absent from most petroleums, they are more frequently found in plant life; the terpene (turpentine) family consists principally of complex ethylenic hydrocarbons, among which isomerism is specially pronounced.

THE ACETYLENE SERIES OF HYDROCARBONS, C_nH_{2n-2}

This series of hydrocarbons can be considered to be derived from ethane (not methane) and its homologues by taking away in each case four atoms of hydrogen, two from each of two neighbouring carbon atoms, and the ethylene hydrocarbons are thus intermediate between the paraffins and the acetylenes, thus:

$$\underset{\text{ethane}}{\begin{array}{c} CH_3 \\ | \\ CH_3 \end{array}} - H_2 \rightarrow \underset{\text{ethylene}}{\begin{array}{c} CH_2 \\ \| \\ CH_2 \end{array} ; \begin{array}{c} CH_2 \\ \| \\ CH_2 \end{array}} - H_2 \rightarrow \underset{\text{acetylene}}{\begin{array}{c} CH \\ ||| \\ CH \end{array}}$$

In acetylene and its homologues there is therefore present a triple linkage or triple bond (p. 27) between two adjacent carbon atoms. Consequently the unsaturation of the acetylenes is twice as great as that of the corresponding olefines.

Acetylene, C_2H_2

Acetylene does not occur in nature. Its presence to a small extent can be detected in coal gas by a test described below.

Pure acetylene can be produced indirectly from ethane by making use of the following series of reactions:

$$C_2H_6 + Br_2 \rightarrow C_2H_5Br + HBr$$

ethyl bromide*

$$\begin{matrix} CH_3 \\ | \\ CH_2Br \end{matrix} + KOH \rightarrow \begin{matrix} CH_2 \\ \| \\ CH_2 \end{matrix} + KBr + H_2O \text{ (p. 27)}$$

$$\begin{matrix} CH_2 \\ \| \\ CH_2 \end{matrix} + \begin{matrix} Br \\ | \\ Br \end{matrix} \rightarrow \begin{matrix} CH_2Br \\ | \\ CH_2Br \end{matrix}$$

ethylene bromide

$$\begin{matrix} CH_2Br \\ | \\ CH_2Br \end{matrix} + KOH \rightarrow \begin{matrix} CH_2 \\ \| \\ CHBr \end{matrix} + KBr + H_2O \text{ (p. 40)}$$

monobromoethylene

$$\begin{matrix} H\text{—}C\text{—}H \\ \| \\ H\text{—}C\text{—}Br \end{matrix} + KOH \rightarrow \begin{matrix} CH \\ \||| \\ CH \end{matrix} + KBr + H_2O$$

Although monobromoethylene (or vinyl bromide) is the intermediate product in the reaction between ethylene bromide and alcoholic potassium hydroxide, acetylene is obtained actually starting with ethylene bromide. The reaction is carried out under similar conditions to those for the preparation of ethylene from ethyl bromide, and the pure gas can be collected over water or in any other suitable way. The method for producing acetylene by the action of water on calcium carbide† was discovered by Wöhler in 1862, but the industrial application was only made possible after Moissan's discovery (1892) of the commercial method of producing calcium carbide.

Calcium carbide is produced in very considerable quantities by heating a mixture of coke and lime to a temperature of 2000° in a closed electric furnace. The reaction taking place is

$$CaO + 3C \rightarrow CaC_2 + 2CO$$

The calcium carbide is a greyish black stony mass. For the preparation in the laboratory the apparatus as indicated in Fig. 1 (p. 14) is used, a steady stream of the gas being obtained as the water is gradually dropped from the tap funnel on to the carbide. The gas

* Ethyl bromide is always, however, prepared from ethyl alcohol.

† Calcium carbide is more systematically described as 'calcium acetylenide' or 'calcium acetylide'. It is a derivative of acetylene and its constitutional formula is, probably, $\begin{matrix} C \\ \||| \\ C \end{matrix}\!\!>\!Ca$. On the other hand aluminium carbide (p. 14) is really a derivative of methane. In view of this, aluminium carbide might be more systematically named aluminium methanide.

is collected over water after the air from the apparatus has first been displaced. The reaction taking place is

$$C_2Ca + \begin{matrix} HO\vdots H \\ HO\vdots H \end{matrix} \rightarrow C_2H_2 + Ca(OH)_2$$

Prepared in this way, the acetylene is far from pure. It frequently contains phosphine and arsine due to the presence of phosphorus and arsenic compounds in the coke.

Berthelot's discovery (1862) that acetylene is formed when the electric arc is struck between carbon terminals in an atmosphere of hydrogen was verified and extended by Bone and Jerdan (1897), who found that methane and acetylene are formed under these conditions and that, under the conditions of the experiment, an equilibrium exists between the hydrogen, methane and acetylene. The synthesis of hydrocarbons from carbon and hydrogen is of fundamental importance, since the hydrocarbons are the starting point for the synthesis of most organic compounds, however complicated they may be.

Acetylene is a colourless gas which is collected over water. When pure it possesses an ethereal odour which is not unpleasant and quite different from the odour of the gas produced by the action of water on calcium carbide. The solubility of acetylene in ethyl alcohol is greater than that of ethylene. Like the latter gas, acetylene mixed with oxygen and some material to mask the smell has been used as a general anaesthetic to a considerable extent. Acetylene was at one time believed to be poisonous, but it is the unpleasant smell of the commercial gas and the risk of explosion which prevent its increased use as an anaesthetic.

When acetylene is ignited in an ordinary gas jar it burns with a sooty flame and carbon is deposited. The gas forms an explosive mixture with air or oxygen and the equation expressing the complete combination of the gas is

$$2C_2H_2 + 5O_2 = 4CO_2 + 2H_2O$$

Using this reaction and a knowledge of the vapour density, the composition of acetylene may be determined.

Acetylene can be liquefied at 0° under a pressure of 26 atmospheres. Liquid acetylene has explosive properties, and even the gas when under a small pressure can be exploded by a detonator. Acetylene is somewhat readily soluble in acetone and its solubility increases rapidly with increasing pressure. It has been found that when acetylene is stored under pressure in the presence of acetone, the gas loses its sensitiveness to shock, and the steel cylinders containing the gas under these conditions can be handled with safety.

When acetylene is burned in such a manner as to admit the mixture of the gas with the requisite quantity of air, i.e. using special burners, a hot dazzling flame is produced, and acetylene is therefore not only used as an illuminant, but also for producing high temperatures. The oxyacetylene blowpipe flame is used in welding and for the cutting of steel plates.

The unsaturated nature of acetylene is much more marked than that of ethylene. It combines with hydrogen in the presence of finely divided platinum (platinum black) at ordinary temperatures producing finally ethane, ethylene being the intermediate product:

$$\begin{array}{c}CH\\ |||\\ CH\end{array} + \begin{array}{c}H\\ |\\ H\end{array} \rightarrow \begin{array}{c}CH_2\\ ||\\ CH_2\end{array};\quad \begin{array}{c}CH_2\\ ||\\ CH_2\end{array} + \begin{array}{c}H\\ |\\ H\end{array} \rightarrow \begin{array}{c}CH_3\\ |\\ CH_3\end{array}$$

Acetylene combines directly with bromine in two stages:

$$\begin{array}{c}CH\\ |||\\ CH\end{array} + \begin{array}{c}Br\\ |\\ Br\end{array} \rightarrow \begin{array}{c}CHBr\\ ||\\ CHBr\end{array}$$

acetylene dibromide*
(*sym*-dibromoethylene)

$$\begin{array}{c}CHBr\\ ||\\ CHBr\end{array} + \begin{array}{c}Br\\ |\\ Br\end{array} \rightarrow \begin{array}{c}CHBr_2\\ |\\ CHBr_2\end{array}$$

acetylene tetrabromide
(*sym*-tetrabromoethane)

When acetylene is brought into contact with chlorine under ordinary conditions, the mixture bursts into flame and finely divided carbon is deposited; the reaction is expressed:

$$C_2H_2 + Cl_2 = 2HCl + 2C$$

The two-stage addition of chlorine acetylene can be effected by the method discovered by Berthelot in 1869. Acetylene is absorbed by antimony pentachloride, forming the compound $C_2H_2.SbCl_5$. This compound on distillation gives antimony trichloride and acetylene dichloride (*sym*-dichloroethylene). When an excess of antimony

* The colourless liquid produced is a mixture of the *cis*- and *trans*- isomers:

$$\begin{array}{c}Br—C—H\\ ||\\ Br—C—H\end{array} \quad \text{and} \quad \begin{array}{c}Br—C—H\\ ||\\ H—C—Br\end{array}$$

Acetylene tetrabromide is a colourless highly refractive liquid with a curious odour resembling that of a mixture of camphor and chloroform. It has b.p. 114° at 12 mm. and its high specific gravity, 2·971 at 18°, renders it useful in determinations of densities by Retger's method.

pentachloride is used, acetylene tetrachloride (*sym*-tetrachloroethane) is produced:

$$C_2H_2.SbCl_5 \rightarrow C_2H_2Cl_2 + SbCl_3$$
$$C_2H_2Cl_2 + SbCl_5 \rightarrow C_2H_2Cl_4 + SbCl_3$$

or

$$\begin{array}{c} CH \\ ||| \\ CH \end{array} + \begin{array}{c} Cl \\ | \\ Cl \end{array} \rightarrow \begin{array}{c} CHCl^* \\ || \\ CHCl \end{array} ; \quad \begin{array}{c} CHCl \\ || \\ CHCl \end{array} + \begin{array}{c} Cl \\ | \\ Cl \end{array} \rightarrow \begin{array}{c} CHCl_2\dagger \\ | \\ CHCl_2 \end{array}$$

acetylene dichloride (*sym*-dichloroethylene) — acetylene tetrachloride (*sym*-tetrachloroethane)

Acetylene also combines with hydrogen halides in two stages. For example, with hydrogen chloride, the first product of the reaction is monochloroethylene (vinyl chloride), a colourless gas, and the second product is *unsym*-dichloroethane (ethylidene chloride, p. 19):

$$\begin{array}{c} CH \\ ||| \\ CH \end{array} + \begin{array}{c} H \\ | \\ Cl \end{array} \rightarrow \begin{array}{c} CH_2 \\ || \\ CHCl \end{array}$$

vinyl chloride

$$\begin{array}{c} CH_2 \\ || \\ CHCl \end{array} + \begin{array}{c} H \\ | \\ Cl \end{array} \rightarrow \begin{array}{c} CH_3 \\ | \\ CHCl_2 \end{array}$$

ethylidene chloride

isomeric with ethylene chloride, $CH_2Cl.CH_2Cl$. The corresponding products, using hydrogen bromide, are monobromoethylene (vinyl bromide, $CH_2:CHBr$) and *unsym*-dibromoethane (ethylidene bromide, $CH_3.CHBr_2$).

Acetylene is absorbed by sulphuric acid, and when the resulting solution is diluted and boiled, acetaldehyde (p. 154) distils over. The reactions which take place may be represented:

$$\begin{array}{c} CH \\ ||| \\ CH \end{array} + \begin{array}{c} H \\ | \\ O.SO_3H \end{array} + \begin{array}{c} H \\ | \\ O.SO_3H \end{array} \rightarrow \begin{array}{c} CH_3 \\ | \\ CH(O.SO_3H)_2 \end{array}$$

$$\begin{array}{c} CH_3 \\ | \\ CH(O.SO_3H)_2 \end{array} + H_2O \rightarrow \begin{array}{l} CH_3 \\ |\;\diagup H \\ C \\ \;\;\diagdown\!\!\diagdown O \end{array} + 2H_2SO_4$$

acetaldehyde

* The acetylene dichloride (or *sym*-dichloroethylene) so produced is a mixture of two isomers:

$$\begin{array}{c} Cl{-}C{-}H \\ || \\ Cl{-}C{-}H \end{array} \quad \text{and} \quad \begin{array}{c} Cl{-}C{-}H \\ || \\ H{-}C{-}Cl \end{array}$$

which boil at 48° and 60° (respectively). It is an important commercial substance, being the best known solvent for rubber and most useful as a solvent for iodine, used for sterilising the skin.

† Acetylene tetrachloride (or *sym*-tetrachloroethane) is a colourless liquid, b.p. 147°. It is an important substance and known commercially under the trade name of 'Westron'. It dissolves oils, fats, resins, waxes, sulphur, phosphorus and chlorine. It is used as a solvent for cellulose acetates and is extensively employed as an insecticide in agriculture (p. 20).

Trichloroethylene, $CCl_2:CHCl$, is known under the trade name of 'Westrosol' and is a colourless liquid, b.p. 88°, used for similar purposes. It is obtained by the action of milk of lime, $Ca(OH)_2$, on tetrachloroethane.

The relative amount of acetylene converted into the important substance acetaldehyde is small, but the reaction is the basis of an important industrial process whereby the percentage conversion of acetylene is very greatly increased. In this process the purified acetylene is passed into a rapidly agitated solution of 20 per cent. sulphuric acid containing 1 per cent. mercuric oxide and 5 per cent. ferric sulphate. The actual mixture of catalysts may vary but mercuric sulphate is always present. To avoid the slowing down of the reaction due to reduction of the mercuric sulphate to metallic mercury, manganese dioxide may be added in small quantities from time to time.

The production of chlorinated solvents and of acetaldehyde are but two examples of the industrial applications of acetylene in the manufacture of organic compounds. These industrial applications are very varied and are continually increasing. To mention only a few compounds which may be manufactured, starting with acetylene we have, apart from the above, acetic acid (p. 198), acetone (p. 179), ethylene, ethyl alcohol (p. 96), hydrocyanic acid, methyl cyanide (p. 199), thiophene (p. 57) and monochloroacetic acid (p. 205).

Probably the most characteristic property of acetylene is its ability to form metallic derivatives in which the hydrogen atoms of the acetylene are replaced by metals. This implies that acetylene possesses a certain acid character. Calcium carbide has already been mentioned as a calcium derivative of acetylene. When acetylene is passed over heated sodium or potassium, hydrogen is liberated and metallic substitution products are formed, such as C_2Na_2 and C_2HNa. Potassium acetylenide (acetylide) was first obtained by Davy (1836) as the product of the action of heat on potassium hydrogen tartrate (p. 397). Davy showed that the product heated with water yielded a gas which he termed 'bicarburet of hydrogen' and which burnt with a brilliant flame. This was the discovery of acetylene.

If acetylene is shaken up with, or passed into, the blue ammoniacal solution of cuprous chloride, a red precipitate is formed. This is *copper* (cuprous) *acetylenide* (acetylide), C_2Cu_2, and its formation constitutes a very delicate test for acetylene. The presence of the small quantity of acetylene in coal gas can be detected by this means, and to make the red precipitate more easily visible, the blue solution may be decolorised by the addition of a little hydroxylamine before shaking up with a large volume of coal gas.* Acetylene may also

* Another very sensitive reagent for acetylene is made by saturating an aqueous solution of copper sulphate with sodium chloride and adding sodium bisulphite until all the green colour disappears. A few drops of ammonia increase the sensitiveness, but are not necessary to secure the precipitation of the C_2Cu_2.

be separated from other gases by means of this copper compound, which may be handled with safety whilst moist but when dry explodes on being struck or heated gently. When copper acetylenide is warmed with a dilute solution of hydrochloric acid, acetylene and monochloroethylene (vinyl chloride) are obtained. If a dilute solution of potassium cyanide be used, pure acetylene may be prepared by this method.

Silver acetylenide (C_2Ag_2) is formed as a colourless precipitate when acetylene is passed into the colourless ammoniacal silver nitrate solution. It is more readily explosive than the copper compound. Mercurous acetylenide is also known; it has the composition C_2Hg_2.

Apart from the formation of metallic derivatives, it has been shown that the characteristic properties of acetylene are its highly unsaturated nature and its tendency to explode. The unsaturated nature of acetylene is at once correlated with the presence of the triple linkage between the two carbon atoms. Its explosive nature is allied with the fact that acetylene is a highly endothermic compound. Ethylene, a less unsaturated compound, is also endothermic, but the formation of ethylene from carbon and hydrogen is not attended by as great an absorption of heat as that of acetylene from carbon and hydrogen. Ethane, a saturated and highly exothermic compound, and ethylene are not explosive, and consequently the explosiveness of acetylene and its highly unsaturated character must both be correlated with its being strongly exothermic and with the presence of the triple linkage between the two carbon atoms.

If acetylene be placed in a tube, the open end of which dips under mercury, and the tube be heated carefully, after some time fumes appear. On allowing the tube and its contents to cool, minute drops of liquid condense and the mercury will be found to have risen in the tube higher than its previous level. The same liquid is formed when acetylene is passed through a porcelain or glass tube heated to dull redness. This colourless liquid formed from acetylene under these conditions contains a compound having the same percentage composition as acetylene (carbon 92·3 per cent.; hydrogen 7·7 per cent.): but its vapour density being 39 ($H=1$), its molecular weight must be 78. The molecular formula of this liquid is therefore C_6H_6, that is, three times the formula of acetylene, whose molecular weight is 26. The liquid is a highly important hydrocarbon known as *benzene*, which can thus be regarded as a polymer of acetylene produced by a change which may be represented:

$$3C_2H_2 \rightarrow C_6H_6$$

It will be shown later (p. 56) that benzene differs very markedly from all the types of hydrocarbons* which have been mentioned up to the present. These hydrocarbons have their carbon atoms arranged in open chains (straight or branched) with single, double or triple linkages between carbon atoms. Benzene, on the other hand, has its carbon atoms in a closed ring, and it will be shown that its constitution is best represented by a formula such as

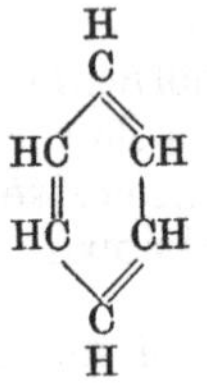

H
C
HC CH
HC CH
C
H

and its formation by polymerisation of acetylene is illustrated thus:

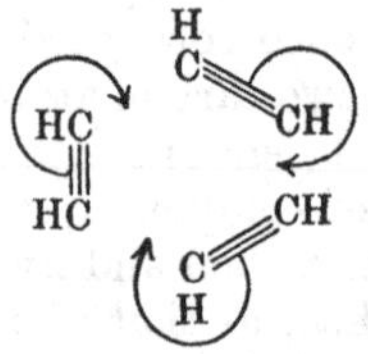

H
C
HC CH
HC CH
C
H

The polymerisation of acetylene to benzene was discovered by Berthelot in 1866.

Other Unsaturated Open Chain Hydrocarbons

Of the hydrocarbons having the general formula C_nH_{2n-2}, those having more than two atoms of carbon in the molecule fall into two classes:

(i) Those which, like acetylene, contain two carbon atoms connected together by a triple linkage. These are the true homologues of acetylene and are derived from acetylene by the replacement of one or both hydrogen atoms by alkyl (methyl, ethyl, propyl, etc.) groups.

* Organic compounds are conveniently classified as follows:

(i) *Open chain* compounds, frequently known as *aliphatic* compounds, the latter name being due to the fact that the fats (p. 222) belong to this class. Aliphatic compounds may also be described as *acyclic*.

(ii) *Closed chain* or *Cyclic compounds*. Of these there are two classes, viz. the polymethylenes (p. 42) and their derivatives, and benzene, its homologues and their derivatives. Benzene, its homologues and their derivatives are usually referred to as *aromatic* compounds (sometimes *benzenoid* compounds). The polymethylene compounds are frequently described generally as *alicyclic* compounds, having some resemblances to the aliphatic or acyclic compounds. When the ring system in a cyclic compound consists of atoms all of one kind, it is described as *homocyclic*, otherwise as *heterocyclic*. Compounds may also contain more than one ring system and are then described as *polycyclic*.

The production of benzene from acetylene is an important example of the conversion of an aliphatic into an aromatic compound.

Methylacetylene (or allylene), $CH_3.C \vdots CH$, (C_3H_4), is the first homologue and a gas very similar to acetylene in properties. It is prepared by heating 1 : 2-dibromopropane (prepared from propylene and bromine) with alcoholic potassium hydroxide:

$$\begin{array}{ccccccc} & & CH_3 & & & & \\ & & | & & & & C.CH_3 \\ KOH & + & H-C-Br & + \; KOH & \rightarrow 2KBr + H_2O + & & \| \! | \\ & & Br-C-H & & & & C.H \\ & & | & & & & \\ & & H & & & & \end{array}$$

It yields copper and silver derivatives which have the formulae $\begin{array}{c} C.CH_3 \\ \| \! | \\ C.Cu \end{array}$ and $\begin{array}{c} C.CH_3 \\ \| \! | \\ C.Ag \end{array}$ respectively, there being only one atom of hydrogen in the molecule of methylacetylene capable of being replaced by metals.*

The next homologue would have the formula C_4H_6, and this might be dimethylacetylene, $CH_3.C \vdots C.CH_3$, or ethylacetylene, $C_2H_5.C \vdots CH$. These two hydrocarbons are known, and while they show the same unsaturation properties as acetylene, only the latter yields metallic derivatives, this property being confined to those hydrocarbons which contain the grouping $—C \vdots CH$.

(ii) Those hydrocarbons which contain two double linkages instead of one triple linkage. It is obvious that the hydrocarbons of this class have the same unsaturation capacity as the isomeric hydrocarbons of class (i). The hydrocarbons of this class are described as *diolefines* and the first member of this series is

Allene, $CH_2 : C : CH_2$, (C_3H_4), a colourless gas which while not important itself is the parent substance of derivatives of theoretical interest (p. 348). The true homologues of allene would be those derived from it by replacing the hydrogen atoms by alkyl groups. The most important diolefines, however, are the hydrocarbons isomeric to the homologues of allene which contain the so-called conjugate double linkages. The simplest of these is

* Methylacetylene is interesting owing to the fact that like acetylene and under similar conditions it undergoes polymerisation. Symmetrical trimethylbenzene (mesitylene) is produced:

$$\begin{array}{ccc} & CH_3 & \\ & C & \\ HC & & CH \\ \| \! | & & \\ CH_3C & & CCH_3 \\ & C & \\ & H & \end{array} \qquad \begin{array}{ccc} & CH_3 & \\ & C & \\ HC & & CH \\ \| & & | \\ CH_3.C & & C.CH_3 \\ & C & \\ & H & \end{array} \quad \text{or} \quad C_6H_3(CH_3)_3$$

mesitylene
b.p. 164·5°

Butadiene, $CH_2:CH.CH:CH_2$, (C_4H_6), usually known as butadiene (1 : 3)* to distinguish it from the isomeric compound $CH_3.HC:C:CH_2$, which is monomethylallene or butadiene (1 : 2).

A homologue of butadiene (1 : 3) is the monomethyl compound $CH_2:C(CH_3).CH:CH_2$, 2-methylbutadiene (1 : 3), usually known as *isoprene*. It has the molecular formula C_5H_8, half the molecular formula $C_{10}H_{16}$ of the important series of hydrocarbons known as terpenes. Isoprene is frequently described as a sesquiterpene, and is a product of the destructive distillation of caoutchouc (rubber). Synthetic rubber has been produced by the polymerisation of both butadiene (1 : 3) and of isoprene, which takes place under a variety of conditions, particularly in the presence of sodium.

The diolefines do not yield metallic derivatives, but have the same unsaturation capacity as the true homologues of acetylene, with which they are isomeric.

It will be realised that a very large number of unsaturated open chain hydrocarbons are capable of existence, but only two need be referred to here. *Diallyl*, C_6H_{10}, is a diolefine possessing the constitution $CH_2:CH.CH_2.CH_2.CH:CH_2$† and therefore belongs to the C_nH_{2n-2} series. It is a liquid, b.p. 59°, produced by the action of sodium on allyl iodide:

$$CH_2:CH.CH_2I + 2Na + ICH_2.CH:CH_2 \rightarrow 2NaI + CH_2:CH.CH_2.CH_2.CH:CH_2$$

Diallyl combines directly with two molecules of bromine forming *diallyl tetrabromide* or 1:2:5:6-*tetrabromohexane* which occurs in two isomeric forms (p. 396). When either or both forms of diallyl tetrabromide are heated with alcoholic potassium hydroxide, the following reaction takes place:

$$CH_2Br.CHBr.CH_2.CH_2.CHBr.CH_2Br + 4KOH \rightarrow 4KBr + 4H_2O + \underset{\text{dipropargyl}}{CH \vdots C.CH_2.CH_2.C \vdots CH}$$

with the formation of dipropargyl, C_6H_6, having the constitution shown. *Dipropargyl*‡ is a colourless liquid, m.p. −6°, b.p. 86–87°, which is a di-acetylene hydrocarbon. It contains two $CH \vdots C$— groups and therefore apart from combining directly four molecules of bromine or chlorine, it yields copper and silver derivatives analogous to those of acetylene. Another point of interest connected

* The numbers written thus indicate the position of the double bonds. Thus, a compound containing the carbon atoms linked $\overset{4}{C}—\overset{3}{C}=\overset{2}{C}=\overset{1}{C}$ is described as the 1 : 2 compound, while a compound containing the grouping $\overset{4}{C}=\overset{3}{C}—\overset{2}{C}=\overset{1}{C}$ is the 1 : 3.

† Its systematic name is hexadiene (1 : 5).

‡ 'Propargyl' is the name given to the radical $CH \vdots C.CH_2$—.

with dipropargyl is its isomerism with benzene, C_6H_6, from which it differs most markedly in chemical and physical properties. Dipropargyl belongs to a series of open chain hydrocarbons of the general formula C_nH_{2n-6}, where n is at least 4. Benzene belongs to a series of closed chain hydrocarbons of the general formula C_nH_{2n-6}, where n is at least 6.

CHAPTER II

THE HYDROCARBONS (*Continued*)

B. Benzene and other Aromatic Hydrocarbons

Benzene, to which reference has already been made (p. 52), was discovered, its molecular formula determined and many of its properties investigated by Faraday in 1825. The inflammable gas produced by the cracking of whale or cod oil was stored under pressure of about 30 atmospheres in portable vessels and the gaseous contents formerly used for illuminating purposes. During the process of compression a liquid was deposited, each 1000 cubic feet of gas yielding nearly a gallon of condensed liquid. Faraday submitted this liquid to careful examination and separated from it a hydrocarbon, benzene, to which he assigned, after careful experiments, the molecular composition C_6H_6. Faraday also determined the melting point of this hydrocarbon as 5·5°, the most recent determinations indicating a melting point of 5·44°. Faraday found that chlorine is without action on benzene in the absence of sunlight, but that when a mixture of chlorine and benzene is exposed to sunlight vigorous action occurs with evolution of hydrogen chloride, and he succeeded in isolating at least one of the chlorination products of benzene. The isolation of benzene by Faraday is one of the fundamentally important discoveries in chemistry of which Berzelius wrote: 'One of the most important chemical investigations which has enriched chemistry during 1825 is without doubt that of Faraday on the oily compounds of carbon and hydrogen obtained by compressing the gases obtained by the decomposition of fatty oils.' Actually, Faraday's discovery of benzene laid the foundation of more than one-half of modern organic chemistry and one of the most important branches of chemical industry.

Although benzene occurs in many natural petroleums, the chief source of benzene and of its homologues as well as of other aromatic hydrocarbons is in coal-tar, in which it was discovered by Hofmann in 1845. When coal is destructively distilled for the manufacture of coal gas, the products at ordinary temperatures are partly gaseous (coal gas), partly liquid (ammoniacal liquor and coal-tar) and partly solid (coke). The gaseous products are separated from the liquid products by cooling, while the coke is left behind in the retorts. The ammoniacal liquor is to a considerable extent separated from the liquid products, which form a separate and lower layer. The coal-tar, which was formerly regarded as an obnoxious by-product in the manufacture of coal gas, is a highly important industrial material.

When the coal-tar is submitted to fractional distillation, a variety of important substances is obtained. That portion of the distillate boiling at about 75° to 100° contains chiefly hydrocarbons, of which benzene and its homologues form the major portion. The higher boiling fractions of the coal-tar distillation will be referred to as the substances contained in them are discussed. The hydrocarbon fraction referred to is washed with dilute acid and then with dilute alkali (which do not affect the hydrocarbons and merely remove basic and acid substances) and finally with water. After drying, the washed hydrocarbon fraction is submitted to more careful fractional distillation. The fraction boiling at from 79° to 82° contains chiefly benzene, from which benzene boiling at 80–81° can be obtained by continued fractional distillation.

This benzene, however carefully distilled, is not quite pure. It contains traces of an organic sulphur compound known as thiophen, C_4H_4S.* Thiophen is much more easily attacked than benzene by cold concentrated sulphuric acid and oxidising agents such as potassium or sodium dichromate. If coal-tar benzene be shaken up repeatedly with cold concentrated sulphuric acid containing sodium dichromate, thiophen is eliminated and pure benzene is then obtained by thorough washing with water, in which it is insoluble, drying with anhydrous calcium chloride and distillation.

Pure benzene is a highly refractive liquid possessing a characteristic odour. It has b.p. 80·4° and m.p. 5·44°; its specific gravity is 0·874 at 20°. By taking advantage of the ease with which benzene solidifies, it can be freed from small quantities of its homologues which may be present in benzene derived from coal-tar. The benzene is cooled in ice water, when it should solidify; any remaining liquid will contain most of the impurities, and this can be separated by rapid filtration while the cooling is maintained. Pure benzene is frequently used for the determination of molecular weights by the cryoscopic method of solutes, soluble in benzene, the convenience of its melting point and its immiscibility with water rendering it particularly suitable for the purpose: the molecular depression constant for benzene is 49. It

* Thiophen has the constitutional formula:

```
HC——CH
‖    ‖
HC   CH
  \ /
   S
```

Its presence in coal-tar benzene is indicated by the indophenin reaction (V. Meyer, 1883). A crystal of isatin (an oxidation product of indigo, p. 337) is dissolved in cold concentrated sulphuric acid. When the solution is shaken with coal-tar benzene, a deep blue colour is produced owing to the presence of thiophen.

should be noted, however, that benzene appears to favour polymerisation of the molecules of certain solutes, and abnormal values for their molecular weights may be obtained.

Benzene is highly inflammable, and when ignited burns with a bright and smoky flame. The small amount of benzene vapour in coal gas is largely responsible for the illuminating power of the latter. Benzene is insoluble in water, but freely soluble in most organic solvents. It is used as a solvent for many different types of organic compounds and mixtures, but the economic importance of benzene is due chiefly to compounds derived from it.

Benzene is a very stable compound and yields simpler compounds (those containing fewer than six carbon atoms in the molecule) only with very great difficulty. It is unattacked by concentrated alkalies and even such strong oxidising agents as chromic acid and potassium permanganate have very little action upon it. The substitution products of benzene are highly important both industrially and scientifically. The most important of these are the compounds derived by the action of concentrated nitric and sulphuric acids, and in its reactions with these acids benzene differs fundamentally from the aliphatic hydrocarbons.

When benzene is added gradually to concentrated nitric acid, or better, to a mixture of equal volumes of concentrated nitric acid and concentrated sulphuric acid, and the temperature of the mixture kept below 40°, a reaction takes place and, when the mixture is poured into an excess of water, a pale yellow oil separates which is heavier than the aqueous solution. This oil can be collected, thoroughly washed with dilute alkali, then with water and after drying with calcium chloride distilled. The oil, b.p. 208°, m.p. 3°, known as *mononitrobenzene*, is a highly important technical product. Its production from benzene by the method indicated above is by the process of *nitration*, the reaction being generally expressed:

$$C_6H_5.\boxed{H + HO}.NO_2 \rightarrow \underset{\text{mononitrobenzene}}{C_6H_5.NO_2} + H_2O$$

In the process of nitration one atom of hydrogen in the benzene molecule is replaced by the monovalent nitro- ($—NO_2$) group. If mononitrobenzene be warmed on the water-bath with a similar nitrating mixture (the mixture of concentrated nitric acid and sulphuric acid), another atom of hydrogen can be readily replaced by the nitro-group and a dinitrobenzene, m.p. 90°, can be easily isolated:

$$C_6H_4\begin{cases} NO_2 \\ \boxed{H + HO}.NO_2 \end{cases} \rightarrow \underset{\text{dinitrobenzene}}{C_6H_4\begin{cases} NO_2 \\ NO_2 \end{cases}} + H_2O$$

Benzene is only very slowly attacked by concentrated sulphuric acid at the ordinary temperature, but if the two substances be heated together, the benzene goes completely into solution, and the reaction is finished when no oil (benzene) separates on pouring the mixture into an excess of water. If this solution be heated with an excess of barium carbonate and then filtered, the excess of sulphuric acid used is got rid of as barium sulphate, which remains on the filter together with the excess of barium carbonate used. The filtrate will be a clear solution, and if evaporated to dryness will yield a crystalline residue of a barium salt of another acid, the barium salt of benzene sulphonic acid, which has been formed by the *sulphonation* of the benzene:

$$C_6H_5 \boxed{H + HO}.SO_3H \rightarrow C_6H_5.SO_3H + H_2O$$

In the process of sulphonation a hydrogen atom of the benzene has been replaced by the monovalent sulphonic acid (—SO_3H) group, the *benzene sulphonic acid* formed being a strong monobasic acid of which the anhydrous barium salt has the formula $(C_6H_5.SO_3)_2Ba$, the acid itself being a very hygroscopic crystalline solid having m.p. 50–51° when anhydrous.

Chlorine and bromine react very slowly with benzene in the absence of sunlight, giving substitution products such as *monochlorobenzene*, b.p. 132°, *monobromobenzene*, b.p. 157°, and at the same time hydrogen halide is evolved:

$$C_6H_6 + Cl_2 \rightarrow C_6H_5Cl + HCl$$

In the presence of suitable catalysts (halogen carriers), such as iron filings or iodine, the reaction goes readily at ordinary temperatures even in the dark. Higher halogenated substitution products are also obtained, up to the compounds C_6Cl_6 and C_6Br_6, *hexachlorobenzene* (m.p. 229°) and *hexabromobenzene* (m.p. higher than 310°) respectively.

In the presence of bright sunlight and in the absence of halogen carriers chlorine and bromine form addition compounds with benzene, *benzene hexachloride*, $C_6H_6Cl_6$, and *benzene hexabromide*, $C_6H_6Br_6$, respectively.

These halogen addition compounds of benzene correspond to the reduction or hydrogenation product of benzene which is formed when benzene is reduced with hydrogen in the presence of finely divided nickel or colloidal platinum. In these circumstances a compound having the formula C_6H_{12} is formed. This is *hexamethylene* or *cyclo*hexane, which has m.p. 6·4° and b.p. 80·8° (p. 44).

Benzene does not form addition compounds with hydrogen halides; it does, however, form a trisozonide, $C_6H_6(O_3)_3$.

The Constitution of Benzene

The stability of benzene towards ordinary reagents and its power of forming substitution products with chlorine and bromine indicate strong resemblances to the paraffins. On the other hand, benzene yields also addition compounds with the above-mentioned halogens and with ozone, and in this respect behaves like an unsaturated aliphatic hydrocarbon. In its behaviour towards concentrated nitric and sulphuric acids, it is different from both the saturated and unsaturated aliphatic hydrocarbons.

The formula of benzene having been established as C_6H_6, it is isomeric with dipropargyl (p. 54), the constitution of which has been established by synthesis to be $HC \vdots C.CH_2.CH_2.C \vdots CH$. In all its reactions, dipropargyl behaves as a highly unsaturated compound, a di-acetylene, and a molecule of the hydrocarbon requires and combines with four molecules of bromine for complete saturation. A molecule of any other unsaturated open chain hydrocarbon isomeric with dipropargyl, i.e. having the formula C_6H_6,* would also combine with four molecules of bromine for complete saturation. When, however, benzene does form addition compounds with chlorine or bromine, the total quantity of halogen with which a molecule of the hydrocarbon will combine is three molecules.

It will be realised later that benzene forms a very large number of derivatives, and when these are submitted to various chemical reactions (oxidation, etc.) they may be converted into simpler derivatives which in almost every case contain a minimum of six carbon atoms. For the above and other reasons, Kekulé† in 1867 concluded (1) that the six carbon atoms in benzene form a closed chain or nucleus, (2) that the molecule of benzene is symmetrical and (3) that each carbon atom is united to one atom of hydrogen. These conclusions are embodied in a formula where all the carbon atoms lie in the same plane and situated at the corners of a regular hexagon, thus:

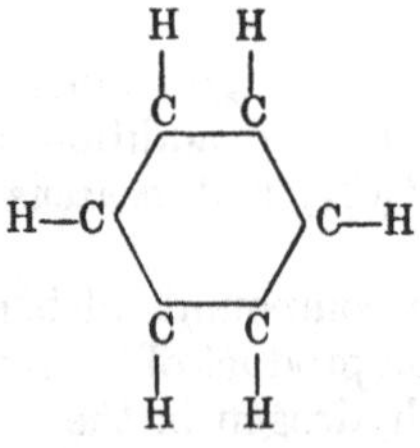

```
        H   H
        |   |
        C   C
       /     \
  H—C <       > C—H
       \     /
        C   C
        |   |
        H   H
```

* Such as the possible hydrocarbons,

$CH_2:C:CH.CH:C:CH_2$ and $CH_3.C \vdots C.C \vdots C.CH_3$.

† Kekulé was the first to enunciate definitely the quadrivalency of carbon and to construct a model of the carbon atom: 'The four units of affinity of the carbon atom, instead of being all placed in one plane, radiate from the sphere representing the atom in the direction of hexahedral axes so that they end in the faces of a tetrahedron....'

This view of the constitution of benzene is universally accepted, but the above formula is incomplete since carbon must be represented as a quadrivalent element.

Kekulé's suggestion together with that of others for the complete benzene formula are as shown:

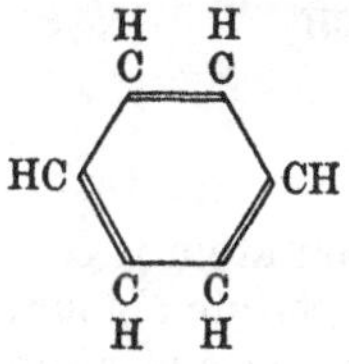

Kekulé

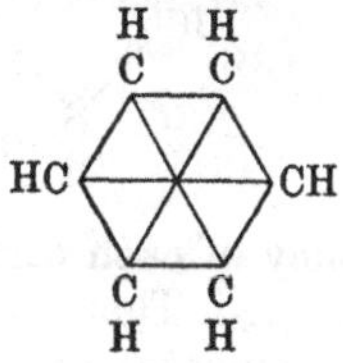

Claus (1867) diagonal formula

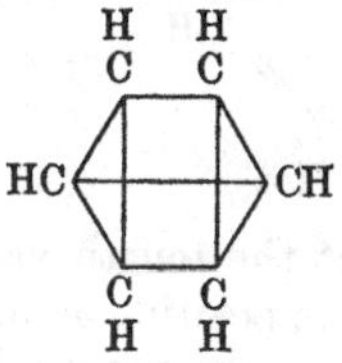

Ladenburg (1869) prism formula

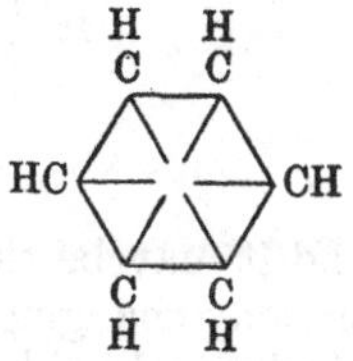

Armstrong (1887) Baeyer (1892) centric formula

Of these, those which represent most closely the chemical behaviour of benzene are the formulae suggested by Kekulé and Armstrong. In the latter formula, the centric bond is considered to be not a real but a potential bond exciting a directive force or pressure towards the centre of the molecule. Under suitable conditions, as in the formation of addition compounds with halogens, the centric bonds can function as ordinary bonds or valencies producing the normal saturated compounds, $C_6H_6X_6$ (X = Cl or Br):

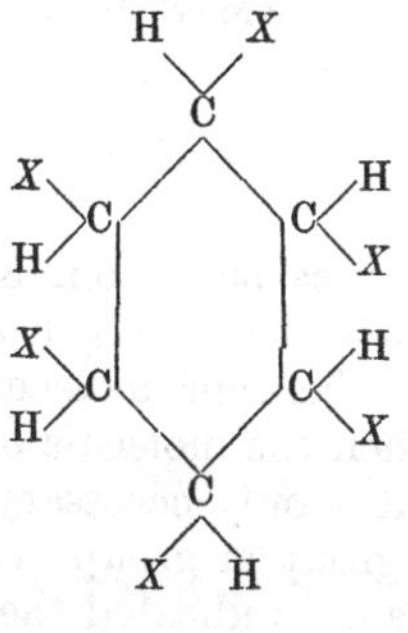

At first sight it would appear that the Kekulé formula suffers from the drawback that it contains three double or ethylene bonds, and therefore one would expect that whenever benzene is brought into contact with chlorine or bromine, addition of the latter must always take place. In actual fact, as has been stated, addition or substitution takes place according to the conditions employed. On the other hand, it can be scarcely expected that the properties due to the presence of a double bond must be the same when the latter linkage is in an open chain compound as in a closed chain compound. It should be realised that there is no difference between the two forms of the Kekulé formula:

```
        H                          H
        C                          C
      //  \                      /  \\
  HC        CH              HC        CH
   |         ||      ⇄       ||        |
  HC        CH              HC        CH
    \\     /                  \     //
        C                          C
        H                          H
```

and it may be that the fourth valency of each carbon atom is continually changing its position as indicated. There is a certain amount of physical evidence which actually indicates the presence of alternate double linkages in the benzene molecule. For most purposes, however, in studying the chemistry of benzene and its derivatives, it is not necessary to include the fourth valency of the carbon atoms in the formula employed, although the fourth valency, whether implying the presence of alternate double bonds or of centric valencies in the benzene molecule, is always understood. It will generally be sufficient to represent benzene as

```
        H
        C                                      H
      /   \                                  /   \
  HC        CH                             H       H
   |         |      or more briefly as     |       |
  HC        CH                             H       H
    \     /                                  \   /
        C                                      H
        H
```

it being understood that a carbon atom is at each corner of the regular hexagon. Frequently the hydrogen atoms are not indicated in the written formula, the benzene molecule there being indicated by a regular hexagon. When the molecule of a substituted benzene derivative is represented it is only necessary to indicate the position of the substituting atom, group or groups, and where a substituting atom, group or groups is not indicated the presence of a hydrogen atom is implied. Strictly speaking, the molecule of benzene should be represented spatially. In the vast majority of cases this is hardly necessary, since an actual model of the benzene molecule, built up on the basis of the tetrahedral configuration of the carbon atom as already indicated, shows that the carbon atoms lie in one plane and that the hydrogen atoms are all symmetrically placed with reference to the molecule as a whole. The actual representation of the benzene molecule employed above is thus an abbreviated way of indicating the spatial configuration.

The symmetry of the benzene molecule necessitates the equivalence in every respect of the six hydrogen atoms in the molecule, and when one atom of hydrogen is replaced by an equivalent monovalent atom or group of atoms the same compound is always obtained. This does not necessarily imply that it is the same atom of hydrogen in the molecule that has been replaced, for it has actually been demonstrated by Ladenburg (1874) that *the six hydrogen atoms in benzene are equivalent to each other*. Monohydroxybenzene (commonly known as phenol), C_6H_5OH, can be produced in a number of ways, and however it is produced it always has identical properties. Similarly, there is only one mononitrobenzene, $C_6H_5NO_2$, one monocarboxybenzene (benzoic acid), C_6H_5COOH, and so on; in other words, *monosubstituted derivatives of benzene exist only in one form*. The method of proving the equivalence of the six hydrogen atoms in benzene is indicated briefly.

In the first place, it is assumed—and there is a large amount of experimental evidence for the assumption—that when one atom or equivalent group in a compound is replaced in a series of normal reactions by an equivalent atom or group of atoms, the replacing atom or group of atoms actually occupies the same position in the molecule of the compound as the replaced atom or group. Thus from phenol it is possible to obtain benzoic acid, and it is assumed that the carboxyl (—CO_2H) group in the latter occupies the same position in the benzene molecule as the hydroxyl (—OH) group of the phenol. If the hydrogen atoms in the benzene molecule are numbered 1, 2, 3, 4, 5, 6, we can represent the formulae of benzene, phenol and benzoic acid thus:

		1	2	3	4	5	6
Benzene	C_6,	—H,	—H,	—H,	—H,	—H,	—H
Phenol	C_6,	—OH,	—H,	—H,	—H,	—H,	—H
Benzoic acid	C_6,	—CO_2H,	—H,	—H,	—H,	—H,	—H

In the second place, when once a hydrogen atom in the benzene molecule has been replaced by an equivalent atom or group of atoms, the remaining five hydrogen atoms are no longer mutually equivalent, because they are now differently placed in the molecule with respect to the substituting group. It has been proved experimentally that there are three isomeric monohydroxybenzoic acids, $C_6H_4(OH)(COOH)$, capable of existence, and each of these compounds has been investigated and its properties examined. The isomerism of these three acids results from the difference in the position in the molecule in each case of the hydroxyl group with respect to the carboxyl group. If the carboxyl group occupies position 1, as in benzoic acid, we can represent the formula of phenol and of the three isomeric hydroxy benzoic acids in a similar way to the above. Now when each of the three isomeric hydroxybenzoic acids is heated with lime* (CaO) it is converted almost quantitatively into phenol identical in every respect with phenol prepared in the ordinary way. As shown below, this proves that the hydrogen atoms in four of the six positions in benzene are mutually equivalent:

	1	2	3	4	5	6		1	2	3	4	5	6
Phenol	C_6, —*OH*,	—H,	—H,	—H,	—H,	—H	→ C_6,	—*OH*,	—H,	—H,	—H,	—H,	—H
(i) $C_6H_4\langle^{OH}_{CO_2H}$	C_6, —CO_2H,	—*OH*,	—H,	—H,	—H,	—H	→ C_6,	—H,	—*OH*,	—H,	—H,	—H,	—H
(ii) $C_6H_4\langle^{OH}_{CO_2H}$	C_6, —CO_2H,	—H,	—*OH*,	—H,	—H,	—H	→ C_6,	—H,	—H,	—*OH*,	—H,	—H,	—H
(iii) $C_6H_4\langle^{OH}_{CO_2H}$	C_6, —CO_2H,	—H,	—H,	—*OH*,	—H,	—H	→ C_6,	—H,	—H,	—H,	—*OH*,	—H,	—H

products all identical

By extending and elaborating the method, the equivalence of all the six hydrogen atoms has been proved.

Regarding the disubstituted products of benzene, it has been shown that in every case, whether the substituting groups be the same or different, there are three and only three isomeric compounds. There are, for example, three dihydroxybenzenes, $C_6H_4(OH)_2$, three hydroxycarboxybenzenes (hydroxybenzoic acids), $C_6H_4\langle^{OH}_{CO_2H}$, three dinitrobenzenes, $C_6H_4(NO_2)_2$, three hydroxynitrobenzenes (nitrophenols), $C_6H_4\langle^{OH}_{NO_2}$, three dicarboxybenzenes, $C_6H_4(CO_2H)_2$, and so on. The carbon atoms in the benzene molecule are systematically numbered in a clockwise direction as shown, and if two hydrogen atoms are substituted by two monovalent atoms or groups, each denoted by *X*, there would appear to be the following possible compounds:

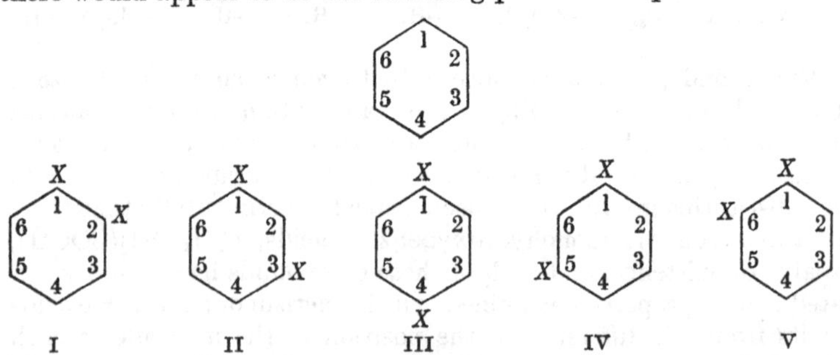

* The reaction between lime and the isomeric hydroxybenzoic acids may be represented:

$$C_6H_4\langle^{OH}_{CO_2H} + CaO \rightarrow C_6H_5OH + CaCO_3$$

It is clear that positions 2 and 3 are respectively similarly placed in the symmetrical configuration as positions 6 and 5; that is, compound IV is the same as compound II and compound V is the same as compound I, and thus there are three and only three isomeric disubstituted products of benzene in every case, and this applies whether the substituting atoms or groups are different from or identical with each other. A disubstituted benzene with the substituting groups on neighbouring carbon atoms, whether these are 1 and 2 or 2 and 3 or 3 and 4 or 4 and 5 or 5 and 6 or 6 and 1—all identical with 1 and 2, on account of the symmetry of the molecule—is known as the *ortho* or 1 : 2 compound: one with the substituting groups on each of two alternate carbon atoms, 1 and 3 (or 2 and 4 or 3 and 5 or 4 and 6 or 5 and 1—all identical with 1 and 3), is known as the *meta* or 1 : 3 compound; the one with the substituting groups on each of two diagonal carbon atoms, 1 and 4 (or 2 and 5 or 3 and 6—each identical with 1 and 4), is known as the *para* or 1 : 4 compound.* Thus there are three dinitrobenzenes:

ortho-dinitrobenzene, $C_6H_4{<}^{NO_2\ 1}_{NO_2\ 2}$, m.p. 117°

meta- „ $C_6H_4{<}^{NO_2\ 1}_{NO_2\ 3}$, m.p. 90°

para- „ $C_6H_4{<}^{NO_2\ 1}_{NO_2\ 4}$, m.p. 172°

The three hydroxynitro- (or nitrohydroxy-) benzenes (nitrophenols) are as follows:

o-nitrophenol, $C_6H_4{<}^{OH\ 1}_{NO_2\ 2}$, m.p. 45°

m- „ $C_6H_4{<}^{OH\ 1}_{NO_2\ 2}$, m.p. 96°

p- „ $C_6H_4{<}^{OH\ 1}_{NO_2\ 4}$, m.p. 114°

and the three dihydroxybenzenes:

o-dihydroxybenzene (catechol), $C_6H_4(OH)_2$ 1 : 2, m.p. 104°
m- „ (resorcinol), $C_6H_4(OH)_2$ 1 : 3, m.p. 119°
p- „ (quinol or hydroquinone), $C_6H_4(OH)_2$ 1 : 4, m.p. 170°

A satisfactory method for determining whether a particular disubstituted benzene derivative is an *ortho*-, *meta*- or *para*- compound

* The following abbreviations explain themselves:

o dinitrobenzene, $C_6H_4(NO_2)_2$ 1 : 2,
m-dinitrobenzene, $C_6H_4(NO_2)_2$ 1 : 3,
p-dinitrobenzene, $C_6H_4(NO_2)_2$ 1 : 4.

was devised by Körner (1874). Körner's method is based on the fact that when a disubstituted derivative of benzene is converted into a trisubstituted derivative by the displacement of another hydrogen atom by an equivalent atom or group of atoms, the number of isomeric compounds which may be obtained from an *ortho-*, *meta-* or *para-* compound is different in all three cases. If a disubstituted benzene designated by the formula $C_6H_4X_2$ be converted into a trisubstituted derivative designated by $C_6H_3X_3$ or $C_6H_3X_2Y$ (X and Y being either the same or different substituting univalent atoms or groups), then two isomeric trisubstituted compounds would be obtained from the *ortho-* compound, three from the *meta-* compound and only one trisubstituted derivative from the *para-* compound. This is shown diagrammatically below. It will be realised, for example, that while

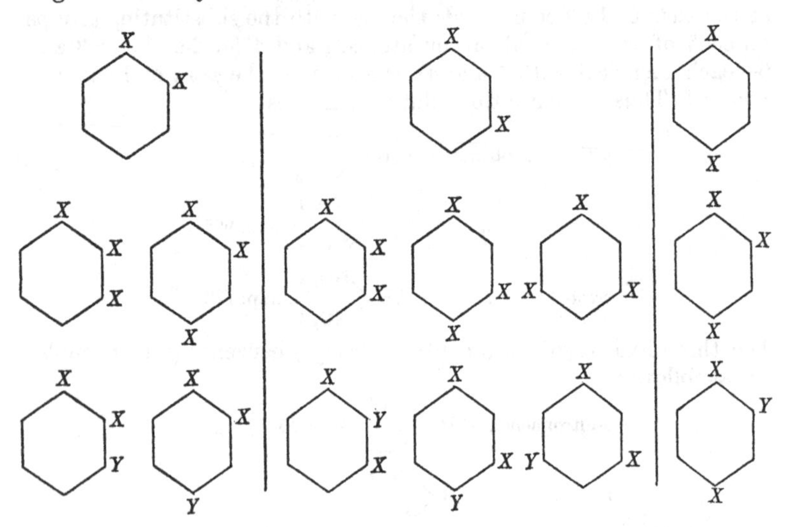

in the *para-* compound, $C_6H_4X_2$, there are four atoms of hydrogen which may be replaced by the third substituting atom or group (X or Y), only one isomer can result, because the hydrogen atoms on the carbon atoms 2, 3, 5 and 6 are symmetrically placed with respect to the 1 and 4 positions. Similar considerations limit the number of isomers of trisubstituted derivatives which can be derived from the corresponding *ortho-* and *meta-* compounds. When once the position of the groups in a particular disubstituted benzene derivative has been decided, the position of the groups in other disubstituted derivatives derived from it by simple or normal reactions is also determined. The determination of the position of the substituting groups in a benzene derivative is described as the *orientation* of the compound.

The isomerism among the trisubstituted derivatives of benzene is more complicated, but the number of isomeric compounds can easily be deduced. If the three substituting groups are identical, the isomeric compounds having the formula $C_6H_3X_3$, the symmetrical structure of the benzene molecule indicates that there are three isomerides possible. The constitutions of these compounds are:

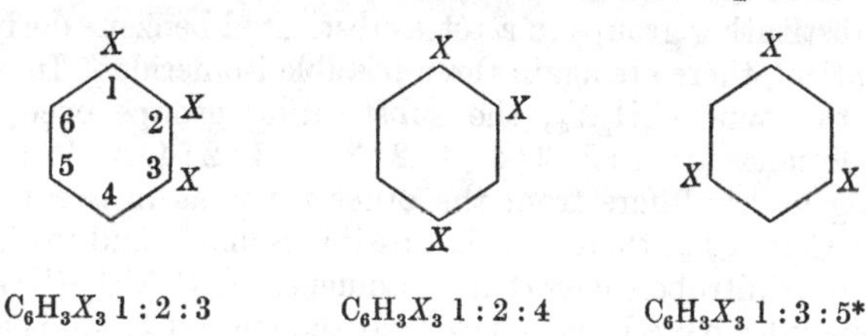

$C_6H_3X_3$ 1 : 2 : 3 $C_6H_3X_3$ 1 : 2 : 4 $C_6H_3X_3$ 1 : 3 : 5*

The three isomeric compounds have been described in a number of cases: for example, the three trihydroxybenzenes, the three tribromobenzenes, the three trimethylbenzenes and the three trinitrobenzenes are all known:

Trihydroxybenzenes	$C_6H_3(OH)_3$ 1 : 2 : 3 (pyrogallol), m.p. 132°
	$C_6H_3(OH)_3$ 1 : 2 : 4 (hydroxyhydroquinone), m.p. 140·5°
	$C_6H_3(OH)_3$ 1 : 3 : 5 (phloroglucinol), m.p. 217°
Tribromobenzenes	$C_6H_3Br_3$ 1 : 2 : 3, m.p. 87°
	$C_6H_3Br_3$ 1 : 2 : 4, m.p. 44°
	$C_6H_3Br_3$ 1 : 3 : 5, m.p. 120°
Trimethylbenzenes	$C_6H_3(CH_3)_3$ 1 : 2 : 3 (hemimellitene), b.p. 175°
	$C_6H_3(CH_3)_3$ 1 : 2 : 4 (pseudocumene), b.p. 170°
	$C_6H_3(CH_3)_3$ 1 : 3 : 5 (mesitylene), b.p. 164°
Trinitrobenzenes	$C_6H_3(NO_2)_3$ 1 : 2 : 3, m.p. 121°
	$C_6H_3(NO_2)_3$ 1 : 2 : 4, m.p. 60°
	$C_6H_3(NO_2)_3$ 1 : 3 : 5, m.p. 122·5°

If the substituting groups in a trisubstituted benzene derivative are not identical, the number of isomerides increases considerably. If two of the substituting groups are identical with each other, as in a compound of the type $C_6H_3XY_2$, there are no less than six possible isomerides, which are designated as follows (X occupying position 1):

1 : 2 : 3, 1 : 2 : 4, 1 : 2 : 5, 1 : 2 : 6, 1 : 3 : 4 and 1 : 3 : 5.

If all the three substituting groups are different from each other, as in a compound of the type C_6H_3XYZ, there are ten possible isomerides,

* In the compound $C_6H_3X_3$ 1 : 2 : 3 the substituting groups are described as being in adjacent or vicinal positions in the molecule.

In the compound $C_6H_3X_3$ 1 : 2 : 4 the substituting groups are described as being in the asymmetrical positions in the molecule.

In the compound $C_6H_3X_3$ 1 : 3 : 5 the substituting groups are described as being in the symmetrical positions in the molecule.

which are designated thus (the groups being mentioned in the order XYZ):

1 : 2 : 3, 1 : 2 : 4, 1 : 2 : 5, 1 : 2 : 6, 1 : 3 : 2,
1 : 3 : 4, 1 : 3 : 5, 1 : 3 : 6, 1 : 4 : 2, 1 : 4 : 3.

In such a complicated case as the latter it is unusual for all the isomerides to be known.

If the substituting groups in a tetrasubstituted benzene derivative are all identical, there are again three possible isomerides. In a compound of the type $C_6H_2X_4$, the substituting groups occupy the positions designated: 1 : 2 : 3 : 4, 1 : 2 : 3 : 5, 1 : 2 : 4 : 5. When one substituting group differs from the other three, as in a compound of the type $C_6H_2XY_3$, there are six possible isomers, and in the case of the methyltrinitrobenzenes (trinitrotoluenes), $C_6H_2(CH_3)(NO_2)_3$, all the six isomerides have been isolated. If the methyl group is always placed in position 1, it is only necessary to signify the positions of the nitro- groups; the compounds are:

2 : 4 : 6	trinitrotoluene,	m.p. 80·8°
2 : 3 : 4	,,	m.p. 112°
2 : 4 : 5	,,	m.p. 104°
3 : 4 : 5	,,	m.p. 137·5°
2 : 3 : 5	,,	m.p. 97·2°
2 : 3 : 6	,,	m.p. 79·5°

The best known of these compounds is the first, 2 : 4 : 6 trinitrotoluene, usually known as T.N.T., the well-known explosive, the constitution of which is

CH_3

O_2N NO_2

NO_2

In a pentasubstituted benzene derivative, when all the substituting groups are the same as in a compound of the type C_6HX_5 only one compound is possible, and the same is true of a hexasubstituted benzene derivative of the type C_6X_6.

So far only the isomerism in substituted benzene derivatives, arising from the position of the substituting groups in what is frequently referred to as the benzene nucleus, has been considered. In the case of the homologues of benzene, however, the isomerism becomes increased on account of the isomerism amongst the hydrocarbon radicals.* There are, for example, three dimethylbenzenes

* When a hydrocarbon radical is a substituting group in benzene, the former is often referred to as a *side chain*. This term is not usually used for substituting groups other than hydrocarbon radicals. In ethyl benzene, the ethyl radical would be designated the side chain, while the C_6H_5— radical would be described as the nucleus.

(*o*-, *m*- and *p*- xylenes), $C_6H_4(CH_3)_2$, and there is another isomeric hydrocarbon, ethylbenzene, $C_6H_5.CH_2.CH_3$. In more complicated cases, the isomerism becomes much greater. In the case of the homologues of benzene of the general formula C_9H_{12}, there are not only the three isomeric trimethylbenzenes already mentioned (p. 67), but there are the three isomeric methylethylbenzenes *o*-, *m*- and *p*- $C_6H_4\langle{}^{CH_3}_{C_2H_5}$ and, in addition, *n*-propylbenzene and *i*-propylbenzene having the formulae

$$C_6H_5.CH_2.CH_2.CH_3 \quad \text{and} \quad C_6H_5{-}CH\langle{}^{CH_3}_{CH_3}$$

respectively, where there is the possibility of isomerism in the side chain as well as in the position which the alkyl groups take up in the nucleus. In the case of substituted derivatives of such homologues of benzene, the number of isomers becomes very considerable. Of a halogen monosubstituted derivative of benzene of the formula $C_9H_{11}X$, there are no less than 39 possible isomerides, due to the substitution of the halogen atom in various positions in the side chains and in the various positions in the nucleus of the isomeric hydrocarbons of the general formula C_9H_{12}.

Homologues of Benzene

The first of the homologues of benzene bears the same relationship to benzene that ethane does to methane. It can be considered to be derived from benzene by the replacement of a hydrogen atom by the methyl group, and the fact that it can be obtained by the action of sodium on a mixture of monobromobenzene and methyl iodide (Fittig's reaction) is quite analogous to the synthesis of ethane from methyl iodide:

$$C_6H_5\,Br + I\,CH_3 + 2Na \rightarrow NaBr + NaI + C_6H_5.CH_3$$

compare $$CH_3\,I + I\,CH_3 + 2Na \rightarrow 2NaI + C_2H_6$$

Methylbenzene, or, as it is usually called, *toluene*, occurs with benzene in the distillate from coal-tar, b.p. from about 75° to 100°. This is the main source of toluene, which industrially is no less important than benzene. Like the latter, toluene also occurs in certain natural petroleums, which constitute a not inconsiderable source of this hydrocarbon. Toluene is isolated from the benzene fractions of the coal-tar distillate by an elaborate fractional distillation. It boils some 30° higher than benzene. When ordinarily isolated from coal-tar, toluene usually contains an organic sulphur compound, thiotolen (a homologue of thiophen), C_5H_6S.* Pure toluene is a colourless highly refractive liquid which has b.p. 110·3°; it solidifies at −94°, which is too low a temperature for it to be conveniently purified, as in the case of benzene, by freezing. It is lighter than and insoluble in water, but it is freely soluble in nearly all dry organic solvents.

Another important reaction for the preparation of toluene and of other homologues of benzene is by the *Friedel and Craft's reaction* (1877). For the preparation of toluene, anhydrous benzene is placed in a suitable flask fitted with a reflux condenser. To the benzene about one-third of its weight of freshly prepared anhydrous aluminium chloride is added and then dry methyl chloride (gas) is passed into the mixture, which is subsequently heated on the water-bath until hydrogen chloride ceases to be evolved. During the reaction care should be taken to prevent access of moisture to the reacting mixture. When the reaction is at an end, powdered ice is added carefully to the resulting mixture to decompose the aluminium compounds, after which the contents of the flask are submitted to steam distillation. The products of this particular reaction are volatile in steam and these float on the surface of the distillate: they are separated, dried with anhydrous calcium chloride and then submitted to careful fractional distillation. That fraction boiling at 109–111° under normal pressure consists almost entirely of toluene. The reaction taking place for the formation of toluene is usually expressed:

$$C_6H_6 + CH_3Cl \rightarrow C_6H_5.CH_3 + HCl$$

Dimethyl- and trimethylbenzenes are formed at the same time:

$$C_6H_6 + 2CH_3Cl \rightarrow C_6H_4(CH_3)_2 + 2HCl$$
$$C_6H_6 + 3CH_3Cl \rightarrow C_6H_3(CH_3)_3 + 3HCl$$

Toluene is, however, easily separated from these by-products and from unchanged benzene by the fractional distillation.†

* Thiotolen has the formula:

```
HC——CH
||    ||
HC    C.CH3
  \  /
   S
```

and like thiophen reacts with isatin in the presence of concentrated sulphuric acid, giving an intense blue colour (indophenin reaction) (compare p. 57).

† It is suggested, with some degree of probability, that aluminium chloride first reacts with benzene to form a compound having the composition $C_6H_5.Al_2Cl_5$ with evolution of hydrogen chloride. Subsequently the aluminium compound is decomposed:

$$C_6H_5.Al_2Cl_5 + ClCH_3 \rightarrow C_6H_5.CH_3 + Al_2Cl_6$$

In one or two cases the intermediate aluminium compound has actually been isolated.

In considering the chemical properties of toluene it is necessary to bear in mind that this hydrocarbon consists of a benzene nucleus, C_6H_5—, and an aliphatic radical, methyl, CH_3—, the side chain. The benzene nucleus will undergo the same reactions as benzene, while the aliphatic portion will behave similarly to methane or any other paraffin.

When toluene is added slowly to concentrated nitric mixed with sulphuric acid under the same conditions as for the preparation of mononitrobenzene, mononitrotoluene is formed:

$$C_6H_5.CH_3 + HO.NO_2 \rightarrow C_6H_4\begin{cases}CH_3\\NO_2\end{cases} + H_2O$$

Two mononitrotoluenes are formed at the same time, these are the *ortho-* and *para-* compounds having the constitution and properties indicated:

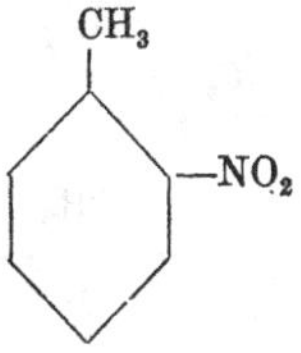

I: benzene ring with CH_3 and —NO_2 in adjacent (ortho) positions

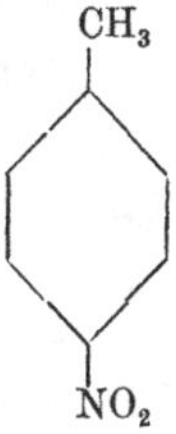

II: benzene ring with CH_3 and NO_2 in opposite (para) positions

I is the *ortho-* or 1 : 2 nitrotoluene, $C_6H_4\begin{cases}CH_3\ 1\\NO_2\ 2\end{cases}$, which occurs in dimorphic forms, m.p. $-9°$ and $-4°$ respectively; b.p. 218°.

II is the *para-* or 1 : 4 nitrotoluene, $C_6H_4\begin{cases}CH_3\ 1\\NO_2\ 4\end{cases}$, which has m.p. 54°, b.p. 230°.

The final product of the nitration of toluene is the symmetrical compound, $C_6H_2(CH_3)(NO_2)_3$ (1 : 2 : 4 : 6), trinitrotoluene (T.N.T.) to which reference has already been made (p. 68).

The action of concentrated sulphuric acid on toluene is similar to the action on benzene, the product of the action in this case being a mixture of the *ortho-* and *para-* toluene sulphonic acids analogous in constitution to the *ortho-* and *para-* mononitrotoluenes:

$$C_6H_5.CH_3 + HO.SO_3H \rightarrow C_6H_4\begin{cases}CH_3\\SO_3H\end{cases} + H_2O$$

The concentrated nitric and sulphuric acids have no action on the aliphatic portion, or methyl group, of the toluene molecule.

The action of halogen (chlorine and bromine) on toluene is more complicated than the action on benzene because the former can act upon both the nucleus or aromatic portion and the side chain or aliphatic portion of the molecule. When chlorine is passed into boiling toluene the three hydrogen atoms in the side chain are replaced:

$$C_6H_5.CH_3 + Cl_2 \rightarrow HCl + C_6H_5.CH_2Cl \text{ (\textit{benzyl chloride}, b.p. 176°)}$$

$$C_6H_5.CH_2Cl + Cl_2 \rightarrow HCl + C_6H_5.CHCl_2 \text{ (\textit{benzal chloride}, b.p. 207°)}$$

$$C_6H_5.CHCl_2 + Cl_2 \rightarrow HCl + C_6H_5.CCl_3 \text{ (\textit{benzotrichloride}, b.p. 213°)}$$

This substitution of chlorine in the side chain takes place in the presence of phosphorus pentachloride and also in the presence of sunlight.

When chlorine acts upon toluene in the cold and in the presence of such halogen carriers as iodine and iron, substitution takes place only in the benzene nucleus or aromatic portion of the molecule:

$$C_6H_5.CH_3 + Cl_2 \rightarrow HCl + C_6H_4\begin{cases} CH_3 \\ Cl \end{cases}$$

$$C_6H_4\begin{cases} CH_3 \\ Cl \end{cases} + Cl_2 \rightarrow HCl + C_6H_3\begin{cases} CH_3 \\ Cl_2 \end{cases}, \text{etc.}$$

The reaction, however, is more complicated than in the case of benzene, owing to the greater amount of isomerism among toluene derivatives. The three monochlorotoluenes are known:

$$\text{o-\textit{chlorotoluene}} \quad C_6H_4\begin{cases} CH_3 \\ Cl\,2 \end{cases}, \text{ b.p. 156°}$$

$$\text{m-\textit{chlorotoluene}} \quad C_6H_4\begin{cases} CH_3 \\ Cl\,3 \end{cases}, \text{ b.p. 150°}$$

$$\text{p-\textit{chlorotoluene}} \quad C_6H_4\begin{cases} CH_3 \\ Cl\,4 \end{cases}, \text{ b.p. 163°}$$

and these are, of course, isomeric with benzyl chloride. The six isomeric dichlorotoluenes, $C_6H_3(CH_3)Cl_2$, are also known and these are isomeric with benzal chloride.*

* The chlorine (or halogen atom generally) is much more easily replaced by hydroxyl and other monovalent groups when in the side chain than when in the nucleus. Benzyl chloride is converted into *benzyl alcohol* under almost the same conditions as the formation of methyl alcohol from methyl chloride or iodide

$$C_6H_5.CH_2Cl + AgOH \rightarrow AgCl + C_6H_5.CH_2OH.$$

It is somewhat difficult to replace chlorine in chlorobenzene or chlorotoluenes.

Like benzene, toluene can be reduced by hydrogen in the presence of reduced nickel or platinum black, the product being methyl*cyclo*hexane or hexahydrotoluene, the first homologue of *cyclo*hexane or hexahydrobenzene:

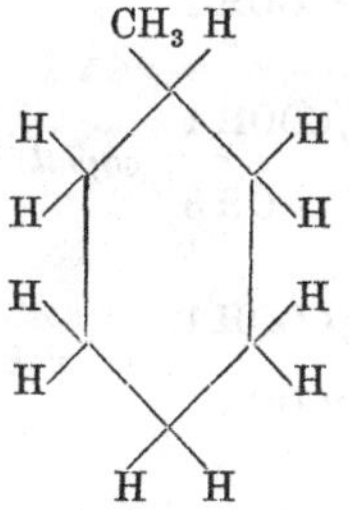

$$\begin{array}{c} CH_3\ \ H \\ \diagdown\ \diagup \\ H\diagdown\ \ C\ \ \diagup H \\ H\diagup C\quad\ C\diagdown H \\ H\diagdown C\quad\ C\diagup H \\ H\diagup\ \ C\ \ \diagdown H \\ \diagup\ \diagdown \\ H\ \ \ H \end{array}$$

methyl*cyclo*hexane, hexahydrotoluene, b.p. 100°

Another important reaction of toluene and other homologues of benzene is their oxidation with an aqueous solution of potassium permanganate. When any aromatic hydrocarbon containing a side chain is boiled with an aqueous solution of potassium permanganate, the nucleus of the hydrocarbon is unaffected and, no matter how long the side chain may be, it is oxidised to the $-C\begin{matrix}\nearrow O \\ \searrow O-H\end{matrix}$, ($CO_2H$), carboxyl group:

$$2C_6H_5.CH_3 + 3O_2 \rightarrow 2H_2O + 2C_6H_5.C\begin{matrix}\nearrow O \\ \searrow O-H\end{matrix}$$

The compound produced from toluene is known as carboxybenzene or *benzoic* acid,* a colourless crystalline substance, m.p. 121°. When toluene is boiled with an aqueous solution of potassium permanganate for a long time, a hydrated oxide of manganese is precipitated and the solution gradually becomes colourless. The solution contains potassium benzoate, and when freed from the oxide by filtration it can be acidified with hydrochloric acid, when the sparingly soluble benzoic acid is precipitated.† The latter after separation by filtration can be recrystallised from hot water. Benzoic acid is sparingly soluble in cold water. Ethylbenzene and the two isomeric propylbenzenes also yield benzoic acid when acted upon by potassium

* All typical organic acids contain the carboxyl or $-CO_2H$ group. The hydrogen of the carboxyl group is capable of being replaced by metals in salt formation. A monobasic organic acid contains one carboxyl group, a dibasic acid contains two carboxyl groups, and so on. Benzoic acid is a monobasic acid. Sodium benzoate is $C_6H_5.COONa$: calcium benzoate is $(C_6H_5COO)_2Ca$ (p. 94).

† If benzyl chloride be used instead of toluene, the oxidation proceeds more rapidly.

permanganate under these conditions. The dimethylbenzenes or xylenes, $C_6H_4(CH_3)_2$, yield dicarboxybenzenes or phthalic acids:

$$C_6H_4\begin{matrix}\diagup CH_3\ 1\\ \diagdown CH_3\ 2\end{matrix} \rightarrow C_6H_4\begin{matrix}\diagup COOH\ 1\\ \diagdown COOH\ 2\end{matrix}$$, *phthalic acid*, m.p. 184°

o-xylene

$$C_6H_4\begin{matrix}\diagup CH_3\ 1\\ \diagdown CH_3\ 3\end{matrix} \rightarrow C_6H_4\begin{matrix}\diagup COOH\ 1\\ \diagdown COOH\ 3\end{matrix}$$, iso*phthalic acid*, m.p. above 300°

m-xylene

$$C_6H_4\begin{matrix}\diagup CH_3\ 1\\ \diagdown CH_3\ 4\end{matrix} \rightarrow C_6H_4\begin{matrix}\diagup COOH\ 1\\ \diagdown COOH\ 4\end{matrix}$$, tere*phthalic acid*, m.p. above 300°

p-xylene

the methylcarboxybenzenes, the monobasic toluic acids being intermediate products in the oxidation:

$$C_6H_4\begin{matrix}\diagup CH_3\ 1\\ \diagdown CH_3\ 2\end{matrix} \rightarrow C_6H_4\begin{matrix}\diagup CH_3\ 1\\ \diagdown COOH\ 2\end{matrix}$$, o-*toluic acid*, m.p. 102°

$$C_6H_4\begin{matrix}\diagup CH_3\ 1\\ \diagdown CH_3\ 3\end{matrix} \rightarrow C_6H_4\begin{matrix}\diagup CH_3\ 1\\ \diagdown COOH\ 3\end{matrix}$$, m-*toluic acid*, m.p. 110°

$$C_6H_4\begin{matrix}\diagup CH_3\ 1\\ \diagdown CH_3\ 4\end{matrix} \rightarrow C_6H_4\begin{matrix}\diagup CH_3\ 1\\ \diagdown COOH\ 4\end{matrix}$$, p-*toluic acid*, m.p. 186°

The types of homologues of benzene so far mentioned can be regarded as substituted derivatives of the paraffins. Toluene is thus described as phenylmethane*

$$\begin{matrix} & C_6H_5 & \\ & | & \\ H— & C & —H \\ & | & \\ & H & \end{matrix}$$

ethylbenzene as phenylethane

$$\begin{matrix} & H & H & \\ & | & | & \\ C_6H_5— & C & —C & —H \\ & | & | & \\ & H & H & \end{matrix}$$

* '*Phenyl*', C_6H_5, is the name given to the monovalent radical theoretically obtained from benzene by the removal of a hydrogen atom and corresponds with methyl, ethyl, etc. Benzene sulphonic acid is actually more accurately described as phenylsulphonic acid: monochlorobenzene might well be described as phenyl chloride corresponding with methyl chloride. The radical theoretically obtained from benzene by the removal of two hydrogen atoms is frequently known as 'phenylene', C_6H_4, and is a divalent radical corresponding with ethylene, although, unlike ethylene, it may be incapable of independent existence.

Of the hydrocarbons having the molecular formula $C_{10}H_{14}$ and which are substituted derivatives of benzene there are 22 possible isomerides. Twenty of these are actually known and have been well characterised. The most important one is 1-*methyl*-4-*isopropyl-benzene* (or *p*-isopropyltoluene), p-*cymene* and frequently re-

n-propylbenzene as α- (or 1-) phenylpropane

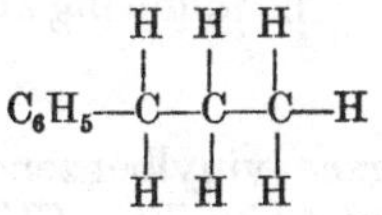

```
          H  H  H
          |  |  |
   C6H5 — C — C — C — H
          |  |  |
          H  H  H
```

and *i*-propylbenzene as β- (or 2-) phenylpropane

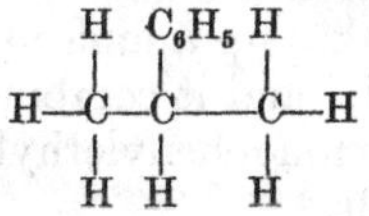

```
      H  C6H5  H
      |   |    |
  H — C — C —— C — H
      |   |    |
      H   H    H
```

Corresponding with these there are phenyl derivatives of unsaturated

ferred to as *cymene* (because it is the best known). This hydrocarbon has the constitution:

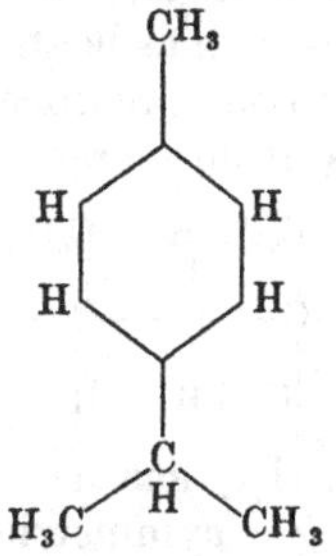

```
            CH3
             |
          /     \
        H         H
        H         H
          \     /
             |
             C
        H3C  H  CH3
```

and occurs in the essential oils (mixtures of terpenes) obtained from a variety of plants. It is closely related to camphor (a terpene derivative), from which it is obtained in good yield by warming with phosphorus pentoxide:

$$\underset{\text{camphor}}{C_{10}H_{16}O} - H_2O \rightarrow \underset{p\text{-cymene}}{C_{10}H_{14}}$$

It is closely related to many terpenes (isomeric hydrocarbons having the formula $C_{10}H_{16}$), from which it can be frequently obtained by oxidation processes.

The constitution of *p*-cymene has been proved by synthesis from *p*-bromoisopropylbenzene and methyl iodide by the action of sodium in the presence of dry ether (Fittig's reaction), the product obtained being identical with the *p*-cymene from natural sources.

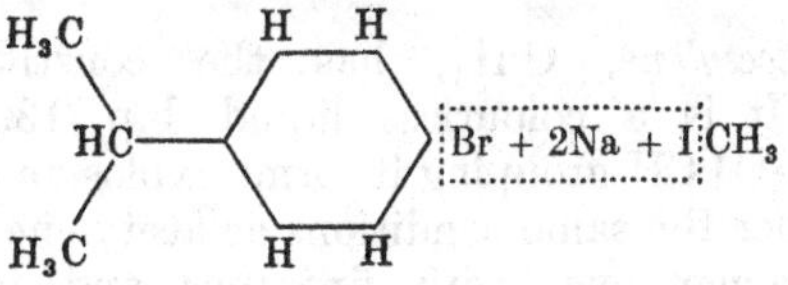

```
  H3C          H   H
     \
      HC ——— (ring) [Br + 2Na + I]CH3
     /
  H3C          H   H
```

p-Cymene is a colourless liquid, b.p. 177·3°, having a sharp odour somewhat reminiscent of carrots. When *p*-cymene is oxidised with chromic acid *p*-phthalic acid (terephthalic acid) is obtained (p. 74), and when heated with somewhat dilute nitric acid *p*-toluic acid (p. 74) is the chief oxidation product. These products of the oxidation of *p*-cymene prove (1) that there are two alkyl groups in the *para*-position to each other in the hydrocarbon, and (2) that one of the alkyl groups is the methyl group.

aliphatic hydrocarbons, such as ethylene and its homologues and acetylene and its homologues. The following are typical hydrocarbons of these two classes.

Monophenylethylene or *styrene* (vinylbenzene) has the formula C_8H_8 and its constitution is represented by $C_6H_5.CH:CH_2$. It is a colourless highly refracting liquid, b.p. 146°, with an odour reminiscent of benzene. When reduced it takes up a molecule of hydrogen forming ethylbenzene, $C_6H_5.CH_2.CH_3$, and it combines readily with a molecule of bromine forming monophenylethylene bromide (styrene bromide), $C_6H_5.CHBr.CH_2Br$.

In the presence of sunlight, styrene rapidly undergoes polymerisation forming 'styrene resins', substances or mixtures of substances of high molecular weight. This polymerisation of unsaturated compounds of many types is by no means uncommon; it may take place under a variety of conditions such as heating, in the presence of light or photochemically and/or in the presence of specific catalysts. The formation of 'styrene resins' from styrene may be represented thus:

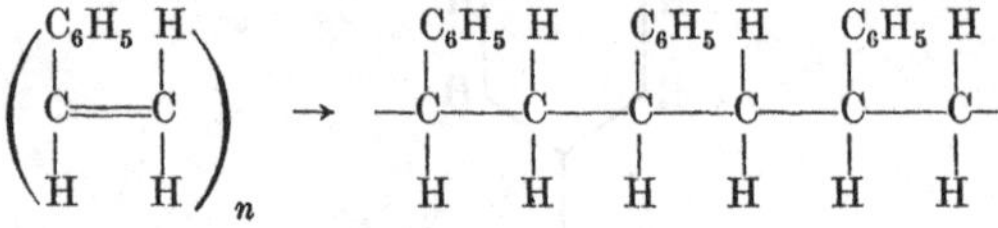

$$\left(\begin{array}{cc} C_6H_5 & H \\ | & | \\ C & = C \\ | & | \\ H & H \end{array}\right)_n \rightarrow \begin{array}{cccccc} C_6H_5 & H & C_6H_5 & H & C_6H_5 & H \\ | & | & | & | & | & | \\ -C- & -C- & -C- & -C- & -C- & -C- \\ | & | & | & | & | & | \\ H & H & H & H & H & H \end{array}$$

sym-*Diphenylethylene*, $C_{14}H_{12}$, has a constitution represented as $C_6H_5.CH:CH.C_6H_5$. Being a symmetrically disubstituted derivative of ethylene, it is capable of existing in two isomeric forms and both of these have been described:

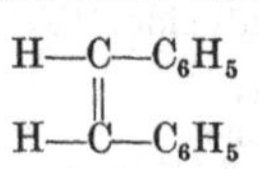

$$\begin{array}{c} H-C-C_6H_5 \\ \| \\ H-C-C_6H_5 \end{array} \qquad\qquad \begin{array}{c} C_6H_5-C-H \\ \| \\ H-C-C_6H_5 \end{array}$$

cis-form	*trans*-form
stilbene	iso*stilbene*
(colourless monoclinic prisms, m.p. 124°)	(colourless liquid, b.p. 143° at 21 mm.)

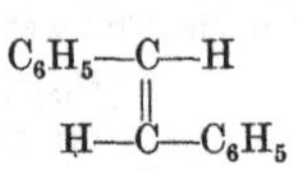

Both of these stereoisomeric forms show the unsaturated properties of ethylene and its derivatives*.

Monophenylacetylene, C_8H_6, has the constitutional formula $C_6H_5.C:CH$. It is a colourless liquid, b.p. 139–140°. Since it possesses the —C:CH grouping it forms explosive copper and silver derivatives under the same conditions as acetylene itself. It readily takes up hydrogen, even with zinc dust and acetic acid, being converted into styrene, $C_6H_5.CH:CH_2$ (*v.* above). Further, like acetylene, it combines with water in the presence of dilute sulphuric

* The configurations of stilbene and *iso*stilbene given above are those assigned by Wislicenus. Other authorities reverse these constitutions and assign to stilbene the *trans*-configuration.

acid, forming the compound *acetophenone* (m.p. 20°, b.p. 202°) (p. 186):

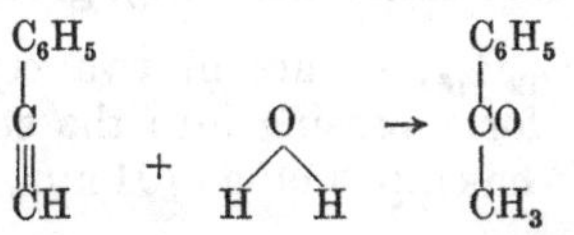

$$\begin{matrix} C_6H_5 & & & & C_6H_5 \\ | & & & & | \\ C & & O & \rightarrow & CO \\ ||| & + & / \backslash & & | \\ CH & & H \quad H & & CH_3 \end{matrix}$$

acetophenone

Diphenylacetylene, $C_{14}H_{10}$, has the constitutional formula $C_6H_5.C:C.C_6H_5$ and is a colourless crystalline compound, m.p. 60°. It cannot form copper and silver derivatives, although it displays the marked unsaturation properties of an acetylene compound. One of the points of interest connected with the compound is its isomerism with two aromatic hydrocarbons of quite a different type, viz. *anthracene* (p. 81) and *phenanthrene* (p. 83), both of which also have the molecular formula $C_{14}H_{10}$.

The following three hydrocarbons are, like benzene and its homologues, fundamental substances from which numerous dyestuffs are derived.

Phenylbenzene or *diphenyl*, $C_{12}H_{10}$, has its constitutional formula represented as

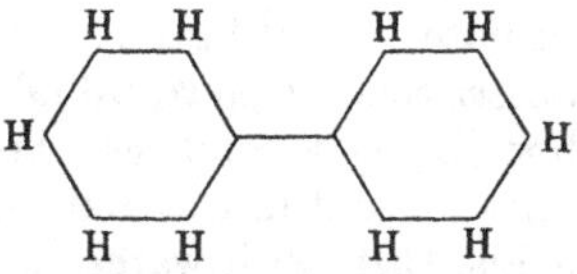

[Structural formula: two benzene rings joined, each ring carbon bearing H]

or $C_6H_5.C_6H_5$

The compound crystallises in colourless plates or prisms, m.p. 70·5°, b.p. 260°, and occurs in the distillate from coal-tar. Its presence in coal-tar is explained by the fact that diphenyl is formed when benzene vapour is passed through a red hot tube. Fittig prepared diphenyl from monobromobenzene and sodium in the presence of ether or benzene:

$$C_6H_5\fbox{Br + 2Na + Br}C_6H_5 \rightarrow 2NaBr + C_6H_5.C_6H_5$$

Diphenylmethane or *benzylbenzene*, $C_{13}H_{12}$, is the next homologue of diphenyl and has the constitutional formula:

[Structural formula: two benzene rings joined through a CH₂ carbon, ring carbons bearing H]

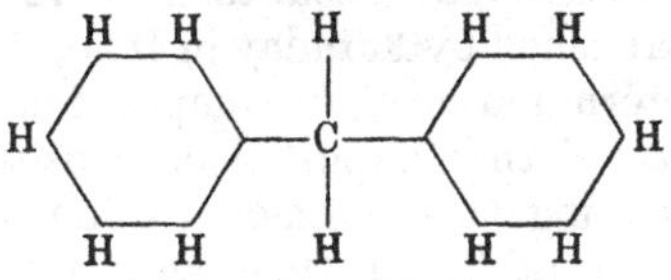

or $(C_6H_5)_2:CH_2$

It crystallises in long prismatic needles, m.p. 26–27°, b.p. 264·7°. If the compound be written as $C_6H_5.CH_2.C_6H_5$, it can be regarded as benzene with one hydrogen atom substituted by the monovalent $C_6H_5.CH_2$. group, usually called the *benzyl* group or radical.

Triphenylmethane, $C_{19}H_{16}$, occurs in two crystalline forms (dimorphism), the labile form passing into the stable form below the melting point, 92·5°. It has b.p. 359° at 704 mm. and its constitution is represented:

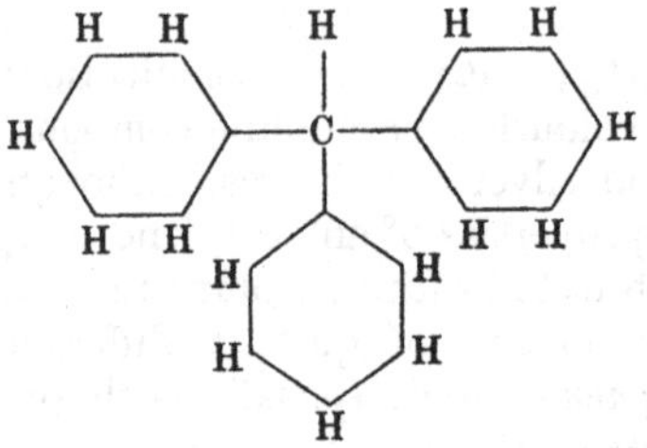

or $(C_6H_5)_3CH$

Diphenylmethane and triphenylmethane are di- and tri- substituted products respectively of methane. The corresponding monosubstituted product of methane is toluene, $C_6H_5.CH_3$.

Other Aromatic Hydrocarbons

From the higher boiling distillates of coal-tar has been isolated a hydrocarbon, having a characteristic and powerful odour, known as *naphthalene*. This hydrocarbon, being a solid, can be readily freed from basic and acidic impurities present in the coal-tar distillate, by treatment first with dilute acids, then with water, then with dilute alkalies and finally with water. After drying, the hydrocarbon can be purified by sublimation. Naphthalene, along with certain of its homologues, also occurs in the higher boiling fraction of certain petroleums. Methane, ethane, ethylene, acetylene, benzene and many other hydrocarbons are converted partly at least into naphthalene when they are passed through a red hot tube. Naphthalene is, of course, not the only product of such a pyrogenic reaction: it has been pointed out (p. 52) that acetylene is converted partly into benzene in these circumstances, and it may be that the benzene is subsequently converted into naphthalene.

Naphthalene, $C_{10}H_8$, was discovered in coal-tar in 1819 by Garden and its composition was established by Faraday in 1826, shortly after the latter's discovery and determination of the composition of benzene. In 1866, Erlenmeyer suggested that naphthalene consisted of two coalesced benzene nuclei having two common carbon atoms, and subsequent work definitely established this fact. Consequently, naphthalene was represented by formula I, which corresponds to the Kekulé formula of benzene, and a centric formula II has also been suggested. As in the case of benzene, it is, however, generally sufficient to represent the naphthalene as in III (without the numbers), it being understood that, although it is not actually portrayed, the fourth valency of the carbon atoms is understood:

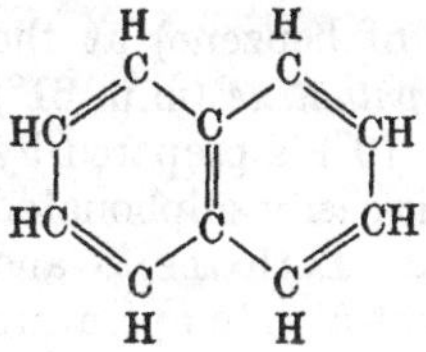

I 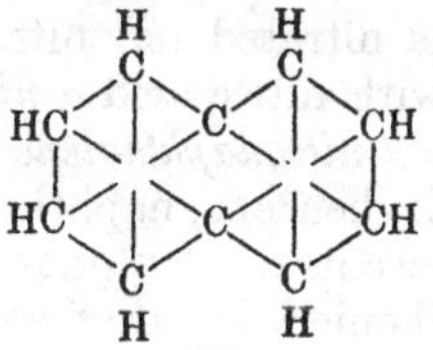 II 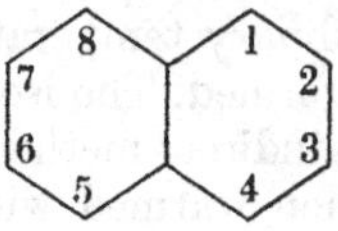 III

The eight carbon atoms to which hydrogen atoms are attached are numbered as shown in III, for convenience in describing substituted derivatives of naphthalene.

Naphthalene crystallises in colourless monoclinic prisms which have a characteristic bluish reflex. It has m.p. 80·98° and b.p. 217·7°. It is insoluble in water and aqueous solutions, but readily soluble in almost all organic solvents.

It will be realised that the naphthalene molecule is highly symmetrical and that positions 1, 4, 5 and 8 are equivalent to each other, and that positions 2, 3, 6 and 7, while equivalent again to each other, are not equivalent to the four former positions. This implies that each monosubstituted derivative of naphthalene exists in two isomeric forms, usually called the α- and the β- derivatives, according to the mode of representation IV:

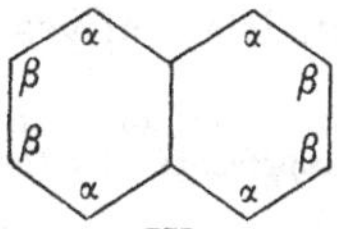

IV 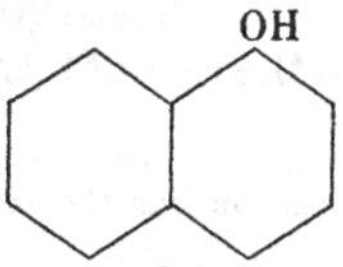 α-naphthol* 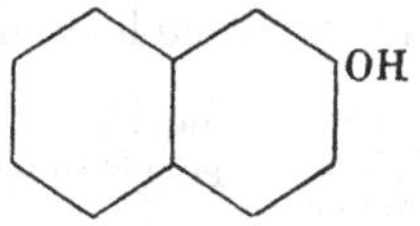 β-naphthol

* That the constitution of α-naphthol is as shown was proved by Fittig. Fittig showed that phenyl*iso*crotonic acid is converted into α-naphthol when heated at 300°:

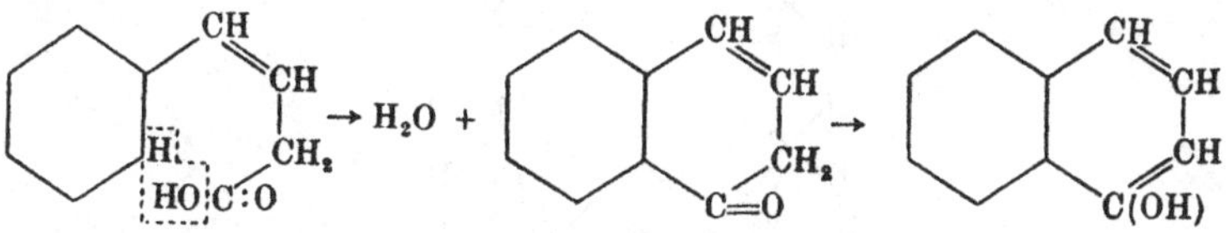

$\rightarrow H_2O +$ $\rightarrow$

This synthesis is fundamentally important. It proves not only the constitution of α-naphthol, that is, it proves the position of the hydroxyl group, but it also proves the constitution of naphthalene.

When naphthalene with excess of hydrogen is passed over reduced nickel at 160°, reduction of the hydrocarbon takes place and *decahydronaphthalene*, $C_{10}H_{18}$,

For example, there are two monohydroxynaphthalenes usually called naphthols. α-*Naphthol*, $C_{10}H_7(OH)$, has the constitutional formula shown and is a colourless crystalline compound, m.p. 94°, b.p. 278–280°. β-*Naphthol* crystallises in colourless plates, m.p. 122°, b.p. 286°.

When naphthalene is nitrated (cf. nitration of benzene) at the ordinary temperature with nitric acid α-*nitronaphthalene* (m.p. 61°) is formed. The isomeric β-*nitronaphthalene* (m.p. 79°) is prepared by an indirect method. Like benzene, naphthalene is easily sulphonated when warmed with concentrated sulphuric acid. Both the α- and the β- naphthalene sulphonic acids are formed, but at a temperature of about 80° the α-acid is produced in the greater quantity, while at about 160° the β-acid is obtained in the larger amount. When the α-acid is heated with sulphuric acid it is converted into the β-acid. α-*Naphthalene sulphonic acid* has m.p. 80°, β-*naphthalene sulphonic acid* has m.p. 124°. The salts of these two isomeric monobasic acids differ usually considerably in solubility and, for example, the difference in solubility in the lead (or calcium) salts is made use of in separating the two acids from each other. α-Naphthalene sulphonic acid and β-naphthalene sulphonic acid yield α-naphthol and β-naphthol respectively when fused with potassium hydroxide (see p. 120):

$$C_{10}H_7\ SO_3H + K\ OH \rightarrow C_{10}H_7OH + KHSO_3$$

It can be shown mathematically that in the case of a disubstituted derivative of naphthalene there are 14 isomerides; 14 isomerides are also possible in the case of the trisubstituted derivative and no less than 22 possible isomerides in the case of tetrasubstituted derivatives;

a colourless liquid, b.p. 187–188° (much lower than the boiling point of naphthalene), is produced. This hydrocarbon has the constitution represented by the formula:

```
         H   H   H   H
          \ /     \ /
           C   H   C
          /  \ | /  \
    H    /     \|/    \    H
     \  /       |       \ /
      C         C        C
     /  |       |       |  \
    H   |       |       |   H
    H   |       |       |   H
     \  |       |       |  /
      C         C        C
     /  \      /|\      /  \
    H    \    / | \    /    H
          \  /  |  \  /
           C    H   C
          / \      / \
         H   H    H   H
```

and it has the same relationship to naphthalene that *cyclo*hexane (hexahydrobenzene) has to benzene.

in the latter two cases the monovalent substituting groups are supposed to be identical, the number of possible isomerides being, of course, much greater if the substituting groups in each compound are different from each other.

Naphthalene can be regarded as the parent substance of a series of aromatic hydrocarbons of the general formula, C_nH_{2n-12}, where n is at least equal to 10.

Anthracene

Anthracene, $C_{14}H_{10}$, was discovered in coal-tar distillate by Dumas and Laurent in 1832. They called the compound 'paranaphthalene', because they at first believed it to be a form of naphthalene. Anthracene was first synthesised in 1886 by Limpricht, and it was then realised that just as naphthalene consisted of two coalesced benzene nuclei, anthracene consisted of three benzene nuclei, coalesced in a similar manner. The constitution of anthracene can be represented by either formula I or II or more briefly by III (the last, of course, without the figures which indicate the numbering of the carbon atoms for the identification of anthracene derivatives):

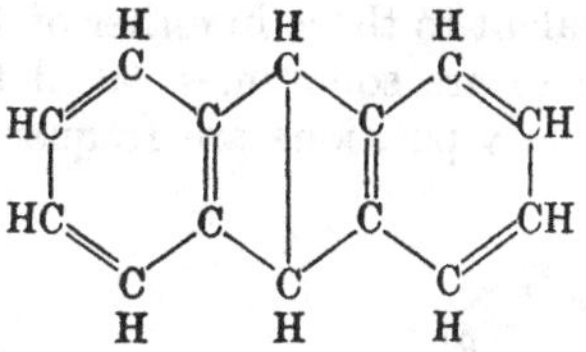

I

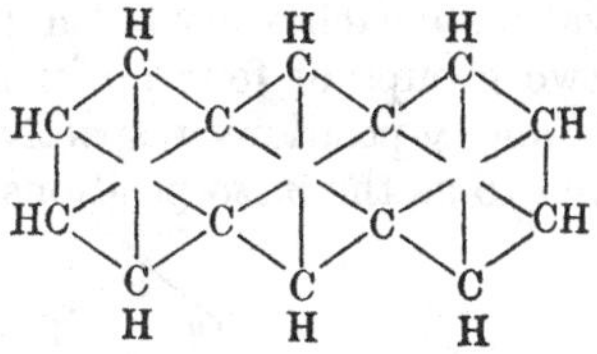

II

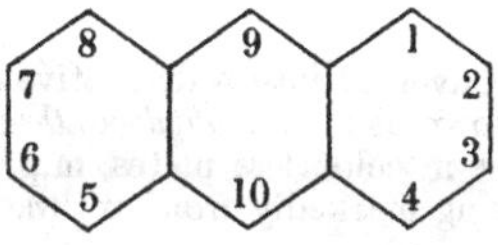

III

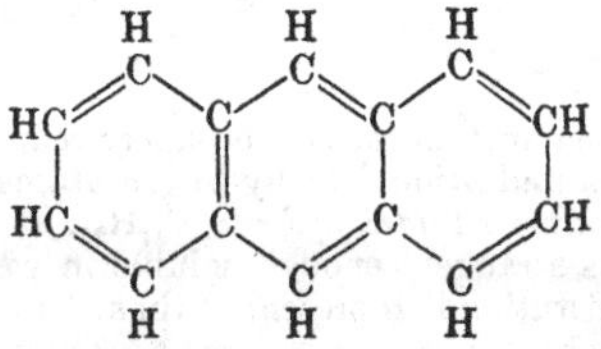

IV

The determination of the disposition of the fourth valency of the carbon atoms is just as difficult in the case of anthracene as in the cases of benzene and naphthalene. The fourth valency of the carbon atoms in anthracene would be satisfied equally well by formula IV, but formula I is preferred as the chemistry of anthracene derivatives indicates strongly that the two central : CH groups (numbered 9 and 10) are different from, or in a different state of combination from, all the other : CH groups. These central or *meso-groups* are usually attacked first by oxidising agents, when the highly stable *anthraquinone* (pale yellow needles, m.p. 285°) is produced. Anthraquinone has been synthesised and shown to have the constitution

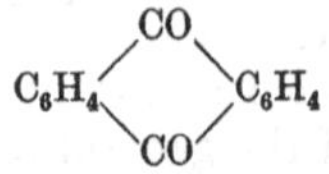

$$C_6H_4\begin{matrix} \diagup CO \diagdown \\ \diagdown CO \diagup \end{matrix}C_6H_4$$

The only source of anthracene is coal-tar distillate, and it is separated from the appropriate fraction in a fashion somewhat similar to that used for the separation of naphthalene. The pure hydrocarbon has m.p. 217° and b.p. 351° and is soluble in most organic solvents. The crystals of anthracene and solutions of anthracene show a characteristic blue fluorescence. The chief use of anthracene is for the production of important dye-stuffs.

Isomerism amongst the substituted derivatives of anthracene is very marked. Any monosubstituted derivative of the general formula $C_{14}H_9X$ (X = a monovalent substituting group) is capable of existing in three isomeric forms. In the above formula the hydrogen atoms in positions 1, 4, 5 and 8 are equivalent to each other, those in positions 2, 3, 6 and 7 are again equivalent to each other but not equivalent to the first four, while those in positions 9 and 10 are again equivalent to each other but not equivalent to those in either of the first two groups of four positions. These are sometimes called the α-, β- and γ-positions respectively. The γ-positions are frequently referred to as the *meso*-positions:

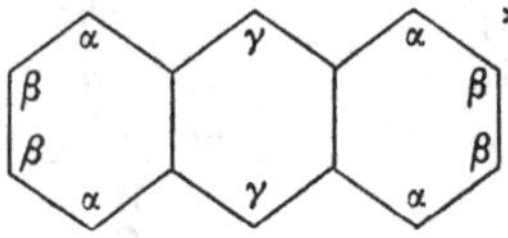

α γ α *
β β
β β
α γ α

* When anthracene is completely reduced, it is converted into a derivative containing an additional 14 hydrogen atoms. It is known as *tetradecahydroanthracene* or *anthracene tetradecahydride*, $C_{14}H_{24}$. It crystallises in colourless plates, m.p. 88°, and it is strangely enough soluble in water, differing markedly from anthracene. Its constitution is represented thus:

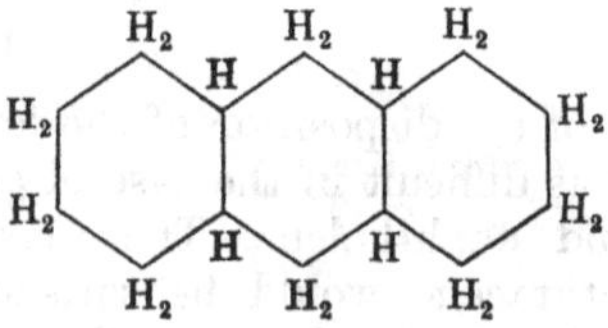

H_2 H_2 H_2
H H
H_2 H_2
H_2 H_2
H H
H_2 H_2 H_2

Phenanthrene

It has been indicated that the three benzene nuclei in the anthracene molecule are joined laterally, and this is the most symmetrical formula which can be constructed. There is, however, another but less symmetrical way of joining the three benzene nuclei to represent the constitution of an aromatic hydrocarbon of the formula $C_{14}H_{10}$. This is represented by Formula I or II:

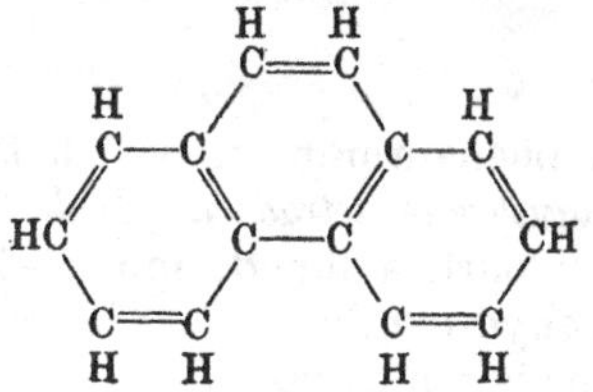

I

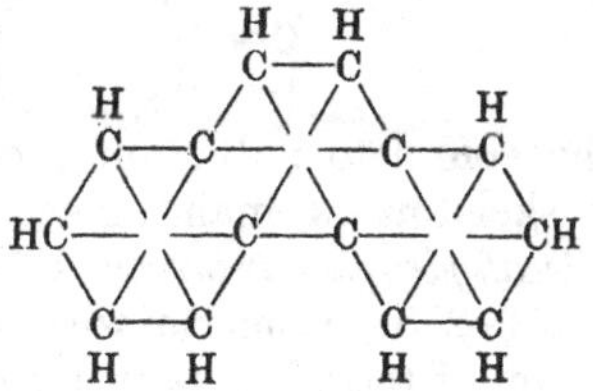

II

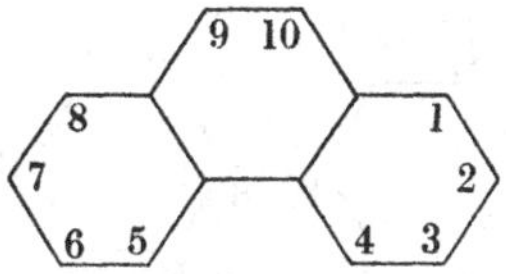

III

Actually I (or II) represents the constitution of a well-known hydrocarbon, *phenanthrene*, $C_{14}H_{10}$.* This hydrocarbon only occurs in coal-tar.

Corresponding to anthraquinone, the oxidation product of anthracene, phenanthrene forms *phenanthrenequinone*,† $C_{14}H_8O_2$, by direct oxidation under similar conditions for the production of

* Confirmatory evidence of the constitution allotted to phenanthrene is afforded by the fact that phenanthrene is formed when stilbene (*sym*-diphenylethylene, p. 76) is passed through a red hot tube. Loss of two hydrogen atoms takes place as indicated:

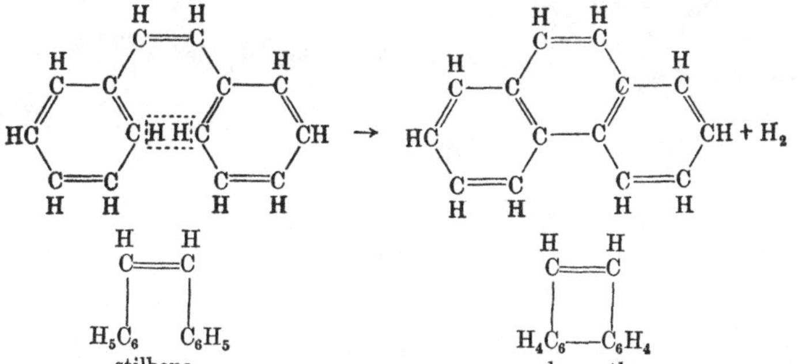

$$\underset{\text{stilbene}}{H_5C_6-CH=CH-C_6H_5} \rightarrow \underset{\text{phenanthrene}}{H_4C_6(CH=CH)C_6H_4} + H_2$$

† or phenanthraquinone.

anthraquinone from anthracene. Phenanthrenequinone crystallises in orange-coloured needles, m.p. 205°. Its constitution is represented thus:

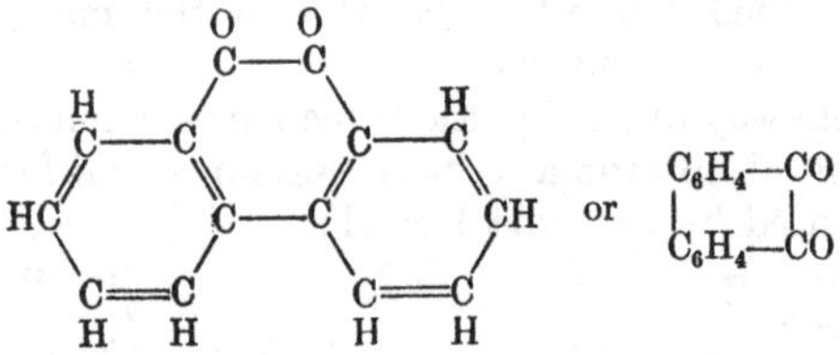

```
           O   O
           C———C
   H      /     \      H
   C———C          C———C
  //      \\    //      \\
HC          C———C          CH   or
  \        /     \        /
   C===C          C===C
   H   H          H   H
```

$$\begin{array}{l} C_6H_4\text{—CO} \\ | \qquad\ \ | \\ C_6H_4\text{—CO} \end{array}$$

The complete reduction product of phenanthrene in which the ring skeleton is maintained is *phenanthrene tetradecahydride* or *tetradecahydrophenanthrene*, $C_{14}H_{24}$, a colourless liquid, m.p. $-3°$, b.p. 270–275°, isomeric with tetradecahydroanthracene. The constitution of tetradecahydrophenanthrene is represented as:

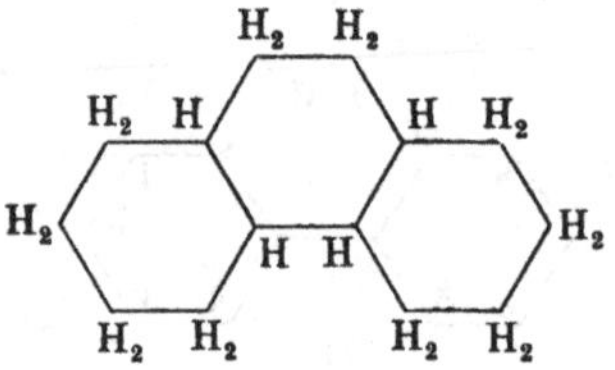

```
                 H2  H2
          H2  H /      \ H  H2
       H2                      H2
                 H    H
          H2  H2         H2  H2
```

CHAPTER III

MONOHYDROXY-DERIVATIVES OF THE PARAFFINS, OF HOMOLOGUES OF BENZENE AND OF UNSATURATED ALIPHATIC HYDROCARBONS

MONOHYDROXY-DERIVATIVES of the hydrocarbons form two important classes of compounds, which comprise (i) those derived by the substitution of a hydroxyl (—OH) group for a hydrogen atom in the aliphatic hydrocarbons, those derived by the same substitution for a hydrogen atom in the side chain of aromatic hydrocarbons together with those similarly derived from the *cyclo*-paraffins, and (ii) those derived by the similar substitution for a hydrogen atom in the nucleus of aromatic hydrocarbons. If the composition of any hydrocarbon be represented by C_xH_y, then the composition of its monohydroxy-derivative is represented by $C_xH_{y-1}OH$, and a monohydroxy-derivative of a hydrocarbon is actually an oxidation product of the latter.

Compounds of class (i) are known as *alcohols*,* and, to distinguish them from those containing more than one hydroxyl group (in place of more than one hydrogen atom in the molecule of the hydrocarbon) in the molecule, as *monohydric alcohols*. Compounds of class (ii) are usually described as *phenols*,† the simplest compound of this class, monohydroxybenzene, C_6H_5OH, being known as phenol.

Being monosubstitution products of hydrocarbons, the number of monohydroxy-derivatives which can be derived from any one hydrocarbon will be the same as the number of any other mono-substituted derivatives to which that hydrocarbon can give rise. Thus, there is one monohydroxymethane, CH_3OH, one mono-hydroxyethane, C_2H_5OH; there are two monohydroxypropanes, C_3H_7OH, four hydroxybutanes, C_4H_9OH—two from each of the two isomeric butanes—one monohydroxybenzene, two monohydroxy-naphthalenes, $C_{10}H_7OH$, etc.

MONOHYDROXY-DERIVATIVES OF THE PARAFFINS

The paraffin hydrocarbons form a homologous series of compounds having the general formula C_nH_{2n+2}, and their monohydroxy-derivatives *or alcohols* form a homologous series having the general formula $C_nH_{2n+1}OH$.

* 'Alcohol' is derived originally from the Arabic *al* = the, *koh'l* = powder for staining eyelids.

† From the Greek φαίνος = shining.

The first member of the series of these alcohols—the most important series of the alcohols—is monohydroxymethane or *methyl alcohol*: frequently known in organic chemistry as *carbinol*. The constitution of methyl alcohol is represented:

```
    H
    |
H—C—O—H
    |
    H
```

a formula which represents the compound as methane in which any one of the four equivalent atoms of the hydrocarbon is replaced by the monovalent hydroxyl, or —OH, group. All the alcohols of the $C_nH_{2n+1}OH$ series can be regarded as substitution derivatives of methyl alcohol or carbinol, just as the homologues of methane are regarded as substitution derivatives of that hydrocarbon. Thus:

```
    H          H H            H H H            H H H H
    |          | |            | | |            | | | |
H—C—H      H—C—C—H        H—C—C—C—H        H—C—C—C—C—H
    |          | |            | | |            | | | |
    H          H H            H H H            H H H H
```

ethane, methylmethane, $CH_3.CH_3$, C_2H_6.

propane, ethylmethane, $CH_3.CH_2.CH_3$, C_3H_8.

n-butane, *n*-propylmethane, or methylethylmethane, $CH_3.CH_2.CH_2.CH_3$, *n*-C_4H_{10}.

```
    H
    |
H—C—H
    |   H
    |   |
H—C—C—H
    |   |
    |   H
H—C—H
    |
    H
```

i-butane, trimethylmethane, or *i*-propylmethane, $(CH_3)_3CH$, *i*-C_4H_{10}.

```
    H            H H              H H H              H O—H H
    |            | |              | | |              | |     |
H—C—O—H      H—C—C—O—H        H—C—C—C—O—H        H—C—C—————C—H
    |            | |              | | |              | |     |
    H            H H              H H H              H H     H
```

methyl alcohol, carbinol, CH_3OH or $H.CH_2OH$.

ethyl alcohol, methylcarbinol, $CH_3.CH_2OH$ or C_2H_5OH.

n-propyl alcohol, ethylcarbinol, $CH_3.CH_2.CH_2OH$ or *n*-$C_3H_7.OH$.

i-propyl alcohol, dimethylcarbinol, $(CH_3)_2CHOH$ or *i*-C_3H_7OH.

$$\begin{array}{ccccccccccc} & & H & & H & & H & & H & & \\ & & | & & | & & | & & | & & \\ H & — & C & — & C & — & C & — & C & — & O—H \\ & & | & & | & & | & & | & & \\ & & H & & H & & H & & H & & \end{array}$$

n-butyl alcohol,
n-propylcarbinol.
$CH_3.CH_2.CH_2.CH_2OH$
or *n*-$C_3H_7.CH_2OH$
or *n*-C_4H_9OH.

$$\begin{array}{ccccccccc} & & H & & H & & O—H & & H \\ & & | & & | & & | & & | \\ H & — & C & — & C & — & C & — & C—H \\ & & | & & | & & | & & | \\ & & H & & H & & H & & H \end{array}$$

methylethylcarbinol,
$$\begin{array}{l} CH_3 \\ C_2H_5 \end{array} \!\!> CHOH.$$

$$\begin{array}{ccccc} & & H & & \\ & & | & & \\ H & — & C—H & & H \\ & & | & & | \\ H & — & C & — & C—OH \\ & & | & & | \\ H & — & C—H & & H \\ & & | & & \\ & & H & & \end{array}$$

i-propylcarbinol,
i-$C_3H_7.CH_2OH$.

$$\begin{array}{ccccc} & & H & & \\ & & | & & \\ & H— & C—H & & H \\ & & | & & | \\ H—O & — & C & — & C—H \\ & & | & & | \\ & H— & C—H & & H \\ & & | & & \\ & & H & & \end{array}$$

trimethylcarbinol,
$(CH_3)_3COH$.

Any alcohol of the $C_nH_{2n+1}OH$ series, no matter how complicated, has its constitution represented as derived from carbinol in which one or more of the three hydrogen atoms attached to the carbon atom is (or are) replaced by alkyl groups.

The alcohols represented above can be divided into three classes according to their formulae:

(*a*) Those alcohols containing the —CH_2OH group united either to hydrogen or to an alkyl group:

Methyl alcohol or carbinol	$H.CH_2OH$
Ethyl alcohol or methylcarbinol	$CH_3.CH_2OH$
n-Propyl alcohol or ethylcarbinol	$C_2H_5.CH_2OH$
n-Butyl alcohol or *n*-propylcarbinol	*n*-$C_3H_7.CH_2OH$
i-Propylcarbinol	*i*-$C_3H_7.CH_2OH$

(*b*) Those alcohols containing the $>CHOH$ group united to two alkyl groups:

i-Propyl alcohol or dimethylcarbinol	$\begin{array}{l} CH_3 \\ CH_3 \end{array}\!\!>C<\!\!\begin{array}{l} H \\ OH \end{array}$
Methylethylcarbinol	$\begin{array}{l} CH_3 \\ C_2H_5 \end{array}\!\!>C<\!\!\begin{array}{l} H \\ OH \end{array}$

(*c*) The alcohol containing the $\geqslant$C—OH group united to three alkyl groups:

Trimethylcarbinol	$(CH_3)_3COH$

The above are representative of the three types of alcohols; those containing the $\cdot CH_2OH$ group known as *primary* alcohols, those containing the $:CHOH$ group known as *secondary* alcohols and those containing the : COH group known as *tertiary* alcohols. In so far as all alcohols contain the hydroxyl group, they all have certain chemical properties in common; but primary, secondary and tertiary alcohols differ in certain of their chemical properties, especially in their behaviour on oxidation. By its behaviour on oxidation, a particular alcohol is classified as a primary, or a secondary, or a tertiary alcohol. However complicated a particular alcohol may be, it must belong to one of the above three classes.

Methyl Alcohol

Methyl alcohol, carbinol, methanol,* CH_3OH, was formerly known as wood spirit, because it was at one time exclusively produced by the destructive distillation of wood. Except for the production of wood charcoal, certain kinds of which have valuable absorptive properties, the wood distillation industry for the production of methyl alcohol and other volatile products (such as acetone and acetic acid, which will be mentioned later) has been almost completely superseded.

The chief modern processes for the manufacture of methyl alcohol employ 'highly purified' water gas as their starting material. *Water gas* is the product of the action of steam on incandescent coke, when the following reactions take place:

$$\text{(i)}\quad C + H_2O \rightleftarrows CO + H_2$$

$$\text{(ii)}\quad CO + 2H_2O \rightleftarrows CO_2 + 2H_2$$

The higher the temperature of the reaction the greater the proportion of carbon monoxide in the product, and for the purpose of the production of methane the gas should consist of one volume of carbon monoxide to two volumes of hydrogen, and the mixture should be as pure as possible. In the production of methyl alcohol, attempts are made to realise the reaction

$$CO + 2H_2 \rightarrow CH_4O \quad (\text{i.e. } CH_3OH)$$

as completely as possible.

The manufacture of methyl alcohol from water gas is now carried out industrially in England, France, Germany and America. The purified mixture of carbon monoxide and hydrogen is passed over certain metallic oxides at pressures of about 200 atmospheres and at various temperatures from 200° to about 400° according to the particular catalyst used. In Petart's (French) process the mixture of gases is passed at 200 atmospheres over zinc oxide at 400–425°. The product is chiefly methyl alcohol and very little methane is

* Methanol is the industrial name for methyl alcohol. 'Methanol' has also been adopted for scientific use. It is particularly suitable as a descriptive name for methyl alcohol; the ending '-ol' connotes alcohol, and the whole word indicates the alcohol derived from methane.

formed. In Audibert's (also French) process, the best catalyst is stated to be uranium suboxide, which is contained in copper-lined steel tubes through which the gases are passed at 200 atmospheres at a temperature of 225–275°. Under these conditions methyl alcohol is produced almost exclusively; at temperatures higher than 325° side reactions set in, methane for example being produced. Under the best conditions, after condensing the methyl alcohol, the issuing gas contains only 1 per cent. of methane and a little carbon dioxide and hydrogen. It is possible that the carbon dioxide is formed as a result of the reaction

$$CO + 2H_2O \rightleftarrows CO_2 + 2H_2$$

the gases having contained a little moisture at the beginning. Almost all possible catalysts which can be employed in the production of methyl alcohol from the carbon monoxide-hydrogen mixtures have been patented for this particular purpose, and it appears that almost any oxide can be used provided that it is not reducible by either hydrogen or carbon monoxide below 550°. For one of the English processes the catalyst consists essentially of a basic zinc chromate having the composition $4ZnO.CrO_3$, and the purified mixture of carbon monoxide and hydrogen is passed over this at high pressures at a temperature of 300–400°. In this particular process it is stated that the catalyst is suitable for the production of methyl alcohol when it is free from soluble alkali salts. If, however, the catalyst contains such salts, it is suitable for the production of higher alcohols, especially when the rate of flow of the gases is slow. The methyl alcohol produced by the high pressure process is, for a commercial product, of a high degree of purity, and the development of the production of this simple fundamentally important compound (and of its oxidation products) is an outstanding achievement of chemical engineering.

The production of methyl alcohol from methane (from natural gas) is, as yet, not developed commercially, but the process has been carefully investigated. The methane is chlorinated under such conditions as will give the maximum yield of methyl chloride in association with methylene chloride, chloroform and carbon tetrachloride (p. 16). The methyl chloride is then converted into methyl alcohol by passing its vapour mixed with steam over hydrated lime, which neutralises the hydrochloric formed in the reaction

$$CH_3Cl + HOH \rightarrow CH_3OH + HCl$$

The relatively high cost of chlorine is a disadvantage in this process in spite of the fact that the important substances, chloroform and carbon tetrachloride, are by-products. These two substances are produced more conveniently however by other methods.

The product obtained by the high pressure process usually only requires to be left for some time over fresh quick lime (to take up moisture) and subsequent careful fractional distillation for complete purification.

Methylated spirit is 'denatured' ethyl alcohol (p. 96) containing up to 90 per cent. ethyl alcohol, up to 9 per cent. wood spirit (crude methyl alcohol), about 0·5 per cent. pyridine (p. 298) and a small quantity of petroleum. The mixing of these substances with the ethyl alcohol is to prevent its use for drinking purposes: methyl alcohol itself is not only an intoxicant but also a dangerous poison.

Methyl alcohol when pure is a colourless liquid with little or no odour. It freezes at −94·7° and has b.p. 64·7°. When ignited it burns with a very pale blue flame producing carbon dioxide and water:

$$2CH_3OH + 3O_2 = 2CO_2 + 4H_2O$$

By analysis and by determination of its vapour density the formula is easily established as CH_4O.

When methyl alcohol is allowed to react with metallic sodium, hydrogen is evolved and the sodium dissolves. One molecular proportion of hydrogen is evolved from two molecular proportions of methyl alcohol, or, when sodium dissolves in methyl alcohol, one atom of hydrogen is replaced by sodium, the product formed being known as sodium methylate (or methoxide). Since only one atom of hydrogen in the methyl alcohol molecule is displaced by sodium, there is reason for believing that one of the hydrogen atoms is differently combined from the other three. This is indicated in the constitutional formula

$$\begin{array}{c} \quad H \\ \quad | \\ H—C—O—H \\ \quad | \\ \quad H \end{array}$$

where three atoms of hydrogen are represented as being joined directly to the carbon atom and the fourth atom of hydrogen—the one capable of being replaced by sodium—is joined to the oxygen atom. The action of sodium on methyl alcohol may be compared with the more vigorous reaction of sodium on water:

$$2H—O—H + 2Na = 2H—O—Na + H_2$$

$$2CH_3—O—H + 2Na = 2CH_3—O—Na^* + H_2$$

* The sodium methylate, CH_3ONa, can be obtained as a colourless crystalline solid by careful evaporation of the resulting liquid under diminished pressure. It is easily soluble in methyl alcohol and is readily decomposed by water:

$$CH_3ONa + H_2O = CH_3OH + NaOH$$

In the presence of methyl alcohol, sodium methylate is frequently used as a reagent in organic chemistry.

The reaction between methyl alcohol and potassium is very similar, potassium methylate having similar properties to those of sodium methylate.

Methyl alcohol is neutral in reaction and is freely soluble in water. Pure methyl alcohol indeed is hygroscopic, and when mixed with water a contraction in volume takes place and heat is developed. In spite of its neutrality towards indicators, methyl alcohol reacts with acids, water being eliminated, the acid radical taking the place of the hydroxyl group in the alcohol. Thus if methyl alcohol be saturated with hydrogen chloride and the mixture heated, methyl chloride (monochloromethane) is given off. The reaction may be compared with the action of (say) hydrochloric acid on sodium hydroxide, when the salt, sodium chloride, is formed:

$$NaOH + HCl = NaCl + H_2O$$

$$CH_3OH + HCl \rightarrow CH_3Cl + H_2O^*$$

Similarly, when methyl alcohol is mixed with concentrated sulphuric acid, heat is developed and a compound known as methyl hydrogen sulphate is produced. If the mixture be distilled under reduced pressure, *dimethyl sulphate*† distils over:

$$CH_3.OH + H_2SO_4 \rightarrow \begin{matrix} CH_3 \\ H \end{matrix}\!\!>SO_4 + H_2O$$

$$CH_3.OH + \begin{matrix} CH_3 \\ H \end{matrix}\!\!>SO_4 \rightarrow \begin{matrix} CH_3 \\ CH_3 \end{matrix}\!\!>SO_4 + H_2O$$

[By distillation, $2\begin{matrix} CH_3 \\ H \end{matrix}\!\!>SO_4 \rightarrow \begin{matrix} CH_3 \\ CH_3 \end{matrix}\!\!>SO_4 + H_2SO_4$]

The formation of dimethyl sulphate does take place to a small extent when sulphuric acid is mixed with methyl alcohol, but dimethyl sulphate is obtained in quantity when the mixture is distilled. The formation of methyl hydrogen sulphate and of dimethyl sulphate

* This reaction affords further proof of the existence of the hydroxyl group in the molecule of methyl alcohol.

† Dimethyl sulphate, the constitution of which should be expressed as

$$\begin{matrix} CH_3{-}O \\ CH_3{-}O \end{matrix}\!\!>SO_2 \quad \text{or} \quad \begin{matrix} CH_3{-}O \\ CH_3{-}O \end{matrix}\!\!>S\!\!<\!\!\begin{matrix} O \\ O \end{matrix}$$

is a colourless liquid, b.p. 188·6°. It is slowly decomposed by water with the formation of methyl alcohol and sulphuric acid:

$$(CH_3)_2SO_4 + 2H_2O = 2CH_3OH + H_2SO_4$$

It is more rapidly decomposed by alkalies, when methyl alcohol and the corresponding sulphates are produced.

Dimethyl sulphate is an important reagent in organic chemistry. It should be handled with care since it is poisonous.

can be compared with the neutralisation of sodium hydroxide and sulphuric acid:

$$NaOH + H_2SO_4 = NaHSO_4 + H_2O$$
$$NaOH + NaHSO_4 = Na_2SO_4 + H_2O$$

The product of the reaction between an alcohol and an acid is known as an *ester*, methyl chloride, methyl hydrogen sulphate and dimethyl sulphate being esters of hydrochloric and sulphuric acids respectively. Similarly methyl bromide (monobromomethane) and methyl iodide (monoiodomethane, CH_3I) are esters respectively of hydrobromic and hydriodic acids.

It will be shown later (p. 213) that the reaction between an alcohol and an acid, whereby an ester is produced, differs from the normal reaction for the formation of a salt by the neutralisation of an acid by an alkali. In a normal case, the reaction between an alkali and an acid proceeds rapidly and completely to neutrality. Under ordinary circumstances the reaction between an alcohol and an acid proceeds with measurable velocity and is never complete. It is more correct to represent the formation of methyl chloride, for example, by

$$CH_3OH + HCl \rightleftarrows CH_3Cl + H_2O$$

and we know that methyl chloride, like dimethyl sulphate, is decomposed by water.

Methyl chloride is also produced in the vigorous reaction which takes place when phosphorus pentachloride, phosphorus trichloride and phosphorus oxychloride react with methyl alcohol. The reactions which take place may be represented:

$$\text{(i)} \quad CH_3OH + PCl_5 \rightarrow CH_3Cl + POCl_3 + HCl$$
$$\text{(ii)} \quad 3CH_3OH + PCl_3 \rightarrow 3CH_3Cl + H_3PO_3$$
$$\text{(iii)} \quad 3CH_3OH + POCl_3 \rightarrow 3CH_3Cl + H_3PO_4$$

It will be noticed in the case of reaction (i) that the hydroxyl group in the methyl alcohol is replaced by a chlorine atom and that hydrogen chloride is evolved. The evolution of hydrogen chloride when phosphorus pentachloride reacts with an organic compound is frequently regarded as a diagnosis of the presence of a hydroxyl group in the latter.

Methyl bromide is produced when the corresponding phosphorus bromides are used. For the production of methyl iodide it is necessary to use iodine and free phosphorus and in this case the reaction is conveniently expressed:

$$6CH_3OH + 2P + 3I_2 \rightarrow 6CH_3I + 2H_3PO_3$$

Methyl iodide (a colourless liquid, frequently slightly coloured owing to the presence of a small trace of free iodine, b.p. 42·3°) is the most important of these three esters and is usually prepared by the direct action of iodine on methyl alcohol containing phosphorus in suspension. It may also be prepared by dropping dimethyl sulphate into a dilute solution of potassium iodide:

$$(CH_3)_2SO_4 + 2KI \rightarrow K_2SO_4 + 2CH_3I^*$$

Methyl alcohol is readily oxidised under a variety of conditions and yields directly two products, first a gas known as *formaldehyde* (p. 144), which is then further oxidised to a colourless liquid known as *formic acid* (p. 192). When the oxidation of methyl alcohol is carried out under controlled conditions, the only other product of

* Methyl iodide is a convenient source for the preparation of pure methane (p. 13). A *copper-zinc couple* is prepared by allowing granulated zinc to remain in a 2 per cent. aqueous solution of copper sulphate until the latter is decolorised, the copper being deposited on the zinc; the solution may then be poured off and the process repeated two or three times. The copper-covered zinc is carefully and well washed with water and then with ethyl alcohol. Some quantity of the couple is placed in a suitable flask as shown and the U-tube is also packed with a further

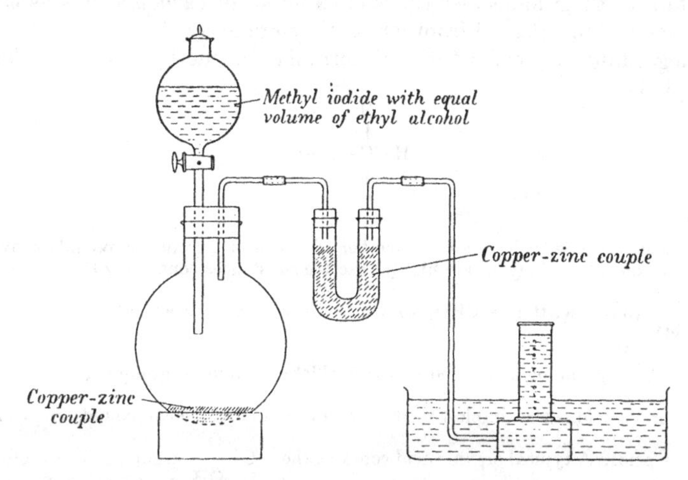

quantity. The couple in the flask is covered with ethyl alcohol (containing some water) and to this is added slowly a mixture of equal volumes of methyl iodide and ethyl alcohol. If a few drops of sulphuric acid have been added to the contents of the flask, the reaction will begin and continue without any external heating being necessary. A steady stream of pure methane is produced and the gas issuing from the U-tube can be collected in any suitable manner. The couple (moistened with ethyl alcohol) in the U-tube serves to reduce any methyl iodide vapour which may be carried over and which would contaminate the methane. Sulphuric acid can also be used to free the methane from any methyl iodide. The reaction for the production of methane by this method is usually expressed:

$$CH_3I + 2H \rightarrow CH_4 + HI$$

the nascent hydrogen, reducing the methyl iodide, being produced by the action of the copper-zinc couple on the aqueous alcohol used.

the reaction is water. The reactions expressing the oxidation of methyl alcohol are

$$\text{(i)}\quad H-\overset{\displaystyle H}{\underset{\displaystyle H}{\overset{|}{\underset{|}{C}}}}-O-H + O \rightarrow H-\underset{\displaystyle H}{\underset{|}{C}}=O + H_2O$$

or H.CHO
formaldehyde*

$$\text{(ii)}\quad H-\overset{\displaystyle H}{\overset{|}{C}}=O + O \rightarrow H-\overset{\displaystyle O-H}{\overset{|}{C}}=O$$

or H.COOH
formic acid†

Methyl alcohol is a primary alcohol: it contains the $-CH_2OH$ grouping. All primary alcohols behave analogously to methyl alcohol on oxidation, i.e. they yield aldehydes and acids both of which contain in their molecules the same number of carbon atoms as the molecule of the alcohol from which they are derived.

Regarding the constitutional formula for methyl alcohol, that adopted

$$H-\overset{\displaystyle H}{\underset{\displaystyle H}{\overset{|}{\underset{|}{C}}}}-O-H$$

* Formaldehyde is the simplest member of a class of organic compounds known as *aldehydes*. The aldehydes are highly reactive substances containing the grouping $-C\begin{smallmatrix}\diagup H\\ \diagdown\!\!\diagdown O\end{smallmatrix}$ (often written $-CHO$) to which their reactivity as aldehydes is due. The $-C\begin{smallmatrix}\diagup H\\ \diagdown\!\!\diagdown O\end{smallmatrix}$ grouping is referred to as the aldehyde group or grouping.

† Formic acid is the simplest member of a class of organic compounds known as organic acids. A typical organic acid contains the $-C\begin{smallmatrix}\diagup\!\!\diagup O\\ \diagdown OH\end{smallmatrix}$ grouping. This is often written as $-COOH$ or $-CO_2H$ and described as the carboxyl group or carboxy grouping. In a typical organic acid it is the hydrogen of the carboxyl group which is replaced by a metal in salt formation. A typical monobasic organic acid therefore contains one carboxyl group, a dibasic acid two carboxyl groups, and so on. The sodium salt of formic acid (sodium formate) has the formula $H-C\begin{smallmatrix}\diagup\!\!\diagup O\\ \diagdown O-Na\end{smallmatrix}$ (or $H.COONa$ or $H.CO_2Na$). Calcium formate is $H-C\begin{smallmatrix}\diagup\!\!\diagup O\\ \diagdown O-Ca-O\diagup\end{smallmatrix}\!\!\begin{smallmatrix}O\diagdown\!\!\diagdown\\ \end{smallmatrix}C-H$ (or $(H.COO)_2Ca$). Except that they are generally weaker, carboxylic acids behave exactly as do inorganic acids, i.e. they undergo salt and ester formation, and, as in the case of the inorganic oxy- acids, they give rise to anhydrides, acid chlorides, etc.

is the only one which can be written for a compound having the molecular formula CH_4O which is in keeping with the accepted valencies of hydrogen, oxygen and carbon. This constitutional formula is in keeping with there being one hydrogen atom differently combined in the molecule from the other three, one hydrogen atom being replaceable by sodium and potassium. It is in keeping with the presence of the hydroxyl group, as indicated by the action of acids and by the action of phosphorus pentachloride, and finally it is in keeping with the direct formation of the two oxidation products, viz. formaldehyde and formic acid.

Methyl alcohol, formaldehyde and formic acid can be regarded as intermediate products in the complete oxidation of methane to carbon dioxide and water; this is illustrated:

$$\left(\begin{array}{l} \begin{array}{c}H\\|\\H—C—H\\|\\H\end{array} + O \rightarrow \begin{array}{c}H\\|\\H—C—OH\\|\\H\end{array} \\ \text{i.e. } CH_4 + O \rightarrow CH_3.OH \text{ (methyl alcohol);} \end{array}\right)$$

$$\begin{array}{c}H\\|\\H—C—OH\\|\\H\end{array} + O \rightarrow \underset{\text{unstable}}{\begin{array}{c}OH\\|\\H—C—OH\\|\\H\end{array}} \xrightarrow{-H_2O} H—C\begin{array}{l}{\nearrow}O\\ \searrow H\end{array}$$

$$\text{i.e. } CH_3OH + O \rightarrow H_2O + H.CHO \text{ (formaldehyde);}$$

$$\begin{array}{c}OH\\|\\H—C—OH\\|\\H\end{array} + O \rightarrow \underset{\text{unstable}}{\begin{array}{c}OH\\|\\H—C—OH\\|\\OH\end{array}} \xrightarrow{-H_2O} H—C\begin{array}{l}{\nearrow}O\\ \searrow OH\end{array}$$

$$\text{i.e. } H.CHO + O \rightarrow H.COOH \text{ (formic acid);}$$

$$\begin{array}{c}OH\\|\\H—C—OH\\|\\OH\end{array} + O \rightarrow \underset{\text{unstable}}{\begin{array}{c}OH\\|\\HO—C—OH\\|\\OH\end{array}} \xrightarrow{-2H_2O} CO_2$$

$$\text{i.e. } H.COOH + O \rightarrow CO_2 + H_2O;$$

the sum of these reactions is:

$$CH_4 + 2O_2 \rightarrow CO_2 + 2H_2O$$

Ethyl Alcohol

*Ethyl alcohol, monohydroxyethane, methylcarbinol, ethanol,** is the next homologue to methyl alcohol and has the formula C_2H_5OH.

The chief source of ethyl alcohol is a sugar known as glucose, which itself may be produced from a large variety of other complex and related substances, such as starch, cane sugar, malt sugar, etc. The process by which alcohol is produced from glucose is known as fermentation (from the Latin *fervere*, to boil). This process will be discussed later (p. 149), and for the present the production of alcohol by the fermentation of dilute solutions of glucose may be represented:

$$\underset{\text{glucose}}{C_6H_{12}O_6} \rightarrow 2CO_2 + 2C_2H_5OH$$

although the changes which take place are far more complicated than that implied here, and moreover other products are formed at the same time. The production of wine depends on the fermentation of solutions of glucose present in grape juice whereby ethyl alcohol, the chief product of the fermentation, is produced. Beer is essentially the product of the fermentation of glucose produced from the starch in barley, and cider is the product of the fermentation of the glucose contained in the juice from apples of particular varieties. The so-called *spirits* are the distilled products of the various fermentation processes, e.g. brandy is essentially the distillate from fermented grape juice, and whisky is the distillate from fermented barley and other cereals. Spirits always have a much higher alcohol content than true fermented liquids, and the particular flavours of both spirits and true fermented liquids may be natural or imparted by the addition of suitable materials. The presence of ethyl alcohol in spirits, wine, beer, etc. can readily be determined by distilling the liquid in question and testing the distillate in the way described below (p. 104). Ethyl alcohol is only one of many substances which can be produced by fermentation, and fermentation processes are now used industrially for the production of a large variety of organic substances; the production of methane by a fermentation process has been referred to previously (p. 14).

The production of ethyl alcohol and carbon dioxide by the fermentation of glucose can be shown in the laboratory. A solution containing 10 grams of glucose (grape sugar) in about 150 c.c. of water is placed in a 2-litre flask provided with a suitable cork through which is passed a tube leading to a suitable vessel containing clear

* Ethanol is now the industrial and scientific name for ethyl alcohol. Ethyl alcohol is also known as 'spirits of wine', especially in pharmacy. Ethyl alcohol is probably the most frequently used substance in organic chemistry, and the word 'alcohol' without any distinguishing term usually connotes 'ethyl alcohol'.

lime water. A small quantity of brewer's yeast is added to the sugar solution and the apparatus is allowed to stand in a warm place (about 30°) for at least 24 hours. After that time, the lime water will have become turbid (proving the evolution of carbon dioxide, care having been taken to prevent access of carbon dioxide from the atmosphere), and the contents of the flask which have been frothing should be filtered and the clear filtrate placed in a suitable distilling flask provided with a condenser and a flask for collecting the distillate. The contents of the distilling flask should be distilled and collected until about one-third has passed over. The distillate will have a distinct vinous odour and consists essentially of an aqueous solution containing almost all the alcohol produced in the fermentation. If anhydrous potassium carbonate be added to the distillate until no more will dissolve, the liquid will separate into two layers, which can be separated by means of a separating funnel (p. 502). The upper layer, which contains chiefly alcohol, can then be dried by the addition of fresh anhydrous potassium carbonate to remove most of the small amount of water still remaining, decanted or filtered from the potassium carbonate and redistilled. It will be found to boil for the most part at about 78°, and the distillate coming over at that temperature can be tested for ethyl alcohol by one or more of the reactions described below. This illustrates essentially the process by which ethyl alcohol is produced industrially. The drying agent on the industrial scale is quick lime, and the distillation is carried out in an elaborate still provided with a long column to tho leading-off pipes for the more complete separation of the ethyl alcohol from the water and higher boiling products of the fermentation process.

When produced under the best conditions in the way indicated, the product, *industrial spirit* as it is called, consists of about 96 per cent. ethyl alcohol, the rest being water and very small quantities of higher boiling alcohols. To obtain *absolute alcohol* (i.e. pure ethyl alcohol), industrial spirit is allowed to stand for some hours with excess of fresh quick lime and then carefully distilled. If this process be repeated, and the resulting alcohol carefully distilled using a long fractionating column, a product containing less than 0·2 per cent. of water can be obtained. To remove the last trace of water is a matter of some difficulty, although various methods for the purpose are known.* In attempting to prepare absolute alcohol it is necessary

* Two of these are the following: distillation in the presence of calcium turnings (the calcium acts upon the water forming calcium hydroxide and hydrogen), distillation in the presence of magnesium amalgam (magnesium hydroxide and hydrogen being formed).

Addition of one part of glycerol (p. 385) to three parts of rectified spirit and subsequent distillation is a rapid method of the dehydration of the latter, leading to production of 99·5 per cent. ethyl alcohol. This is stated to be more efficient than the use of calcium oxide. The glycerol residue can be recovered, dried and used again.

to prevent access of atmospheric moisture to the apparatus, since absolute alcohol is highly hygroscopic.

When alcoholic liquids, such as wine, beer, spirits, etc., are distilled, the distillate contains the ethyl alcohol free from any colouring matter. If the distillation is carried on until the volume of the distillate is one-third that of the liquid distilled, it contains practically all the ethyl alcohol formerly present in the latter. Pure ethyl alcohol can be obtained from this distillate in the manner indicated.

In spite of the fact that a large number and variety of natural starchy and sugar-containing materials have been proved to be capable of yielding ethyl alcohol by fermentation processes, the high cost of fermentation ethyl alcohol has made it desirable to investigate other means of production of this fundamentally important compound. Ethyl alcohol may be described as almost the basic substance of the organic chemical industry, and apart from this its uses in industry generally and its possible future use in internal combustion engines render additional sources of supply absolutely essential. Three chemical processes for the industrial manufacture of ethyl alcohol have been investigated. The first of these uses ethylene (in coal gas) as the starting material. The gas from the gas works and coke-oven plants is freed from sulphur and passed into concentrated sulphuric acid at 70–80°. Under these conditions the ethylene is absorbed and ethyl hydrogen sulphate (p. 37) is produced:

$$\begin{matrix} CH_2 \\ \| \\ CH_2 \end{matrix} + \begin{matrix} H \\ | \\ OSO_2OH \end{matrix} \rightarrow \begin{matrix} CH_3 \\ | \\ CH_2—OSO_2OH \end{matrix}$$

The mixture of ethyl hydrogen sulphate and excess sulphuric acid is diluted considerably with water and subjected to distillation, when a 15 per cent. solution of ethyl alcohol in water collects as the distillate, owing to the hydrolysis of the ester:

$$\begin{matrix} CH_3 \\ | \\ CH_2—OSO_2OH \end{matrix} + H_2O \rightarrow \begin{matrix} CH_3 \\ | \\ CH_2OH \end{matrix} + H_2SO_4$$

The ethyl alcohol can be obtained from the distillate by careful fractional distillation, drying and redistillation. The sulphuric acid in a diluted condition remains in the 'still'; it can be concentrated and used again. Theoretically the process is an economical one, but there are difficulties connected with the necessity of using acid-resisting materials for the plant, the necessity of obtaining rapid absorption of the ethylene and the economic factor connected with the recovery of the sulphuric acid in such a condition that it can be used indefinitely.

The second process which may be used for the production of ethyl alcohol by chemical methods starts with acetylene (from calcium carbide). This is converted into ethyl alcohol in the following stages:

$$\text{(i)}\quad \begin{matrix}CH\\ \|\!|\\ CH\end{matrix} + \begin{matrix}H_2\\ \|\\ O\end{matrix} \rightarrow \begin{matrix}CH_3\\ |\\ C\end{matrix}\!\!\begin{matrix}/H\\ \\ \backslash\!\backslash O\end{matrix} \qquad \text{(p. 154)}$$

$$\text{(ii)}\quad \begin{matrix}CH_3\\ |\\ C\end{matrix}\!\!\begin{matrix}/H\\ \\ \backslash\!\backslash O\end{matrix} + H_2 \rightarrow \begin{matrix}CH_3\\ |\\ CH_2OH\end{matrix}$$

The conversion of acetylene into acetaldehyde is effected in the presence of a mercury catalyst, and the reduction of the acetaldehyde to ethyl alcohol is effected in the presence of a copper or nickel catalyst. (Incidentally, acetaldehyde from acetylene is used in the industrial manufacture of acetic acid, p. 198.) The manufacture of ethyl alcohol from acetylene involves cheap electric power for the manufacture of calcium carbide and for the production of hydrogen, and, although at present it cannot compete economically with the fermentation processes, there are great possibilities in the process. Alcohols higher in the series than methyl alcohol are also produced from the water-gas reaction (p. 88) and among these is ethyl alcohol. This is another highly important industrial process for the production of ethyl alcohol.

Pure ethyl alcohol is a colourless liquid which freezes at −117° and has b.p. 78·4°.* When ignited it burns with a pale blue flame forming carbon dioxide and water. The combustion or complete oxidation of alcohol is expressed by the equation:

$$C_2H_5OH + 3O_2 = 2CO_2 + 3H_2O$$

Pure ethyl alcohol is soluble in almost all organic solvents and in water in all proportions; it is hygroscopic and absorbs moisture from the atmosphere. When it is mixed with water, heat is developed and a contraction in volume occurs. The greatest contraction occurs when three molecular proportions of water are mixed with one molecular proportion of ethyl alcohol. The specific gravity of pure ethyl alcohol is 0·78945 at 20° (compared with water = 1 at 4°), and the specific gravity of alcohol-water mixtures increases from this figure according to the water content. These specific gravities have been very carefully determined and the so-called 'alcohol tables' render the determination of the alcohol content of any particular alcohol-water mixture from its specific gravity a mere matter of reference.

The determination of the percentage of alcohol in any alcoholic liquid consists of the distillation of the liquid under standard conditions until one-third of the original volume has been collected in

* The low freezing point of ethyl alcohol and the regularity of its volume changes with temperature make it a suitable liquid for low temperature thermometers.

the distillate. The distillate is then made up to the original volume with water, the specific gravity of this mixture taken under standard conditions, and then by reference to the tables the percentage of alcohol noted. This is taken as the percentage of alcohol in the original liquid.

Like methyl alcohol, ethyl alcohol reacts vigorously with sodium and potassium, hydrogen being evolved and sodium or potassium ethylate formed. The reaction is less vigorous than in the case of water, the metal rarely catching fire:

$$2C_2H_5OH + 2Na = 2C_2H_5ONa + H_2$$

The *sodium ethylate* is usually left dissolved in the excess of ethyl alcohol, but it can be obtained as a colourless crystalline and hygroscopic solid by carefully distilling the solution in ethyl alcohol under reduced pressure. The sodium ethylate is easily decomposed by water, sodium hydroxide being produced:

$$C_2H_5ONa + H_2O = C_2H_5OH + NaOH$$

Sodium in the presence of ethyl alcohol is very frequently employed as a reducing agent in organic chemistry when the presence of an alkali is not contra-indicated.

Like methyl alcohol, ethyl alcohol is neutral in reaction towards indicators, and it behaves similarly towards acids forming esters.

Ethyl chloride (monochloroethane), C_2H_5Cl, a colourless liquid, b.p. 12·5°—used as a local anaesthetic on account of the cold produced during its evaporation and also as a general anaesthetic, especially in dental operations, particularly for children—is prepared by the action of hydrogen chloride on ethyl alcohol. The reaction is expressed:

$$C_2H_5OH + HCl \rightleftarrows C_2H_5Cl + H_2O$$

In the actual preparation ethyl alcohol is saturated with hydrogen chloride and the mixture very carefully warmed. The ethyl chloride is thus removed from the sphere of the reaction which so tends to go more to completion, the equilibrium indicated above being shifted towards the right. Zinc chloride is often mixed with the reactants, and this on account of its affinity for water also tends to shift the equilibrium in the same direction.

The action of hydrogen bromide on ethyl alcohol is similar to that of hydrogen chloride. In the preparation of *ethyl bromide* (monobromoethane), C_2H_5Br—a colourless liquid, b.p. 38·4°, often coloured somewhat brown owing to the presence of a very small amount of free bromine—the hydrogen bromide is best developed *in situ*, and the ethyl bromide is distilled off continuously, thus destroying the equilibrium and increasing the relative amount of ethyl bromide obtained. The following illustrates the preparation of ethyl bromide, and is based on a published description:

60 c.c. of concentrated sulphuric acid are added carefully to 64 c.c. of rectified industrial spirit, the mixture being well cooled during the process. To the cold mixture contained in a 500 c.c. distillation flask, powdered sodium bromide (105 grams) is added. The flask suitably stoppered and attached to an efficient condenser is warmed in a large water-bath. The outlet from the condenser is usefully provided with an adapter, the end of which just dips under a quantity of water in a receiver which is kept cold. The water-bath is slowly warmed to such a temperature that the ethyl bromide gently drops from the end of the condenser and falls under the water in the receiver, the temperature of the water-bath being gradually raised, finally to the boiling point, as the distillation of the ethyl bromide slackens and until finally no more ethyl bromide distils. The contents of the receiver are transferred to a separating funnel, the ethyl bromide separated, washed with dilute sodium carbonate solution to remove any free acid and finally with water and then dried with anhydrous (fused) calcium chloride. The dried ethyl bromide should be finally redistilled, the distillation flask being placed in a warmed water-bath, the portion distilling between 37·5° and 39·5° being collected as moderately pure ethyl bromide. A further distillation, using a fractionating column, is necessary for more complete purification. Working carefully, a yield of 80 per cent. based on the ethyl alcohol used may be obtained. The reaction taking place under the conditions indicated may be expressed

$$C_2H_5OH + H_2SO_4 + NaBr \rightarrow C_2H_5Br + NaHSO_4 + H_2O$$

Ethyl iodide, an important reagent in organic chemistry, is prepared by the action of iodine on alcohol in the presence of phosphorus (*v.* below).

As far as the formation of its ethyl esters is concerned the reaction between sulphuric acid and ethyl alcohol is analogous to that with methyl alcohol. It has already been observed that when ethylene is absorbed by concentrated sulphuric acid *ethyl hydrogen sulphate* is formed, and the same ester is formed when concentrated sulphuric acid is mixed with ethyl alcohol according to the equilibrium reaction:

$$C_2H_5OH + H_2SO_4 \rightleftarrows C_2H_5O.SO_2.OH + H_2O$$

When equimolecular quantities of the alcohol and acid are taken and the mixture allowed to stand on a heated water-bath for 4 hours, 50 per cent. of the sulphuric acid is converted into the ethyl hydrogen sulphate; if three molecular proportions of the alcohol are used to one molecular proportion of sulphuric acid, 77·4 per cent. of the latter is converted into ethyl hydrogen sulphate. Many salts of ethyl hydrogen sulphate are known and the free acid or ester-acid has been isolated through the lead and barium salts, which have the formulae $(C_2H_5O.SO_2.O)_2Pb$ and $(C_2H_5O.SO_2.O)_2Ba$ respectively. For ordinary purposes, there is little need to isolate the very hygroscopic ester-acid.

When ethyl hydrogen sulphate mixed with sulphuric acid and alcohol is heated, ethylene is formed:

$$C_2H_5O.SO_2.OH \rightarrow C_2H_4 + H_2SO_4 \quad \text{(p. 29)}$$

When ethyl hydrogen sulphate (i.e. the mixture containing it, alcohol and sulphuric acid) is diluted largely with water and submitted to distillation, ethyl alcohol is formed owing to the hydrolysis of the ester-acid (p. 98):

$$C_2H_5O.SO_2.OH + H.OH \rightleftarrows C_2H_5.OH + H_2SO_4$$

If the alcohol-sulphuric acid mixture is heated with excess of ethyl alcohol, diethyl ether (p. 30) is produced and distils over:

$$C_2H_5O.SO_2.OH + C_2H_5.OH \rightarrow C_2H_5O.C_2H_5 + H_2SO_4$$

When the mixture of one volume of absolute ethyl alcohol and two volumes of concentrated sulphuric acid (containing, of course, ethyl hydrogen sulphate) is submitted to distillation under greatly reduced pressure, *diethyl sulphate*, $(C_2H_5)_2SO_4$ or $(C_2H_5O)_2SO_2$, the diethyl (or normal ethyl) ester of sulphuric acid, slowly distils over:

$$\begin{array}{l} C_2H_5O.SO_2.OH \\ C_2H_5O.SO_2.OH \end{array} \rightarrow \begin{array}{l} C_2H_5O \\ C_2H_5O \end{array}\!\!>SO_2 + H_2SO_4$$

Diethyl sulphate is a colourless oil possessing a characteristic odour. It has b.p. 98° at 15 mm. It undergoes decomposition when distilled at ordinary pressure. Like dimethyl sulphate it is an important synthetic reagent in organic chemistry. When it is mixed with an aqueous solution of potassium iodide, ethyl iodide is produced:

$$(C_2H_5)_2SO_4 + 2KI \rightarrow 2C_2H_5I + K_2SO_4$$

The ethyl esters of most of the known acids have been prepared and those of organic acids are frequently used for the identification of those acids. Many of the ethyl esters of organic acids are important as solvents and as synthetic reagents.

The reactions of ethyl alcohol with the phosphorus halides are precisely analogous to the reactions between the latter and methyl alcohol. The products of the reactions are ethyl chloride, ethyl bromide, etc., the monohalogenoethanes or ethyl esters of hydrochloric acid, hydrobromic acid, etc.

The following method for the preparation of ethyl iodide was described by Walker in 1892, and illustrates generally the method of the replacement of the alcoholic hydroxyl group by the iodine atom. The quantities stated are one-fifth those mentioned in the original description.

50 grams ethyl alcohol, 6 grams yellow phosphorus and 6 grams red phosphorus together with two or three pieces of porous tile are placed in the round-bottom

(300 c.c.) flask, set up as shown. The tube A is to contain 20 grams iodine. The tube is provided with a loose plug of glass wool at B and the end of the tube C is constricted somewhat. The iodine is packed into A, and the rest of the apparatus fitted up as indicated. The flask is heated on the water-bath, and the alcohol in it boiled. The alcohol vapour passes up the tube D and into the adapter E, where it is condensed. The liquid alcohol drips down at E and percolates through the iodine in A dissolving some of the iodine. When all the iodine has dissolved, heating is continued until there is no iodine colour in the liquid in the flask. At this stage, and after cooling the contents of the flask, another 20 grams of iodine are introduced into A and the process repeated until finally 100 grams of iodine have been used. The heating should be gentle throughout the operations and towards the end less heat should be used because the ethyl iodide, which is accumulating in the flask, is more volatile than ethyl alcohol and a better solvent for iodine. When the operations are over, a little water is poured down the condenser to destroy any phosphonium compound present, the upper part of the apparatus removed and replaced by an ordinary V-shaped tube to which is attached a condenser and the contents of the flask distilled, using the water-bath. The distillate is washed thoroughly with water in a separating funnel, dried with fused calcium chloride and then redistilled, when the final distillate consists of practically pure ethyl iodide. This method can be successfully used for the preparation of other alkyl iodides from the corresponding alcohols, the arrangements being suitably modified according to the boiling point of the alcohol and alkyl iodide concerned.* It is not difficult to obtain a yield of 90 per cent. of ethyl iodide based on the alcohol used; the reaction taking place may be represented:

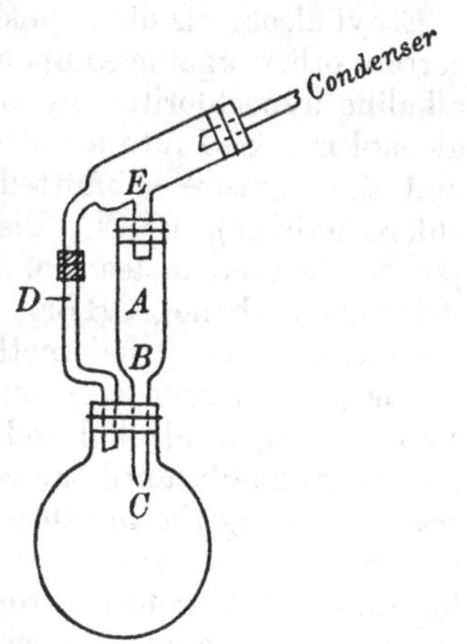

$$6C_2H_5OH + 2P + 3I_2 \rightarrow 6C_2H_5I + 2H_3PO_3$$

Ethyl iodide is a colourless liquid (generally coloured reddish owing to its containing a trace of free iodine), b.p. 72·4°. It is an important organic compound and used in a considerable variety of reactions in organic chemistry. In this and in all organic compounds containing halogen united directly to carbon, the halogen is not ionised in the ordinary sense in aqueous solution. Halogen ions in these circumstances only manifest themselves after hydrolysis: e.g.

$$C_2H_5I + H_2O \rightleftarrows C_2H_5OH + HI$$

On the other hand, the halogen in this and analogous compounds can be readily replaced, especially when heated with silver compounds under appropriate conditions: e.g.

$$C_2H_5I + AgOH \rightarrow C_2H_5OH + AgI$$

$$2C_2H_5I + Ag_2SO_4 \rightarrow (C_2H_5)_2SO_4 + 2AgI$$

* The method can be modified for the preparation of alkyl bromides, using the necessary alcohol, red phosphorus and bromine. In this case, the bromine is added from a tap funnel to the alcohol and red phosphorus, the latter mixture being kept cold during the addition on account of the very energetic reaction between bromine and phosphorus.

The ease with which such reactions proceed is explained largely by the affinity of silver for halogen and by the insolubility of silver halides.

Ethyl alcohol is distinguished from methyl alcohol, but not from certain other organic compounds, by the way in which it reacts with alkaline hypochlorites, hypobromites and hypoiodites. When ethyl alcohol is mixed into a paste with fresh alkaline bleaching powder and the mixture submitted to gentle distillation, *chloroform* (trichloromethane), $CHCl_3$, distils over. When 'bromine' bleaching powder is used instead of ordinary bleaching powder, *bromoform* (tribromomethane), $CHBr_3$, is obtained. The analogous reaction to produce *iodoform* (triodomethane), CHI_3, is carried out by dissolving iodine (a relatively large quantity is necessary compared with the alcohol used) in alcohol and then adding sodium hydroxide or carbonate gradually until the colour of the iodine just disappears. On *gently* warming the mixture iodoform is formed and in many cases separates in the crystalline form even from the hot solution. The formation of chloroform, bromoform and iodoform from ethyl alcohol involves the change of a molecule containing two carbon atoms to molecules containing one carbon atom:

$$CH_3.CH_2OH \rightarrow CHCl_3, CHBr_3, CHI_3$$

and the reactions concerned are somewhat complicated. They will be discussed later (p. 164), and in the meantime it may be observed that ethyl alcohol is actually the material from which the important anaesthetic chloroform and the, formerly used, antiseptic iodoform are chiefly made.

Ethyl alcohol like methyl alcohol is a primary alcohol and on oxidation it yields first *acetaldehyde* and then *acetic acid*, both of which contain like ethyl alcohol itself two carbon atoms in the molecule. The reactions expressing the formation of acetaldehyde and acetic acid from ethyl alcohol are exactly analogous to those indicating the formation of formaldehyde and formic acid from methyl alcohol:

$$\text{(i)}\quad CH_3-\underset{\displaystyle H}{\overset{\displaystyle H}{\underset{|}{\overset{|}{C}}}}-O-H + O \rightarrow CH_3-\overset{\displaystyle H}{\overset{|}{C}}{=}O + H_2O$$

acetaldehyde

$$\text{(ii)}\quad CH_3-\overset{\displaystyle H}{\overset{|}{C}}{=}O + O \rightarrow CH_3-\overset{\displaystyle H}{\overset{|}{\overset{\displaystyle O}{\overset{|}{C}}}}{=}O$$

acetic acid

The production of acetaldehyde (p. 154) by the oxidation of ethyl alcohol by means of sulphuric acid and potassium (or sodium) dichromate is one of the conclusive tests for ethyl alcohol. The test is easily carried out as follows. To a small quantity of concentrated sulphuric acid diluted with about twice its volume of water and cooled add a small quantity of sodium or potassium dichromate dissolved in an equal weight of water. Then add a small amount of ethyl alcohol and warm the mixture. If the reaction be carried out in a test-tube the characteristic odour of acetaldehyde can be detected, and if a piece of filter paper moistened with an ammoniacal solution of silver oxide* be placed at the mouth of the test-tube a stain due to metallic silver will appear on the filter paper. The reactions involved are:

$$K_2Cr_2O_7 + 4H_2SO_4 \rightarrow K_2SO_4 + Cr_2(SO_4)_3 + 4H_2O + 3O$$

$$3CH_3.CH_2OH + 3O \rightarrow 3CH_3.CHO + 3H_2O$$

$$CH_3.\overset{\displaystyle H}{\overset{|}{C}}{=}O + Ag_2O \rightarrow CH_3.\overset{\displaystyle O{-}H}{\overset{|}{C}}{=}O + 2Ag$$

It will be noticed that the solution in the test-tube becomes green owing to the formation of chromium sulphate. If care be taken chrome alum, $K_2SO_4.Cr_2(SO_4)_3.24H_2O$, can actually be crystallised out from the aqueous solution provided that all the dichromate has been reduced. It is thus seen that one of the characteristic properties of an aldehyde (in this case acetaldehyde) is its reducing power, whereby it tends to be converted into the corresponding acid (in this case, acetic acid).

Ethyl alcohol is not the only alcohol formed during the fermentation of glucose. The higher boiling fraction obtained during the distillation of the alcoholic products of the fermentation is known as *fusel oil*, and is a mixture of alcohols having a somewhat indefinite boiling point (85–140°). This mixture contains alcohols which have three, four and five atoms of carbon in their molecules. These higher alcohols of the $C_nH_{2n+1}OH$ series are not normal products of the fermentation of glucose but owe their origin to chemical changes produced in the proteins present in the yeast, these chemical changes proceeding at the same time as the conversion of glucose into alcohol and carbon dioxide. Fusel oil constitutes an important source for some of these higher alcohols, which to some extent can be separated by careful fractional distillation of the crude product.

One of the constituents of fusel oil is the next primary alcohol to ethyl alcohol, viz. *n*-propyl alcohol, and this can be separated during the fractional distillation of the fusel oil.

* This solution is made from the ordinary silver nitrate (bench reagent). To about 3 c.c. of the solution of silver nitrate add aqueous sodium hydroxide solution drop by drop until no further precipitate is formed. Then add aqueous ammonia drop by drop until on shaking the precipitate of silver hydroxide just dissolves leaving a clear solution.

Normal (n-)*propyl alcohol*, ethylcarbinol, propanol(1), has the constitutional formula $CH_3.CH_2.CH_2OH$, and this may be abbreviated to $n\text{-}C_3H_7OH$. Another name which expresses its constitution is 1-hydroxypropane. It is a colourless liquid, b.p. 97·2°, which burns with a luminous flame and which has a similar odour and very similar properties generally to those of ethyl alcohol. The similarity of its properties to those of ethyl alcohol makes it unnecessary to describe them in detail. On oxidation it yields first an aldehyde (propionaldehyde) and then an acid (propionic acid), both of which like *n*-propyl alcohol itself contain three atoms of carbon in the molecule:

$$\text{(i)}\quad CH_3.CH_2.\overset{\boxed{H}}{\underset{H}{C}}\!-\!O\!-\!\boxed{H + O} \rightarrow CH_3.CH_2.C\begin{matrix}\nearrow O \\ \searrow H\end{matrix} + H_2O$$

propionaldehyde

$$\text{(ii)}\quad CH_3.CH_2.C\begin{matrix}\nearrow O \\ \searrow H\end{matrix} + O \rightarrow CH_3.CH_2.C\begin{matrix}\nearrow O \\ \searrow O\!-\!H\end{matrix}$$

propionic acid

Propionaldehyde and propionic acid are analogous in their properties to acetaldehyde and acetic acid respectively.

Iso*propyl alcohol*, i-*propyl alcohol*, 2-hydroxypropane, propanol(2), is isomeric with *n*-propyl alcohol and has the constitutional formula $(CH_3)_2CH(OH)$. Its relationship to methyl alcohol is expressed by the designation dimethylcarbinol.

i-Propyl alcohol is only rarely found in fusel oil, and even then the fusel oil originates from the fermentation of sugar derived from particular starch materials and the fermentation has to be carried out under special conditions. *i*-Propyl alcohol is prepared almost entirely from acetone (p. 179), which has the constitutional formula

$$\begin{matrix} CH_3 \\ | \\ C\!=\!O \\ | \\ CH_3 \end{matrix}$$

When one volume of acetone is mixed with five volumes of water in which it is freely soluble and sodium amalgam added gradually to the mixture, acetone is reduced by the hydrogen produced by the action of the sodium in the amalgam on the water and *iso*propyl alcohol is produced:

$$\begin{matrix} CH_3 \\ | \\ C\!=\!O \\ | \\ CH_3 \end{matrix} + H_2 \rightarrow \begin{matrix} & CH_3 & \\ & | & \\ H\!-\! & C & \!-\!OH \\ & | & \\ & CH_3 & \end{matrix}$$

After the reaction the aqueous solution, which contains sodium hydroxide, is submitted to distillation and the products boiling below 105° collected. Since acetone has b.p. 56·5°, any acetone which has escaped reduction is easily separated. The higher boiling distillate (75–105°), which contains chiefly *i*-propyl alcohol and water, can be saturated with anhydrous potassium carbonate until two layers form. The upper layer, chiefly *i*-propyl alcohol, is then dried separately with anhydrous potassium carbonate and redistilled. On the industrial scale freshly calcined quick lime is used as the dehydrating agent.

Acetone can also be reduced catalytically to *i*-propyl alcohol by passing its vapour mixed with hydrogen over freshly reduced nickel at 115–125°. Processes have been patented for the electrolytic reduction of acetone to *i*-propyl alcohol.*

*Iso*propyl alcohol is a colourless liquid, b.p. 82·4°, possessing a faint alcoholic odour. It is soluble in water and forms a constant boiling mixture, b.p. 80·4°, containing 87·9 per cent. of the alcohol and 12·1 per cent. of water. It is soluble in most organic solvents.

As an alcohol it behaves like all other alcohols, because of the presence of the hydroxyl group. Thus, it reacts with sodium, the hydrogen of the hydroxyl group being replaced, and it reacts with acids, forming esters under analogous conditions to methyl, ethyl and

* *i*-Propyl alcohol is not the only product of the reduction of acetone, especially when the reduction is carried out either by the sodium or electrolytic method. The by-product which is formed under these conditions is known as *pinacol* and its formation is illustrated:

$$\begin{array}{c} CH_3 \\ | \\ O{=}C \\ | \\ CH_3 \end{array} + 2H + \begin{array}{c} CH_3 \\ | \\ C{=}O \\ | \\ CH_3 \end{array} \rightarrow \begin{array}{c} \quad CH_3 \quad CH_3 \\ \quad | \qquad | \\ HO{-}C{-}{-}C{-}OH \\ \quad | \qquad | \\ \quad CH_3 \quad CH_3 \end{array}$$

pinacol
(tetramethylethylene glycol)

Pinacol, when free from water, is a colourless crystalline substance, m.p. 38°, b.p. 171–172° at 739 mm. It belongs to the class of dihydric alcohols or glycols (p. 351). When pinacol is heated with dilute sulphuric or hydrochloric acid, it loses a molecule of water and at the same time a remarkable intramolecular change takes place with the formation of *pinacolone* (a colourless liquid with a characteristic odour; it has b.p. 105°). The formation of pinacolone from pinacol is illustrated as follows:

$$\begin{array}{c} \quad O\,\boxed{H \quad OH} \\ \quad | \qquad | \\ CH_3{-}C{-}{-}C{-}CH_3 \\ \quad | \; \nearrow | \\ \quad \boxed{CH_3} \; CH_3 \end{array} \rightarrow H_2O + \begin{array}{c} \qquad O \quad CH_3 \\ \qquad \| \quad\; | \\ CH_3{-}C{-}C{-}CH_3 \\ \qquad\quad | \\ \qquad\quad CH_3 \end{array}$$

pinacolone

n-propyl alcohols. Likewise, when acted upon by phosphorus halides, the hydroxyl group is replaced by halogen and the esters *i*-propyl chloride (2-chloropropane), *i*-propyl bromide (2-bromopropane) and *i*-propyl iodide (2-iodopropane) are formed according to the phosphorus halide used.

*Iso*propyl alcohol, however, differs from methyl, ethyl and *n*-propyl alcohols in that it is a *secondary alcohol* while the other three are primary alcohols. A secondary alcohol behaves differently from a primary alcohol on oxidation; the difference in behaviour is explained by there being the $—CH_2OH$ group (united either to hydrogen or to an alkyl group) present in a primary alcohol and the $\rangle C\langle^{H}_{OH}$ group (united to two alkyl groups) in a secondary alcohol.

Under the conditions for the conversion of ethyl alcohol into acetaldehyde or of *n*-propyl alcohol into propionaldehyde, *i*-propyl alcohol readily yields the substance acetone (p. 179). The actual mechanism of the oxidation of the primary alcohol as far as the first stage of the process is concerned is similar to that of the oxidation of a secondary alcohol, a molecule of water being eliminated in both cases, thus:

$$\begin{array}{c} CH_3 \\ | \\ H—C—O\boxed{H + O} \\ | \\ \boxed{H} \end{array} \rightarrow H_2O + \begin{array}{c} CH_3 \\ | \\ C{=}O \\ | \\ H \end{array}$$

$$\begin{array}{c} CH_3 \\ | \\ CH_3—C—O\boxed{H + O} \\ | \\ \boxed{H} \\ \text{i-propyl alcohol} \end{array} \rightarrow H_2O + \begin{array}{c} CH_3 \\ | \\ C{=}O \\ | \\ CH_3 \\ \text{acetone*} \end{array}$$

* Acetone is the first member of a series of compounds known as *ketones*. Acetone is systematically described as dimethyl ketone. The ketones are compounds whose molecules contain the : C=O group united to two monovalent radicals derived from the hydrocarbons by loss of one hydrogen atom from each molecule of the latter. Thus we have, for example:

$$\begin{array}{cccc} CH_3 & CH_3 & C_6H_5 & C_6H_5 \\ | & | & | & | \\ C{=}O & C{=}O & CO & C{=}O \\ | & | & | & | \\ CH_3 & C_2H_5 & CH_3 & C_6H_5 \end{array}$$

dimethyl ketone acetone; methyl ethyl ketone; methyl phenyl ketone (acetophenone); diphenyl ketone (benzophenone), etc.

A ketone may be isomeric with an aldehyde containing the same number of carbon

In the case of a primary alcohol, the oxidation can proceed further and an acid is produced by the addition of an oxygen atom to a molecule of the aldehyde. In the case of a secondary alcohol, the oxidation cannot proceed further without breaking up the molecule, yielding a mixture of products each of which contains fewer carbon atoms than the original alcohol. In the present case, for instance, acetone is not a reducing agent and is an extremely stable compound. It can actually be distilled from potassium permanganate without being affected. If, however, acetone be submitted to the action of an oxidising agent under such extreme conditions that oxidation does take place, the molecule is disrupted and carbon dioxide and acetic acid are produced:

$$\begin{array}{l} CH_3 \\ | \\ C{=}O \\ | \\ CH_3 \end{array} + 2O_2 \rightarrow CO_2 + CH_3.COOH + H_2O$$

Such vigorous oxidation conditions are seldom used in organic chemistry for preparative purposes, but if they are employed there is no reason why the reaction should not proceed further with the production of simply carbon dioxide and water, thus:

$$\begin{array}{l} CH_3 \\ | \\ CO \\ | \\ CH_3 \end{array} + 4O_2 \rightarrow 3CO_2 + 3H_2O$$

Such a vigorous oxidation of an organic compound is uncontrollable, and the distinguishing feature of a secondary alcohol is that under controllable conditions of oxidation it yields a ketone containing the same number of carbon atoms as the original alcohol. Further oxidation (under 'uncontrollable' conditions) leads to substances containing fewer carbon atoms than the alcohol or ketone.

There are four isomeric monohydric alcohols of the formula C_4H_9OH, and these are derived from the two isomeric butanes of the formula C_4H_{10} by the replacement of one hydrogen in the molecule by the hydroxyl group.

atoms in the molecule. Thus, acetone is isomeric with propionaldehyde produced by the oxidation of *n*-propyl alcohol:

$$CH_3.CH_2.\overset{H}{\underset{H}{C}}{-}O\,H + O \rightarrow H_2O + CH_3.CH_2.\overset{H}{C}{=}O$$

propionaldehyde

Apart from other reactions, a ketone can be readily distinguished from an aldehyde, because a ketone is not a reducing agent.

n-*Butyl alcohol, n*-propylcarbinol, butanol(1), 1-hydroxybutane, has the constitutional formula $CH_3.CH_2.CH_2.CH_2.OH$. It has now become an industrially important organic compound and is produced by other processes which will be described later (p. 470–2). It is a colourless liquid, b.p. 117·0°. It is soluble in most organic solvents, but its solubility in water is less than that of the propyl alcohols. It is a primary alcohol, and by direct oxidation yields first *n*-butyraldehyde and then *n*-butyric acid:

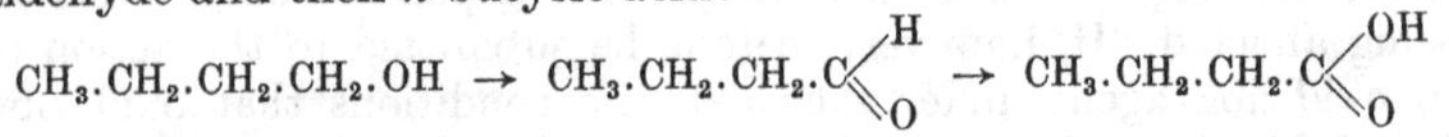

$$CH_3.CH_2.CH_2.CH_2.OH \rightarrow CH_3.CH_2.CH_2.C\begin{matrix}\diagup H\\ \diagdown\!\!\diagdown O\end{matrix} \rightarrow CH_3.CH_2.CH_2.C\begin{matrix}\diagup OH\\ \diagdown\!\!\diagdown O\end{matrix}$$

The other monohydric alcohol derived from *n*-butane is *methylethylcarbinol*, having the constitution formula $CH_3.CH_2.CH(OH).CH_3$. This secondary alcohol is also known as 2-hydroxybutane, and butanol(2). It is a synthetic compound and is of little interest except from a theoretical point of view. It is a colourless liquid, b.p. 99·9° at 756 mm. Being a secondary alcohol it yields a ketone, *methylethyl ketone* (b.p. 78·6°), a homologue of acetone, by direct oxidation:

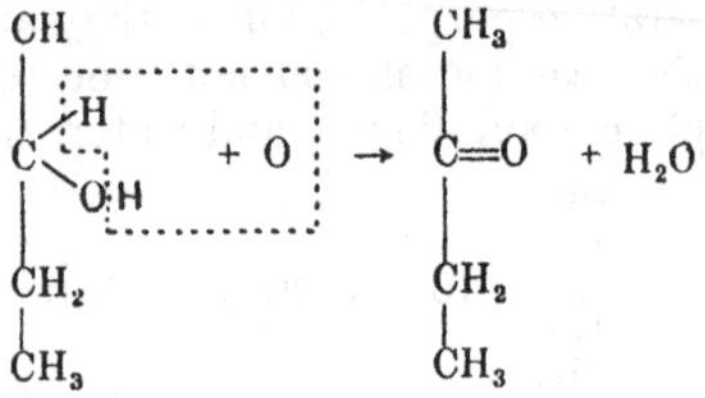

$$\begin{matrix} CH \\ | \\ C\begin{matrix}\diagup H\\ \diagdown OH\end{matrix} \\ | \\ CH_2 \\ | \\ CH_3 \end{matrix} + O \rightarrow \begin{matrix} CH_3 \\ | \\ C{=}O \\ | \\ CH_2 \\ | \\ CH_3 \end{matrix} + H_2O$$

The monohydric alcohols derived from *iso*-butane by the replacement of a hydrogen atom by a hydroxyl group have the following constitutional formula respectively:

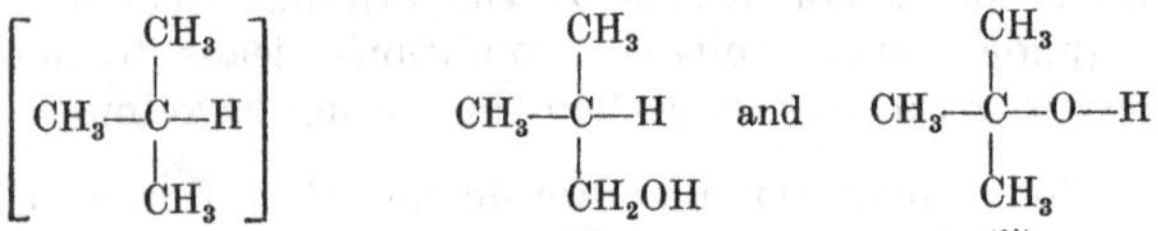

$$\left[\begin{matrix} & CH_3 & \\ & | & \\ CH_3— & C & —H \\ & | & \\ & CH_3 & \end{matrix}\right] \qquad \begin{matrix} & CH_3 & \\ & | & \\ CH_3— & C & —H \\ & | & \\ & CH_2OH & \\ & \text{(i)} & \end{matrix} \quad \text{and} \quad \begin{matrix} & CH_3 & \\ & | & \\ CH_3— & C & —O—H \\ & | & \\ & CH_3 & \\ & \text{(ii)} & \end{matrix}$$

The first of these is:

Iso-*propylcarbinol*, 1-hydroxy-2-methylpropane, and occurs in certain varieties of fusel oil. It is a colourless liquid, b.p. 108·4°, soluble in most organic solvents, but, like the other isomeric alcohols, somewhat sparingly soluble in water. Being a primary alcohol it yields by direct oxidation, first an aldehyde (*iso*-butyraldehyde) and then an acid (*iso*-butyric acid):

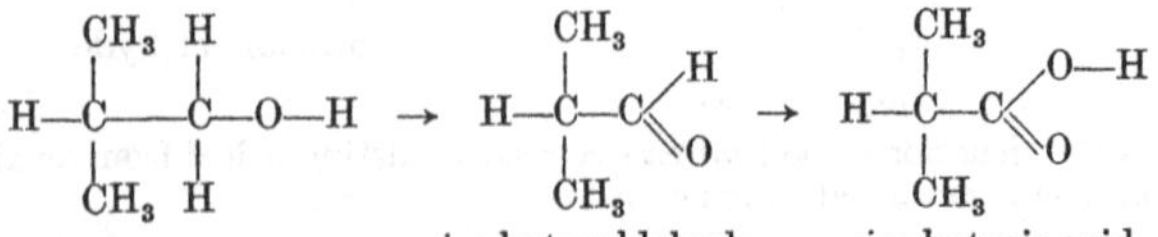

$$\begin{matrix} & CH_3 & & H & \\ & | & & | & \\ H— & C & — & C & —O—H \\ & | & & | & \\ & CH_3 & & H & \end{matrix} \rightarrow \begin{matrix} & CH_3 & \\ & | & \\ H— & C & —C\begin{matrix}\diagup H\\ \diagdown\!\!\diagdown O\end{matrix} \\ & | & \\ & CH_3 & \end{matrix} \rightarrow \begin{matrix} & CH_3 & \\ & | & \\ H— & C & —C\begin{matrix}\diagup O—H\\ \diagdown\!\!\diagdown O\end{matrix} \\ & | & \\ & CH_3 & \end{matrix}$$

iso-butyraldehyde *iso*-butyric acid

The second of these alcohols is:

Tertiary butyl alcohol (*tert*-butyl alcohol), trimethylcarbinol, 2-hydroxy-2-methylpropane, which does not occur in fusel oil and is prepared synthetically from acetone (p. 179). *tert*-Butyl alcohol forms colourless rhombohedral plates and prisms, m.p. 25·5°, b.p. 82·55°. It dissolves freely in water and in organic solvents. As an alcohol it yields a sodium derivative (sodium *tert*-butylate: $C_3H_7.C.ONa$) and other metallic derivatives. It also forms esters like other alcohols, on account of the presence of the hydroxyl group. It has a distinctive property, viz. a tendency to lose a molecule of water forming *iso*-butylene (*asym.*-dimethylethylene):

$$\begin{array}{c} CH_3 \\ | \\ CH_3-C-\boxed{O-H} \\ | \\ H-C-\boxed{H} \\ | \\ H \end{array} \rightarrow H_2O + \begin{array}{c} CH_3-C-CH_3 \\ \| \\ CH_2 \\ \textit{iso}\text{-butylene} \end{array}$$

When *tert*-butyl alcohol is heated to a high temperature (i.e. above its boiling point) *iso*butylene (*asymm.*-dimethylethylene) is formed, and this same compound is formed when tert-*butyl chloride* is strongly heated:

$$\begin{array}{c} CH_3 \\ | \\ CH_3-C-\boxed{Cl} \\ | \\ H-C-\boxed{H} \\ | \\ H \end{array} \rightarrow HCl + \begin{array}{c} CH_3-C-CH_3 \\ \| \\ CH_2 \end{array}$$

Such behaviour is characteristic of all analogous compounds containing a tertiary carbon atom, united to at least one alkyl group. (A tertiary carbon atom is one united directly to three other carbon atoms: thus

$$\begin{array}{c} \cdots \\ C \\ | \\ :C-C-X \\ | \\ C \\ \cdots \end{array}$$

where X is a monovalent group which is not united to the tertiary carbon atom through a carbon atom.)

tert-Butyl alcohol being a tertiary alcohol behaves differently from primary and secondary alcohols on oxidation. It gives no direct oxidation products, and if oxidation does take place it can only yield a mixture of products each containing fewer carbon atoms in its molecule than the original *tert*-butyl alcohol. All other tertiary alcohols, i.e. those containing the grouping $\geqslant C-O-H$ united to three other monovalent radicals derived from the hydrocarbons by the loss of one hydrogen atom from each molecule, behave in an analogous fashion; and the tertiary alcohols are thus distinguished from the primary and secondary alcohols.

The eight isomeric alcohols of the formula $C_5H_{11}OH$, the so-called *amyl alcohols*, are all known. They are derived theoretically from the three isomeric pentanes by substitution of a hydroxyl group for a hydrogen atom in the molecule of the hydrocarbon. From *n*-pentane are thus derived:

(*a*) $CH_3.CH_2.CH_2.CH_2.CH_2OH$ n-*amyl alcohol*, *n*-butylcarbinol (primary) (colourless liquid, b.p. 137·9°)

(*b*) $$\begin{matrix} CH_3.CH_2.CH_2 & & H \\ & C & \\ CH_3 & & OH \end{matrix}$$ *methyl*-n-*propylcarbinol* (secondary) (colourless liquid, b.p. 119·5°)

(*c*) $$\begin{matrix} CH_3.CH_2 & & H \\ & C & \\ CH_3.CH_2 & & OH \end{matrix}$$ *diethylcarbinol* (secondary) (colourless liquid, b.p. 114–115° at 744 mm.)

From 2-methyl-*n*-butane (dimethylethylmethane) are derived:

(*d*) $$\begin{matrix} CH_3.CH_2 & & H \\ & C & \\ CH_3 & & CH_2OH \end{matrix}$$ 2-*methyl*-n-*butyl alcohol* (primary) (colourless liquid, b.p. 128·7°)

(*e*) $$\begin{matrix} H & & CH_2.CH_2OH \\ & C & \\ CH_3 & & CH_3 \end{matrix}$$ 3-*methyl*-n-*butyl alcohol* (primary) (also known as *iso*amyl alcohol) (colourless liquid, b.p. 131°)

(*f*) $$\begin{matrix} CH_3 & & H & & \\ & C & & & H \\ CH_3 & & & C & \\ & CH_3 & & & OH \end{matrix}$$ *methyl*-i-*propylcarbinol* (secondary) (colourless liquid, b.p. 113–114°)

(*g*) $$\begin{matrix} CH_3.CH_2 & & CH_3 \\ & C & \\ CH_2 & & OH \end{matrix}$$ *dimethylethylcarbinol* (tertiary) (also known as amylene hydrate) (colourless liquid, b.p. 102° at 762 mm.)

The remaining monohydric alcohol of this series can be considered to be derived from the third isomeric pentane, tetramethylmethane, and this is

(*h*) $$\begin{matrix} CH_3 & & CH_3 \\ & C & \\ CH_3 & & CH_2OH \end{matrix}$$ 2 : 2-*dimethyl*-n-*propyl alcohol* (primary) (colourless crystalline substance, m.p. 52–53°, b.p. 113–114°)

All these alcohols are much less soluble in water than any of the alcohols previously mentioned, and they react much more sluggishly with sodium. Commercial amyl alcohol is obtained from *fusel oil*—the higher boiling mixture of alcohols produced as by-products in the fermentation of glucose under various conditions and obtained by distillation after the ethyl alcohol has been separated. The boiling point of commercial amyl alcohol is 125–143°. Its two chief constituents are 3-methyl-*n*-butyl alcohol (*e*) and 2-methyl-*n*-butyl alcohol (*d*), of which the former is usually present in the greater proportion. Commercial amyl alcohol is usually optically active (p. 113) and slightly *laevo*-rotatory. This property is due to the presence of

2-methyl-*n*-butyl alcohol (*d*), which as produced under fermentation conditions is always *laevo*-rotatory.*

Certain higher alcohols of this $C_nH_{2n+1}OH$ series are important in that they occur as esters in certain waxes. Spermaceti Cetaceum from the head of the sperm whale consists chiefly of the cetyl ester of palmitic acid (p. 222), which has the formula $C_{15}H_{31}.CO_2C_{16}H_{33}$.

Cetyl alcohol, $C_{16}H_{33}OH$, is known also as hexadecanol, being theoretically derived from the normal paraffin hexadecane, $C_{16}H_{34}$, the constitution of which is written in abbreviated form $CH_3.(CH_2)_{14}.CH_3$. Cetyl alcohol is a primary alcohol and its constitution is therefore written $CH_3(CH_2)_{14}.CH_2OH$. It crystallises from ethyl alcohol in colourless leaflets, m.p. 49·5°, and is obtained from spermaceti by boiling it with an ethyl alcoholic solution of sodium hydroxide for a long time and then pouring the product into water. The cetyl alcohol then crystallises out.

Myricyl alcohol (also known as melissyl alcohol) contains 30 carbon atoms in the molecule and, belonging to the $C_nH_{2n+1}OH$ series, has the formula $C_{30}H_{61}OH$. It occurs as myricyl palmitate, $C_{15}H_{31}.CO_2C_{30}H_{61}$, in bee's wax. It crystallises in colourless needles, m.p. 88°. It yields melissic acid on oxidation. The formula of melissic acid is $C_{30}H_{60}O_2$ or $C_{29}H_{59}.COOH$, and hence myricyl or melissyl alcohol is a primary alcohol.

The simplest monohydric aromatic alcohols can be considered to be derived from those aromatic hydrocarbons in which one hydrogen atom in the saturated side chain is replaced by a hydroxyl group: or, they may be considered to be derived from methyl alcohol or carbinol by the substitution of a hydrogen atom directly attached to carbon by a monovalent aromatic hydrocarbon radical. Two of the simplest examples are:

Benzyl alcohol, $C_6H_5.CH_2OH$, monohydroxytoluene (—OH group

* The property which certain compounds possess of rotating the plane of plane polarised light either to the right or left is known as optical activity. The possession of this property by compounds in the liquid state or in solution or in the gaseous condition is always associated with a deficiency of symmetry in the molecule. In the case of 2-methyl-*n*-butyl alcohol (*d*), the carbon atom written in heavy type is united to four different groups:

$$\begin{array}{ccc} CH_3.CH_2 & & H \\ & \mathbf{C} & \\ CH_3 & & CH_2OH \end{array}$$

and the molecule as a whole regarded spatially therefore possesses no single principle of symmetry, and is described as being asymmetric. The optical activity of certain petroleums is due to their containing compounds the molecules of which are asymmetric. The optical activity of these petroleums is strong evidence of their having been produced by biochemical processes.

substituted for a hydrogen atom in the CH_3— side chain of toluene) or phenylcarbinol (phenyl group, C_6H_5—, substituted for a hydrogen atom attached to carbon in methyl alcohol), and phenylethyl alcohol, $C_6H_5.CH_2.CH_2.OH$, monohydroxyphenylethane or benzylcarbinol, indicating the replacement of a hydrogen atom in methyl alcohol by the benzyl group, $C_6H_5.CH_2$—.

Benzyl alcohol occurs free and in the form of esters in certain ethereal oils. It is a colourless liquid, b.p. 206°, and is soluble in most organic solvents but only slightly soluble in water at the ordinary temperature.

Its synthesis from benzyl chloride proves its constitution:

$$C_6H_5.CH_2\boxed{Cl + Ag}OH \rightarrow C_6H_5.CH_2OH + AgCl$$

A process for the manufacture is based on the reaction for the replacement of the chlorine atom in benzyl chloride by the hydroxyl group. It is generally manufactured from benzaldehyde by a reaction discussed on p. 170.

That benzyl alcohol is a primary alcohol is proved by its yielding first an aldehyde, *benzaldehyde* (p. 165), and then an acid, *benzoic acid* (p. 235), by direct oxidation:

$$\text{(i)}\quad C_6H_5\text{—}\overset{H}{\underset{H}{C}}\text{—O—H} + O \rightarrow C_6H_5.C\begin{matrix}\diagup H\\ \diagdown\!\!= O\end{matrix} + H_2O$$

benzaldehyde

$$\text{(ii)}\quad C_6H_5\text{—}C\begin{matrix}\diagup H\\ \diagdown\!\!= O\end{matrix} + O \rightarrow C_6H_5.C\begin{matrix}\diagup O\text{—}H\\ \diagdown\!\!= O\end{matrix}$$

benzoic acid

The hydrogen of the hydroxyl group is capable of being replaced by sodium, and benzyl alcohol like other alcohols already mentioned forms esters both with phosphorus halides and with acids. When benzyl alcohol is heated with hydriodic acid in the presence of phosphorus at 140° it is reduced to toluene. This reaction may be expressed:

$$C_6H_5\text{—}\overset{H}{\underset{H}{C}}\text{—O—H} + \begin{matrix}H\text{—}I\\ H\text{—}I\end{matrix} \rightarrow C_6H_5\text{—}\overset{H}{\underset{H}{C}}\text{—H} + I_2 + H_2O$$

Phenylethyl alcohol, benzylcarbinol, is a colourless liquid, b.p. 219°, the chief constituent of natural rose perfume. It is a primary alcohol and yields by direct oxidation with sodium or potassium dichromate and dilute sulphuric acid first *phenylacetaldehyde* and then *phenylacetic acid*:

$$\text{(i)}\quad C_6H_5.CH_2.\overset{\displaystyle H}{\underset{\displaystyle H}{C}}\text{—O—}\boxed{H + O \ (H)} \rightarrow H_2O + C_6H_5.CH_2.C\begin{smallmatrix}\diagup H\\ \diagdown\!\!\!= O\end{smallmatrix}$$

phenylacetaldehyde, b.p. 194°

$$\text{(ii)}\quad C_6H_5.CH_2.C\begin{smallmatrix}\diagup H\\ \diagdown\!\!\!= O\end{smallmatrix} + O \rightarrow C_6H_5.CH_2.C\begin{smallmatrix}\diagup O\text{—}H\\ \diagdown\!\!\!= O\end{smallmatrix}$$

phenylacetic acid, m.p. 76·7°

Since the molecule of phenylethyl alcohol contains a two-carbon-atom side chain, it may yield benzoic acid on more vigorous oxidation. When, for example, it is oxidised with potassium permanganate the chief oxidation product is benzoic acid, which contains fewer carbon atoms (one less) in the molecule than the phenylethyl alcohol itself although the alcohol is primary:

$$2C_6H_5.CH_2.CH_2OH + 5O_2 \rightarrow 2C_6H_5.COOH + 2CO_2 + 4H_2O$$

The simplest secondary aromatic monohydric alcohols may be considered to be derived from carbinol or methyl alcohol by the replacement of one hydrogen atom by an aromatic hydrocarbon radical (or aryl group) and a second hydrogen may be replaced either by an aliphatic hydrocarbon radical (alkyl group) or by an aryl group. As typical of the first class may be mentioned

Methylphenylcarbinol, $\begin{matrix}C_6H_5\\CH_3\end{matrix}\!\!>C<\!\!\begin{matrix}H\\OH\end{matrix}$, which is obtained by the reduction of acetophenone (p. 186), just an *iso*propyl alcohol (dimethylcarbinol) is obtained from acetone. It is a colourless liquid, b.p. 205°, which is directly oxidised by chromic acid to acetophenone:

$$\begin{matrix}C_6H_5\\CH_3\end{matrix}\!\!>C<\!\!\begin{matrix}H\\OH\end{matrix} + O \rightarrow H_2O + \begin{matrix}C_6H_5\\CH_3\end{matrix}\!\!>C=O$$

Typical of the second class is:

Diphenylcarbinol, $\begin{matrix}C_6H_5\\C_6H_5\end{matrix}\!\!>C<\!\!\begin{matrix}H\\OH\end{matrix}$, which is obtained by the reduction of benzophenone (p. 187), as well as by other methods. It is obtained in colourless needles, m.p. 67·5–68°, and is almost insoluble in water. Its chief direct oxidation product is benzophenone.

The simplest monohydric tertiary aromatic alcohols may be exemplified by *dimethylphenylcarbinol* and by *triphenylcarbinol*.

Dimethylphenylcarbinol, $\begin{matrix} C_6H_5 \\ CH_3 \\ CH_3 \end{matrix} \!\!> C{-}OH$, is obtained from acetone and from acetophenone using a general method for the preparation of tertiary alcohols (p. 499). It is a colourless crystalline substance (prisms), m.p. 23°, b.p. 89° at 10 mm.

Triphenylcarbinol, $(C_6H_5)_3COH$, is prepared by an analogous reaction as dimethylphenylcarbinol, starting with benzophenone. It crystallises in hexagonal plates, m.p. 162·5°.

Apart from their properties as primary, secondary and tertiary alcohols, these aromatic alcohols also possess properties which differentiate them from the analogous aliphatic compounds, in consequence of the presence in the molecule of at least one aromatic nucleus. In considering the properties of these aromatic alcohols the chemical properties of the aromatic nucleus should be borne in mind.

Monohydroxy Derivatives of Unsaturated Hydrocarbons

The simplest of these is *monohydroxyethylene* or *vinyl alcohol*, $H_2C:C\!<\!\begin{matrix} H \\ OH \end{matrix}$. This compound would be isomeric with acetaldehyde, $CH_3.CHO$, and reactions which one would imagine would lead to the production of vinyl alcohol always yield acetaldehyde, probably because of the tendency of the grouping $=C{=}C\!<\!\begin{matrix} H \\ OH \end{matrix}$ to be converted into $=\overset{H}{\underset{|}{C}}{-}C\!<\!\begin{matrix} H \\ O \end{matrix}$. If it did exist, vinyl alcohol would of course be a secondary alcohol; but only certain derivatives are known.

The most important alcohol derived from open chain unsaturated hydrocarbons is the one derived theoretically from propylene, $CH_3.CH:CH_2$, by the substitution of a hydroxyl group for a hydrogen atom in the methyl radical. This compound is known as *allyl alcohol* or vinylcarbinol, $CH_2:CH.CH_2OH$, and is obtained from the trihydric alcohol, glycerol, $CH_2OH.CHOH.CH_2OH$, by a reaction discussed on p. 389. It is a colourless liquid, miscible with water in all proportions, possessing a pungent odour and a burning taste. It has b.p. 96·5°, and is much more toxic than the corresponding saturated primary alcohol, viz. *n*-propyl alcohol. In addition to its properties as a primary alcohol, yielding, for example, the corresponding aldehyde, *acrylic aldehyde* or *acrolein*, $CH_2:CH.CHO$, on oxidation it is also an unsaturated compound.

For example, with bromine, it combines directly yielding 1-hydroxy-2 : 3-dibromopropane:

$$\begin{array}{l} CH_2 \\ \| \\ CH \\ | \\ CH_2OH \end{array} + Br_2 \rightarrow \begin{array}{l} CH_2Br \\ | \\ CHBr \\ | \\ CH_2OH \end{array}$$

which is also, of course, a primary alcohol.

In view of its unsaturated character, the oxidation of allyl alcohol is somewhat more complicated than that of a primary alcohol derived from the saturated hydrocarbons: and there is a strong tendency during oxidation for the compound to be decomposed at the double linkage, so that although it is a primary alcohol, it may yield products on oxidation containing fewer carbon atoms than the original alcohol. In spite of this, the fact that allyl alcohol does yield acrylic aldehyde or acrolein on oxidation is a practical proof of its being a primary alcohol.

The analogue of allyl alcohol among the aromatic alcohols is phenylallyl alcohol or, as it is more commonly named, *cinnamic* or *cinnamyl alcohol*, $C_6H_5.CH:CH.CH_2OH$. It is a primary alcohol which occurs combined in certain esters in storax and in Peru balsam. It is obtained from these esters by boiling with alkali (hydrolysis). Cinnamic alcohol crystallises in colourless needles, m.p. 33°, b.p. 257·5°. Its unsaturated character is shown by its combining directly with bromine forming the dibromo-addition compound having the constitution $C_6H_5.CHBr.CHBr.CH_2OH$, 2 : 3-*dibromo*-3-*phenyl*-n-*propyl alcohol* or 1-*hydroxy*-2 : 3-*dibromo*-3-*phenylpropane*. As would be expected from its being a primary alcohol, cinnamic alcohol on oxidation yields *cinnamic acid*:

$$C_6H_5.CH:CH.\overset{\overset{\displaystyle H}{|}}{\underset{\underset{\displaystyle H}{|}}{C}}-O-H + \begin{array}{c} O \\ \| \\ \| \\ O \end{array} \rightarrow H_2O + C_6H_5.CH:CH.C\begin{array}{l} \nearrow O \\ \searrow O-H \end{array}$$

cinnamic acid, m.p. 133°

As an acetylene alcohol, may be mentioned *acetylenylcarbinol* or *propiolic alcohol*, having the constitution $HC:C.CH_2OH$, i.e. carbinol, having one hydrogen substituted by the acetylenyl ($HC:C-$) radical.

It is obtained by the action of alkali on the dibromo-addition product of allyl alcohol:

$$\begin{array}{c} H \\ | \\ H\text{—}C\text{—}Br \\ | \\ Br\text{—}C\text{—}H \\ | \\ CH_2OH \end{array} + 2KOH \rightarrow 2KBr + 2H_2O + \begin{array}{c} CH \\ ||| \\ C \\ | \\ CH_2OH \end{array}$$

This alcohol is an agreeable smelling liquid, b.p. 114–115°. It is both a primary alcohol and an acetylene compound. It combines directly with halogens, and it forms a silver derivative as would be expected from its having a replaceable hydrogen atom in the acetylenic part of the molecule.

The aromatic analogue is *phenylpropiolic alcohol*, $C_6H_5.C:C.CH_2OH$, a colourless liquid, b.p. 139° at 16 mm. Unlike propiolic alcohol it does not yield metallic derivatives, having no replaceable hydrogen in the acetylenic part of the molecule; it is, of course, an unsaturated compound. On the other hand, it has the ordinary properties of an alcohol and forms esters.

CHAPTER IV

HYDROXY DERIVATIVES OF CYCLIC HYDROCARBONS

THE compounds obtained theoretically from the aromatic hydrocarbons by the substitution of at least one hydrogen atom in the aromatic nucleus by a hydroxyl group are described generally as phenols, from the name given to the simplest compound of this type. These compounds have many reactions in common: they all yield characteristic colours (blue, violet, red)* with a neutral solution of ferric chloride, and this is often employed as a rough method of detecting the presence of a phenolic substance, provided other substances which may be present do not interfere with the reaction; they usually undergo nitration and sulphonation in the aromatic nucleus much more easily than the corresponding hydrocarbons, the presence of the hydroxyl group favouring these reactions; and similarly, they form substitution products in the aromatic nucleus with halogens much more easily than do the corresponding hydrocarbons.

PHENOL

Phenol, monohydroxybenzene, C_6H_5OH, was discovered by Runge in coal-tar in 1834. Runge applied to it the name 'carbolic acid' (acid from coal (*carbo*) oil or oleum), and the material examined by him contained certain homologues now known as cresols or methylphenols. It was first obtained in a state of purity by Laurent in 1841 and was analysed and identified by him and, in the same year, Gerhardt proposed the name 'phenol' for the substance.

The chief source of phenol is coal-tar. The so-called 'middle oil', b.p. 160–250°, contains from 25–40 per cent. phenols (phenol and cresols) together with naphthalene and small quantities of organic basic substances. After further fractional distillation, that portion of the 'middle oil', b.p. 160–205°, is allowed to stand in the 'cold house', when most of the naphthalene crystallises out. The naphthalene is separated from the liquid portion, which latter is first washed with dilute acid to remove basic substances, and the remaining oil again submitted to careful fractionation. The cresols boil at higher temperatures than phenol itself and are found in the higher boiling distillate. The phenol itself can be further purified by

* An aqueous solution of phenol itself yields a deep violet colour.

dissolving it in sodium hydroxide solution, and then passing carbon dioxide into this, when the phenol again separates as an oil. This can be distilled and that portion boiling at 181–182° is practically pure phenol. Fractional distillation is the best way to purify the substance.

The so-called *synthetic* phenol—merely given this name to distinguish it from phenol obtained from coal and from which it differs in no respect—is made, starting with benzene. The benzene is converted into benzene monosulphonic acid (p. 59) and the latter into its sodium salt. This sodium salt is mixed with dry sodium hydroxide (or a mixture of sodium and potassium hydroxides) and the mixture heated to 300–350° in silver or nickel pots, the mixture being mechanically stirred during the fusion. The reactions taking place consist in the formation of sodium sulphite (or a mixture of potassium and sodium sulphite) and the conversion of phenol into its metallic derivative:

$$C_6H_5.\boxed{SO_3Na + Na}OH \rightarrow C_6H_5OH + Na_2SO_3$$

$$C_6H_5OH + NaOH \rightarrow C_6H_5ONa + H_2O$$

At the end of the reaction (about 3 hours), the contents (strongly alkaline) of the fusion pot are allowed to cool and mixed with water in such quantity so that the sulphite for the greater part remains undissolved. The aqueous alkaline solution is separated and acidified with sulphuric acid (or sulphur dioxide is passed in). Under these conditions the phenol with a certain amount of water separates as an oil, which after being freed from the supernatant liquid is submitted to distillation.

Phenol can also be obtained from benzene by another method, described on p. 275.

Phenol crystallises in colourless needles, m.p. 43°, b.p. 181·4°. Its odour is characteristic and easily recognisable. At ordinary temperatures it is partially miscible with water; at 20°, a saturated aqueous solution of phenol contains 8·2 per cent. by weight of phenol and a saturated solution of water in phenol contains 72·2 per cent. by weight of phenol. Such a saturated two-liquid system becomes homogeneous at 68·8° at which temperature it contains 34·0 per cent. by weight of phenol; this is the solubility of phenol in water at that temperature. Phenol is freely soluble in diethyl ether and in ethyl alcohol.

Many suggestions have been made to explain the turning pink or brown of colourless phenol crystals when kept for some time under ordinary conditions. It has been found that specimens of phenol which have not been completely purified become coloured much more readily than purer specimens, and specimens of pure phenol have kept free from colour during several years under ordinary laboratory conditions. When phenol becomes coloured, surface oxidation probably has taken place, since in the presence of sulphur dioxide it remains colourless almost indefinitely.

Phenol is somewhat hygroscopic, but with precautions it forms a suitable solvent for the determination of molecular weights by the cryoscopic method, the melting point of pure phenol being convenient for this purpose. The molecular constant for the lowering of the freezing point of phenol is 72.

Regarding the constitution of phenol, chemical analysis and molecular weight determinations show that its molecular formula is C_6H_6O. When phenol vapour is passed over heated zinc dust, benzene is produced, $C_6H_6O + Zn \rightarrow C_6H_6 + ZnO$, indicating that it is a benzene derivative and the only possible constitutional formula of phenol is therefore

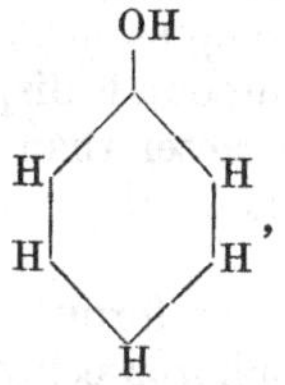

OH
H H
H H , C_6H_5OH
H

i.e. benzene with one hydrogen atom substituted by a hydroxyl group*. Accepting the recognised formulae for benzene, the constitutional formula of phenol becomes

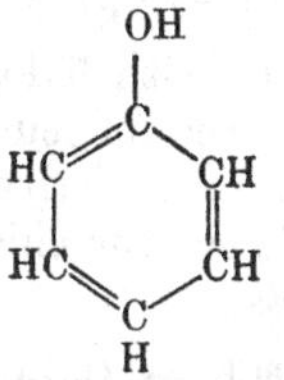

OH
C
HC CH
HC CH
C
H

There thus appears to be in phenol a potential tertiary alcohol ($\equiv$C—O—H) grouping. Indeed, in many respects, phenol does have the properties of a tertiary alcohol. On vigorous oxidation it yields compounds whose molecules contain fewer than six carbon atoms. It has one hydrogen atom in the molecule capable of being directly replaced by sodium or potassium, and it forms esters just like any other tertiary alcohol.

On the other hand, an aqueous solution of phenol is distinctly acidic in character and reacts as an acid towards indicators. *Sodium*

* The recognition of phenols as oxidation products of aromatic hydrocarbons is theoretically important. Phenol, itself, has been prepared by the catalytic oxidation of benzene under various conditions.

and *potassium phenates*,* which, as has been stated, are formed when sodium and potassium react directly with phenol:

$$2C_6H_5OH + 2Na \rightarrow 2C_6H_5ONa + H_2$$

are really salts of the weak acid, phenol. Phenol dissolves readily in aqueous solutions of sodium and potassium hydroxides forming the same phenates:

$$C_6H_5OH + NaOH \rightarrow C_6H_5ONa + H_2O$$

The aqueous solutions of these phenates, however, are strongly alkaline, due to the hydrolysis of the sodium salt of the weak acid, phenol:

$$C_6H_5ONa + H_2O \rightarrow C_6H_5OH + NaOH$$

Phenol is precipitated from these solutions by means of carbon dioxide, and phenol would therefore appear to be, and is, a weaker acid than carbonic acid. It is reasonable to ascribe the definite though weakly acid character of phenol to the presence of the phenyl group, which in phenol and many other compounds displays a more electro-negative or less electro-positive character than the hydrogen atom or alkyl groups, indicated by the neutral properties of water and alcohols respectively.

Additional confirmation of the above formula for phenol is given by its reaction on heating with phosphorus pentachloride, a reaction analogous to that taking place more easily in the case of alcohols. In this case, monochlorobenzene or phenyl chloride is formed, due to the replacement of the hydroxyl group of chlorine:

$$C_6H_5OH + PCl_5 \rightarrow C_6H_5Cl + POCl_3 + HCl$$

Phenol reacts very readily with halogens, producing compounds in which hydrogen atoms in the nucleus are substituted by halogen atoms. Various halogen substituted derivatives are produced according to the conditions employed, but the chief ones are 2 : 4 : 6-*trichloro-* and 2 : 4 : 6-*tribromo- phenols*:

$$C_6H_5OH + 3Cl_2 \rightarrow C_6H_2Cl_3OH + 3HCl$$

The constitutional formulae of these compounds are

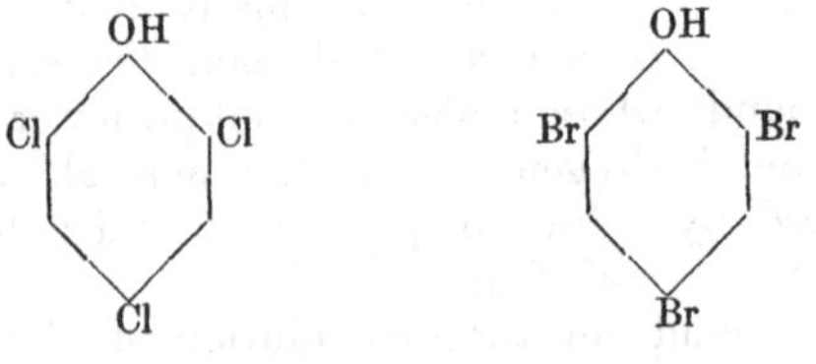

OH, Cl, Cl, Cl (2 : 4 : 6-trichlorophenol ring) OH, Br, Br, Br (2 : 4 : 6-tribromophenol ring)

One of the most satisfactory methods for the identification of phenol,

* Other metallic phenates are also known, e.g. $(C_6H_5O)_2Ca$, $(C_6H_5O)_2Ba$, $(C_6H_5O)_3Al$.

even in dilute aqueous solution, is to add to the latter bromine water until the colour of the bromine just persists. The white precipitate is filtered off, and recrystallised from very dilute aqueous alcohol. The tribromophenol crystallises in colourless needles, m.p. 96°.* Trichlorophenol is very similar in appearance and has m.p. 67–68°.

Even *triiodophenol* (2 : 4 : 6), m.p. 157°, analogous to the above compounds, is formed by the direct action of a solution of iodine in potassium iodide on an aqueous solution of phenol, especially in the presence of a little alkali. The ease with which these substitutions by halogen take place is proof of the hydrogen atoms in the benzene nucleus being rendered more active by the initial presence of the hydroxyl group than they are in benzene itself.

The reduction of phenol to benzene when distilled in the presence of zinc dust has already been mentioned. When a mixture of phenol vapour and hydrogen is passed over finely divided nickel at a temperature of 140–150°, the phenol is reduced to the hexahydro-derivative, cyclo*hexanol* (having properties resembling those of a typical secondary alcohol):

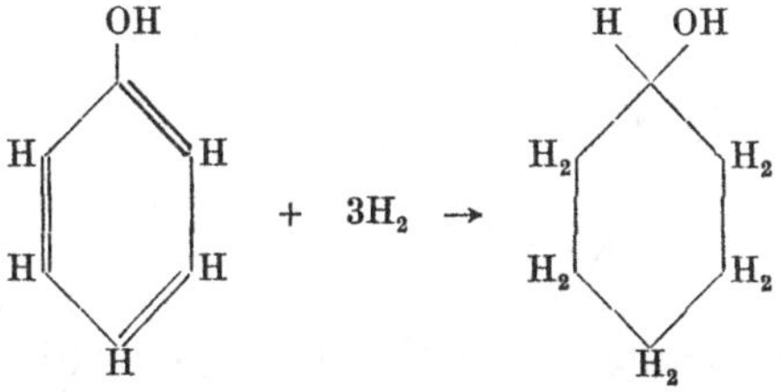

OH; H, H, H, H, H + $3H_2$ → H OH; H_2, H_2, H_2, H_2, H_2

$C_6H_{11}OH$, m.p. 24°, b.p. 160–161°

Phenol is readily acted upon by acids such as sulphuric, nitrous and nitric, yielding respectively sulphonic acid, nitroso, and nitro derivatives. When phenol is mixed with an equal quantity of concentrated sulphuric acid at ordinary temperature, a mixture of *ortho*- and *para*- sulphonic acids of phenol is produced. When the reaction is carried out at higher temperatures (90–100°), chiefly the *para*-sulphonic acid is produced. *Phenol-o-sulphonic acid* and *phenol-p-sulphonic acid* have respectively the following formulae:

OH, SO_3H (ortho)   OH, SO_3H (para)

* Although it is not the most accurate method, the weight of tribromophenol produced from a given solution may be used to estimate the amount of phenol originally present.

and the reaction by which they are produced is represented:

$$C_6H_5OH + HO.SO_3H \rightarrow C_6H_4\begin{cases} OH \\ SO_3H \end{cases} + H_2O$$

Sulphonation takes place by the mere solution of phenol in concentrated sulphuric acid, i.e. very much more readily than the sulphonation of benzene.

When phenol is mixed with sulphuric acid containing sodium nitrite and gently warmed, a compound, known as *p*-nitrosophenol is produced by the action of nitrous acid on the phenol. This reaction, generally known as *Liebermann's nitroso reaction* for phenols, is carried out as follows:

A few crystals of sodium nitrite are added to 2 or 3 c.c. of concentrated sulphuric acid and the mixture gently warmed (water-bath) until the crystals dissolve. To the mixture is added a small quantity of phenol and again the mixture is warmed. The solution at first changes to a deep blue colour. If the solution is diluted with water it becomes red, and if this be made alkaline it becomes blue.

p-*Nitrosophenol* has the constitutional formula I:

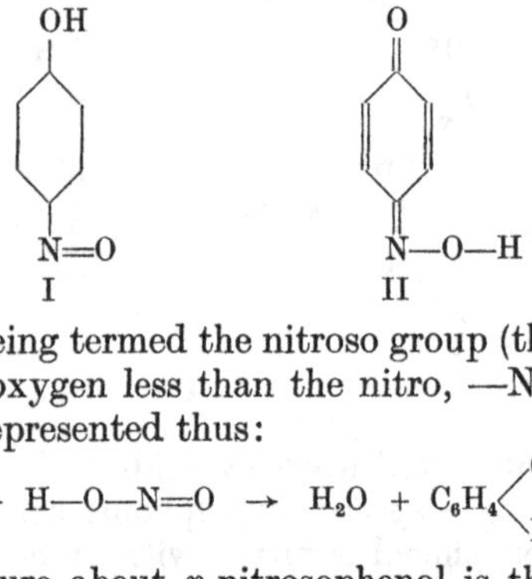

OH O

N=O N—O—H

I II

the —N=O group being termed the nitroso group (the nitroso group contains one atom oxygen less than the nitro, $—NO_2$, group). The formation may be represented thus:

$$C_6H_5.OH + H—O—N=O \rightarrow H_2O + C_6H_4\begin{cases} OH \\ NO \end{cases}$$

The interesting feature about *p*-nitrosophenol is that a compound having the above constitution is not actually isolated from the reaction, but a compound having the constitution represented by II. This compound is known as *benzoquinonemonoxime* (pale yellow needles, m.p. 126° with decomposition) and belongs to a class of substances known as oximes, which are referred to later (p. 159). The conversion of *p*-nitrosophenol into benzoquinonemonoxime is referred to as a tautomeric change and benzoquinonemonoxime and *p*-nitrosophenol are frequently spoken of as dynamic isomerides:

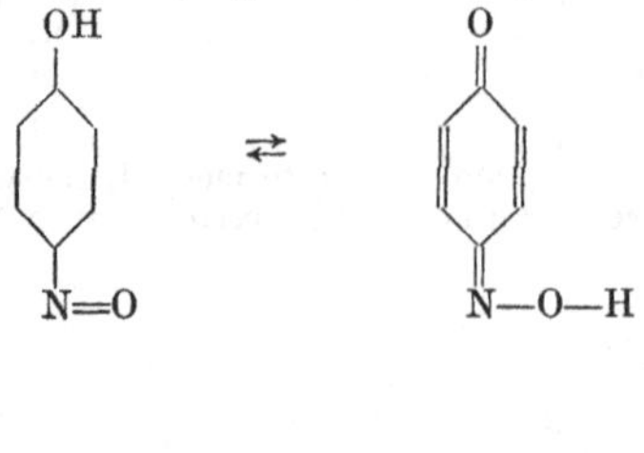

OH O

$\rightleftarrows$

N=O N—O—H

It is usual to adopt the oxime formula for the substance itself, which, however, is capable of reacting as if it possessed either of the two constitutions according to the conditions. Thus, it is feebly basic as would be expected from the oxime formula, and, on the other hand, it is oxidised to *p*-nitrophenol (*v.* below), which is what would be anticipated if it possessed the nitrosophenol constitution.

Action of Nitric Acid on Phenol

When nitric acid acts on phenol, the latter is nitrated, and according to the conditions the following compounds are obtained:

(i) *Mononitrophenols*:

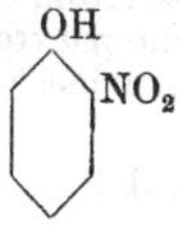

OH, NO_2 — *o*-nitrophenol

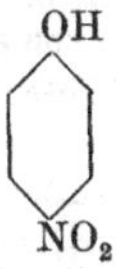

OH, NO_2 — *p*-nitrophenol

(ii) *Dinitrophenols*:

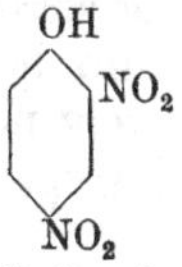

OH, NO_2, NO_2 — 2 : 4-dinitrophenol

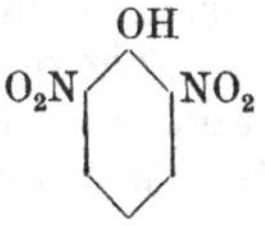

OH, O_2N, NO_2 — 2 : 6-dinitrophenol

and finally

(iii) *Trinitrophenol* or *picric acid*:

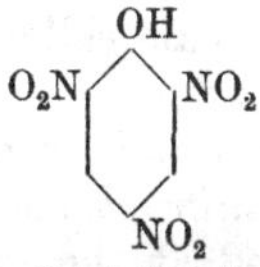

OH, O_2N, NO_2, NO_2 — 2 : 4 : 6-trinitrophenol

The above compounds are formed by the direct action of nitric acid on phenol under various conditions, the reaction being similar to that of nitric acid on benzene, but taking place much more readily, and even with dilute nitric acid.

o- and *p-Nitrophenols* are conveniently prepared in the laboratory as follows:

80 grams of sodium nitrate are dissolved in 200 c.c. of warm water, and to this solution is added very slowly 100 grams of concentrated sulphuric acid in such a way that the temperature of the mixture does not rise above 25°. This solution is kept stirred and a mixture of 50 grams of phenol with 5 grams of alcohol slowly added, the temperature of the mixture during the reaction never being allowed to rise above 25–30°. After all the phenol has been added, the stirring is continued

for about 2 hours. Water (200 c.c.) is now added, when a dark oil, often containing crystalline material, separates. The aqueous solution is poured off and the oil washed several times by decantation with water. Finally water (about 500 c.c.) is added and the mixture distilled in steam. The *o*-nitrophenol is volatile in steam while the *p*-nitrophenol remains in the distillation flask. *o*-Nitrophenol may crystallise in the condenser from the steam distillation flask, but any blocking of the condenser is easily avoided by cutting off the cooling water from time to time. The distillate contains solid, pure *o*-nitrophenol which can be filtered off and dried at ordinary temperature. The residue in the distilling flask containing the *p*-nitrophenol is poured out while hot into a suitable vessel, boiled up with dilute hydrochloric acid and the solution poured off from the tarry residue. This is repeated several times. The united extracts are heated to boiling and then boiled up further with decolorising charcoal (to absorb extraneous coloured and resinous material) and the solution filtered while hot. The filtrate is then made just alkaline by addition of an aqueous solution of sodium hydroxide and evaporated to small bulk on the water-bath. Finally, the concentrated solution is made just acid by the careful addition of hydrochloric acid, when the *p*-nitrophenol crystallises out and can be collected in the usual way. It may be recrystallised from boiling dilute hydrochloric acid or from boiling water, the solution being filtered from any resinous material which may have escaped separation during the previous operations.

o-Nitrophenol crystallises from alcohol in sulphur-yellow prisms, m.p. 44·5°. It has a distinct odour somewhat resembling that of burnt sugar.

p-Nitrophenol is less strongly coloured than *o*-nitrophenol and possesses little odour. It is dimorphic, occurring in two crystalline forms, the α- or metastable form passing into the β-form at about 63°. It has m.p. 114°, and is almost completely non-volatile in steam. It is readily soluble in hot water and very sparingly soluble in cold water.

Both these nitrophenols are much stronger acids than phenol and, for example, decompose sodium carbonate which is not affected by phenol. They are, however, relatively weak acids. On account of the striking colour of the solutions of their sodium salts (*o*-nitrophenol—yellowish red, *p*-nitrophenol—pale yellow), these substances may be used as indicators in acidimetry.

The next direct nitration products of phenol are the isomeric 2 : 4-*dinitrophenol* and 2 : 6-*dinitrophenol*. These are not very important; they are both produced at the same time, but the latter is formed only in very small amount.

2 : 4-Dinitrophenol crystallises from water in almost colourless plates, m.p. 113°. 2 : 6-Dinitrophenol crystallises from water in bright yellow fine needles, m.p. 64°. These two substances are again stronger acids than the mononitrophenols, and many metallic salts have been prepared. These are salts of monobasic acids, e.g. $Na(C_6H_3O_5N_2)$, $Ba(C_6H_3O_5N_2)_2$, etc. The salts are all brightly coloured and crystalline.

The final direct nitration product of phenol is *picric acid*, 2 : 4 : 6-*trinitrophenol*, and it has long been known that it is produced by the action of nitric acid on a number of other substances as well. For example, in 1771 it was shown by Peter Woulfe that a yellow-coloured substance is produced by the action of nitric acid on indigo (p. 333); and in 1799 it was observed that the same substance was produced in a similar manner from silk. Its preparation in the laboratory from phenol may be carried out as follows:

A mixture of 10 grams of phenol and 10 grams of concentrated sulphuric acid is added in very small quantities to 30 grams of concentrated nitric acid. The mixture is warmed on the water-bath while the nitric acid reacts gently. It will be observed that a dark-coloured oil (dinitrophenol) separates. After about 2 hours fuming nitric acid is added and the heating continued until the oil disappears, the solution becomes clear, and on cooling and diluting a small test portion with water, yellow crystals separate which completely dissolve when the diluted solution is heated to boiling. The main bulk of the solution is then diluted considerably with water after cooling, the yellow crystals filtered off and recrystallised from boiling water. The picric acid crystallises from the aqueous solution in bright yellow plates, m.p. 122·5°.

The name was applied to this substance on account of its bitter taste (*πικρός* = bitter) and it was the first artificial dye; animal fabrics, wool and silk, are dyed permanently yellow by simply immersing them in a dilute aqueous solution of picric acid, but picric acid is not now used as a dye-stuff to any great extent. Picric acid is only sparingly soluble in cold water, but this solution is bright yellow and stains the skin. A 1 per cent. aqueous solution or a suitable picric acid ointment is frequently applied to the skin for the treatment of burns and certain infectious skin diseases. Its use, however, for such purposes is deprecated, for, although an antiseptic, picric acid, like all aromatic nitro compounds, is poisonous and poisoning may occur by absorption from an open wound.

Picric acid under ordinary circumstances is quite safe to handle and small portions may be heated to decomposition. When detonated picric acid explodes violently, and it is still used as an explosive, trinitrotoluene being now used to a much greater extent. Picric acid is a relatively strong monobasic acid and forms well-defined crystalline salts. The sparing solubility of potassium picrate, $K(C_6H_2O_7N_3)$, may be employed as a qualitative test for the potassium ion. Many metallic salts of picric acid have been described. In the dry state these picrates are dangerous, being extremely sensitive to shock and liable to explode when heated. Another remarkable feature about picric acid (and, to a less degree, other nitro derivatives of benzene and phenols) is its property of forming compounds with certain types of neutral substances. Thus it will form compounds with naphthalene, anthracene, phenanthene, benzene and other aromatic hydrocarbons. These compounds are formed when equimolecular quantities of the particular hydrocarbon and picric acid are brought into solution in a suitable medium, the compounds being formed by the combination of one molecule of each constituent. Picric acid is frequently employed for the separation and identification of organic bases (p. 245), the picrates of these bases being generally well-crystallised compounds stable and safe to handle. The compound, whether a hydrocarbon or an organic base, combined with picric acid, can readily be obtained from the picrate by treatment of the latter with a strong inorganic base in aqueous solution, e.g. sodium hydroxide. In these circumstances soluble sodium picrate is formed and the substance previously combined with the picric acid can be isolated by methods depending on the properties of the particular compound.

When picric acid is warmed with an aqueous solution of bleaching powder a somewhat remarkable reaction takes place with the formation of a colourless liquid, b.p. 112°, known as *chloropicrin*, $C(NO_2)Cl_3$, having marked lachrymatory properties. Chloropicrin is chloroform with the one hydrogen atom in the molecule of the latter replaced by the monovalent nitro group. *Bromopicrin*, $C(NO_2)Br_3$, is a similar compound produced by the action of an aqueous solution of calcium hypobromite on picric acid.*

In the preparation of picric acid, the nitric acid reacts with a solution of phenol in sulphuric acid. This latter solution, as has been pointed out, at ordinary temperatures contains both *o*- and *p*-phenolsulphonic acids, and therefore the nitric acid not only nitrates the phenol in the ordinary sense but also reacts in such a way as to replace the sulphonic group by the nitro- group, thus:

$$C_6H_4(OH)(SO_3H) + 3HNO_3 \rightarrow C_6H_2(OH)(NO_2)_3 + 2H_2O + H_2SO_4$$

(OH at top of ring, SO_3H para; product: OH at top, O_2N and NO_2 ortho, NO_2 para)

Several examples have been given to show that when once a hydrogen atom in the benzene nucleus has been replaced by a monovalent atom or group of atoms, such as $—CH_3$, —Cl, —OH, $—SO_3H$, further substitution takes place more readily than in the first place. Considering, for the present, only the case of the first nitration products of monosubstituted benzenes, well-marked differences are observed in the position in the molecule which the entering nitro- group takes relative to the group already present in different compounds. Thus, mononitrobenzene yields *m*-dinitrobenzene, benzenesulphonic acid yields *m*-nitrobenzenesulphonic acid; on the other hand, monochloro- and monobromo- benzenes yield in

* Chloropicrin and bromopicrin may be described as *nitrochloroform* and *nitrobromoform* respectively. They are aliphatic nitro-compounds (trihalogenonitromethanes).

each case a mixture of the *o*- and *p*-nitro-compounds, toluene yields a mixture of *o*- and *p*-nitro-compounds and phenol yields a mixture of *o*- and *p*-nitrophenols. These are the main products of the mononitration in each case; small quantities of the other possible mononitration product or products may be formed in each case, but the amount of these is relatively insignificant and frequently does not appear after isolation and purification of the product or products of the reaction. Further, the identification of a particular disubstituted benzene as an *ortho*-, *meta*- or *para*- compound has been carried out by converting the compound into other compounds which have been definitely orientated by (say) Körner's method (p. 166), or which has been prepared by a method which leaves no doubt as to its constitution.

The above examples are by no means exhaustive, since there is a very large number of monosubstituted benzene derivatives; but whatever be the nature of a monosubstituted benzene derivative the substituting group may be described as a *meta-directing* or as an *ortho-para-directing* group when a second substituting group is introduced into the nucleus. This classification of the substituting groups applies not only to the mononitration of such compounds but also to halogenation and to any monovalent atom or group which can be substituted directly into the nucleus. Thus, the —OH group in phenol is *ortho-para*-directing not only with regard to the position which an entering nitro- group takes up in the benzene nucleus, but also with regard to the position which a halogen atom (chlorine, bromine or iodine) takes up, the first chlorination products of phenol being mixtures of *o*-chloro- and *p*-chloro- phenols and similarly for the bromination and iodination products. On account of the experimental convenience, the classification and attempted explanation of the phenomenon has been largely based on the results of nitration experiments.

In 1892, Crum Brown and Gibson published a paper entitled "A Rule for determining whether a given Benzene Mono-derivative shall give a *Meta*-di-derivative or a mixture of *Ortho*- and *Para*-di-derivatives". One way of expressing the rule of Crum Brown and Gibson is the following: If the hydrogen derivative of the substituting group in a monosubstituted benzene derivative is capable of being *directly* oxidised by addition of oxygen, the second substituting atom or group will enter the *meta*-position in the nucleus; if the hydrogen derivative of the first substituent group be not capable of direct oxidation to the corresponding hydroxy-derivative, the second substituting atom or group will enter the *ortho*- and *para*-positions, a mixture of the two isomeric disubstituted compounds resulting. The results so far described can be expressed in a table somewhat similar to that given by Crum Brown and Gibson:

Monosubstituted benzene derivative	Substituting group	Hydrogen compound of substituting group	Oxidation product	Disubstituted derivative or derivatives produced
C_6H_5Cl	—Cl	HCl	HOCl	*o*-, *p*-
C_6H_5Br	—Br	HBr	HOBr	*o*-, *p*-
$C_6H_5CH_3$	$—CH_3$	HCH_3	$HOCH_3$	*o*-, *p*-
C_6H_5OH	—OH	HOH	HO.OH	*o*-, *p*-
$C_6H_5NO_2$	$—NO_2$	HNO_2	HNO_3	*m*-
$C_6H_5SO_2OH$	$—SO_2OH$	HSO_2OH	H_2SO_4	*m*-

It will be appreciated that HCl, HBr, CH_4 and H_2O cannot be converted directly into the corresponding oxidation products, HOCl, HOBr, $HOCH_3$ and H_2O_2 respectively, and C_6H_5Cl, C_6H_5Br, $C_6H_5CH_3$ and C_6H_5OH all yield *ortho*- and *para*- disubstituted derivatives. On the other hand, HNO_2 and H_2SO_3 are converted directly into HNO_3 and H_2SO_4, and $C_6H_5NO_2$ and $C_6H_5SO_3H$ yield chiefly *meta*- disubstituted derivatives. Although not infallible, Crum Brown and Gibson's rule applies in the majority of cases, only a few of which are actually mentioned above.

Crum Brown and Gibson were careful to emphasise that the rule is not a law, because it has no visible relation to any mechanism by which the substitution is carried out in one way rather than in another way. It must be acknowledged that, in spite of an ever-increasing knowledge of the subject since it was formulated, the rule provides still a useful working system.

More recently, theories as distinct from rules to explain as well as predict substitution effects have been put forward chiefly by Lapworth, Robinson, Flürscheim and Ingold; but it would be undesirable as well as unnecessary from the present point-of-view to give any detailed account of these, since no finality has yet been reached on the subject. The very considerable amount of work done by the above-mentioned authors on the subject endeavours to explain the substitution effects observed and to predict others by taking into account the distribution of affinity in the molecule or the relative electrical character of different parts of the molecule (so-called polarity effects). It seems unlikely, however, that we shall arrive at any fully adequate explanation of such phenomena as substitution effects until the difficulties attaching to the determination of the fundamental physical nature of the molecules of complicated organic substances have been overcome.

Derivatives of phenol in which the hydrogen of the hydroxyl group is replaced by alkyl radicals are obtained by relatively simple reactions. These compounds belong to the class of substances known as ethers (p. 135). *Methylphenyl ether* or *anisole* is prepared by heating potassium phenate in alcoholic solution with methyl iodide:

$$C_6H_5.O\boxed{K + I}CH_3 \rightarrow KI + \underset{\text{anisole}}{C_6H_5.O.CH_3}$$

Similarly, *ethylphenyl ether* or *phenetole* is obtained by an analogous reaction using ethyl iodide. Anisole is a colourless fragrant liquid, b.p. 154°: phenetole has similar properties and has b.p. 172°. Both of these substances, which are typical of a large number of analogous compounds, when heated with hydriodic acid (the constant boiling aqueous solution, b.p. 126°, is usually employed for such a purpose) are reconverted into phenol and the alkyl iodide.

$$C_6H_5O\boxed{R + I}H \rightarrow RI + C_6H_5OH$$

With suitable precautions the reaction is quantitative, and if the alkyl iodide evolved be absorbed in an alcoholic solution of silver nitrate, silver iodide is precipitated and can be weighed. The weight of silver iodide corresponds to that of the alkyl iodide evolved. This reaction is frequently employed for the determination of methoxyl (CH_3O—) and ethoxyl (C_2H_5O—) groups in organic compounds; it is known as Zeisel's method.

Apart from its use as a source of such compounds as have been described, phenol is very largely employed as a powerful antiseptic and, when applied locally, as an anaesthetic. Phenol is highly poisonous, but in carefully regulated doses it is an important medicament.

One of the most important derivatives of phenol is *salicylic acid* (p. 132). This acid has been shown to have the following constitution:

$$\text{(benzene ring)}\begin{cases}\text{OH}\\ \text{COOH}\end{cases} \qquad \left(C_6H_4\begin{cases}\text{OH (1)}\\ \text{COOH (2)}\end{cases}\right)$$

and is therefore *o*-hydroxybenzoic acid (or, as showing its relation to phenol, *o*-carboxyphenol). Formerly, salicylic acid was obtained from natural sources, but it is now prepared almost exclusively from phenol by Kolbe's (1874) process. In this process, sodium phenate is treated with carbon dioxide under pressure, the gas is absorbed by the phenate and sodium phenylcarbonate is formed:

$$C_6H_5ONa + CO_2 \rightarrow C_6H_5O{-}C\begin{cases}{=}O\\ \text{ONa}\end{cases} \quad \text{or} \quad C\begin{cases}\text{OC}_6\text{H}_5\\ {=}O\\ \text{ONa}\end{cases}$$

When this sodium phenylcarbonate is heated at 120–140° under pressure, an intra-molecular change takes place with the production of sodium salicylate:

$$\text{(ring: H, H, H, H, H)}{-}O{-}C\begin{cases}{=}O\\ O{-}Na\end{cases} \longrightarrow \text{(ring: H, H, H, H)}\begin{cases}O{-}H\\ {-}C\begin{cases}{=}O\\ O{-}Na\end{cases}\end{cases}$$

The salicylic acid is precipitated by dissolving the sodium salicylate in water and acidifying the solution with hydrochloric acid:

$$C_6H_4\left\langle\begin{matrix}OH\\COONa\end{matrix}\right. + HCl \rightarrow C_6H_4\left\langle\begin{matrix}OH\\COOH\end{matrix}\right. + NaCl$$

Salicylic acid crystallises from hot water in colourless needles, m.p. 159°. Being a phenol, it gives in neutral solution with ferric chloride an intense violet coloration. When heated with lime it is converted into phenol, calcium carbonate being formed:

$$C_6H_4\left\langle\begin{matrix}OH\\COOH\end{matrix}\right. + CaO \rightarrow CaCO_3 + C_6H_5OH$$

Acetylsalicylic acid (*aspirin*), $C_6H_4{:}(O.OC.CH_3)(COOH)$, is prepared from salicylic acid as described on p. 209. It is an easily hydrolised 'ester' (p. 213) of acetic acid.

Of the homologues of phenol, the three methylhydroxybenzenes or *cresols* occur with phenol in coal-tar and are obtained in the distillate along with the latter. These compounds are isomeric with methylphenyl ether (anisole) and benzyl alcohol and have the following constitutional formulae:

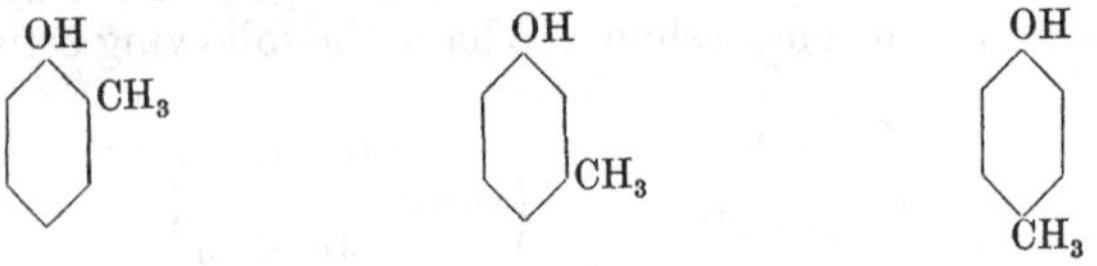

o-*cresol*, m.p. 31°, b.p. 191° | m-*cresol*, m.p. 3–4°, b.p. 202° | p-*cresol*, m.p. 36°, b.p. 202°

These compounds are not readily obtained pure from coal-tar on account of the difficulty of separating them from each other; they are prepared, however, in a state of purity by other methods. One of these, which is analogous to the method of preparing phenol, is by the fusion of the corresponding toluene sulphonic acids (or their sodium salts) with sodium hydroxide:

$$C_6H_4\left\langle\begin{matrix}CH_3\\SO_3Na\end{matrix}\right. + NaOH \rightarrow C_6H_4\left\langle\begin{matrix}CH_3\\OH\end{matrix}\right. + Na_2SO_3$$

These three cresols resemble phenol in most properties. They yield sodium and potassium derivatives and are soluble in solutions of sodium and potassium hydroxides. From these solutions, the cresols are precipitated by carbon dioxide.

The cresols give a bluish coloration with ferric chloride and all yield toluene when distilled with zinc dust, just as phenol yields benzene under similar conditions:

$$C_6H_4\left\langle\begin{matrix}OH\\CH_3\end{matrix}\right. + Zn \rightarrow ZnO + C_6H_5.CH_3$$

The corresponding methyl ethers (methyl cresols or methoxy-toluenes) are prepared by methods analogous to those for obtaining methylphenyl ether, and these methyl ethers, $C_6H_4\langle{}^{OCH_3}_{CH_3}$, when oxidised, under the conditions for converting toluene into benzoic acid, are converted into the corresponding methoxybenzoic acids, $C_6H_4\langle{}^{OCH_3}_{COOH}$. The cresols, themselves, cannot be oxidised directly to the corresponding hydroxybenzoic acids, $C_6H_4\langle{}^{OH}_{COOH}$.

Another important homologue of phenol is *thymol*, which has been shown to have the following constitution:

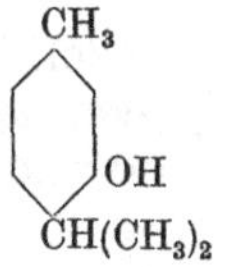

CH_3

OH

$CH(CH_3)_2$

3-hydroxy-1-methyl-4-*iso*propyl benzene

It is obtained from the essential oils of *Thymus vulgaris*, *Monarda punctata* and *Carum copticum*. The oil from the last named, generally referred to as Ajowan oil, contains not less than 40 per cent. thymol. The Ajowan oil is shaken with sodium hydroxide solution which dissolves the thymol, and the alkaline solution after separation from the insoluble portion is treated with hydrochloric acid when the thymol is precipitated.

Thymol has the characteristic odour of thyme and crystallises in large colourless plates, m.p. 51·5°, b.p. 233·5°. Although it is a phenol it gives no coloration with ferric chloride. It is easily soluble in organic solvents but very sparingly soluble in water. The constitution of thymol is proved by its reactions with zinc dust and phosphorus pentoxide respectively.

When thymol is distilled with zinc dust, cymene (p. 75) is produced; compare the action of zinc dust on phenol:

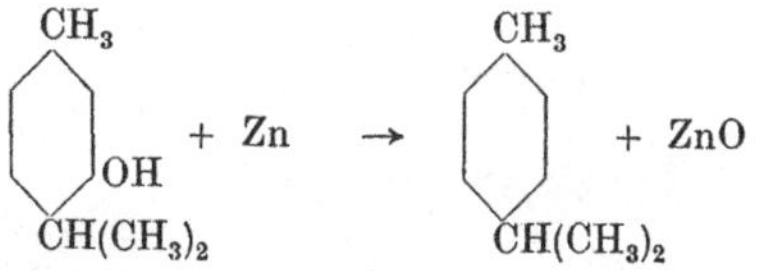

CH_3 CH_3

OH   + Zn →   + ZnO

$CH(CH_3)_2$ $CH(CH_3)_2$

When thymol is heated with phosphorus pentoxide, *m*-cresol (p. 132) and propylene are produced:

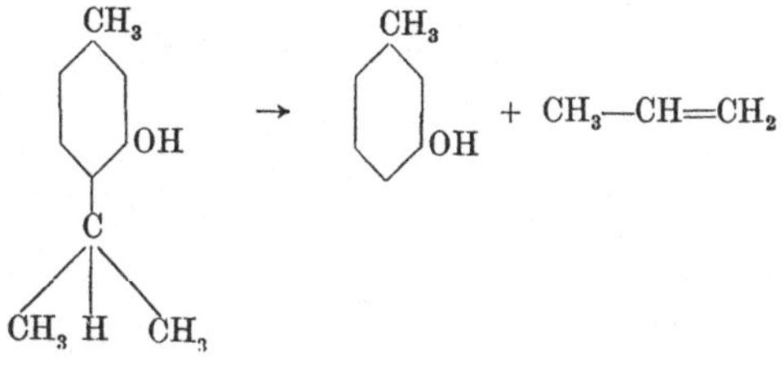

CH_3 CH_3

OH → OH + $CH_3—CH=CH_2$

C

CH_3 H CH_3

The latter reaction can be considered to consist in the transference of a hydrogen atom from the isopropyl radical to the nucleus in the position formerly occupied by the radical.

Thymol is an important antiseptic and anthelmintic and is used largely in the treatment of ankylostomiasis.

Isomeric with thymol is a closely related phenol known as *carvacrol*, which occurs in the oil from *Origanum hirtum* and other Origanum oils, from which it is isolated. It may be obtained by heating camphor with iodine:

$$C_{10}H_{16}O + I_2 \rightarrow \underset{\text{carvacrol}}{C_{10}H_{14}O} + 2HI$$

Carvacrol gives a green coloration with ferric chloride. The facts that on distillation with zinc dust it yields *p*-cymene, and when heated with phosphorus pentoxide it is decomposed into *o*-cresol and propylene, prove its constitution to be

CH_3 (top of ring), OH (adjacent position), $CH(CH_3)_2$ (bottom of ring) — 2-hydroxy-1-methyl-4-isopropyl benzene.

The use of carvacrol as an internal antiseptic and antiparasitic as supplementary to thymol has been suggested.

CHAPTER V

ETHERS

THE monohydroxy-derivatives of the hydrocarbons of the general formula, R—O—H [where R=a monovalent hydrocarbon radical either alkyl (aliphatic radical, methyl-, ethyl-, etc.) or aryl (aromatic radical, phenyl-, etc.)], have been considered to be derived theoretically from water, H—O—H, by the replacement of a hydrogen atom by the hydrocarbon radical. If R is an alkyl group, the monohydroxy-derivative is a monohydric alcohol which like water is a neutral substance, whereas if it be an aryl group, the monohydroxy-derivative is a monohydric phenol which is feebly acidic in character, probably on account of aryl radicals being potentially more electronegative in character than alkyl radicals. An analogy has been drawn between the monohydric alcohols on the one hand and the monobasic alkali hydroxides (e.g. NaOH) on the other, particularly with regard to their reaction with acids.

The class of substances which can be considered to be derived theoretically from water by the replacement of both hydrogen atoms by equivalent alkyl or aryl groups are known as *ethers*, which thus have the general formula R—O—R', the oxygen atom being united to two carbon atoms, and the radicals R and R' being the same as, or different from, each other. In the former case, the compound has the general formula R—O—R, and is usually described as a *simple ether*, whereas, in the latter case, the compound has the general formula R—O—R', and is described as a *mixed ether*.

Since the monohydric aliphatic alcohols derived from the paraffins constitute a homologous series having the general formula $C_nH_{2n+1}OH$, the ethers containing two saturated aliphatic or alkyl radicals also constitute a homologous series of compounds having the general formula $C_nH_{2n+1}.O.C_nH_{2n+1}$ or $(C_nH_{2n+1})_2O$, n (the number of carbon atoms in the radical) being the same or different in the two radicals. These may be termed saturated aliphatic ethers and are the most important members of this class of organic compounds. These ethers may be compared with the oxides of the alkali metals (e.g. Na_2O) in inorganic chemistry; thus

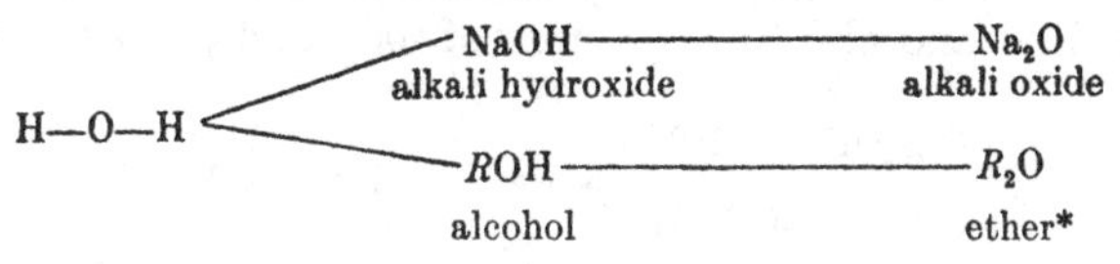

* Ethers have frequently been termed alkyl or aryl oxides.

Ethers and monohydric alcohols may be isomeric; thus, the simplest ether, dimethyl ether, $CH_3.O.CH_3$, is isomeric with ethyl alcohol, C_2H_5OH, both compounds having the molecular formula C_2H_6O, and methylethyl ether, $CH_3.O.C_2H_5$, is isomeric with *n*- and *i*-propyl alcohols, C_3H_7OH.

The constitution of any ether can be proved by a simple method of synthesis, which consists in allowing an appropriate alkyl iodide (CH_3I, C_2H_5I, etc.) to react with the sodium derivative of the necessary alcohol, in a suitable medium. The reaction can be generally expressed:

$$R.O.Na + IR' \rightarrow NaI + R.O.R'$$

it usually takes place readily and can be applied not only for the formation of the aliphatic ethers but also for that of the aromatic ethers. This is the best method for obtaining mixed ethers and has already been described for the preparation of methylphenyl ether or anisole.

Aliphatic Ethers

Dimethyl ether is a colourless gas, b.p. −23·65°, and is only of theoretical importance. It may be obtained by the action of methyl iodide on sodium methylate:

$$CH_3.ONa + I.CH_3 \rightarrow NaI + CH_3.O.CH_3$$

and its formation by this reaction proves its constitution to be satisfactorily represented by the formula

```
    H     H
    |     |
H—C—O—C—H
    |     |
    H     H
```

Its preparation by the action of concentrated sulphuric acid or of syrupy phosphoric acid on methyl alcohol is analogous to that of the preparation of diethyl ether (see below). Having the molecular formula, C_2H_6O, dimethyl ether is isomeric with ethyl alcohol, from which it is readily distinguished by its physical properties and by the fact that it does not react with sodium. In its chemical behaviour, dimethyl ether closely resembles the most important member of this class, diethyl ether.

Methylethyl ether is the next homologue and is also a colourless gas, b.p. 10·8°. It is prepared by the action of sodium ethylate on methyl iodide and also by the action of sodium methylate on ethyl iodide:

$$\left.\begin{array}{l} C_2H_5ONa + ICH_3 \\ CH_3ONa + IC_2H_5 \end{array}\right\} \rightarrow NaI + CH_3.O.C_2H_5$$

These methods of preparation establish its constitutional formula

Diethyl ether (generally known as ether). This is the next homologue and, from the scientific and industrial points of view, the most important compound of the series.

Although it had been known for a long time previously, its constitutional formula

$$\begin{array}{ccccccccc} & & H & & H & & H & & H & & \\ & & | & & | & & | & & | & & \\ H & — & C & — & C & — & O & — & C & — & C & — & H \\ & & | & & | & & | & & | & & \\ & & H & & H & & H & & H & & \end{array}$$

was finally established by Williamson in 1851, who prepared the compound by heating sodium or potassium ethylate with ethyl iodide in alcoholic solution:

$$C_2H_5O\boxed{Na + I}C_2H_5 \rightarrow NaI + C_2H_5.O.C_2H_5$$

It has been suggested (p. 29) that ethylene and diethyl ether may be regarded as dehydration products of ethyl alcohol:

$$C_2H_5OH \rightarrow C_2H_4 + H_2O$$

$$2C_2H_5OH \rightarrow (C_2H_5)_2O + H_2O$$

Whenever it is possible, such apparently simple reactions which lead to the production of industrially important substances are nowadays carried out by catalytic processes. It is known that those catalytic processes which lead to the production of ethylene from alcohol always yield diethyl ether as a by-product, and that by modifying the conditions (usually only change of temperature is necessary) the production of ethylene may be diminished and that of the ether increased. In spite, however, of the apparent simplicity of such a process as passing alcohol over a dehydrating catalyst (e.g. aluminium oxide, kaolin, etc.) at a particular temperature (about 250°), diethyl ether is still produced by a process whereby it was first obtained by Cordus in 1540. This process consisted in heating alcohol with sulphuric acid (hence the old names for ether, vinic ether, sulphuric ether, etc.) and was thoroughly investigated by Williamson (1850–1856) and is frequently known as Williamson's 'Continuous Process'.

It has already been indicated that the reaction between ethyl alcohol and sulphuric acid leading to the production of diethyl ether takes place in two stages, the ester, ethyl hydrogen sulphate, being the intermediate product:

$$\text{(i)}\quad C_2H_5OH + H_2SO_4 \rightleftarrows \begin{matrix} C_2H_5O \\ HO \end{matrix}\!\!>SO_2 + H_2O$$

$$\text{(ii)}\quad \begin{matrix} C_2H_5O \\ HO \end{matrix}\!\!>SO_2 + C_2H_5OH \rightarrow C_2H_5.O.C_2H_5 + H_2SO_4$$

Theoretically, it would appear that a fixed quantity of sulphuric acid is sufficient to convert an indefinitely large quantity of alcohol into ether. If it were not for the impossibility of preventing side reactions from taking place to a certain extent this would be the case. In the first place, the conditions of carrying out the reaction provide largely for eliminating most of the water formed in reaction (i), the water distilling over with the ether, so that reaction (i) approaches completion. Now the temperature of carrying out the preparation of ether by this method is 135–145°, but at a slightly higher temperature 160–180° and in presence of a larger amount of sulphuric acid ethylene is produced:

$$\begin{matrix} C_2H_5O \\ HO \end{matrix} \!\!>\! SO_2 \rightarrow C_2H_4 + H_2SO_4$$

and in this side reaction some of the alcohol is therefore converted into ethylene and in practice it is found impossible to prevent its taking place to some extent. Finally there are the side reactions, impossible to prevent completely, due to the oxidising action of concentrated sulphuric acid on the comparatively easily oxidised substance ethyl alcohol, whereby oxidation products (acetaldehyde and acetic acid) are produced, the sulphuric acid being at the same time reduced to sulphurous acid or sulphur dioxide at the temperature of the reaction.* It might be mentioned that diethyl sulphate produced by the reaction

$$2C_2H_5O.SO_2OH \rightarrow (C_2H_5)_2SO_4 + H_2SO_4$$

is scarcely likely to be formed under the present conditions, since the distillation is carried out at the ordinary pressure. If care is taken to adhere as rigidly as possible to the conditions of the reaction, i.e. the proper proportion of alcohol and sulphuric acid and the correct temperature, the production of by-products, ethylene, acetaldehyde, sulphur dioxide, etc., is reduced to a minimum and the more nearly Williamson's process approaches to a 'continuous' one.

Diethyl ether is a colourless liquid possessing a characteristic and not unpleasant odour. It has m.p. −117·6°, and b.p. 34·5°. It is much ligher than water, having a specific gravity of 0·720 at 15°. Ether is soluble in water; 6·6 grams of ether dissolve in 100 grams of

* It has been suggested that sulphuric acid may be replaced in the continuous process by certain sulphonic acids, particularly benzene sulphonic acid, the reaction being

$$C_6H_5.SO_2.\boxed{OH + H}OC_2H_5 \rightarrow C_6H_5.SO_2.OC_2H_5 + H_2O$$

ethyl benzene sulphonate

$$C_6H_5.SO_2.\boxed{OC_2H_5 + C_2H_5}OH \rightarrow C_6H_5.SO_2.OH + (C_2H_5)_2O$$

In this country this method is usually not employed.

water at 19°. Water will also dissolve in ether; 1·24 grams of water dissolve in 100 grams of ether at 20°. Ether and water form a binary mixture of constant boiling point, 34·14°.

Ether is highly inflammable and its vapour, which is $2\frac{1}{2}$ times as heavy as air, forms explosive mixtures with air and oxygen. On account of its low boiling point ether is very volatile at ordinary temperatures and because of its inflammability should not be brought near a flame. Ether burns when ignited with a pale blue flame, carbon dioxide and water being the products of combustion or complete oxidation:

$$C_4H_{10}O + 6O_2 \rightarrow 4CO_2 + 5H_2O$$

Extra precautions are necessary in manipulating ether in the laboratory. For safety, when ether is being distilled either in its preparation or from solutions containing it, the receiver provided with a side tube should be attached by a suitable cork to the water-cooled condenser (p. 508). To the side tube of the receiver is attached a long piece of rubber tubing which leads as far as possible below the surface of the working bench so that any uncondensed ether vapour is not only led away from flames on the bench but also to some distance below them.

Ether is much more soluble in an aqueous solution of hydrochloric acid than in water. The reason for the greater solubility of ether in hydrochloric acid than in water lies probably in the fact that ether forms unstable and soluble compounds with hydrogen chloride, and one of these has the composition $(C_2H_5)_2O.HCl$. (This is a so-called 'oxonium' compound which is a salt analogous to, but less stable than, the ammonium compounds. In the latter, the nitrogen atom in the ammonium ion is quadricovalent and in the oxonium ion the oxygen atom is tercovalent. Electronically, the constitution of diethylether hydrochloride is represented:

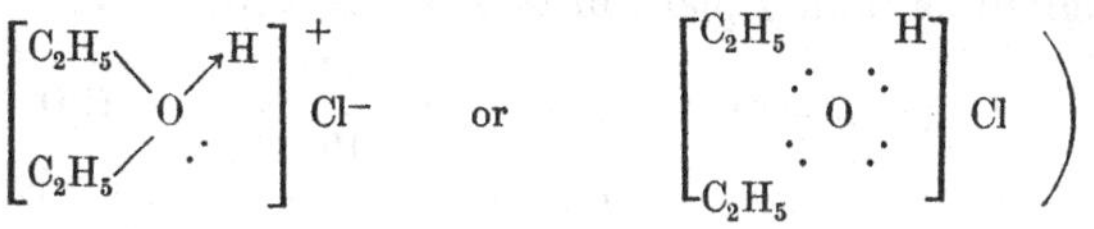

$$\left[\begin{array}{ccc} C_2H_5 & & H \\ & \overset{\searrow\ \nearrow}{O} & \\ C_2H_5 & \nearrow \ \ \colon & \end{array}\right]^+ Cl^- \quad \text{or} \quad \left(\left[\begin{array}{ccc} C_2H_5 & & H \\ & \colon O \colon & \\ C_2H_5 & & \end{array}\right] Cl\right)$$

Ether is also soluble in concentrated sulphuric acid, and it is liberated from this solution by the addition of ice.

Chemically, ether is a very stable substance. It is unattacked by sodium, and this metal is used for the removal of the last traces of water from ether, which has to be dry for use as a medium in certain chemical reactions. Ether forms substitution compounds when submitted to the action of halogens (chlorine and bromine). The reaction with chlorine is vigorous, and at 90° tetrachloro- and pentachloro- diethyl ethers are formed with the liberation of a corresponding amount of hydrogen chloride.

When ether is heated with concentrated hydriodic acid ethyl iodide is formed and can be identified:

$$(C_2H_5)_2O + 2HI \rightarrow H_2O + 2C_2H_5I$$

This reaction of hydriodic acid is a general one with nearly all ethers and it is largely employed in their detection and estimation. Zeisel's method (p. 131) depends on this reaction, and the method of determination of methoxyl ($—OCH_3$) or ethoxyl ($—OC_2H_5$) groups is by the action of hydriodic acid on methyl and ethyl ethers respectively.

Ether is largely used in the laboratory and industry as a solvent and for the extraction of substances, which are soluble in ether, from their solutions with which ether is not miscible or only partly so. One of its most important uses is as an anaesthetic, for which purpose the ether has to be specially pure and as free as possible from even minute traces of sulphur dioxide, acetaldehyde and from the products which are formed, although even only in small quantities, when ether is exposed to air and bright sunlight.* The use of ether as a general anaesthetic was introduced into this country about the middle of last century after the discovery of the use of chloroform for the same purpose in 1848 by Sir James Simpson, Professor of Midwifery in the University of Edinburgh.

Ether having the molecular formula, $C_4H_{10}O$, is not only isomeric with the four butyl alcohols but also with the two isomeric mixed ethers, *methyl*-n-*propyl ether* and *methyl*-i-*propyl ether*, having the constitutional formulae $CH_3.O.CH_2.CH_2.CH_3$ and $CH_3.O.CH(CH_3)_2$ respectively. Both of these ethers have been prepared by the action of methyl iodide on the sodium derivative of the corresponding alcohol.

On theoretical grounds it would appear possible to prepare mixed aliphatic ethers by the sulphuric acid method, since the reaction is represented as taking place in two stages, thus:

$$\text{(i)}\quad R—OH + H_2SO_4 \rightleftarrows \begin{matrix} RO \\ HO \end{matrix}\!\!>SO_2 + H_2O$$

$$\text{(ii)}\quad R'—OH + \begin{matrix} RO \\ HO \end{matrix}\!\!>SO_2 \rightarrow \begin{matrix} R \\ R' \end{matrix}\!\!>O + H_2SO_4$$

* When ether is exposed to bright sunlight and at the same time is in free contact with air, a peroxide is formed, the presence of which can be demonstrated by the ethereal solution liberating iodine from potassium iodide in aqueous solution. This same product is formed when ozonised oxygen is passed through ether. This peroxide is highly explosive.

Divinyl ether, $(CH_2:CH)_2O$, b.p. 39°, may be present in some specimens of ether, which are then unsuitable for anaesthetic use. Its presence in some specimens of ether has not been clearly explained: we do not know whether it is due to the action of oxygen in bright sunlight or to the presence of acetaldehyde, which is isomeric with vinyl alcohol ($CH_2:CHOH$), and the conversion of the latter into the corresponding ether.

Ether made from industrial methylated spirit is known as 'Methylated Ether'. It contains some dimethyl ether and while adapted for spraying to produce local anaesthesia on account of its low boiling point, it is not suitable for producing general anaesthesia.

The reversibility of reaction (i) to some extent, even under the conditions described above, makes it impossible to prevent the formation of the simple ethers *R*—O—*R* and *R'*—O—*R'*. Consequently, although some mixed ether, *R*—O—*R'*, is formed, this would have to be separated, by distillation or other suitable means, from the two simple ethers. As this may be a matter of some difficulty owing to the closeness of the boiling points of the three compounds and the possibility of their forming constant boiling binary and ternary mixtures, the best and simplest way of producing a given aliphatic mixed ether is by the general reaction:

$$R\text{—O—Na} + \text{I}R' \rightarrow R\text{—O—}R' + \text{NaI}$$

Aromatic Ethers

Diphenyl ether, or as it is still sometimes called *diphenyl oxide,* $(C_6H_5)_2O$, is an example of a true aromatic simple ether. This is formed by means of the reaction:

$$\underset{\text{sodium phenate}}{C_6H_5ONa} + \underset{\text{phenyl iodide}}{IC_6H_5} \rightarrow C_6H_5.O.C_6H_5 + NaI$$

This reaction, however, does not take place anything like so easily as the corresponding reaction with aliphatic compounds and requires finely divided copper as a catalyst.

Diphenyl ether forms colourless slender plates or prisms, m.p. 28°, b.p. 259°, and has a characteristic geranium-like odour. Unlike the aliphatic ethers it is not acted upon by concentrated hydriodic acid even at 250°. In their behaviour towards hydriodic acid the mixed aliphatic-aromatic ethers, anisole and phenetole (p. 130), are intermediate between the true aliphatic and true aromatic ethers. Anisole and phenetole are both decomposed by heating with concentrated hydriodic acid, with the production of phenol and methyl and ethyl iodides respectively:

$$\underset{\text{anisole}}{CH_3\boxed{O.C_6H_5 + H}I} \rightarrow CH_3I + C_6H_5OH$$

$$\underset{\text{phenetole}}{C_2H_5.O.C_6H_5} + HI \rightarrow C_2H_5I + C_6H_5OH$$

CHAPTER VI

ALDEHYDES AND KETONES

It has been indicated that when a primary alcohol is submitted to oxidation, the first product is known as an *aldehyde*.* Similarly, the direct product of the oxidation of a secondary alcohol is a *ketone*, both substances being derived from the corresponding alcohol by the removal of two hydrogen atoms from the molecule of the latter by oxidation, thus:

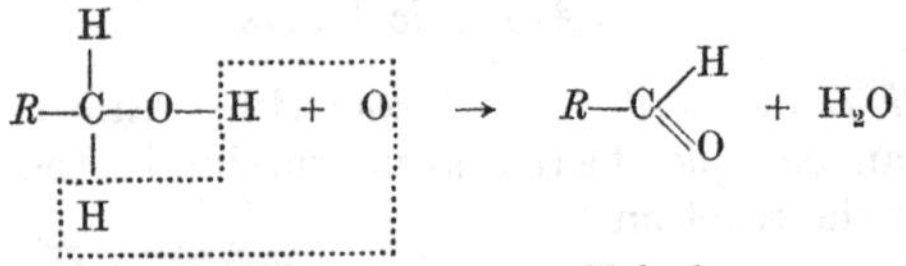

$$R{-}CH_2{-}O{-}H + O \rightarrow R{-}C\genfrac{}{}{0pt}{}{H}{O} + H_2O$$

primary alcohol — aldehyde

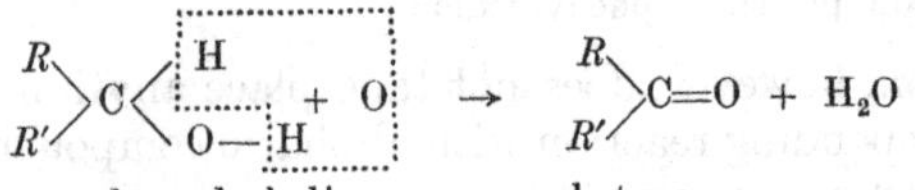

$$\genfrac{}{}{0pt}{}{R}{R'}\!>C\genfrac{}{}{0pt}{}{H}{O{-}H} + O \rightarrow \genfrac{}{}{0pt}{}{R}{R'}\!>C{=}O + H_2O$$

secondary alcohol† — ketone

The monohydric alcohols derived from the saturated aliphatic hydrocarbons or paraffins have the general formula $C_nH_{2n+1}OH$, and hence the aldehydes and ketones, derived from them by the removal of two hydrogen atoms from the molecule in each case, may be considered to have the general formula $C_nH_{2n}O$. This indicates that aldehydes and ketones may be isomeric with each other. On the other hand, it should be emphasised that the characteristic grouping in an aldehyde is the monovalent $-C\genfrac{}{}{0pt}{}{H}{O}$ group, generally abbreviated to $-CHO$, united either to a hydrogen atom or to a univalent hydrocarbon radical. The general formula for aldehydes derived from the primary alcohols of the general formula $C_nH_{2n+1}OH$ is $C_nH_{2n+1}.CHO$, and for the ketones derived from the secondary alcohols having the same general formula is $(C_nH_{2n+1})_2{:}C{=}O$.

Aldehydes and ketones can be reduced again to the corresponding primary and secondary alcohols respectively, and both aldehydes and ketones consequently have certain *unsaturation* properties in common.

* Aldehyde is an abbreviation of the Latin *alcohol dehydrogenatum* (deprived of hydrogen).

† The alkyl groups in the secondary alcohol may be the same as, or different from, each other.

On the other hand, aldehydes differ markedly from ketones in that the former are reducing agents and tend to yield the corresponding acid on further oxidation. Ketones do not yield direct oxidation products and are therefore not reducing agents. These properties may be summed up:

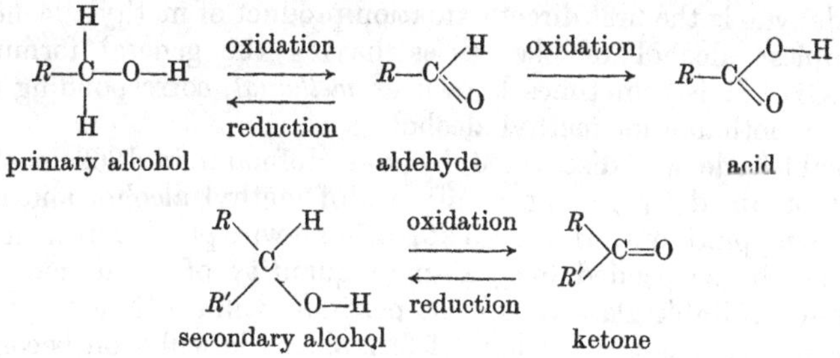

$$R\text{—}\overset{\displaystyle H}{\underset{\displaystyle H}{C}}\text{—O—H} \underset{\text{reduction}}{\overset{\text{oxidation}}{\rightleftarrows}} R\text{—C}\langle^{H}_{O} \xrightarrow{\text{oxidation}} R\text{—C}\langle^{O—H}_{O}$$

primary alcohol — aldehyde — acid

$$\begin{matrix} R & & H \\ & C & \\ R' & & O\text{—H} \end{matrix} \underset{\text{reduction}}{\overset{\text{oxidation}}{\rightleftarrows}} \begin{matrix} R \\ R' \end{matrix}\rangle C{=}O$$

secondary alcohol — ketone

While the aldehydes of the general formula, $R\text{—C}\langle^{H}_{O}$, and the ketones of the general formula, $\begin{matrix} R \\ R' \end{matrix}\rangle C{=}O$, display unsaturation properties, it should be remembered that the unsaturation of these compounds is of a different kind from that of compounds containing the double or triple linkage between carbon atoms. The unsaturation properties of aldehydes and ketones arise from the presence of the $\rangle C{=}O$ grouping in the molecule in each case. When this grouping is attached either to two hydrogen atoms (as in formaldehyde, $H\text{—C}\langle^{H}_{O}$), or to one hydrogen atom and one univalent hydrocarbon radical (as in other aldehydes, $R\text{—C}\langle^{H}_{O}$), or to two univalent hydrocarbon radicals (as in a ketone, $\begin{matrix} R \\ R' \end{matrix}\rangle C{=}O$), the compound has the property of forming additive compounds of certain types. Aldehydes and ketones, however, do not combine directly with halogens (bromine and chlorine) at the $\rangle C{=}O$ grouping, as do the ordinary unsaturated compounds containing either the carbon-carbon double linkage ($\rangle C{=}C\langle$) or the carbon-carbon triple bond (—C≡C—), which are present in what are regarded as typically unsaturated organic substances.

ALDEHYDES OF THE GENERAL FORMULA, C_nH_{2n+1}, $C\langle{}^{H}_{O}$

Formaldehyde, $H{-}C\langle{}^{H}_{O}$, H.CHO

This aldehyde is the first direct oxidation product of methyl alcohol, the simplest alcohol of the series having the general formula $C_nH_{2n+1}OH$. It is sometimes known as *methanal*, corresponding to the name methanol for methyl alcohol.

Formaldehyde was discovered by von Hofmann in 1868 in the product obtained by passing a mixture of methyl alcohol and air over glowing platinum. If a spiral of platinum wire previously heated to redness be suspended over a small quantity of warm methyl alcohol in a suitable glass vessel, the platinum will continue to glow and the penetrating sharp odour of formaldehyde will soon become evident. Bone and his co-workers have shown that formaldehyde is an intermediate product in the combustion of methane, ethylene, acetylene and other hydrocarbons under certain conditions, and the regulated oxidation of ethylene according to the equation

$$\begin{matrix} CH_2 \\ \Vert \\ CH_2 \end{matrix} + \begin{matrix} O \\ \Vert \\ O \end{matrix} \rightarrow 2H{-}C\langle{}^{H}_{O}$$

has been suggested as a possible process for the production of formaldehyde on the commercial scale.

Formaldehyde is manufactured in very large quantities by processes all of which depend on the flameless combustion of methyl alcohol vapour in air in the presence of a catalyst. The methyl alcohol is oxidised according to the reaction:

$$H{-}\overset{\displaystyle H}{\underset{\displaystyle H}{\overset{|}{\underset{|}{C}}}}{-}O{:}H + O \rightarrow H{-}C\langle{}^{H}_{O} + H_2O$$

and the formaldehyde produced is absorbed in water, in which it is highly soluble. For this contact oxidation process, platinum, copper and silver are the chief catalysts used; the temperature varies in the different processes and may be as low as 100°.

Ever since Chapman and Holt (1905) showed that mixtures of hydrogen, carbon dioxide and steam and also mixtures of carbon monoxide and steam (with or without hydrogen) when heated by incandescent platinum produce formaldehyde in considerable quantity, it has been realised that the commercial production of formaldehyde by the reduction of carbon dioxide is a possibility. It has been shown that the reduction of carbon monoxide and dioxide to methyl alcohol (p. 88) has been realised industrially and formaldehyde may be the intermediate product:

$$O{=}C{=}O + 2H_2 \rightarrow H{-}C(H){=}O + H_2O$$

Carbon dioxide may be the intermediate product when carbon monoxide and steam are employed:

$$CO + H_2O \rightleftarrows CO_2 + H_2$$

Special interest attaches to Fenton's classical experiment (1907) on the reduction of carbon dioxide to formaldehyde. Fenton showed that when a stream of carbon dioxide is passed to saturation for 18 to 24 hours through water in contact with several rods of amalgamated magnesium, the solution obtained gives rise to slight but unmistakable indications of the presence of formaldehyde. It was also shown that in the presence of certain basic substances, among which may be mentioned ammonia, aniline and inorganic substances such as aluminium hydroxide, ferric hydroxide and calcium carbonate, the production of formaldehyde becomes more marked. This reduction of carbon dioxide to formaldehyde in aqueous solution acquires special significance, as it may have a bearing on the problem of assimilation in plants.

Formaldehyde is a colourless gas, b.p. $-21°$; when it is cooled in liquid air it forms a colourless crystalline mass having a melting point of about $-92°$.

By chemical analysis and molecular weight determination the formula of formaldehyde is CH_2O, and the only constitutional formula for the substance which can be devised, having in mind the normal valency of the constituent elements, is $H{-}C(H){=}O$, and this constitutional formula is in keeping with its properties and with its formation by the oxidation of methyl alcohol:

$$H{-}C(H)(H){-}O{-}H + O \rightarrow H{-}C(H){=}O + H_2O$$

Formaldehyde is readily soluble in water, and under special conditions an aqueous solution containing as much as 55 per cent. of formaldehyde can be obtained. 'Formalin' is the commercial aqueous solution of formaldehyde; it always contains methyl alcohol and is frequently acid to litmus on account of the presence of formic acid; the solution should contain not less than 36 and not more than

38 grams of formaldehyde in 100 c.c., and it is only rarely that it contains 40 grams in that volume. 'Formalin' is colourless and possesses the characteristic pungent odour of formaldehyde. This solution considerably diluted with water may be used for the following experiments.

On account of its tendency to be converted by oxidation into formic acid:

$$H{-}C\begin{matrix}\diagup H \\ \diagdown\!\!\diagdown O\end{matrix} + O \rightarrow H{-}C\begin{matrix}\diagup O{-}H \\ \diagdown\!\!\diagdown O\end{matrix}$$

formaldehyde is a strong reducing agent. It will reduce an ammoniacal solution of silver oxide (p. 105) to metallic silver in the cold. On adding an aqueous solution of formaldehyde, silver is precipitated, partly in the form of a bright metallic film on the sides of the test-tube. The reaction taking place is represented

$$H{-}C\begin{matrix}\diagup H \\ \diagdown\!\!\diagdown O\end{matrix} + Ag_2O \rightarrow H{-}C\begin{matrix}\diagup O{-}H \\ \diagdown\!\!\diagdown O\end{matrix} + 2Ag$$

Similarly, formaldehyde will reduce a solution of cupric hydroxide to cuprous oxide, which is formed as a red precipitate. The solution of cupric hydroxide is known as Fehling's solution. Usually, for Fehling's solution, two solutions are made up and mixed as required. The first solution is a solution of copper sulphate very slightly acidified with dilute sulphuric acid. The second solution contains Rochelle salt (sodium potassium tartrate, p. 399) and sodium hydroxide. When these aqueous solutions are mixed a clear deep blue solution is obtained; the Rochelle salt is used to prevent the precipitation of cupric hydroxide. When the formaldehyde solution is added to the Fehling's solution, and the mixture warmed, it becomes turbid and a red precipitate of cuprous oxide is produced: the reaction may be represented

$$H{-}C\begin{matrix}\diagup H \\ \diagdown\!\!\diagdown O\end{matrix} + 2CuO \rightarrow H{-}C\begin{matrix}\diagup O{-}H \\ \diagdown\!\!\diagdown O\end{matrix} + Cu_2O$$

A strong solution of formaldehyde may reduce the cuprous oxide further to metallic copper. Similarly, mercury and bismuth salts are reduced in presence of alkalis by strong solutions of formaldehyde, the metal being deposited in each case. In neutral solution, iron salts are reduced, the metal being deposited in the crystalline condition. Metallic gold is also deposited when acid solutions of gold are mixed with formaldehyde solution, and the latter has been used for the preparation of colloidal solutions of gold.

Formaldehyde, like most aldehydes, gives the Schiff reaction. The reagent for this is a dilute aqueous solution of a dye-stuff known as magenta (also as fuchsine or rosaniline) which has been decolorised by means of sulphurous acid. If such a solution be warmed, the pink colour of the dye-stuff returns. When a dilute solution of formaldehyde is added to the *cold* reagent a pink colour somewhat resembling that of the dye-stuff appears. With very few exceptions, all aldehydes give this reaction, although in some cases the action is slow.*

Rimini's test is probably the most delicate and most satisfactory colour reaction for formaldehyde. To the dilute aqueous solution of formaldehyde is added a few drops each of dilute aqueous solution of phenylhydrazine hydrochloride (p. 279), sodium nitroprusside p. 5) and sodium hydroxide; the result is a bright but transient blue colour.†

When an aqueous solution of formaldehyde is warmed with sodium or potassium hydroxide no visible change takes place (compare this behaviour of formaldehyde with that of its homologues), but the solution now contains methyl alcohol and sodium or potassium formate. This is known as Cannizzaro's reaction, and the two substances have been formed by the mutual reduction and oxidation of two molecules of the formaldehyde, thus:

$$2HCHO + KOH \rightarrow CH_3.OH + H.COOK$$

```
     H                     H
     |                     |                  O
  H—C=O ←——— H   →   H—C—O—H  +  H—C⟨
     ↑       |             |                  O—K
     H   +   |             H
     |       |
  H—C ←———— OK
     ‖
     O
```

It will be seen (p. 170) that in this reaction formaldehyde resembles benzaldehyde, which under similar conditions gives rise to benzyl alcohol and sodium or potassium benzoate.

In its reaction with ammonia, formaldehyde differs from its homologues and somewhat resembles benzaldehyde. When formalin is mixed with an excess of diluted ammonia solution, the solution allowed to stand for some hours and then evaporated to dryness on the water-bath, ammonia being kept in excess during the process, a white crystalline residue is obtained. This may be recrystallised from alcohol. The same substance is obtained when an excess of ammonium carbonate is treated with formalin. The am-

* The Schiff reaction has not yet been adequately explained. The pink colour given by aldehydes in the cold is not identical with the pink colour obtained by warming the Schiff reagent by itself.

† Another colour reaction which is often used as a test for formaldehyde is the deep red coloration produced when a small quantity is added to about 5 c.c. of sulphuric acid in which a little salicylic acid has been dissolved.

monium carbonate dissolves readily with effervescence owing to the evolution of carbon dioxide. The solution is conveniently evaporated to dryness under reduced pressure and the product recrystallised from alcohol.

The substance is known as *hexamethylenetetramine* and has the composition $(CH_2)_6N_4$ and is produced according to the equation:

$$6CH_2O + 4NH_3 \rightarrow (CH_2)_6N_4 + 6H_2O$$

which indicates that hexamethylenetetramine is a condensation product of formaldehyde and ammonia. Its constitution may be

```
              N
           /  |  \
       H2C   CH2   CH2
        |     |     |
        |     N     |
        |   /   \   |
        | CH2   CH2 |
        N           N
           \     /
              C
              H2
```

It is a colourless crystalline substance which sublimes at 263° without melting. It is readily soluble in water giving an alkaline (to litmus) solution. It is less soluble in alcohol. Hexamethylenetetramine is a very important substance medicinally, and is used under various names—such as *hexamine, urotropine, cystamine,* etc.—as a urinary antiseptic and in the treatment of cholecystitis. When hexamethylenetetramine is heated with dilute sulphuric acid or dilute hydrochloric acid formaldehyde is evolved, and if the resulting solution is cooled and made alkaline with sodium hydroxide, ammonia is liberated:

$$(CH_2)_6N_4 + 6H_2O + 2H_2SO_4 \rightarrow 6CH_2O + 2(NH_4)_2SO_4$$

The reaction between ammonia and formaldehyde is made use of in determining the strength of solutions of the latter.

With certain substances, formaldehyde reacts similarly to other aldehydes. When a concentrated solution of formaldehyde is mixed with a saturated solution of sodium bisulphite direct combination takes place and formaldehyde sodium bisulphite (highly soluble in water) is formed:

$$\text{H—C(H)=O} + \text{S(ONa)(=O)OH} \rightarrow \text{H—C(H)(OH)—O.SO}_2\text{Na}$$

The property of combining directly with sodium bisulphite is common to aldehydes and ketones and illustrates the unsaturated nature of these two classes of compounds.

The reaction of formaldehyde with ammonia has been described as one of condensation, and formaldehyde forms a number of other condensation products. One of these is produced by interaction with an aqueous solution of hydroxylamine:

$$H{-}C\begin{matrix}{\nearrow}H\\ {\searrow}O\end{matrix} + H_2N.OH \rightarrow H{-}C\begin{matrix}{\nearrow}H\\ {\searrow}N.OH\end{matrix} + H_2O$$

The substance produced is described as *formaldoxime*, which is only known in aqueous solution.

Another example of the formation of a condensation product by formaldehyde is its reaction with methyl alcohol. The reaction is carried out by mixing formalin with one and a half times its volume of methyl alcohol containing 2 per cent. of hydrogen chloride and an equal weight of fused calcium chloride. The mixture is allowed to stand for 15 hours and then fractionally distilled. The product of the reaction is known as *methylal*, formed according to the reaction:

$$H{-}C\begin{matrix}{\nearrow}H\\ {\searrow}O\end{matrix} + \begin{matrix}H.O.CH_3\\ H.O.CH_3\end{matrix} \rightarrow H{-}\overset{H}{\underset{O.CH_3}{C}}{-}O.CH_3 + H_2O$$

Methylal is a colourless liquid, b.p. 45·5°, possessing a pleasant odour. When distilled with dilute sulphuric acid it is decomposed, giving an aqueous solution of methyl alcohol and formaldehyde, the reaction being the reverse of that represented above. Formaldehyde forms similar compounds by condensation with ethyl alcohol and higher homologues.

Polymerisation of Formaldehyde

Formaldehyde has a very considerable tendency to yield polymerisation products which have the general formula $(CH_2O)_n$. These polymerisation products are of two types; the one type includes paraformaldehyde and the polyoxymethylenes from both of which formaldehyde or simple derivatives of formaldehyde can be obtained, and the other type includes the sugars, especially some of those having the formula $C_6H_{12}O_6$. The latter do not regenerate formaldehyde by any reaction and therefore, in spite of their molecular weight and their containing carbon, hydrogen and oxygen in the same proportions as formaldehyde should not be considered as ordinary polymerides of that substance (compare acetylene and benzene).

When an aqueous solution of formaldehyde is concentrated, a white amorphous powder separates. This is known as *paraformaldehyde*; it melts somewhat indefinitely between 150–160°, and its composition is usually written as $(CH_2O)_n + xH_2O$; apart from the water, it has at least three times the molecular weight of gaseous formaldehyde. When it is distilled, especially in a current of nitrogen, it is converted into the monomolecular and gaseous formaldehyde, and this constitutes a convenient method for preparing the latter substance in a state of purity. Tablets of paraformaldehyde are used in vaporisers for disinfection of rooms, and others containing small quantities of the substance are employed for internal administration as mouth and throat antiseptics.

By the action on aqueous formaldehyde solution of about one-tenth of its volume of concentrated sulphuric acid various so-called *polyoxymethylenes* are produced which are colourless, indefinitely crystalline substances and less soluble in water than *paraformaldehyde*. This mixture of polyoxymethylenes does not yield formaldehyde so easily as paraformaldehyde, but formaldehyde is always produced by the action of heat. If gently warmed in a current of dry ammonia, hexamethylenetetramine is produced. This mixture of polyoxymethylenes has not been investigated completely, but no less than four polyoxymethylenes with slightly differing properties have been isolated. At present, the formula of each of them can only be represented as $(CH_2O)_n$, and although they have all practically the same melting point (163–168° in a closed capillary tube) they have other distinguishing properties. From the ordinary dried mixture of polyoxymethylenes a substance, *α-trioxymethylene* (m.p. 60–61°), has been obtained by the action of a trace of sulphuric acid in a closed tube at 115°. To this compound has been assigned the constitutional formula:

```
        H   H
         \ /
          C
         / \
        O   O
   H    |   |    H
    \  |   |  /
     >C     C<
    /   \ / \
   H     O    H
```

It is a colourless volatile crystalline substance: its vapour density indicates a molecular weight corresponding to $(CH_2O)_3$, and the above constitutional formula is assigned to the compound, because it does not give any reactions of an aldehyde and unlike the other polyoxymethylenes does not yield formaldehyde on being heated.

The formation of these polymerisation products is undoubtedly connected with the unsaturated nature of the formaldehyde molecule. If the formaldehyde molecule I passes transiently into the form II:

```
        H                    H
       /                    /
  H—C                  H—C—O—
       \\                   \
        O
     I                    II
```

this suggests the possibility of the formation of an endless series of polymeric compounds of which the trimeric *α-trioxymethylene* might very well be the most stable. Thus, the polymerisation of four molecules of formaldehyde might lead to the formation of an unstable compound having the constitution:

```
        H          H
        |          |
    H—C——O——C—H
        |          |
        O          O
    H   |          |  H
     \ C——O——C /
    H /            \ H
```

Such large ring compounds may not be stable and would be expected to yield monomolecular formaldehyde or its derivatives under comparatively mild conditions.

In the formation of these polymerisation products it is generally assumed that the carbon atoms of the different molecules of formaldehyde are not directly united, that is, there is a repetition of the $-\overset{|}{\underset{|}{C}}-O-\overset{|}{\underset{|}{C}}-$ grouping. On the other hand, in the other type of polymerisation products, viz. the sugars, there is ample evidence that the carbon atoms are directly united to each other. From the properties of carbon compounds generally it is known that the —C—C—C— grouping, for example, is much more stable than the —C—O—C— grouping, and this is in keeping with the general properties of the two types of condensation products of formaldehyde.

When a dilute solution of formaldehyde was shaken with an excess of an aqueous suspension of calcium hydrate and after half an hour the filtered solution allowed to stand for several days, the latter was found by Loew (1886) to contain a mixture of sugars, known as *formose*. This confirmed the earlier observation of Butlerow (1861). From this mixture Emil Fischer (1887–1890) isolated a pure product to which he gave the name of *α-acrose* and from this he obtained *fructose*, identical with the naturally occurring sugar. These sugars belong to a large class of so-called *carbohydrates** and possess the molecular formula $C_6H_{12}O_6$. It is known that the carbon atoms in

* The carbohydrates include a very large number of compounds, many, but by no means all, of which occur in nature as products of plant and animal physiological processes. They include sugars, starches and celluloses. The name was applied to express the numerical relation between 'carbon' and 'water' in the molecule, and the general formula of these substances regarded from this point of view may be written $C_x(H_2O)_y$, where x and y are numbers which may be the same as, or different from, each other. Of the commonest sugars, *glucose* has the molecular formula $C_6H_{12}O_6$, *cane sugar* or *sucrose* has the formula $C_{12}H_{22}O_{11}$, while the starches, owing to their molecular weight not being yet accurately known, are represented by the formula $(C_6H_{10}O_5)_n$ (p. 456).

these sugar molecules are in a straight chain, —C—C—C—C—C—C—, and this polymerisation of formaldehyde

$$6CH_2O \rightarrow C_6H_{12}O_6$$

is of a different nature from that which takes place in the production of the polyoxymethylenes.

Other sugars can be produced from formaldehyde or polyoxymethylenes. For example, by boiling a 2 per cent. formaldehyde solution with one-hundredth of its weight of calcium carbonate for at least 10 hours, a condensation product is obtained the chief constituent of which is known as *arabinoketose* and which has the formula $C_5H_{10}O_5$. This must also be regarded as a polymerisation product of formaldehyde:

$$5CH_2O \rightarrow C_5H_{10}O_5$$

and again the polymerisation takes place by the joining together of the carbon atoms in a chain.

It was von Baeyer (1870) who put forward the suggestion that formaldehyde formed by the reduction of carbon dioxide is the first product of assimilation by plants, the formaldehyde then undergoing subsequent polymerisation to carbohydrates (sugars, starch, cellulose). Fenton's proof of the reduction of carbon dioxide to formaldehyde by magnesium in the presence of water, and the proof by Irvine that a six carbon sugar along with other products is present in the syrup produced by the exposure of formaldehyde solution to ultraviolet light, would appear to be evidence in favour of Baeyer's hypothesis of the photosynthetic production of carbohydrates by plants. Formaldehyde has been proved to be present in small quantities in certain plants. On the other hand, it may be that the formaldehyde detected is the result of the breaking down in the dark of the photosynthesised carbohydrates, and, in that case, it is not an intermediate product in the building up of carbohydrates from carbon dioxide and water.

Baly and his collaborators have shown that the action of ultraviolet light on carbon dioxide in the presence of water is to establish a state of equilibrium:

$$6H_2CO_3 \rightleftarrows C_6H_{12}O_6 + 6O_2$$

The amount of carbohydrate present in the equilibrium mixture is very small and, in the presence of oxidisable impurities, the carbohydrate formed is decomposed to formaldehyde. Although no measurable reaction takes place when pure carbon dioxide in aqueous solution and free from all suspended matter is exposed to light, a very definite action occurs when a surface which can absorb carbonic acid is present in the solution. Aluminium oxide, barium sulphate, freshly precipitated aluminium hydroxide, the basic carbonates of aluminium, zinc and magnesium all furnish suitable surfaces and, in these circumstances, the above equilibrium is shifted from left to right, complex organic compounds and not formaldehyde being formed. Since, however, the plant only makes use of visible light, the above experiments do not give much information concerning the natural photosynthetic processes. A distinct advance was made when Baly and his co-workers showed that the photosynthesis can take place in visible light provided that coloured surfaces are used in place of white ones. The first experiments were carried out using a suspension of nickel carbonate in water maintained by a stream of carbon dioxide—the green colour (of the nickel carbonate) being suggested by the green colour of the plant pigment, chlorophyll. Under these conditions, when the apparatus was exposed to the light of a tungsten filament lamp, a larger yield of carbohydrates resulted than when white powders and ultraviolet light were employed. There is no special virtue in the green colour, and equally good results were obtained using cobalt carbonate as the surface catalyst. When the aqueous solution resulting from the reaction is evaporated, there remains a syrup which is a complex mixture, but which undoubtedly contains glucose or fructose, or both, and other more complex carbohydrates. Although much more investigation is needed, it is clear that Baly's photosynthesis of carbohydrates *in vitro* is somewhat similar to that occurring in the living leaf. They are both surface actions which take place in visible light. The light must not be too strong. In the natural process, this results in the cessation of the reaction or reactions followed by slow recovery in the dark. In the laboratory process, it results in the poisoning of the catalytic surface by the oxygen set free, and here again slow recovery proceeds in the absence of light.

Uses of formaldehyde.

In the gaseous form, formaldehyde, usually obtained by the action of heat on paraformaldehyde, is used as a disinfectant for the destruction of bacteria and vermin. In dilute aqueous (1 per cent.) solution, formaldehyde is used in medicine as an internal antiseptic, while in stronger solutions it is used for sterilising instruments. Owing to its power of precipitating proteins, formaldehyde in aqueous solution is used in the preservation by fixing of anatomical specimens and in the preparation of histological specimens.

Formaldehyde is employed industrially in the preparation of certain types of synthetic resins. The phenol-formaldehyde resins are produced by the interaction of formaldehyde and phenol or its homologues in aqueous solutions under prescribed conditions. Uncrystallisable resins are also produced by the action of formaldehyde and urea (p. 476), etc. The chemical reactions whereby these resins are formed are highly complex. In the case of those produced from phenol and its homologues, the first products are simple substitution products:

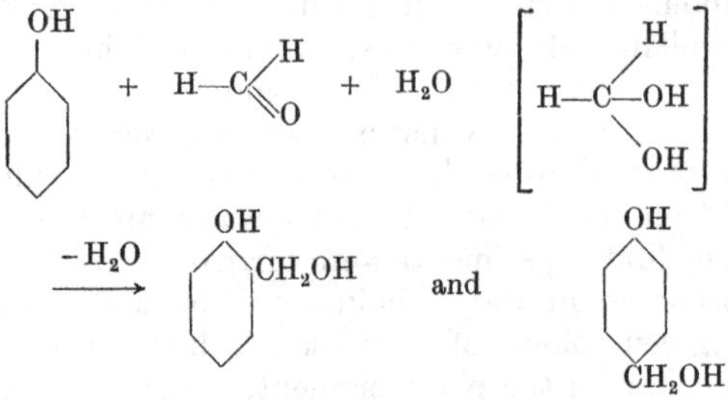

$$C_6H_5OH + H{-}C{\overset{H}{\underset{O}{\lessgtr}}} + H_2O \quad \left[H{-}C(H)(OH){-}OH \right]$$

$$\xrightarrow{-H_2O} \; o\text{-}HOC_6H_4CH_2OH \quad \text{and} \quad p\text{-}HOC_6H_4CH_2OH$$

o-hydroxybenzyl alcohol (saligenin) — *p*-hydroxybenzyl alcohol

such as *o*- and *p*- hydroxybenzyl alcohols, which, while being crystalline and very reactive, easily form much more complicated compounds by combining again with phenol and with formaldehyde. The result is a highly complex uncrystallisable resin setting to a hard mass which can be moulded and coloured. These resins find considerable and increasing applications in the arts and in industry, especially in the electrical industries on account of their unique electrical properties.*

Acetaldehyde, $H{-}\underset{H}{\overset{H}{C}}{-}C{\overset{H}{\underset{O}{\lessgtr}}}$, $CH_3.CHO$

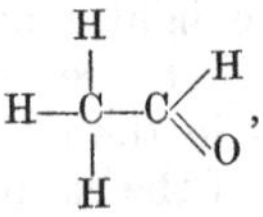

This is the next homologue to formaldehyde and, being the most important and typical member of this series of compounds, is often referred to simply as *aldehyde*. It is also known as *ethanal*, the aldehyde derived from ethyl alcohol or ethanol.

The material obtained in the manner described below can be used without further purification for illustrating the chief and simpler reactions of acetaldehyde.

A mixture containing concentrated sulphuric acid (30 c.c.) and water (100 c.c.) is placed in a distilling flask fitted with a dropping funnel and attached to a good condenser leading to an ice-cooled receiver. The mixture is heated just to boiling and the flame removed. From the tap funnel is added slowly a mixture containing sodium dichromate (100 grams), water (100 c.c.) and ethyl alcohol or rectified spirit (60 c.c.). The aldehyde is formed without any further heating and collects in the receiver together with some unchanged alcohol and acetal (*v.* below). If it is desired to obtain the aldehyde pure, the product can be carefully distilled, the aldehyde collected in ethereal solution and aldehyde ammonia (*v.* below) precipitated from this by means of dry ammonia. The aldehyde may be obtained from this by distillation with diluted sulphuric acid.

* See 'The Chemistry of Synthetic Resins' by Ellis (1935).

The oxidation of ethyl alcohol has previously been referred to, and acetaldehyde is the first direct oxidation product:

$$\begin{array}{c} \text{H}\ \ \text{H} \\ | \quad | \\ \text{H—C—C—O—H} + \text{O} \\ | \quad | \\ \text{H}\ \ \text{H} \end{array} \rightarrow \begin{array}{c} \text{H} \\ | \\ \text{H—C—C}\!\begin{array}{l} \diagup \text{H} \\ \diagdown\!\!\diagdown \text{O} \end{array} \\ | \\ \text{H} \end{array} + H_2O$$

The complete reaction, omitting by-products, is expressed by the equation

$$3C_2H_5OH + Na_2Cr_2O_7 + 4H_2SO_4 \rightarrow 3CH_3.CHO + Na_2SO_4 + Cr_2(SO_4)_3 + 7H_2O$$

The aqueous solution left in the distillation flask is dark green in colour, and from it by careful evaporation can be obtained sodium chrome alum, $Na_2SO_4.Cr_2(SO_4)_3.24H_2O$.

Acetaldehyde is prepared on the large scale by the catalytic oxidation of ethyl alcohol. In one process the alcohol vapour is passed through a tube heated to 300° containing a mixture of copper and magnesium oxides deposited on pumice. The products from the reaction are led up a condenser maintained at 30° and the aldehyde which is gaseous at this temperature passes on and is collected in a specially cooled receiver. Unchanged alcohol is returned so as to be passed again over the heated catalyst. The aldehyde obtained is in concentrated alcoholic solution, and any which escapes condensation is finally passed up an ice tower and obtained in aqueous solution. From the former, it is easy to obtain a fairly pure (97 per cent.) acetaldehyde by careful distillation.

The preparation of acetaldehyde from acetylene has already been referred to (p. 49).

Acetaldehyde is a colourless liquid possessing a characteristic sharp but not unpleasant odour. It has b.p. 20·8° and m.p. −120·6°. It is isomeric with ethylene oxide (p. 38) and with the unknown or unstable vinyl alcohol (p. 116).

Acetaldehyde is freely soluble in organic solvents and in water. From its aqueous solutions it may be obtained by addition of calcium chloride, since it is much less soluble in a strong aqueous solution of this salt. The crude mixture obtained in the above-described method of preparation can be used to illustrate the typical reactions of an aliphatic aldehyde.

Analysis and molecular weight determinations prove that the molecular formula of aldehyde is C_2H_4O. Its constitutional formula follows from its being the primary oxidation product of ethyl alcohol, as indicated above. Its properties also show clearly that it possesses the aldehydic or $\text{—C}\!\begin{array}{l} \diagup \text{H} \\ \diagdown\!\!\diagdown \text{O} \end{array}$ grouping. The only other stable compound having the molecular formula C_2H_4O is ethylene oxide which has physical and chemical properties entirely different from those of acetaldehyde.

When a small quantity of an aqueous solution of acetaldehyde is warmed with an aqueous solution of sodium or potassium hydroxide, the liquid remains clear for a little time and then becomes yellow and cloudy and a curious resinous material is deposited. This is known as *aldehyde-resin*, and although little is known about its actual nature, its production is typical of this and other aliphatic aldehydes (compare, however, formaldehyde, which behaves quite differently, p. 147). Another portion of the above distillate can be used to illustrate the reducing action of acetaldehyde on Fehling's solution or an ammoniacal solution of silver oxide, carried out as described for formaldehyde. In the latter case the acetaldehyde is oxidised to acetic acid, while the silver oxide is reduced to metallic silver:

$$H-\underset{H}{\overset{H}{\underset{|}{\overset{|}{C}}}}-C\left\langle\begin{matrix}H\\ O\end{matrix}\right. + Ag_2O \rightarrow H-\underset{H}{\overset{H}{\underset{|}{\overset{|}{C}}}}-C\left\langle\begin{matrix}O-H\\ O\end{matrix}\right. + 2Ag$$

Acetaldehyde, like formaldehyde, shows the Schiff reaction (production of a pink colour in a cold solution of magenta which has been decolorised by sulphurous acid).

When dry ammonia is passed into an ethereal solution of acetaldehyde a white crystalline precipitate is formed which rapidly becomes somewhat yellow on exposure to air. This precipitate is known as aldehyde-ammonia (cf. reaction between formaldehyde and ammonia):

$$CH_3-C\left\langle\begin{matrix}H\\ O\end{matrix}\right. + NH_3 \rightarrow CH_3-C\begin{matrix}/H\\ -OH\\ \backslash NH_2\end{matrix}$$

Aldehyde-ammonia can be filtered off, and when it is distilled with dilute sulphuric acid acetaldehyde is regenerated. The formation, isolation and subsequent decomposition of aldehyde-ammonia in the manner indicated form a convenient method of purifying acetaldehyde.

When a concentrated solution of acetaldehyde is mixed with a concentrated aqueous solution of sodium bisulphite, an addition compound, colourless and crystalline, is formed:

$$CH_3-C\left\langle\begin{matrix}H\\ O\end{matrix}\right. + NaHSO_3 \rightarrow CH_3-C\begin{matrix}/H\\ -OH\\ \backslash O.SO_2Na\end{matrix}$$

from which the aldehyde is regenerated on warming with dilute acid.

Acetaldehyde combines directly with hydrogen cyanide to form *acetaldehyde cyanhydrin*, a colourless liquid, b.p. 103° at 26 mm.:

$$CH_3{-}C\begin{matrix}\diagup H\\ \diagdown\!\diagdown O\end{matrix} + HCN \rightarrow CH_3{-}C\begin{matrix}\diagup H\\ -OH\\ \diagdown C:N\end{matrix}$$

This is an important compound theoretically, because from it lactic acid can be easily obtained (p. 362), thereby proving not only the relationship between acetaldehyde and lactic acid but also the constitution of the latter compound.

When acetaldehyde is acted upon by phosphorus pentachloride, the oxygen atom is replaced by the equivalent two chlorine atoms and ethylidene chloride (p. 19) isomeric with ethylene chloride is produced:

$$\begin{matrix}CH_3\\ |\\ C\begin{matrix}\diagup H\\ \diagdown\!\diagdown O\end{matrix}\end{matrix} + PCl_5 \rightarrow \begin{matrix}CH_3\\ |\\ CHCl_2\end{matrix} + POCl_3$$

Like formaldehyde, acetaldehyde readily forms polymerisation products. When hydrogen chloride or sulphur dioxide is passed into acetaldehyde cooled in a freezing mixture and allowed to stand for about 2 hours, a white solid separates and can be filtered off. This is known as *metaldehyde*, the composition of which is not known with certainty. Molecular weight determinations indicate that it has a formula of either $(C_2H_4O)_4$ or $(C_2H_4O)_6$. It does not possess the properties of an aldehyde, since it has no reaction with Fehling's solution, with potassium permanganate or with chromic acid. When it is heated, metaldehyde sublimes and at the same time is partly reconverted into acetaldehyde. On being heated in a sealed tube it is completely reconverted into acetaldehyde.

A more important polymerisation product is known as *paraldehyde*, which is formed almost explosively and with considerable evolution of heat when two or three drops of concentrated sulphuric acid are mixed with acetaldehyde. It is a colourless liquid, b.p. 124°. It does not possess the properties of an aldehyde, but is readily reconverted into acetaldehyde by distillation with dilute sulphuric acid. Its formation and constitution may be represented as follows:

$$3\,CH_3{-}CHO \underset{\text{dil. } H_2SO_4}{\overset{\text{conc. } H_2SO_4}{\rightleftharpoons}} \begin{matrix} & CH_3\ \ H & \\ & \diagdown\diagup & \\ & C & \\ & \diagup\quad\diagdown & \\ & O\qquad O & \\ H\diagdown & |\qquad | & \diagup CH_3\\ & C\qquad C & \\ CH_3\diagup & \diagdown\ \diagup & \diagdown H\\ & O & \end{matrix}$$

and its molecular weight has been proved to be three times that of acetaldehyde. Paraldehyde has important uses in medicine as a sedative and hypnotic.

Another polymerisation product of a different type is formed when acetaldehyde is allowed to stand in the cold with dilute hydrochloric acid or with potassium carbonate; the latter is the condensing agent more usually employed. This condensation product has twice the molecular weight of acetaldehyde, and the constitution of the substance known as *aldol*, a colourless liquid, b.p. 83° at 20 mm., is as shown

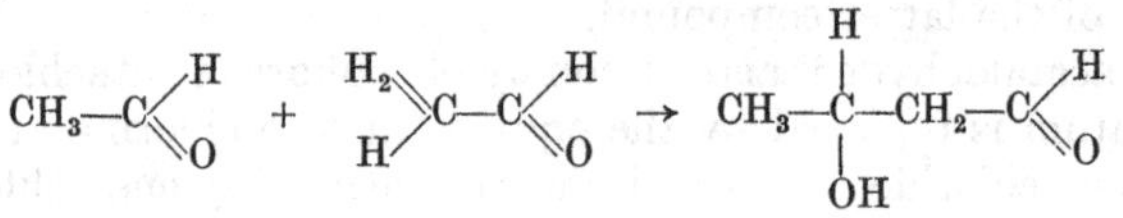

$$CH_3{-}C{\scriptstyle\begin{matrix}\nearrow H\\ \searrow O\end{matrix}} + {\scriptstyle\begin{matrix}H_2\searrow\\ H\nearrow\end{matrix}}C{-}C{\scriptstyle\begin{matrix}\nearrow H\\ \searrow O\end{matrix}} \rightarrow CH_3{-}\underset{OH}{\overset{H}{C}}{-}CH_2{-}C{\scriptstyle\begin{matrix}\nearrow H\\ \searrow O\end{matrix}}$$

or $CH_3.CHOH.CH_2.CHO$

Aldol is a typical aliphatic aldehyde; it yields an ammonia addition compound and reduces an ammoniacal solution of silver oxide, becoming converted into the corresponding acid,

$$CH_3.CHOH.CH_2.COOH$$

(β-hydroxybutyric acid, p. 367). When aldol is distilled under atmospheric pressure it is partly reconverted into acetaldehyde, but mostly it is converted into an unsaturated aldehyde, *crotonaldehyde*,* by loss of a molecule of water (p. 472):

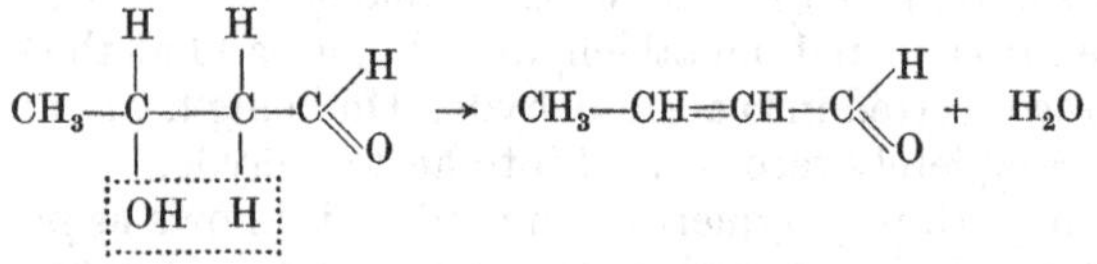

$$CH_3{-}\underset{OH}{\overset{H}{C}}{-}\underset{H}{\overset{H}{C}}{-}C{\scriptstyle\begin{matrix}\nearrow H\\ \searrow O\end{matrix}} \rightarrow CH_3{-}CH{=}CH{-}C{\scriptstyle\begin{matrix}\nearrow H\\ \searrow O\end{matrix}} + H_2O$$

The polymerisation of acetaldehyde in the presence of aluminium ethylate, $Al(OC_2H_5)_3$, is again of quite a different type. When acetaldehyde is allowed to react with aluminium ethylate, the chief product of the reaction (aldol is also formed in small quantities) is a well-known substance, *ethyl acetate* (p. 214). The constitution of this compound has long been established beyond doubt, and the combination of two molecules of acetaldehyde to form this compound

* Crotonaldehyde is a colourless liquid, b.p. 104–105°. From its constitution, it is theoretically capable of existing in stereoisomeric forms, thus:

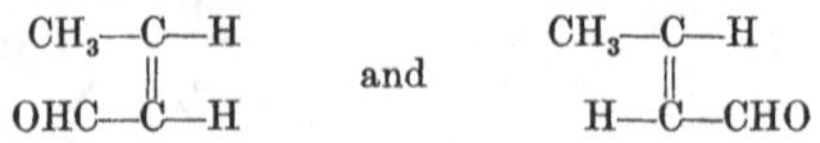

$$\begin{matrix}CH_3{-}C{-}H\\ \Vert\\ OHC{-}C{-}H\end{matrix} \quad \text{and} \quad \begin{matrix}CH_3{-}C{-}H\\ \Vert\\ H{-}C{-}CHO\end{matrix}$$

follows a different course from any of the polymerisation processes already mentioned:

$$CH_3{-}C(H){=}O + O{=}C(H){-}CH_3 \rightarrow CH_3{-}C({=}O){-}O{-}CH_2{-}CH_3$$

ethyl acetate

The other typical reactions of acetaldehyde may be described as condensation reactions. By a *condensation** is usually meant the reaction of one substance with another to form a third substance by elimination of a simple compound, usually water. Thus crotonaldehyde may be regarded as a condensation product of two molecules of acetaldehyde:

$$CH_3.CHO + CH_3.CHO \rightarrow CH_3.CH{:}CH.CHO + H_2O$$

In this case, aldol is the intermediate product in the reaction.

Acetaldehyde, like all carbonyl compounds (aldehydes and ketones), reacts readily with hydroxylamine, water being eliminated, forming a compound known as *acetaldehyde oxime* (usually abbreviated to acetaldoxime).

The usual method of preparation of an oxime is simply to allow one molecular proportion of hydroxylamine hydrochloride and half a molecular proportion of sodium carbonate to react in aqueous solution with one molecular proportion of the aldehyde at the ordinary temperature. The oxime is generally isolated by extracting the aqueous solution with ether, washing the extract with a little water, drying the ethereal solution with anhydrous sodium sulphate and distilling off the ether. If the original aldehyde is insoluble in water, the reaction is generally carried out in aqueous ethyl alcohol solution. The oximes of ketones are prepared in a similar way. In some cases the reaction takes place on simple mixing of hydroxylamine and the aldehyde or ketone in aqueous or alcoholic solution; in other cases it is necessary to heat the reaction mixture on the water-bath for a somewhat longer period, and, in a few cases, a higher temperature is employed.

* Condensation is a somewhat misleading term which has come into general use. It is unfortunate that it has been adopted, since the same term has long been used to describe other processes, particularly the passage from the gaseous phase to the liquid or solid phase of the same substance.

Acetal is a condensation product of acetaldehyde and ethyl alcohol (cf. methylal). It is always produced when ethyl alcohol is oxidised to acetaldehyde in the presence of acids,

$$CH_3.C(H){=}O + H{\cdot}OC_2H_5 + H{\cdot}OC_2H_5 \rightarrow CH_3.CH(OC_2H_5)(OC_2H_5) + H_2O$$

acetal

Acetal may be systematically described a 1-1-diethoxyethane. It has b.p. 104° and it is probably an intermediate product in the preparation of chloral from ethyl alcohol.

The equation expressing the formation of acetaldoxime is

$$CH_3{-}C{\leq}^{H}_{O} + H_2N{-}OH \rightarrow CH_3{-}C{\leq}^{H}_{N-OH} + H_2O$$

acetaldoxime

Acetaldoxime is a colourless crystalline substance, m.p. 47°. Although two isomeric forms of acetaldoxime have not been definitely isolated, there is strong evidence that two isomeric forms of this oxime and the oximes of other aldehydes are capable of existence. These are undoubtedly stereoisomers, the isomerism being similar to that existing in compounds of the type $XYC:CXY$ (p. 41). Detailed investigations have shown that the nitrogen atom in tervalent nitrogen compounds should be regarded as being at the apex of of a flattened tetrahedron, and the stereoisomeric forms of the oximes of an aldehyde are conveniently represented:

$$\begin{array}{c} X{-}C{-}H \\ \| \\ N{-}OH \end{array} \quad \text{and} \quad \begin{array}{c} X{-}C{-}H \\ \| \\ HO{-}N \end{array}$$

syn-oxime *anti*-oxime

Methods of distinguishing between the two stereoisomers will be described later (p. 168).

Semicarbazide, m.p. 96°, a compound* having the formula

$$H_2N.NH.CO.NH_2$$

generally used in the form of its hydrochloride—reacts similarly to hydroxylamine with aldehydes and ketones, and yields with acetaldehyde a colourless crystalline compound, m.p. 163°, known as *acetaldehyde semicarbazone*:

$$CH_3{-}C{\leq}^{H}_{O} + H_2N.NH.CO.NH_2 \rightarrow CH_3{-}C{\leq}^{H}_{N.NH.CO.NH_2} + H_2O$$

The preparation of such simple crystalline derivatives possessing definite melting points affords not only a rigorous method of identifying a particular aldehyde (or ketone) but also a method of obtaining the aldehyde (or ketone) in a state of purity, since an oxime when warmed with mineral acids yields hydroxylamine and the aldehyde (or ketone) by hydrolysis:

$$>C{=}N{-}OH + H_2O \rightarrow >CO + H_2N{-}OH$$

* Prepared by heating hydrazine hydrochloride with potassium cyanate:

$$(NH_2.NH_3)Cl + KOCN \rightarrow KCl + (NH_2.NH_3)OCN \rightarrow NH_2.CO.NH.NH_2$$

(compare p. 476).

Other condensation products of a similar type are formed by a reaction carried out in much the same way with phenylhydrazine, $C_6H_5.NH.NH_2$ (p. 279), and certain substitution products of the latter, chiefly *p*-nitrophenylhydrazine, $O_2N\langle\quad\rangle NH.NH_2$. The reaction taking place, in the case of acetaldehyde, is represented:

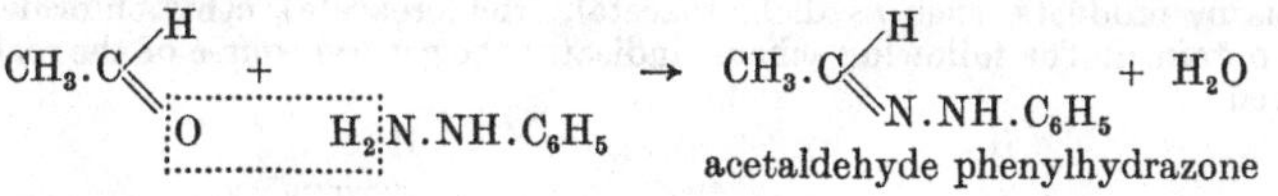

$$CH_3.C\genfrac{}{}{0pt}{}{\diagup H}{\diagdown O} + H_2N.NH.C_6H_5 \rightarrow CH_3.C\genfrac{}{}{0pt}{}{\diagup H}{\diagdown N.NH.C_6H_5} + H_2O$$

acetaldehyde phenylhydrazone

Laws and Sidgwick (1911) have investigated the reaction between phenylhydrazine and acetaldehyde, and they not only confirmed previous work proving the existence of two acetaldehyde phenylhydrazones, the β-compound having m.p. 56°, and the α-compound having m.p. 98·6°, but they showed that the two forms are interconvertible, thus:

$$\alpha \underset{\text{alkalis}}{\overset{\text{acids}}{\rightleftarrows}} \beta$$

Laws and Sidgwick conclude that the two forms of acetaldehyde phenylhydrazone are stereoisomerides, possessing the constitutions:

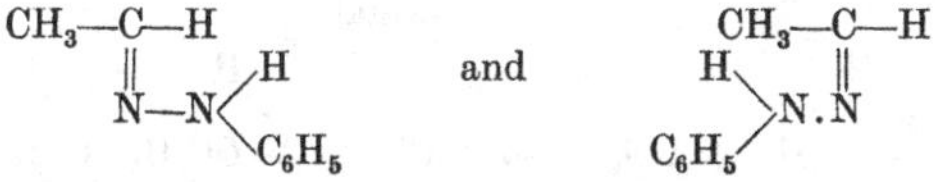

$$\begin{matrix} CH_3\text{—}C\text{—}H \\ \| \\ N\text{—}N\genfrac{}{}{0pt}{}{\diagup H}{\diagdown C_6H_5} \end{matrix} \quad \text{and} \quad \begin{matrix} CH_3\text{—}C\text{—}H \\ \| \\ \genfrac{}{}{0pt}{}{H\diagdown}{C_6H_5\diagup}N.N \end{matrix}$$

similar to those of the analogous oximes.

The pale yellow crystalline p-*nitrophenylhydrazone* of acetaldehyde is very easily prepared. It has m.p. 128·5°, and is frequently used for the identification of the aldehyde.

Chloral, trichloroacetaldehyde, $CCl_3.C\genfrac{}{}{0pt}{}{\diagup H}{\diagdown O}$, is an important derivative of acetaldehyde, from which it can be prepared by direct chlorination in the presence of calcium carbonate:

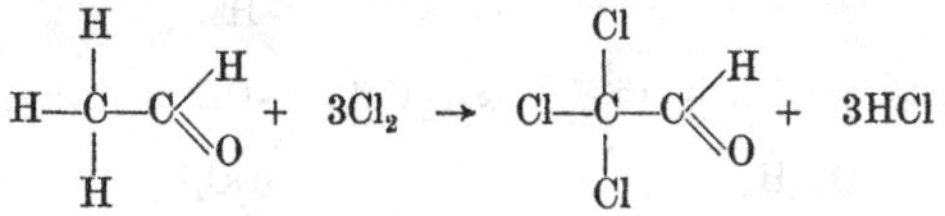

$$H\text{—}\underset{H}{\overset{H}{\underset{|}{\overset{|}{C}}}}\text{—}C\genfrac{}{}{0pt}{}{\diagup H}{\diagdown O} + 3Cl_2 \rightarrow Cl\text{—}\underset{Cl}{\overset{Cl}{\underset{|}{\overset{|}{C}}}}\text{—}C\genfrac{}{}{0pt}{}{\diagup H}{\diagdown O} + 3HCl$$

The calcium carbonate is necessary to neutralise the hydrogen chloride (hydrochloric acid) as it is formed; the presence of the latter would otherwise lead to the production of polymerisation products. Chloral was discovered by von Liebig in 1831, who obtained it by the long-continued chlorination of absolute alcohol, and this method of preparation is used on the manufacturing scale to-day.

Chlorine is passed into absolute alcohol containing ferric chloride as a catalyst for several days. At first it is necessary to cool the mixture and then as the reaction slackens it is heated to 60° and later to 100°. When chlorine ceases to be absorbed, the mixture on cooling solidifies almost completely to a white mass of chloral alcoholate (see below). This is then carefully mixed with concentrated sulphuric acid and the mixture distilled up to a temperature of 100°, and the distillate is finally refractionated, that portion distilling at 97–98° being practically pure chloral. The reaction taking place is somewhat complicated, various by-products, such as dichloroacetal, trichloroacetal, ethyl chloride, etc., being obtained. The following scheme indicates the general course of the reactions involved:

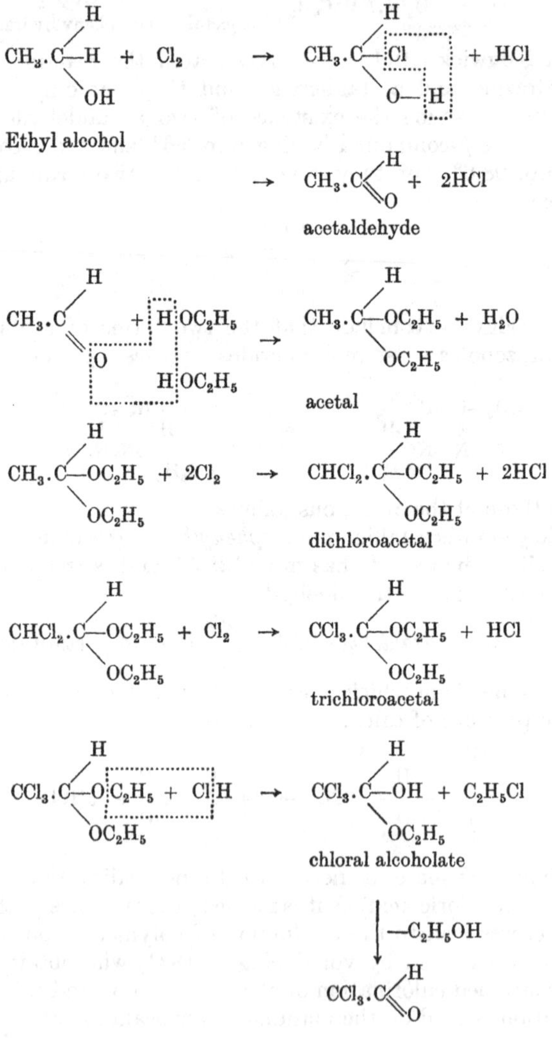

$$\underset{\text{Ethyl alcohol}}{CH_3.CH_2(OH)} + Cl_2 \rightarrow CH_3.CH(Cl)(OH) + HCl \rightarrow \underset{\text{acetaldehyde}}{CH_3.CHO} + 2HCl$$

$$CH_3.CHO + 2\,HOC_2H_5 \rightarrow \underset{\text{acetal}}{CH_3.CH(OC_2H_5)_2} + H_2O$$

$$CH_3.CH(OC_2H_5)_2 + 2Cl_2 \rightarrow \underset{\text{dichloroacetal}}{CHCl_2.CH(OC_2H_5)_2} + 2HCl$$

$$CHCl_2.CH(OC_2H_5)_2 + Cl_2 \rightarrow \underset{\text{trichloroacetal}}{CCl_3.CH(OC_2H_5)_2} + HCl$$

$$CCl_3.CH(OC_2H_5)(OC_2H_5) + ClH \rightarrow \underset{\text{chloral alcoholate}}{CCl_3.CH(OH)(OC_2H_5)} + C_2H_5Cl$$

$$\downarrow -C_2H_5OH$$

$$CCl_3.CHO$$

Chloral is a colourless liquid with a sweetish taste and an extremely pungent odour. It has m.p. $-57{\cdot}5^\circ$ and b.p. $97{\cdot}7^\circ$. As indicated above, it is a stable compound, being unaffected by concentrated sulphuric acid even at 100°. It possesses most of the general properties of a typical aliphatic aldehyde. It will reduce an ammoniacal solution of silver oxide, a silver mirror being formed, and the chloral is oxidised to *trichloroacetic acid* (p. 206), the reaction being analogous to the oxidation of acetaldehyde to acetic acid:

$$CCl_3.C\begin{matrix}\diagup H \\ \diagdown\!\!\diagdown O\end{matrix} + O \rightarrow CCl_3.C\begin{matrix}\diagup O\text{—}H \\ \diagdown\!\!\diagdown O\end{matrix}$$

Chloral is also directly oxidised with fuming nitric acid to trichloroacetic acid. Like formaldehyde and acetaldehyde, it reacts as a typical aldehyde towards Schiff's reagent in the cold. Like acetaldehyde, it combines directly with ammonia and potassium bisulphite forming chloral ammonia and chloral potassium bisulphite respectively:

$$CCl_3.C\begin{matrix}\diagup H \\ \text{—}OH \\ \diagdown NH_2\end{matrix} \qquad CCl_3.C\begin{matrix}\diagup H \\ \text{—}OH \\ \diagdown OSO_2K\end{matrix}$$

When chloral is shaken up with water, considerable heat is evolved, and a crystalline compound, *chloral hydrate*, is formed:

$$CCl_3.C\begin{matrix}\diagup H \\ \diagdown\!\!\diagdown O\end{matrix} + H_2O \rightarrow \underset{\text{chloral hydrate}}{CCl_3.C\begin{matrix}\diagup H \\ \text{—}OH \\ \diagdown OH\end{matrix}}$$

Chloral hydrate crystallises in massive colourless monoclinic prisms, m.p. $51{\cdot}7^\circ$, b.p. $97{\cdot}5^\circ$, dissociation into chloral and water taking place to some extent during the distillation.

Chloral hydrate dissolves to the extent of about 1 part in 4 parts of water: it is more soluble in alcohol and is moderately soluble in many other organic solvents. The name chloral hydrate must not be regarded as implying that the water is combined with the chloral as water of crystallisation, since so-called chloral hydrate does not give the Schiff reaction and therefore presumably the compound is not an aldehyde. Chloral hydrate (so-called) is an outstanding exception to the general rule that two hydroxyl groups are not attached to the same carbon atom in a stable compound, and chloral hydrate is more correctly represented by the formula indicated above rather than by such a formula as $CCl_3CHO.H_2O$.

Chloral hydrate undergoes interesting photochemical reactions. In the absence of oxygen and in presence of sunlight the following decomposition takes place:

$$CCl_3.CH(OH)_2 = 3HCl + 2CO$$

In the presence of oxygen the photochemical decomposition is represented by the following equation:

$$CCl_3.CH(OH)_2 + O_2 = 3HCl + 2CO_2$$

Chloral hydrate is a very important and, in normal cases, safe hypnotic; it is usually given admixed with potassium bromide.

Chloral and chloral hydrate when heated with aqueous sodium or potassium hydroxide are decomposed into chloroform, already referred to as a substitution product of methane, a salt of formic acid being produced at the same time:

$$\begin{matrix}Cl\\Cl\\Cl\end{matrix}\!\!>\!C-C\!\!\begin{matrix}\diagup H\\ \diagdown\!\!\diagdown O\end{matrix} + NaO\vdots H \rightarrow \begin{matrix}Cl\\Cl\\Cl\end{matrix}\!\!>\!C-H + H-C\!\!\begin{matrix}\diagup O-Na\\ \diagdown\!\!\diagdown O\end{matrix}$$

sodium formate

This is a convenient method for obtaining *pure* chloroform (p. 16)*.

When chloral is reduced with zinc and hydrochloric acid, acetaldehyde is produced:

$$CCl_3.C\!\!\begin{matrix}\diagup H\\ \diagdown\!\!\diagdown O\end{matrix} + 3H_2 \rightarrow CH_3.C\!\!\begin{matrix}\diagup H\\ \diagdown\!\!\diagdown O\end{matrix} + 3HCl$$

Chloral forms a number of polymerisation products analogous to those of formaldehyde and acetaldehyde. When chloral is warmed with aluminium chloride to 60°, and the mixture then precipitated with water, an oil separates which boils at 239·5–240°. This is a liquid polymeric form of chloral which has a pronounced odour of that substance. When chloral is admixed with small quantities of various substances, especially sulphuric acid, a white amorphous substance insoluble in water is slowly formed. This is known as *metachloral*, which behaves towards alkalies and nitric acid just as chloral does. It is converted into chloral on heating to 180°. Both the liquid and solid polymerised forms of chloral have a composition which is, at present, unknown. They are usually given the formula $(C_2HOCl_3)_x$.

* From a knowledge of the action of chlorine on ethyl alcohol and of the hydrolysis of chloral by alkali, the preparation of chloroform (p. 16) from ethyl alcohol and bleaching power (oxidising agent, chlorinating agent and alkali) in the presence of water obviously consists in the oxidation of the alcohol to acetaldehyde, chlorination of the latter to chloral and the hydrolysis of the chloral to chloroform and calcium formate, the reactions being

$$CH_3.CH_2OH + Ca(OCl)Cl \rightarrow CH_3.CHO + H_2O + CaCl_2$$

$$2CH_3.CHO + 6Cl_2 + 3Ca(OH)_2 \rightarrow 2CCl_3.CHO + 3CaCl_2 + 6H_2O$$

$$2CCl_3.CHO + Ca(OH)_2 \rightarrow 2CHCl_3 + Ca(OOC.H)_2$$

Acetaldehyde is the typical aliphatic aldehyde and the higher homologues have analogous properties, so that it is unnecessary to describe these compounds in detail.

The aromatic aldehydes fall into two classes: (*a*) those which contain the aldehyde (—CHO) group united to a carbon atom of the nucleus, and (*b*) those which contain the aldehyde group united to a carbon atom of the side chain. The aldehydes of the latter class behave generally as aliphatic aldehydes and do not require detailed description. The aromatic aldehydes of the first class are the more important and include such compounds as *benzaldehyde*, $C_6H_5.C\langle{}^{H}_{O}$,

and *salicylaldehyde*, $C_6H_4\langle{}^{OH\,(1)}_{CHO\,(2)}$ (*o*-hydroxybenzaldehyde), which may be regarded as typical aromatic aldehydes. These typical aromatic aldehydes resemble in many respects formaldehyde, of which they may be considered substitution products. The simplest aromatic aldehyde of the (*b*) class is *phenylacetaldehyde*,

$C_6H_5.CH_2.C\langle{}^{H}_{O}$, which resembles acetaldehyde and its higher homologues in general properties. Both classes of aromatic aldehydes, of course, bear the same relation to the primary aromatic alcohols as do the aliphatic aldehydes to the primary aliphatic alcohols; thus: benzaldehyde and salicylaldehyde are the first direct oxidation products of benzyl alcohol, $C_6H_5.CH_2OH$, and salicyl alcohol,

$C_6H_4\langle{}^{OH(1)}_{CH_2OH\,(2)}$, respectively, whereas phenylacetaldehyde is the direct oxidation product of phenylethyl alcohol or benzylcarbinol, $C_6H_5.CH_2.CH_2OH$.

Benzaldehyde, $C_6H_5.C\langle{}^{H}_{O}$

Benzaldehyde has long been known as the 'Essential Oil of Bitter Almonds', and it was formerly prepared from either bitter almonds or cherry kernels. Bitter almonds and cherry kernels contain a glucoside (p. 449), *amygdalin*, and when they are macerated with water an enzyme (p. 446), *emulsin*, is liberated from certain cells and this enzyme has the power of facilitating the hydrolysis of the amygdalin whereby hydrocyanic acid, benzaldehyde and the sugar, glucose (p. 423), are produced:

$$\underset{\text{amygdalin}}{C_{20}H_{27}O_{11}N} + 2H_2O = \underset{\text{benzaldehyde}}{C_6H_5.CHO} + HCN + \underset{\text{glucose}}{2C_6H_{12}O_6}$$

After the hydrolysis is complete the benzaldehyde may easily be separated by steam distillation, the aldehyde being volatile in steam.

Benzaldehyde can be obtained by the direct oxidation of benzyl alcohol with nitric acid:

$$C_6H_5.C(H)(H)(OH) + O \rightarrow C_6H_5.C\begin{smallmatrix}H\\ \Vert O\end{smallmatrix} + H_2O$$

It is also prepared by the hydrolysis of benzal chloride with dilute sulphuric acid or calcium hydroxide:

$$C_6H_5.C(H)(Cl)(Cl) + \begin{matrix}H.OH\\H.OH\end{matrix} \rightarrow C_6H_5.C(H)(O\text{—}H)(O\text{—}H) + 2HCl$$

unstable intermediate 'hydrate'

$$C_6H_5.C\begin{smallmatrix}H\\ \Vert O\end{smallmatrix} + H_2O$$

The preparation of benzaldehyde from benzyl chloride can be carried out by heating the chloride with an aqueous solution of either lead or copper nitrate; the reaction probably goes in two stages: (i) the conversion of the benzyl chloride by hydrolysis into benzyl alcohol, and (ii) the oxidation of the alcohol to benzaldehyde, reduction of the nitrate with production of oxides of nitrogen taking place at the same time.

A mixture of benzyl chloride (25 grams), water (250 c.c.) and copper nitrate (20 grams) is boiled for 8 hours in a round-bottom flask using a reflux condenser (p. 513), carbon dioxide being passed through the liquid to drive off the oxides of nitrogen and so prevent any oxidation of the benzaldehyde to benzoic acid. When the contents of the flask have cooled to the ordinary temperature, they are extracted with ether, which dissolves the benzaldehyde; the ethereal extract is separated, washed with a little water, and then shaken with a concentrated solution of sodium bisulphite*. The addition compound, benzaldehyde sodium bisulphite, $C_6H_5.C(H)(OH)(SO_3Na)$, separates out in colourless crystals which are filtered off and washed with a little ether. The crystals are placed in a suitable flask together with a solution of sodium carbonate, the solution showing a strongly alkaline reaction, and the mixture is steam distilled until no more oil passes over into the distillate. The cooled distillate is then extracted with ether, the ethereal extract dried with fused calcium chloride and submitted to distillation, using a water-bath for the removal of the ether and finally in the ordinary manner. The yield of the benzaldehyde should be between 75–80 per cent. of the theoretical amount. The reactions above referred to may be expressed:

(i) $2C_6H_5.CH_2Cl + 2H_2O = 2C_6H_5.CH_2OH + 2HCl$

(ii) $2C_6H_5.CH_2OH + Cu(NO_3)_2 + 2HCl = 2C_6H_5.CHO + CuCl_2 + N_2O_3 + 3H_2O$

* Prepared by passing sulphur dioxide into an aqueous solution of sodium carbonate containing sodium carbonate crystals. The latter substance dissolves and the gas is passed until the solution acquires a greenish tint. Free sulphurous acid should be avoided, as this dissolves the bisulphite compound.

Benzaldehyde is produced on the commercial scale by Gattermann's method, which is an application of the Friedel-Craft reaction. A mixture of carbon monoxide and hydrogen chloride is passed into benzene containing anhydrous cuprous chloride and anhydrous aluminium chloride. It is possible that an unstable compound of carbon monoxide and hydrogen chloride, $H{-}C\begin{smallmatrix}{/\!\!/}O\\ \diagdown Cl\end{smallmatrix}$ (formyl chloride, compare acetyl chloride, p. 208), is first produced and this immediately reacts with the benzene as follows:

$$C_6H_6 + ClCO.H = C_6H_5.CHO + HCl$$

Benzaldehyde is a colourless highly refractive liquid having the characteristic odour of bitter almonds. It has b.p. 179°, is very sparingly soluble in water and readily soluble in most organic solvents.

It is readily oxidised to benzoic acid even on exposure to air; crystals of the acid are always formed round the stopper of a bottle containing benzaldehyde, which is opened from time to time.* It reduces an ammoniacal solution of silver oxide, benzoic acid again being formed:

$$C_6H_5.C\begin{smallmatrix}\diagup H\\ \diagdown\!\!\diagdown O\end{smallmatrix} + Ag_2O \rightarrow C_6H_5.C\begin{smallmatrix}\diagup OH\\ \diagdown\!\!\diagdown O\end{smallmatrix} + 2Ag$$

On reduction benzaldehyde is readily converted into benzyl alcohol:

$$C_6H_5.C\begin{smallmatrix}\diagup H\\ \diagdown\!\!\diagdown O\end{smallmatrix} + H_2 \rightarrow C_6H_5.CH_2OH$$

and when treated with phosphorus pentachloride it yields benzal chloride:

$$C_6H_5.C\begin{smallmatrix}\diagup H\\ \diagdown\!\!\diagdown O\end{smallmatrix} + PCl_5 \rightarrow C_6H_5.C\begin{smallmatrix}\diagup H\\ -Cl\\ \diagdown Cl\end{smallmatrix} + POCl_3$$

by a reaction entirely analogous to the production of ethylidene chloride from acetaldehyde.

When benzaldehyde is shaken with a concentrated aqueous solution of sodium bisulphite a solid crystalline compound separates with marked rise of temperature. This is the benzaldehyde sodium

* The oxidation of benzaldehyde when exposed to air can be considerably delayed by mixing it with small quantities of certain easily oxidisable substances, such as hydroquinone (p. 65) and pyrogallol (p. 67). To such substances, when employed in this manner, Moureu (1920) has given the name *antioxygènes*, which act as, what may be described as, *negative catalysts*.

bisulphite mentioned above and is analogous to the corresponding compound of acetaldehyde:

$$C_6H_5.C\begin{matrix}\diagup H\\ \diagdown\!\!\diagdown O\end{matrix} + NaHSO_3 \rightarrow C_6H_5.C\begin{matrix}\diagup H\\ -OH\\ \diagdown SO_3Na\end{matrix}$$

From this compound benzaldehyde may be recovered in the manner indicated.

The unsaturated nature of the aldehyde group in benzaldehyde is also shown by its combining directly with hydrogen cyanide, forming a compound, benzaldehyde cyanohydrin, analogous to acetaldehyde cyanohydrin:

$$C_6H_5.C\begin{matrix}\diagup H\\ \diagdown\!\!\diagdown O\end{matrix} + HCN \rightarrow C_6H_5.C\begin{matrix}\diagup H\\ -OH\\ \diagdown CN\end{matrix}$$

Benzaldehyde cyanohydrin is a colourless oil, which when heated to 170° decomposes into hydrogen cyanide and benzaldehyde. Like the corresponding acetaldehyde compound, it is hydrolysed when boiled with dilute mineral acids, producing, in this case, *mandelic acid* analogous to lactic acid:

$$C_6H_5.C\begin{matrix}\diagup H\\ -OH\\ \diagdown C\vdots N\end{matrix} + \begin{matrix}H\vdots OH\\ H\vdots O\vdots H\end{matrix} \rightarrow C_6H_5.C\begin{matrix}\diagup H\\ -OH\\ \diagdown COOH\end{matrix} + NH_3$$

The condensation reactions of benzaldehyde with hydroxylamine, semicarbazide and phenylhydrazine are analogous to the corresponding reactions of aliphatic aldehydes. When benzaldehyde reacts with hydroxylamine directly, the usual reaction takes place:

$$C_6H_5.C\begin{matrix}\diagup H\\ \diagdown\!\!\diagdown \boxed{O + H_2}N.OH\end{matrix} \rightarrow C_6H_5.C\begin{matrix}\diagup H\\ \diagdown\!\!\diagdown N.OH\end{matrix} + H_2O$$

The compound formed under these conditions has m.p. 35° and is known as *α-benzaldoxime*. When this α-compound is submitted to the action of hydrochloric acid or sulphuric acid or bromine, unstable salts or addition compounds are formed from which an isomeric compound, *β-benzaldoxime*, having m.p. 125°, is liberated. This same β-compound is obtained when benzaldehyde reacts with hydroxylamine hydrochloride in alcoholic solution, and when the β-compound is distilled under diminished pressure it is converted into the α-compound. These two compounds are undoubtedly stereoisomerides and the explanation of their isomerism is based on the Hantzsch-Werner hypothesis (1890), by which an analogy is traced between those cases of isomerism of the general type $CXY:CWZ$ among ethylene compounds and those cases of compounds containing tervalent nitrogen doubly linked to carbon. The β-compound loses a molecule of water more readily than the α-compound, yielding phenyl cyanide when submitted to the action of certain reagents. Formerly, it was supposed that the isomeride (having the higher melting point) which yielded phenyl cyanide the more readily must have the H atom and the —OH group on the same side of the double bond. This assumption in this case and in those of analogous compounds is now known to be incorrect and there is chemical evidence to prove that the more reactive isomeride must have the *anti-* configuration. A satisfactory explanation of the greater reactivity of the isomeride having the *anti-* configuration based on the electronic configuration of the nitrogen and carbon atoms has been given by Mills*. The relationships between the two stereoisomeric oximes may be briefly indicated:

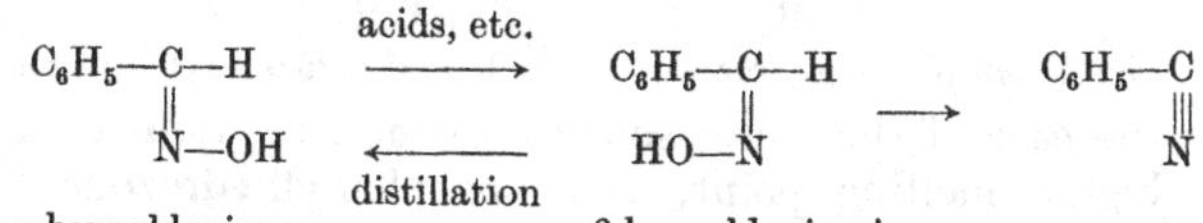

$$\begin{array}{c} C_6H_5\text{—}C\text{—}H \\ \| \\ N\text{—}OH \end{array} \underset{\text{distillation}}{\overset{\text{acids, etc.}}{\rightleftarrows}} \begin{array}{c} C_6H_5\text{—}C\text{—}H \\ \| \\ HO\text{—}N \end{array} \longrightarrow \begin{array}{c} C_6H_5\text{—}C \\ ||| \\ N \end{array}$$

α-benzaldoxime m.p. 35° *syn-* configuration

β-benzaldoxime† m.p. 125° *anti-* configuration

The two benzaldoximes are weak bases and each yield a number of isomeric derivatives. Many aromatic aldehydes in the ordinary way yield only one oxime, although two isomeric oximes are

* "Some aspects of stereochemistry," being the presidential address to the Chemistry Section of the British Association for the Advancement of Science, 1932, by W. H. Mills.

† β-Benzaldoxime when warmed with acetyl chloride is converted into the corresponding acetyl derivative:

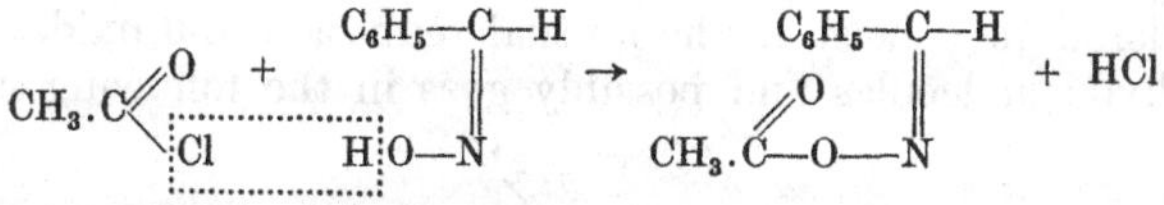

$$CH_3.C\begin{cases}O \\ Cl\end{cases} + \begin{array}{c} C_6H_5\text{—}C\text{—}H \\ \| \\ H\text{—}O\text{—}N \end{array} \rightarrow \begin{array}{c} C_6H_5\text{—}C\text{—}H \\ \| \\ CH_3.C(=O)\text{—}O\text{—}N \end{array} + HCl$$

The acetyl derivative when warmed with sodium carbonate is decomposed quantitatively and phenyl cyanide is produced:

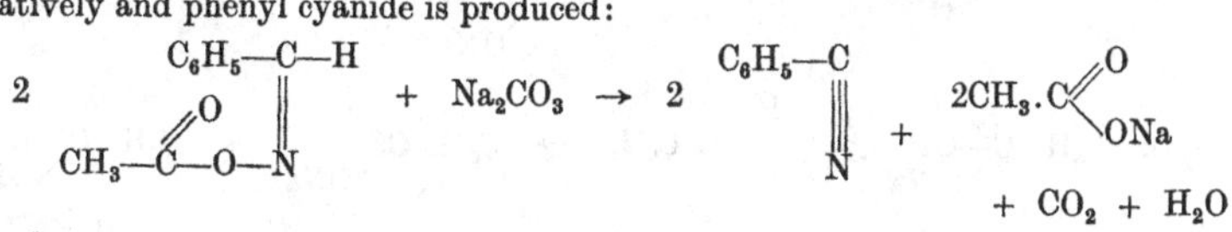

$$2\ \begin{array}{c} C_6H_5\text{—}C\text{—}H \\ \| \\ CH_3\text{—}C(=O)\text{—}O\text{—}N \end{array} + Na_2CO_3 \rightarrow 2\ \begin{array}{c} C_6H_5\text{—}C \\ ||| \\ N \end{array} + 2CH_3.C\begin{cases}O \\ ONa\end{cases} + CO_2 + H_2O$$

The latter reaction can be regarded as consisting in the removal of acetic acid, by the neutralising action of sodium carbonate:

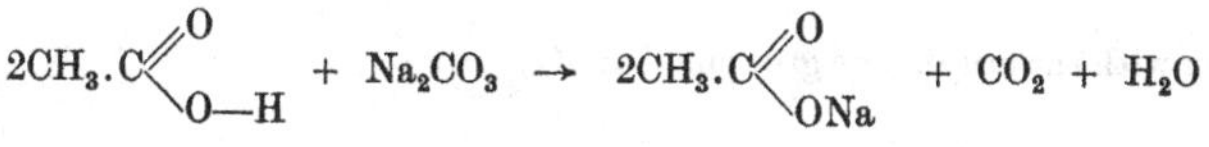

$$2CH_3.C\begin{cases}O \\ O\text{—}H\end{cases} + Na_2CO_3 \rightarrow 2CH_3.C\begin{cases}O \\ ONa\end{cases} + CO_2 + H_2O$$

theoretically possible. In such cases we must assume that the oxime produced has generally the *anti-* configuration, which is generally more stable than the one having the *syn-* configuration.

Benzaldehyde phenylhydrazone (benzylidene phenylhydrazone* or benzal phenylhydrazone) is very readily obtained by simply mixing equimolecular quantities of benzaldehyde and phenylhydrazine in alcoholic solution. The compound crystallises out almost at once and when pure (by recrystallisation from alcohol) yields almost colourless small prisms, m.p. 154–155°. The phenylhydrazone so obtained can be converted into an isomeric compound, m.p. 136°, by heating with acetic anhydride or dilute sulphuric acid. The two benzaldehyde phenylhydrazones are also stereoisomerides, the isomerism being analogous to that of the oximes:

$$C_6H_5—C—H \;\; \| \;\; N—N\begin{matrix}H\\C_6H_5\end{matrix} \qquad\qquad C_6H_5—C—H \;\; \| \;\; \begin{matrix}H\\C_6H_5\end{matrix}N—N$$

benzal *syn*-phenylhydrazone   benzal *anti*-phenylhydrazone

If, as in the case of the stereoisomeric oximes, the *anti-* compound has the higher melting point, then the phenylhydrazone formed directly by mixing benzaldehyde and phenylhydrazine in alcoholic solution should be described as benzal *anti*-phenylhydrazone.

Although benzaldehyde is a reducing agent and reduces an ammoniacal solution of silver oxide, unlike the aliphatic aldehydes, it does not reduce Fehling's solution (alkaline solution of cupric oxide). It gives the Schiff reaction rather more slowly than the aliphatic aldehydes which are generally more soluble in water. When benzaldehyde is shaken with a concentrated aqueous solution of sodium or potassium hydroxide and the emulsion so formed allowed to stand for several hours, a mixture of sodium (potassium) benzoate and benzyl alcohol is produced. This reaction (Cannizzaro, 1854) may be considered to consist in the mutual reduction and oxidation of benzaldehyde molecules and possibly goes in the following stages:

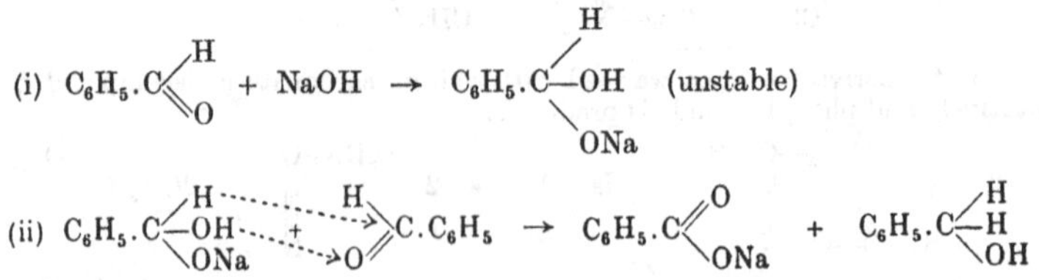

$$\text{(i)}\quad C_6H_5.C\begin{matrix}H\\ \|O\end{matrix} + NaOH \rightarrow C_6H_5.C\begin{matrix}H\\—OH\\ONa\end{matrix} \quad \text{(unstable)}$$

$$\text{(ii)}\quad C_6H_5.C\begin{matrix}H\\—OH\\ONa\end{matrix} + \begin{matrix}H\\O\end{matrix}\!\!>\!C.C_6H_5 \rightarrow C_6H_5.C\begin{matrix}\|O\\ONa\end{matrix} + C_6H_5.C\begin{matrix}H\\—H\\OH\end{matrix}$$

* The name *benzylidene* or *benzal* is given to the grouping, $C_6H_5.C\begin{matrix}H\\ \|\end{matrix}$, corresponding with the ethylidene grouping $CH_3.C\begin{matrix}H\\ \|\end{matrix}$.

Towards sodium (potassium) hydroxide, therefore, benzaldehyde behaves in an analogous manner to formaldehyde.

The behaviour of benzaldehyde towards ammonia is again different from that of a typical aliphatic aldehyde and is somewhat similar to that of formaldehyde. In this case, the chief product of the reaction is a colourless substance, *hydrobenzamide*, m.p. 110°, which is formed according to the equation:

$$3C_6H_5.CHO + 2NH_3 = (C_6H_5.CH:)_3N_2 + 3H_2O$$

It has been shown that one of the characteristic properties of the aliphatic aldehydes is their tendency to polymerise under suitable conditions. This tendency is not nearly so marked in the case of the typical aromatic aldehydes like benzaldehyde, but, in the presence of an aqueous solution of potassium cyanide, they form polymerisation products of a different type. In the case of benzaldehyde—and its behaviour towards an aqueous solution of potassium cyanide is generally typical of aromatic aldehydes of this class—a substance known as *benzoin* is produced, the formation of which may be briefly indicated:

$$C_6H_5.C\begin{matrix}\diagup H\\ \diagdown O\leftarrow\cdots\end{matrix} + \begin{matrix}O\diagdown\\ \cdots H\diagup\end{matrix}C.C_6H_5 \rightarrow C_6H_5-\overset{\displaystyle H}{\underset{\displaystyle OH}{\overset{|}{\underset{|}{C}}}}-\overset{\displaystyle O}{\overset{\|}{C}}-C_6H_5$$

benzoin

The reaction, however, probably takes place in two stages:

$$\text{(i)}\quad C_6H_5.C\begin{matrix}\diagup H\\ \diagdown O\end{matrix} + H.C\vdots N \text{ (from hydrolysis of KCN)} \rightarrow C_6H_5.C\begin{matrix}\diagup H\\ -OH\\ \diagdown CN\end{matrix}$$

benzaldehyde cyanohydrin

$$\text{(ii)}\quad C_6H_5.\overset{\displaystyle H}{\underset{\displaystyle OH}{\overset{|}{\underset{|}{C}}}}-\boxed{C\vdots N} + \boxed{H}\begin{matrix}\diagdown\\ O\diagup\end{matrix}C.C_6H_5 \rightarrow C_6H_5-\overset{\displaystyle H}{\underset{\displaystyle OH}{\overset{|}{\underset{|}{C}}}}-\overset{\displaystyle O}{\overset{\|}{C}}-C_6H_5 + HCN$$

The preparation of benzoin is carried out in the following manner: Benzaldehyde (20 grams) is mixed with potassium cyanide (2 grams, as pure as possible) dissolved in water (40 c.c.) and ethyl alcohol (40 c.c.), and the mixture heated for one hour on the boiling water-bath in a suitable flask using a water-cooled condenser. The benzoin crystallises from the cooling solution and is filtered off; a further quantity may be obtained by heating the mother liquor with fresh potassium cyanide. The benzoin is recrystallised from ethyl alcohol and obtained in colourless needles, m.p. 137°.

Benzoin has interesting properties; it has one asymmetric carbon in its molecule and is both a secondary alcohol and a ketone. When it is reduced with sodium amalgam, *hydrobenzoin* or sym.-*diphenylglycol* (m.p. 139°) (compare p. 107) is obtained:

$$C_6H_5-\underset{OH}{\overset{H}{C}}-\overset{O}{\overset{\|}{C}}-C_6H_5 + 2H \rightarrow C_6H_5-\underset{OH}{\overset{H}{C}}-\underset{OH}{\overset{H}{C}}-C_6H_5$$

Hydrobenzoin contains two asymmetric carbon atoms in the molecule, and on account of the configuration of identical groups attached to the asymmetric carbon atoms its stereoisomerism is similar to that of the tartaric acids (p. 393).

When benzoin is heated with concentrated nitric acid for two hours on the boiling water-bath it is oxidised to *benzil*, which crystallises from alcohol in yellow prisms, m.p. 95°:

$$C_6H_5-\underset{OH}{\overset{H}{C}}-\overset{O}{\overset{\|}{C}}-C_6H_5 + O \rightarrow C_6H_5-\overset{O}{\overset{\|}{C}}-\overset{O}{\overset{\|}{C}}-C_6H_5$$

Benzil is described as an α-diketone (having two ketonic groups adjoining).

Similar to the formation of crotonaldehyde from two molecules of acetaldehyde, benzaldehyde condenses with acetaldehyde in the presence of sodium hydroxide to form an unsaturated aromatic aldehyde known as *cinnamic aldehyde* (cinnamylaldehyde):

$$C_6H_5.CHO + H-\underset{H}{\overset{H}{C}}.CHO \rightarrow C_6H_5.CH:CH.CHO + H_2O$$

Cinnamic aldehyde is a colourless liquid, b.p. 251°. It has the characteristic odour of cinnamon, and it is to the presence of this substance in oil of cinnamon and oil of cassia that the characteristic odour of these oils is due. As an unsaturated compound, cinnamic aldehyde combines directly with chlorine and bromine forming compounds which may be represented thus: $C_6H_5.CHX.CHX.CHO$; as an aldehyde it yields a sodium bisulphite compound

$$C_6H_5.CH:CH.\underset{SO_3Na}{\overset{H}{C}}-OH$$

and is directly oxidised on exposure to air to *cinnamic acid*, which like cinnamic aldehyde is capable of existing in two (*cis-* and *trans-*) stereoisomeric forms:

$$\begin{array}{c} C_6H_5—C—H \\ \| \\ CO_2H—C—H \end{array} \qquad \begin{array}{c} C_6H_5—C—H \\ \| \\ H—C—CO_2H \end{array}$$

Cinnamic aldehyde readily forms a *phenylhydrazone*, m.p. 168°, by condensation with phenylhydrazine; four stereoisomers of this phenylhydrazone are theoretically possible.

$$C_6H_5.CH:CH.C\begin{array}{l} H \\ O \end{array} + H_2N.NH.C_6H_5 \rightarrow C_6H_5.CH:CH.C\begin{array}{l} H \\ N.NH.C_6H_5 \end{array} + H_2O$$

In its reaction with ammonia, cinnamic aldehyde behaves similarly to benzaldehyde and yields *hydrocinnamide*

$$(C_6H_5.CH:CH.CH:)_3N_2, \text{ m.p. } 131°.$$

The reaction of benzaldehyde with acetaldehyde to form cinnamic aldehyde is typical of all aldehydes belonging to the same class of aromatic aldehydes as benzaldehyde.

Another reaction of benzaldehyde and other aldehydes of the same class is exemplified by the reaction of benzaldehyde with sodium acetate in the presence of acetic anhydride (p. 210). This reaction, which in the case of benzaldehyde leads to the formation of cinnamic acid, was discovered by Sir W. H. Perkin (1877) and is generally referred to as Perkin's reaction:

$$C_6H_5.C\begin{array}{l} H \\ O \end{array} + \begin{array}{c} H \\ H—C.CO_2Na \\ H \end{array} \rightarrow C_6H_5.CH:CH.CO_2Na + H_2O$$

sodium salt of cinnamic acid

The sodium salt is converted into free cinnamic acid in the presence of the acetic acid formed from the acetic anhydride by the action of the water liberated during the course of the reaction. The preparation of cinnamic acid is carried out as follows:

Benzaldehyde (20 grams), acetic anhydride (30 grams) and finely powdered recently fused sodium acetate (10 grams) are heated in a flask provided with an air condenser for 8 hours at 180° (temperature of the heating metal or oil bath). Whilst still hot the product is poured into water and the mixture submitted to steam distillation until free from benzaldehyde. The residue in the steam distillation flask is boiled with decolorising charcoal and filtered whilst still hot. The cinnamic acid crystallises in colourless glistening plates, m.p. 133°, from the cooling solution.

As an unsaturated compound, cinnamic acid combines directly with halogens, as does ethylene, forming, for example, with bromine, *cinnamic acid dibromide* (β-phenyl-$\alpha\beta$-dibromopropionic acid), a compound containing two asymmetric carbon atoms in the molecule.

$C_6H_5.CHBr.CHBr.CO_2H$. It has already been remarked that cinnamic acid exhibits *cis-* and *trans-* isomerism; ordinary cinnamic acid is believed to have the *trans-* configuration and the other isomeride exhibits the unusual property of trimorphism, i.e. it occurs in three different crystalline modifications which are interconvertible.

As an aromatic compound, benzaldehyde undergoes nitration. When it is treated with a mixture of concentrated nitric and sulphuric acids (compare the nitration of benzene) the chief product of the reaction is m-*nitrobenzaldehyde*, $C_6H_4\left\langle\begin{matrix}CHO\,(1)\\ NO_2\,(3)\end{matrix}\right.$, thin needles, m.p. 58°, and the isomeric o-*nitrobenzaldehyde*, $C_6H_4\left\langle\begin{matrix}CHO\,(1)\\ NO_2\,(2)\end{matrix}\right.$, colourless needles, m.p. 44·5°, is produced in small quantity at the same time. These nitrobenzaldehydes react in general very similarly to benzaldehyde; for example, they undergo Perkin's reaction whereby nitrocinnamic acids are obtained.

There exist three isomeric monohydroxybenzaldehydes; of these the most important is o-*hydroxybenzaldehyde*, usually known as *salicylaldehyde* (salicylic aldehyde). This compound and its two isomerides may have their constitutions represented:

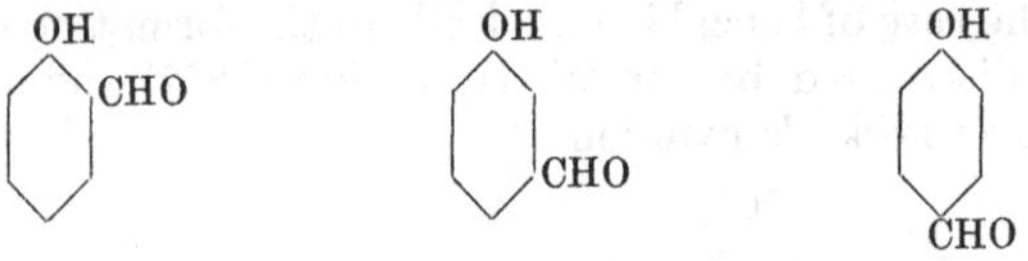

OH CHO | OH CHO | OH CHO

m-hydroxybenzaldehyde *p*-hydroxybenzaldehyde

In so far as these compounds have the aldehyde group directly attached to a carbon atom of the nucleus, their properties will obviously resemble those of benzaldehyde. At the same time these compounds are also phenols and hence possess properties and undergo reactions of these substances.

Salicylaldehyde occurs in the volatile oil of certain species of *Spirea* and particularly in that of *Spirea ulmaria*. It may be obtained by the oxidation of saligenin (*o*-hydroxybenzyl alcohol, p. 175) with chromic acid, but it is usually prepared by the Reimer-Tiemann reaction (1876), a general reaction for the preparation of certain types of aromatic aldehydes and which, in this particular case, is carried out as described below.

Phenol (50 grams) is dissolved in water (170 c.c.) containing sodium hydroxide (100 grams) in a suitable flask fitted with a reflux condenser and a tap funnel (p. 513). Chloroform (75 grams) is added very slowly from the tap funnel while the contents of the flask are kept at 50–60° and frequently shaken vigorously during the addition. After all the chloroform has been added the contents of the flask are boiled for 1–2 hours. The excess chloroform is distilled off and, after making the contents of the flask acid by the addition of sulphuric acid (dilute), they are submitted to steam distillation. Unchanged phenol and salicylaldehyde are volatile under these conditions, and the residue remaining in the steam-distillation flask contains *p*-hydroxybenzaldehyde (see below). The distillate is extracted with ether and the ethereal extract shaken with a concentrated aqueous solution of sodium bisulphite, when the sparingly soluble salicylaldehyde sodium bisulphite compound,

$$HO.C_6H_4\!\!\begin{array}{l}\diagup H\\ -OH\\ \diagdown O.SO_2Na\end{array}$$

, crystallises out. This is separated by filtration, washed with a little ether and then decomposed by warming with dilute sulphuric acid. The salicylaldehyde separates as an oil, which, after cooling, is extracted with ether, the ethereal extract washed with water and dried with anhydrous calcium chloride. The ether is distilled off and the residue, salicylaldehyde, carefully distilled. The salicylaldehyde (10–14 grams) is obtained as a colourless oil, b.p. 196°.

In the mononitration of phenol it has been shown that a mixture of *ortho-* and *para-* nitrophenols is obtained. Similarly, the aldehyde group introduced into the phenol molecule enters the *ortho-* and (to a less extent) *para-* positions and both *o-* and *p-* hydroxybenzaldehydes are obtained in the above reaction (see below).

Salicylaldehyde has a small solubility in water, and this aqueous solution gives a violet coloration with ferric chloride on account of the presence of the phenolic hydroxy group. Like benzaldehyde, salicylaldehyde is easily oxidised, the oxidation product being *salicylic acid* (*o*-hydroxybenzoic acid):

$$C_6H_4\!\!\begin{array}{l}\diagup OH(1)\\ \diagdown C\!\!\begin{array}{l}\diagup H\\ \diagdown\!\!\diagdown O\end{array}(2)\end{array} + O \rightarrow C_6H_4\!\!\begin{array}{l}\diagup OH(1)\\ \diagdown C\!\!\begin{array}{l}\diagup OH\\ \diagdown\!\!\diagdown O\end{array}(2)\end{array}$$

Salicylaldehyde reduces an ammoniacal solution of silver oxide, but it does not reduce Fehling's solution. When salicylaldehyde is reduced, *salicylic alcohol* (*o*-hydroxybenzyl alcohol, saligenin) is produced:

$$C_6H_4\!\!\begin{array}{l}\diagup OH(1)\\ \diagdown C\!\!\begin{array}{l}\diagup H\\ \diagdown\!\!\diagdown O\end{array}(2)\end{array} + 2H \rightarrow C_6H_4\!\!\begin{array}{l}\diagup OH(1)\\ \diagdown C\!\!\begin{array}{l}\diagup H\\ -H\\ \diagdown OH\end{array}(2)\end{array}$$

As a phenol, salicylaldehyde is soluble in aqueous solutions of sodium and potassium hydroxides, and when such a solution of the sodium or potassium derivative is allowed to react with methyl or ethyl iodide or with dimethyl or diethyl sulphate the corresponding *ethers* are obtained:

$$C_6H_4\!\!\begin{array}{l}\diagup ONa(1)\\ \diagdown C\!\!\begin{array}{l}\diagup H\\ \diagdown\!\!\diagdown O\end{array}(2)\end{array} + ICH_3 \rightarrow C_6H_4\!\!\begin{array}{l}\diagup OCH_3(1)\\ \diagdown C\!\!\begin{array}{l}\diagup H\\ \diagdown\!\!\diagdown O\end{array}(2)\end{array} + NaI$$

As an aldehyde, salicylaldehyde reacts similarly to benzaldehyde with hydroxylamine and forms an *oxime*, $C_6H_4\langle^{OH}_{C\langle^{H}_{NOH}}$, m.p. 57°.

Towards phenylhydrazine, it also reacts in a similar way, yielding a *phenylhydrazone*, m.p. 143°. This phenylhydrazone is always obtained together with a small quantity of an isomeride, m.p. 104–105°, and there is evidence that these two substances are the two possible stereoisomerides.

Para-*hydroxybenzaldehyde* is obtained along with salicylaldehyde in the Reimer-Tiemann reaction described above.

It remains in the steam distillation flask after distilling unchanged phenol and salicylaldehyde. This residue is heated to boiling and allowed to cool. The cold mixture is extracted with ether and the ethereal extract evaporated. The residue remaining is crystallised from hot water using decolorising charcoal. The *p*-hydroxybenzaldehyde crystallises in colourless needles, m.p. 115–116°.

In many of its properties, *p*-hydroxybenzaldehyde resembles salicylaldehyde. It yields an *oxime*, m.p. 65°, and a *phenylhydrazone*, m.p. 178°, having the formulae

$$C_6H_4\langle^{OH(1)}_{CH:NOH(4)} \quad \text{and} \quad C_6H_4\langle^{OH(1)}_{CH:N.NHC_6H_5(4)}$$

respectively. Only one form of each of these two compounds has been so far described and both should theoretically exist in *cis*- and *trans*- forms.

p-Hydroxybenzaldehyde is soluble in aqueous solutions of sodium and potassium hydroxides, and from these solutions has been obtained, by the reaction mentioned above, *anisaldehyde*, or *p*-methoxybenzaldehyde, $C_6H_4\langle^{OCH_3(1)}_{CHO(4)}$, a colourless liquid, b.p. 248°. This compound is generally prepared by the oxidation of *anethole* (which occurs in oil of aniseed) with potassium dichromate and sulphuric acid. Anethole is systematically described as 1-methoxy-4-propenylbenzene, and the relationship between it and anisaldehyde is indicated:

$$C_6H_4\langle^{OCH_3(1)}_{CH:CH.CH_3(4)} \xrightarrow{\text{oxidation}} C_6H_4\langle^{OCH_3(1)}_{CHO(4)}$$

Anisaldehyde behaves similarly to benzaldehyde, having no phenolic properties. Possibly, connected with the suppression or absence of phenolic properties, is the fact that the two stereoisomeric *oximes* are easily obtained, which in the light of our present information have the constitutions:

$$CH_3O.C_6H_4—\underset{\underset{N—OH}{\parallel}}{C}—H \qquad CH_3O.C_6H_4—\underset{\underset{HO—N}{\parallel}}{C}—H$$

syn-anisaldoxime, m.p. 61–62° *anti*-anisaldoxime, m.p. 130·5°

Apart from their stereoisomerism, it is curious that the *syn*-anisaldoxime has a distinctly sweet taste while the *anti*- compound is tasteless. The *anti*- oxime is converted into its stereoisomeride by simply boiling its ethereal solution.

When anisaldehyde is reduced it is converted into *anisyl alcohol* (*p*-methoxybenzyl alcohol, $C_6H_4\langle^{OCH_3(1)}_{CH_2OH(4)}$), m.p. 25°; and it is very easily oxidised to *anisic acid* (*p*-methoxybenzoic acid, $C_6H_4\langle^{OCH_3(1)}_{COOH(4)}$), m.p. 185°. The last-named compound when heated with hydriodic acid is converted into *p*-hydroxybenzoic acid, m.p. 210°, isomeric with salicylic acid; its formation by this method may be expressed:

$$C_6H_4\langle^{OCH_3}_{CO_2H} + HI \rightarrow C_6H_4\langle^{OH}_{CO_2H} + CH_3I$$

Anisic acid is converted into anisole (p. 130) by heating with lime:

$$C_6H_4\langle^{OCH_3(1)}_{CO_2H(4)} + CaO \rightarrow C_6H_5.OCH_3 + CaCO_3$$

The relationships between these various substances may be summarised as follows:

$$\underset{\text{anethole}}{C_6H_4\langle^{OCH_3(1)}_{CH:CH.CH_3(4)}} \xrightarrow{\text{oxidation}} \underset{\text{anisaldehyde}}{C_6H_4\langle^{OCH_3(1)}_{CHO(4)}} \xrightarrow{\text{reduction}} \underset{\text{anisyl alcohol}}{C_6H_4\langle^{OCH_3(1)}_{CH_2OH(4)}}$$

anisaldehyde → (oxidation) → $C_6H_4\langle^{OCH_3(1)}_{CO_2H(4)}$ anisic acid

$$\underset{p\text{-hydroxybenzoic acid}}{C_6H_4\langle^{OH}_{CO_2H}} \xleftarrow{\text{HI}} \underset{\text{anisic acid}}{C_6H_4\langle^{OCH_3(1)}_{CO_2H(4)}}$$

p-hydroxybenzoic acid → phenol; anisic acid $\xrightarrow{\text{CaO}}$ anisole

$$\left[\underset{\text{salicylaldehyde}}{C_6H_4\langle^{OH(1)}_{CHO(2)}}\right] + \underset{p\text{-hydroxybenzaldehyde}}{C_6H_4\langle^{OH(1)}_{CHO(4)}} \xleftarrow[+CHCl_3]{NaOH} \underset{\text{phenol}}{C_6H_5OH} \underset{HI}{\overset{NaOH+CH_3I}{\rightleftarrows}} \underset{\text{anisole}}{C_6H_5.OCH_3}$$

p-hydroxybenzaldehyde $\xrightarrow{NaOH+CH_3I}$ anisaldehyde

The third isomeric hydroxybenzaldehyde, m-*hydroxybenzaldehyde*, $C_6H_4\left\langle\begin{matrix}OH\,(1)\\CHO\,(3)\end{matrix}\right.$, is obtained from *m*-nitrobenzaldehyde (produced by the nitration of benzaldehyde) by reactions which will be discussed later (pp. 265, 275). *m*-Hydroxybenzaldehyde crystallises in colourless needles, m.p. 108°. It forms an *oxime*, m.p. 87·5°, and a *phenylhydrazone*, m.p. 130°, having the formulae $C_6H_4\left\langle\begin{matrix}OH\,(1)\\CH:NOH\,(3)\end{matrix}\right.$ and $C_6H_4\left\langle\begin{matrix}OH\,(1)\\CH:N.NHC_6H_5\,(3)\end{matrix}\right.$ respectively. Each of these compounds is theoretically capable of existing in *cis*- and *trans*- isomerides.

Of these aromatic aldehydes, from an industrial standpoint, probably the most important is benzaldehyde, which is employed in the manufacture of certain important dye-stuffs. The dye-stuff known as *malachite green* is typical of this class. Benzaldehyde is also used extensively as a flavouring essence and the odour of these aromatic aldehydes is a characteristic property. Belonging to the same class are *vanillin* and *piperonal*; vanillin is the sweet-smelling constituent of the vanilla bean and its constitution* is

OH, OCH_3, CHO (ring) or CHO, OCH_3, OH (ring)

It is a colourless crystalline substance, m.p. 83°. Piperonal is also a colourless crystalline substance, m.p. 37°, b.p. 263°. It possesses a very pleasant odour resembling that of heliotrope and it is sold as a perfume under the name of ***heliotropin***. The constitution of piperonal is represented:

CHO, O, O—CH_2 (ring) or $C_6H_3\begin{cases}CHO(1)\\O\\O\end{cases}\!\!>CH_2\begin{matrix}(3)\\(4)\end{matrix}$

A very large number of derivatives of vanillin and piperonal are known.

* Usually this is written briefly as $C_6H_3\begin{cases}CHO\,(1)\\OCH_3\,(3)\\OH\,(4)\end{cases}$. Since the characteristic or most reactive grouping is the aldehyde grouping, the systematic name of vanillin is usually given as 3-methoxy-4-hydroxybenzaldehyde, it being assumed that the aldehyde grouping occupies position (1).

Ketones

Acetone, $CH_3.CO.CH_3$, $(CH_3)_2CO$, Me_2CO

Acetone (dimethylketone) occurs in very small quantities as a normal constituent in the urine and blood; in larger quantities, it occurs in these fluids in the abnormal condition known as *acetonuria*, and along with acetoacetic acid (p. 368) and β-hydroxybutyric acid (p. 367) in the condition known as *diabetes mellitus*. The three substances, acetone, acetoacetic acid and β-hydroxybutyric acid, are frequently referred to as the acetone bodies.

Formerly, acetone was exclusively obtained from the distillation of wood and by the action of heat on calcium acetate (p. 203). Modern processes for the manufacture of this important substance consist essentially in passing acetic acid vapour (acetic anhydride, p. 210, can also be used) over catalysts, which bring about the elimination of carbon dioxide and water and the formation of acetone:

$$\begin{array}{l} CH_3 \\ | \\ C(=O)-O-H \\ C(=O)-O-H \\ | \\ CH_3 \end{array} \rightarrow CO_2 + H_2O + \begin{array}{c} CH_3 \\ | \\ C{=}O \\ | \\ CH_3 \end{array}$$

The catalysts employed are oxides of aluminium, thorium, zinc and cadmium at 300–400°. Finely divided metallic copper is also employed as a catalyst in the same reaction at 400°. The production of acetone from ethyl alcohol is described on p. 472.

Another method which is likely to become of considerable importance for the large scale production of acetone is that of fermentation. Fernbach (1910) discovered that certain starchy materials, including maize and potatoes, could be fermented by what may be termed the *acetone bacillus*, the chief products being acetone and *n*-butyl alcohol. Essentially, the process consists in the fermentation of the prepared starchy material under strictly aseptic conditions and the separation of the two main products by fractional distillation. This process (p. 470) may be briefly summarised:

$$\underset{\text{starch}}{C_6H_{10}O_5} \rightarrow \underset{\text{sugar}}{C_6H_{12}O_6} \rightarrow \underset{\text{acetone}}{C_3H_6O} + \underset{n\text{-butyl alcohol}}{C_4H_{10}O}$$

Carbon dioxide and hydrogen are evolved during the reaction.

Although acetone is the source of *iso*propyl alcohol, the latter can be oxidised catalytically to the former:

$$CH_3\text{—}CH(OH)\text{—}CH_3 + O \rightarrow CH_3\text{—}CO\text{—}CH_3 + H_2O$$

This may be carried out by passing the *iso*propyl alcohol vapour with oxygen over finely divided copper at 250–400° or over platinum black at 250–300°. When *iso*propyl alcohol vapour mixed with air is passed over a glowing platinum spiral the same reaction occurs.

Acetone is a colourless liquid having a characteristic and not unpleasant odour and a burning taste. It has m.p. −94·6° and b.p. 56·5°. It mixes with water and most organic solvents in all proportions. It is a very convenient liquid for the determination of molecular weights by the ebullioscopic method, its constant for the raising of the boiling point being 17·1 per 100 grams of acetone.

Acetone is isomeric with propionaldehyde. It is not a reducing agent and is, in reality, a very stable substance. It resists the action of neutral potassium permanganate, and a routine method for the preliminary purification of acetone is to distil the crude material mixed with that substance. When, however, it is oxidised under energetic conditions, such as the long continued action of chromic acid or alkaline permanganate, acetone yields a mixture of carbon dioxide and acetic acid, the molecule being disrupted as shown:

$$CH_3\text{—}CO\text{—}CH_3 + 2O_2 \rightarrow CO_2 + H_2O + CH_3.COOH$$

A typical aldehyde under these conditions, or on catalytic oxidation, is converted into an acid containing the same number of carbon atoms as the aldehyde itself.

Like an aldehyde, acetone is reduced to the corresponding alcohol; in this case, the product of the reduction is *iso*propyl alcohol, a secondary alcohol. The reduction can be carried out by passing a mixture of hydrogen and *iso*propyl alcohol vapour over reduced nickel at 120°:

$$CH_3\text{—}CO\text{—}CH_3 + 2H \rightarrow CH_3\text{—}CH(OH)\text{—}CH_3$$

Under similar conditions, an aldehyde would be converted into the corresponding primary alcohol. The reduction of acetone, however, is not quite so simple as it appears from the above equation, since pinacol (p. 107) is also produced as a by-product.

When submitted to the action of halogens, acetone undergoes substitution just as aldehyde does. The final product of the chlorination of acetone being *hexachloroacetone*, $CCl_3.CO.CCl_3$; the final product of the bromination of acetone appears to be *pentabromoacetone*, $CHBr_2.CO.CBr_3$, and of iodination symmetrical *diiodoacetone*, $CH_2I.CO.CH_2I$. The first product of the chlorination of acetone is a colourless liquid, *monochloroacetone*, $CH_2Cl.CO.CH_3$, b.p. 119°, which is a powerful lachrymator.

When acetone is allowed to react with phosphorus pentachloride it undergoes an analogous reaction to acetaldehyde, the oxygen of the :CO grouping being replaced by its equivalent of two chlorine atoms:

$$CH_3.CO.CH_3 + PCl_5 \rightarrow CH_3.CCl_2.CH_3 + POCl_3$$

The substance produced, *ββ-dichloropropane*, a colourless liquid, b.p. 70°, when treated with alkali easily loses hydrogen chloride in two stages, giving first chlorinated propylene and finally methylacetylene:

$$CH_3.CCl_2.CH_3 \xrightarrow{-HCl} CH_3.CCl{=}CH_2 \xrightarrow{-HCl} CH_3.C{\equiv}CH$$

When either acetone or ethyl alcohol is mixed with an aqueous paste of bleaching powder and the mixture distilled, chloroform is obtained. This is the basis of the industrial production of *chloroform*. The bleaching powder may be considered to act as an oxidising agent, chlorinating agent and hydrolytic agent. When ethyl alcohol is used the intermediate product is chloral, which is then decomposed by the alkali of the bleaching powder (p. 164). When acetone is employed, the intermediate product is trichloroacetone produced by chlorination, the trichloroacetone then undergoing hydrolysis as shown:

$$2\,CCl_3.CO.CH_3 + Ca(OH)_2 \rightarrow 2CHCl_3 + Ca(OOC.CH_3)_2$$

calcium acetate

Chloroform is a colourless liquid, b.p. 61·5°, having an agreeable odour and a sweetish taste which it imparts to water, in which, however, it is very sparingly soluble. It is a very important anaesthetic; its use for this purpose in surgical operations was introduced by Sir James Simpson of Edinburgh in 1847. When chloroform is exposed to sunlight in the presence of air or oxygen, the highly poisonous gas, *phosgene* (carbonyl chloride), is produced, the presence of which in anaesthetic chloroform should be rigorously avoided:

$$CHCl_3 + O \rightarrow \underset{\text{phosgene}}{COCl_2} + HCl$$

To minimise the production of phosgene by this reaction about 1 per cent. of ethyl alcohol is usually added to anaesthetic chloroform.

When chloroform is boiled with an alcoholic solution of potassium hydroxide, hydrolysis takes place slowly with the production of potassium formate:

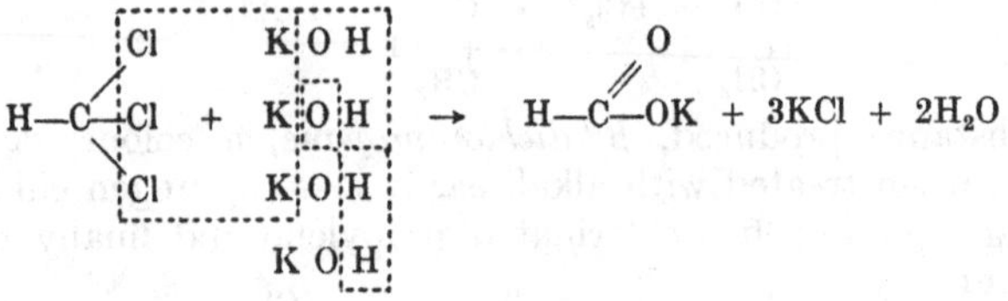

$$H{-}CCl_3 + 4KOH \rightarrow H{-}C({=}O){-}OK + 3KCl + 2H_2O$$

Apart from its being an anaesthetic, chloroform is a useful solvent for many organic compounds and for iodine. It is also important in the preparation of a variety of organic compounds.

The reaction of acetone and of ethyl alcohol with iodine in the presence of alkali is analogous to the reaction of these two substances with bleaching powder; the product of the reaction in this case is *iodoform*, CHI_3. The preparation of iodoform from acetone is carried out as follows:

A small quantity of acetone (1 c.c.) is diluted with water (10 c.c.) and to this is added a solution of iodine in potassium iodide (iodine, 12 grams, potassium iodide, 60 grams, water, 180 c.c.). To the mixture is added slowly and with shaking a concentrated aqueous solution of sodium hydroxide until the colour of the iodine disappears and, at the same time, a yellow crystalline precipitate of iodoform is produced. The reaction taking place can be considered to go in two stages:

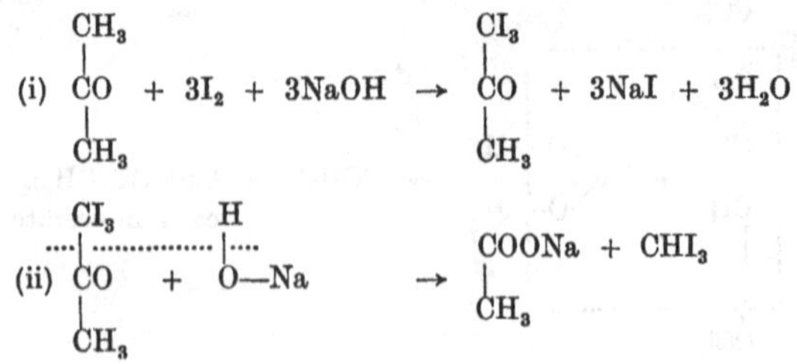

$$\text{(i)}\quad CH_3{-}CO{-}CH_3 + 3I_2 + 3NaOH \rightarrow CI_3{-}CO{-}CH_3 + 3NaI + 3H_2O$$

$$\text{(ii)}\quad CI_3{-}CO{-}CH_3 + H{-}O{-}Na \rightarrow CH_3{-}COONa + CHI_3$$

Ethyl alcohol, acetaldehyde, *iso*propyl alcohol and certain other compounds, but *not* methyl alcohol, also yield iodoform, but in these cases the reaction takes place either on warming or when the mixture is allowed to stand for some time. The reaction takes place much more rapidly in the case of acetone. Moreover, iodoform is produced from acetone when the sodium hydroxide is replaced by an aqueous solution of ammonia in the above experiment. Ethyl alcohol does not yield iodoform under these conditions. This constitutes a method of differentiating between ethyl alcohol and acetone. It appears that the hydrolysis of triiodoacetone, $CI_3.CO.CH_3$, is effected by ammonium hydroxide, whereas triiodoacetaldehyde is not hydrolysed under these conditions.*

Iodoform crystallises from alcohol or aqueous alcohol in yellow hexagonal plates, m.p. 120°, which are insoluble in water and readily soluble in ethyl alcohol. It possesses a characteristic odour and is volatile in steam. It is an important antiseptic, although it has now ceased to be used as a surgical dressing.

Acetone, like acetaldehyde, forms an addition compound when shaken with a concentrated aqueous solution of sodium bisulphite (*v.* below). The bisulphite compound can be filtered off and so separated from other materials which may be originally present and which do not form such a compound. The acetone may then be liberated from the compound by distillation with sodium carbonate. This is the best method of purifying crude acetone:

$$\begin{matrix} CH_3 \\ | \\ CO \\ | \\ CH_3 \end{matrix} + NaHSO_3 \rightarrow \begin{matrix} CH_3 \\ | \\ C \\ | \\ CH_3 \end{matrix}\!\!\begin{matrix} \diagup OH \\ \\ \diagdown O.SO_2Na \end{matrix}$$

$$2(CH_3)_2C\!\begin{matrix} \diagup OH \\ \\ \diagdown O.SO_2Na \end{matrix} + Na_2CO_3 \rightarrow 2(CH_3)_2CO + 2Na_2SO_3 + CO_2 + H_2O$$

* The production of iodoform from ethyl alcohol is illustrated:

(i) $CH_3.CH_2OH + I_2 + 2NaOH \rightarrow CH_3.CHO + 2NaI + 2H_2O$

(ii) $CH_3.CHO + 3I_2 + 3NaOH \rightarrow CI_3.CHO + 3NaI + 3H_2O$
triiodoacetaldehyde

(iii) $$\begin{matrix} CI_3 \\ | \\ C \end{matrix}\!\!\begin{matrix} \diagup H \\ \\ \diagdown\!\!\diagdown O \end{matrix} + \begin{matrix} H \\ | \\ ONa \end{matrix} \rightarrow NaOOC.H + CHI_3$$
sodium formate

Iodoform is also conveniently prepared by the electrolysis of an aqueous solution of potassium iodide and sodium carbonate in the presence of ethyl alcohol, the last named being contained in an inner cell surrounding the positive electrode (anode). Acetone may be substituted for the ethyl alcohol.

Acetone combines directly, as acetaldehyde does, with hydrogen cyanide forming *acetone cyanohydrin*, a colourless liquid, b.p. 82° at 23 mm. This compound is more conveniently prepared by the action of sodium cyanide on acetone sodium bisulphite:

$$CH_3COCH_3 + HCN \rightarrow (CH_3)_2C(OH)CN$$

acetone cyanohydrin

$$(CH_3)_2C(OH)O.SO_2Na + NaCN \rightarrow (CH_3)_2C(OH)CN + Na_2SO_3$$

Like acetaldehyde cyanohydrin, which undergoes hydrolysis to lactic acid, acetone cyanohydrin undergoes hydrolysis when heated with dilute mineral acid (or alkali) to the corresponding acid, *α-hydroxy-iso*butyric acid*:

$$(CH_3)_2C(OH)CN + 2H_2O \rightarrow (CH_3)_2C(OH)COOH + NH_3$$

Acetone condenses readily with hydroxylamine even on mixing aqueous solutions of the two substances and still more easily in the presence of alkali, forming *acetoxime*, a colourless crystalline substance, m.p. 59–60°:

$$(CH_3)_2C{=}O + H_2NOH \rightarrow (CH_3)_2C{=}NOH + H_2O$$

Since the two alkyl groups in acetone are identical, there is no possibility of stereoisomerism in this case. The compound is readily soluble in water and crystallises from ether. When heated with dilute acids, it is hydrolysed acetone being regenerated and a salt of hydroxylamine produced:

$$(CH_3)_2C{=}NOH + H_2O + HCl \rightarrow (CH_3)_2CO + H_2NOH.HCl$$

Similarly, acetone condenses with phenylhydrazine forming *acetone phenylhydrazone*, colourless crystals, m.p. 16°, b.p. 165° at 93 mm.:

$$(CH_3)_2C{=}O + H_2N.N(H)C_6H_5 \rightarrow (CH_3)_2C{=}N.N(H)C_6H_5 + H_2O$$

In this case also there is no possibility of stereoisomerism.

Unlike acetaldehyde and other aliphatic aldehydes, acetone does not undergo polymerisation. On the other hand, it readily yields condensation products similar to crotonaldehyde, the condensation product of acetaldehyde. In the latter case, the intermediate polymerisation product, aldol, is known, but no corresponding substance is known in the case of the formation of condensation products of aliphatic ketones in general and of acetone in particular.

When acetone is saturated with hydrogen chloride, the mixture allowed to stand for two days, then precipitated with water and finally neutralised with sodium hydroxide, an oil is obtained which is a mixture of two condensation products known as *mesityl oxide* (colourless liquid, b.p. 130°, having a peppermint-like odour) and *phorone* (m.p. 28°, b.p. 196°), which can be separated by fractional distillation. These compounds are formed according to the following reactions:

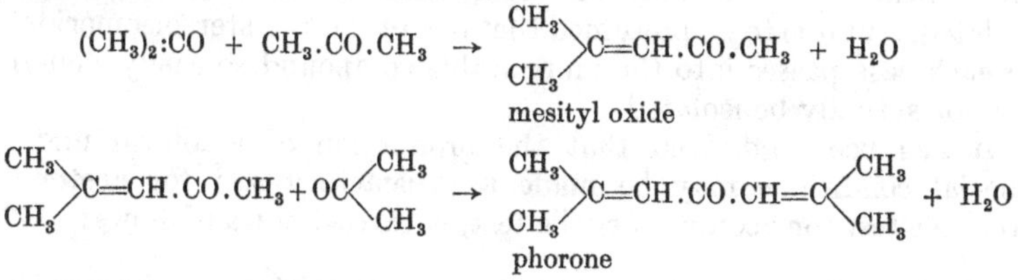

$$(CH_3)_2{:}CO + CH_3.CO.CH_3 \rightarrow (CH_3)_2C{=}CH.CO.CH_3 + H_2O$$

mesityl oxide

$$(CH_3)_2C{=}CH.CO.CH_3 + OC(CH_3)_2 \rightarrow (CH_3)_2C{=}CH.CO.CH{=}C(CH_3)_2 + H_2O$$

phorone

Another condensation product is formed when acetone is distilled with concentrated sulphuric acid. This condensation product is the aromatic hydrocarbon, mesitylene (*sym*-trimethylbenzene, p. 67), and its formation may be represented:

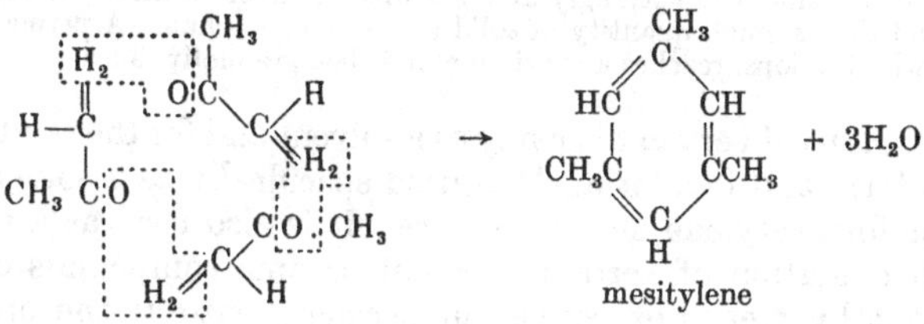

$$3\,CH_3.CO.CH_3 \rightarrow C_6H_3(CH_3)_3 + 3H_2O$$

mesitylene

Or it may be, just as in the case of the polymerisation of acetylene to benzene, that methylacetylene is first formed and then it polymerises as indicated:

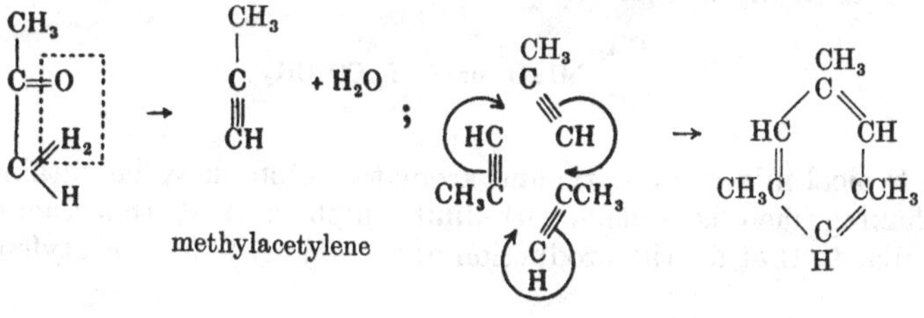

$$CH_3.CO.CH_3 \rightarrow CH_3.C{\equiv}CH + H_2O\ ;\quad 3\,CH_3.C{\equiv}CH \rightarrow C_6H_3(CH_3)_3$$

methylacetylene

Acetone also yields condensation products with aldehydes; one of the most important and typical is that with benzaldehyde with which it condenses in the presence of dilute aqueous sodium hydroxide as indicated:

$$C_6H_5.C\begin{matrix}\diagup H\\ \diagdown O\end{matrix} + \begin{matrix}H\diagdown\\ H_2\diagup\end{matrix}C.CO.CH_3 \rightarrow C_6H_5.\overset{H}{\overset{|}{C}}:\overset{H}{\overset{|}{C}}.CO.CH_3 + H_2O$$

The substance produced, *benzal acetone* or benzylidene acetone, forms colourless crystals, m.p. 42°, b.p. 262°, and is an unsaturated ketone which yields an *oxime*, m.p. 115° and a *phenylhydrazone*, m.p. 156°. Although two stereoisomeric oximes and phenylhydrazones are theoretically possible only one compound in each case has been isolated, and it is to be presumed that one of the two stereoisomerides in each case passes into the more stable compound so easily that it cannot actually be isolated.

It has been indicated that the production of iodoform under special conditions may be made a satisfactory test for acetone. Another test for acetone, Legal's test, is carried out as follows:

A few drops of a freshly prepared dilute aqueous solution of sodium nitroprusside is added to a dilute aqueous solution of acetone and then a few drops of an aqueous solution of sodium hydroxide; a ruby-red colour is produced which fades to a yellow colour on standing.

Rothera's test, which is based on Legal's test, is also frequently employed for the identification of acetone; a few drops of a freshly prepared aqueous solution of sodium nitroprusside made strongly alkaline with ammonia is added to the dilute acetone and then a small quantity of solid ammonium sulphate. A permanganate colour slowly develops, reaches a maximum and then gradually fades.

Acetone is used very extensively as a solvent and for the production of chloroform and iodoform. It is used specifically as a solvent and stabiliser for acetylene under pressure. It is also used as a solvent in which oxidation of certain types of organic compounds can be carried out by means of potassium permanganate at the ordinary temperature.

Acetophenone (methylphenylketone, acetylbenzene)

$$\begin{matrix}CH_3\diagdown\\ C_6H_5\diagup\end{matrix}C:O \quad \text{or} \quad C_6H_5.CO.CH_3$$

This typical mixed aliphatic and aromatic ketone may be obtained by shaking phenylacetylene with dilute sulphuric acid; this reaction is similar to that for the production of acetaldehyde from acetylene:

$$\begin{matrix}CH\\ \|\| \\ C.C_6H_5\end{matrix} + \begin{matrix}H \quad H\\ \diagdown\diagup\\ O\end{matrix} \rightarrow \begin{matrix}CH_3\\ |\\ O:C.C_6H_5\end{matrix}$$

It is prepared* by the Friedel-Craft reaction, using one molecular proportion each of benzene and anhydrous aluminium chloride and rather more than one molecular proportion of acetyl chloride (p. 208). The reagents must be as dry as possible, the reaction taking place being usually represented:

$$C_6H_6 + Cl.CO.CH_3 \rightarrow C_6H_5.CO.CH_3 + HCl$$

This, however, does not take into account the part played by the aluminium chloride in the reaction. It has been found that the reaction is influenced greatly by the amount of aluminium chloride employed, and since the relative amount of this reagent is large it is not a true catalyst. There is definite evidence of the formation of an intermediate compound between the aluminium chloride and the hydrocarbon which is subsequently decomposed by the acid chloride. Homologues of acetophenone (*propiophenone*, $C_6H_5.CO.C_2H_5$, *butyrophenone*, $C_6H_5.CO.C_3H_7$, etc.) can be prepared in an analogous manner.

When pure, acetophenone crystallises in colourless plates, m.p. 20·5°, b.p. 94·5° at 20 mm. and 201–202° at 748 mm. It has an odour somewhat resembling that of benzaldehyde (bitter almonds) and is used to a small extent as a hypnotic under the name of *hypnone*. When it is oxidised with an alkaline solution of potassium perman-

* One of the products of the dry distillation of a mixture of calcium benzoate and calcium acetate is acetophenone:

$$(C_6H_5.CO.O)_2Ca + (CH_3.CO.O)_2Ca \rightarrow 2CaCO_3 + 2C_6H_5.CO.CH_3$$

The other products of this reaction are *benzophenone*, and acetone produced by the action of heat on calcium benzoate and on calcium acetate respectively:

$$(C_6H_5.CO.O)_2Ca \rightarrow CaCO_3 + C_6H_5.CO.C_6H_5;$$

benzophenone (m.p. 49°)

$$(CH_3.CO.O)_2Ca \rightarrow CaCO_3 + CH_3.CO.CH_3$$

ganate, *benzoylformic acid* and benzoic acid are obtained, only the side chain being oxidised:

$$C_6H_5.CO.CH_3 \rightarrow \underset{\text{benzoylformic acid}}{C_6H_5.CO.COOH} \rightarrow C_6H_5.COOH$$

Acetophenone reacts readily with chlorine and bromine, halogenation taking place in the side chain. The first product of the action of chlorine is to produce *ω-chloroacetophenone* (sometimes referred to as phenacyl chloride, the radical, $C_6H_5.CO.CH_2$—, being termed 'phenacyl'), $C_6H_5.CO.CH_2Cl$, a colourless crystalline substance, m.p. 58–59°, possessing powerful lachrymatory properties.

In so far as it is a ketone and possesses the acetyl grouping, acetophenone undergoes reactions analogous to those of acetone. When it is reduced with sodium in the presence of ethyl alcohol the corresponding secondary alcohol, *methylphenylcarbinol*, or 1-phenylethanol, b.p. 202–204°, is produced. This compound has one asymmetric carbon atom in its molecule. At the same time, some of the corresponding pinacol or ditertiary alcohol, *acetophenone pinacol*, *βγ*-dihydroxy-*βγ*-diphenylbutane or *sym*-dimethyldiphenylethylene glycol, m.p. 120°, is also produced. The latter compound has two asymmetric carbon atoms in its molecule, the stereoisomerism, in this case, being analogous to that of the tartaric acids (p. 393):

$$\begin{array}{c} C_6H_5 \\ | \\ CO \\ | \\ CH_3 \end{array} + 2H \rightarrow \begin{array}{c} C_6H_5 \\ | \\ C\!\!\begin{array}{l} H \\ OH \end{array} \\ | \\ CH_3 \end{array} ; \quad CH_3.CO.C_6H_5 + 2H + C_6H_5.CO.CH_3 \rightarrow \begin{array}{c} CH_3 \\ | \\ HO\text{—}C\text{—}C_6H_5 \\ | \\ HO\text{—}C\text{—}C_6H_5 \\ | \\ CH_3 \end{array}$$

Corresponding to the condensation of acetone to mesityl oxide and mesitylene, acetophenone even when boiled for a long time under a reflux condenser undergoes condensation to *dypnone* and 1 : 3 : 5-*triphenylbenzene*, thus:

$$C_6H_5.CO.CH_3 + OC\!\!\begin{array}{l} C_6H_5 \\ CH_3 \end{array} \rightarrow C_6H_5.CO.CH:C\!\!\begin{array}{l} C_6H_5 \\ CH_3 \end{array} + H_2O$$

dypnone, b.p. 225° at 22 mm.

$$\begin{array}{ccc} & C_6H_5 & \\ & | & \\ & [O]C & \\ & & \\ [H_2]HC & & CH[H_2] \\ | & & | \\ C_6H_5[O]C & & C[O]\ C_6H_5 \\ & [H_2]C & \\ & H & \end{array} \rightarrow \begin{array}{ccc} & C_6H_5 & \\ & | & \\ & C & \\ HC & & CH \\ \| & & | \\ C_6H_5.C & & C.C_6H_5 \\ & C & \\ & H & \end{array} + 3H_2O$$

1 : 3 : 5-triphenylbenzene, m.p. 169–170°

Similarly, acetophenone condenses with benzaldehyde in the presence of sodium ethylate, giving *benzylidene acetophenone* (benzal acetophenone), m.p. 58°:

$$C_6H_5.CO.CH_3 + O:C\begin{matrix} C_6H_5 \\ H \end{matrix} \rightarrow C_6H_5.CO.CH:C\begin{matrix} C_6H_5 \\ H \end{matrix} + H_2O$$

It also reacts, similarly to acetone, with phosphorus pentachloride, yielding by replacement of the ketonic oxygen with two chlorine atoms β-phenyl-$\beta\beta$-dichloroethane:

$$C_6H_5.CO.CH_3 + PCl_5 \rightarrow C_6H_5.CCl_2.CH_3 + POCl_3$$

Although it does not combine with sodium bisulphite,* in almost all other respects it possesses ketonic properties similar to those of acetone. For example, it combines directly with hydrogen cyanide forming a cyanohydrin which on hydrolysis is converted into *α-phenyllactic acid* (atrolactinic acid), which like lactic acid contains an asymmetric carbon atom:

$$\begin{matrix} C_6H_5 \\ | \\ C:O \\ | \\ CH_3 \end{matrix} + HCN \rightarrow \begin{matrix} C_6H_5 \\ | \\ C\begin{matrix} OH \\ CN \end{matrix} \\ | \\ CH_3 \end{matrix} ; \quad \begin{matrix} C_6H_5 \\ | \\ C\begin{matrix} OH \\ CN \end{matrix} \\ | \\ CH_3 \end{matrix} + 2H_2O \rightarrow \begin{matrix} C_6H_5 \\ | \\ C\begin{matrix} OH \\ COOH \end{matrix} \\ | \\ CH_3 \end{matrix}$$

Similarly to acetone, acetophenone condenses readily with hydroxylamine and with phenylhydrazine, yielding *acetophenoneoxime* (colourless needles, m.p. 59°) and *acetophenonephenylhydrazone* (m.p. 105°) respectively:

$$\begin{matrix} C_6H_5 \\ | \\ C:N.OH \\ | \\ CH_3 \end{matrix} \qquad \begin{matrix} C_6H_5 \\ | \\ C:N.N\begin{matrix} N \\ C_6H_5 \end{matrix} \\ | \\ CH_3 \end{matrix}$$

These two compounds are theoretically capable of existing in two stereoisomeric forms. It is somewhat remarkable that while the two stereoisomeric oximes of benzaldehyde and of other aromatic aldehydes are well authenticated, stereoisomeric oximes are apparently not obtained in the case of the mixed aromatic-aliphatic ketones such as acetophenone. This is not necessarily an objection to the Hantzsch-Werner hypothesis of the stereoisomerism of the oximes; it indicates that one of the stereoisomers is so unstable that it changes spontaneously into the other stable and isolable oxime. When acetophenoneoxime is warmed with concentrated sulphuric acid, or allowed to stand with acetic acid saturated with hydrogen chloride or heated at 100° with acetyl chloride, it undergoes the Beckmann

* Ketones having an aromatic nucleus attached to the :CO group do not combine with sodium bisulphite.

(1887) rearrangement or transformation and yields acetanilide (N-phenylacetamide, p. 270) isomeric with acetophenoneoxime. Reasons have been advanced (p. 169) indicating that reacting groups in an oxime are probably in the *anti* position to each other; the *anti* (phenyl) configuration is here assigned to acetophenone-oxime. According to our present knowledge, if it possessed the *syn*-configuration, the product of the Beckmann rearrangement would be N-methylbenzamide, $C_6H_5.CONH(CH_3)$, also isomeric with the oxime. The Beckmann rearrangement or isomerisation may be outlined:

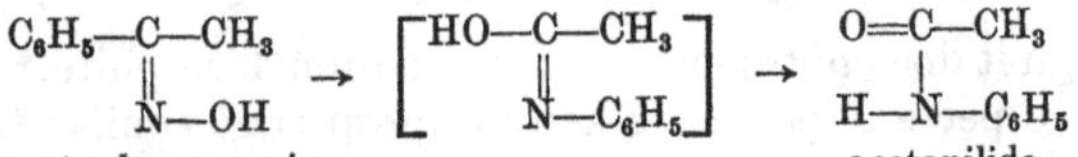

$$\begin{array}{c} C_6H_5\text{—}C\text{—}CH_3 \\ \| \\ N\text{—}OH \end{array} \rightarrow \left[\begin{array}{c} HO\text{—}C\text{—}CH_3 \\ \| \\ N\text{—}C_6H_5 \end{array}\right] \rightarrow \begin{array}{c} O{=}C\text{—}CH_3 \\ | \\ H\text{—}N\text{—}C_6H_5 \end{array}$$

anti-acetophenoneoxime acetanilide

When acetophenone is dissolved in dry ether and treated with one molecular proportion of finely divided sodium, the latter reacts with the acetophenone, liberating an equivalent amount of hydrogen and producing the sodio-derivative of acetophenone. This reaction is explained by acetophenone being capable of reacting as if it had an *enol*-form (compare p. 370) so that the sodium derivative is formed by replacing the hydrogen of the hydroxyl group:

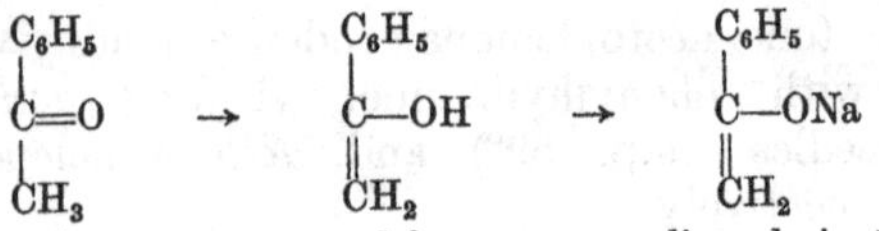

$$\begin{array}{c} C_6H_5 \\ | \\ C{=}O \\ | \\ CH_3 \end{array} \rightarrow \begin{array}{c} C_6H_5 \\ | \\ C\text{—}OH \\ \| \\ CH_2 \end{array} \rightarrow \begin{array}{c} C_6H_5 \\ | \\ C\text{—}ONa \\ \| \\ CH_2 \end{array}$$

keto-form *enol*-form sodium derivative

The *enol*-form of acetophenone has not been isolated.

Being an aromatic compound, acetophenone will undergo nitration and sulphonation under conditions similar to those employed in the case of benzene and aromatic compounds generally. Nitration of acetophenone takes place very readily when the ketone is slowly mixed with fuming nitric acid at 0°:

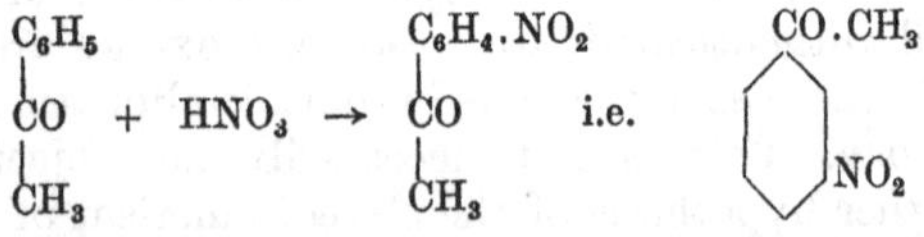

$$\begin{array}{c} C_6H_5 \\ | \\ CO \\ | \\ CH_3 \end{array} + HNO_3 \rightarrow \begin{array}{c} C_6H_4.NO_2 \\ | \\ CO \\ | \\ CH_3 \end{array} \quad \text{i.e.} \quad \text{(benzene ring with } CO.CH_3 \text{ and } NO_2 \text{ in meta position)}$$

The chief product of the reaction is m-*nitroacetophenone*, pale yellow prisms, m.p. 80–81°, and when this has been separated an oil remains which, while still containing some of the *meta*-compound, consists chiefly of the o-*nitroacetophenone*, b.p. 159° at 16 mm.

All the above ketonic derivatives of acetophenone form oximes, phenylhydrazones, etc., similarly to acetophenone itself.

CHAPTER VII

SIMPLE MONOBASIC ACIDS

It has been shown that the direct oxidation of a primary alcohol leads to the formation first of an aldehyde and then of an acid,

$$R.CH_2OH \xrightarrow{-2H} R.C\begin{smallmatrix}H\\ \\ O\end{smallmatrix} \xrightarrow{+O} R.C\begin{smallmatrix}OH\\ \\ O\end{smallmatrix}$$

where '*R*' represents hydrogen or a monovalent hydrocarbon radical. The final direct oxidation product of a primary alcohol may be described as a carboxylic acid since it contains the monovalent $—C\begin{smallmatrix}OH\\ \\ O\end{smallmatrix}$ or 'carboxyl' group.* The acid properties of a carboxylic acid are due to the carboxyl group and the basicity of the acid is, of course, the same as the number of carboxyl groups present in the molecule. As a class, carboxylic acids are weak acids when compared with typical mineral acids.

In the above general formula, if '*R*' = H or a monovalent saturated aliphatic hydrocarbon radical (one derived from a paraffin by loss of a hydrogen atom, e.g. $CH_3—$, $C_2H_5—$, $C_3H_7—$, etc.), the acid is described as a saturated monocarboxylic acid belonging to a homologous series having the general formula $C_nH_{2n+1}.COOH$. Similarly, '*R*' may represent a radical derived from an olefine hydrocarbon, and in that case the acid belongs to a homologous series having the general formula $C_nH_{2n-1}.COOH$; and '*R*' may represent a radical similarly derived from an acetylene or diolefine (C_nH_{2n-2}), when the acid belongs to a homologous series having the general formula $C_nH_{2n-3}.COOH$. '*R*' may also be derived similarly from a cycloparaffin or aromatic hydrocarbon.

The simplest monocarboxylic acid must consequently have the formula $H—C\begin{smallmatrix}OH\\ \\ O\end{smallmatrix}$, which represents the constitution of *formic acid.* Apart from this compound, all the monocarboxylic acids of the types mentioned can be considered to be derived from hydrocarbons by substituting the carboxyl group for a hydrogen atom in the

* Another type of organic acid is the sulphonic acid, containing the monovalent $—SO_2(OH)$ or 'sulphonic' group. Some sulphonic acids (the sulphonation products of benzene and other aromatic compounds) have already been mentioned. As determined by conductivity and other physical methods, sulphonic acids are stronger than carboxylic acids and they approach more nearly in strength the typical mineral acids.

hydrocarbon, just as the alcohols (also phenols) and aldehydes can be considered to be derived from the hydrocarbons by substituting a hydroxyl or aldehyde group respectively for an atom of hydrogen, thus:

$CH_4 \rightarrow CH_3.COOH$ (acetic acid), $CH_2:CH_2 \rightarrow CH_2:CH.COOH$ (acrylic acid)

$CH \vdots CH \rightarrow CH \vdots C.COOH$ (propiolic acid), $C_6H_6 \rightarrow C_6H_5.COOH$ (benzoic acid)

$C_6H_5.CH_3 \rightarrow C_6H_5.CH_2.COOH$ (phenylacetic acid)

$C_6H_5.CH_3 \rightarrow C_6H_5\langle^{CH_3}_{COOH}$ (*o*-, *m*- and *p*- toluic acids)

$C_{10}H_8 \rightarrow C_{10}H_7.COOH$ (α- and β- naphthoic acids)

Acids of the series, $C_nH_{2n+1}.COOH$

This series is generally known as that of the 'Fatty Acids', because certain of the higher members are contained as esters in animal and vegetable fats, from which they were first obtained. It is also known as the 'Acetic Acid Series' since, although formic acid is the first or lowest member, its properties are not typical of those of the series as a whole.

Formic Acid, $H—C\langle^{OH}_{O}$

The constitutional formula—the only possible one for a compound having the molecular formula, CH_2O_2, assigning the normal valencies to the constituent elements—shows the presence of the aldehyde group, $—C\langle^{H}_{O}$. Consequently the reducing properties of an aldehyde are superimposed on the normal acid properties of the compound. In this respect, formic acid differs from all other monocarboxylic acids, which must have a hydrocarbon radical attached to the carboxyl group and therefore cannot contain the aldehyde group. It is a much stronger acid than all the other acids of the same series, being some twelve times stronger than acetic acid, the next higher homologue.

Formic acid occurs in the excretions of certain insects (particularly of ants, Latin *formica*) and animals (perspiration). The often quoted statement that it occurs in the stinging nettle has been denied. Together with formaldehyde and methyl alcohol, it can be obtained by the oxidation of methane by air in the presence of bark shavings at a temperature of 30–50°. The direct formation of formic acid by the oxidation of methane is a justification for describing formic acid as 'methane acid'. Along with formaldehyde, it is produced by passing a mixture of methyl alcohol vapour and air over platinum black at the ordinary temperature, when the two reactions take place:

$$\text{H—}\underset{\text{H}}{\overset{\text{H}}{\text{C}}}\text{—O—H} + \text{O} \rightarrow \text{H—C}\begin{smallmatrix}\nearrow\text{O} \\ \searrow\text{H}\end{smallmatrix} + H_2O; \quad \text{H—C}\begin{smallmatrix}\nearrow\text{O} \\ \searrow\text{H}\end{smallmatrix} + \text{O} \rightarrow \text{H—C}\begin{smallmatrix}\nearrow\text{O} \\ \searrow\text{O—H}\end{smallmatrix}$$

Hydrocyanic acid is the nitrile* of formic acid, and ammonium formate is always present in aqueous solutions of hydrocyanic acid which have been allowed to stand for some time:

$$\text{H—C}\equiv\text{N} + \begin{matrix}\text{H O H} \\ \text{H O H}\end{matrix} \rightarrow \text{H—C}\begin{smallmatrix}\nearrow\text{O} \\ \searrow\text{ONH}_4\end{smallmatrix}$$

The direct combination of carbon monoxide and water to give formic acid has not been effected, but potassium hydroxide and carbon monoxide combine directly at 100°, producing potassium formate:

$$\text{CO} + \text{H—OK} \rightarrow \text{H—C}\begin{smallmatrix}\nearrow\text{O} \\ \searrow\text{OK}\end{smallmatrix}$$

A sealed tube containing a small piece of potassium hydroxide and filled with carbon monoxide is heated for some hours in a boiling water-bath. On opening the tube, little, if any, carbon monoxide is left and potassium formate is found mixed with excess of potassium hydroxide.

The more rapid absorption of carbon monoxide by soda lime at 200° is actually a process of manufacture of sodium formate:

$$\text{CO} + \text{H—ONa} \rightarrow \text{H—C}\begin{smallmatrix}\nearrow\text{O} \\ \searrow\text{ONa}\end{smallmatrix}$$

Since formic acid can easily be obtained from sodium and potassium formate, carbon monoxide may be regarded as the anhydride of formic acid.

When chloroform is boiled with sodium or potassium hydroxide in alcoholic solution, sodium or potassium formate is produced;

* Hydrocyanic acid or hydrogen cyanide is often described as 'formonitrile', i.e. the nitrile yielding formic acid. A 'nitrile' is a compound containing the univalent —C≡N group. Such a compound is also a 'cyanide' and the term 'nitrile' is now falling into disuse. Methyl cyanide (*v.* below) may be termed 'acetonitrile'.

orthoformic acid may be an unstable intermediate product in the reaction:

$$H\text{—}C(Cl)_3 + 3NaOH \rightarrow 3NaCl + H\text{—}C(O\text{—}H)_3;$$

orthoformic acid

$$H\text{—}C(O\text{—}H)_3 + NaOH \rightarrow H\text{—}C\begin{smallmatrix}O\text{—}Na\\ \| O\end{smallmatrix} + 2H_2O$$

Formic acid is prepared from the simplest dicarboxylic acid, oxalic acid, $H_2C_2O_4$, whose constitutional formula must be

$$\begin{array}{l} O{=}C\text{—}O\text{—}H \\ \quad | \\ O{=}C\text{—}O\text{—}H \end{array}$$

When oxalic acid is heated, it decomposes in two ways:

$$(a)\ \begin{array}{c} COO{:}H \\ | \\ COOH \end{array} \rightarrow \begin{array}{c} H \\ | \\ COOH \end{array} + CO_2; \quad (b)\ \begin{array}{c} COO{:}H \\ | \\ CO{:}OH \end{array} \rightarrow CO + CO_2 + H_2O$$

of which (*b*) predominates. When, however, oxalic acid is heated with glycerol (p. 385), a series of reactions takes place which will be discussed later, the ultimate effect being to convert oxalic acid through intermediate compounds into formic acid by loss of carbon dioxide as indicated by reaction (*a*). By this method formic acid of 95–98 per cent. purity is obtained, the impurity being water. Anhydrous formic acid is obtained by distilling a mixture of dry sodium formate and sodium hydrogen sulphate and collecting the distillate:

$$H\text{—}C\begin{smallmatrix}{=}O\\ O{:}Na\end{smallmatrix} + Na{:}H{:}SO_4 \rightarrow H\text{—}C\begin{smallmatrix}{=}O\\ O\text{—}H\end{smallmatrix} + Na_2SO_4$$

It can also be obtained by heating dry lead formate in a current of hydrogen sulphide at 100°, the formic acid which distils being collected in a suitable receiver:

$$\begin{array}{l} H\text{—}C({=}O)\text{—}O \\ H\text{—}C({=}O)\text{—}O \end{array}\!\!>Pb + \begin{array}{l} H \\ H \end{array}\!\!>S \rightarrow PbS + 2H\text{—}C\begin{smallmatrix}{=}O\\ O\text{—}H\end{smallmatrix}$$

Formic acid is a colourless liquid, m.p. 8·6°, b.p. 100·8°. It has a characteristic penetrating odour and blisters the skin. It is soluble in all proportions in water, alcohol, ether and organic solvents generally. When mixed with concentrated sulphuric acid even at the ordinary temperature, formic acid and the formates are decomposed, producing carbon monoxide:

$$H{-}C(=O){-}O{-}H \rightarrow CO + H_2O$$

This reaction is employed as a method of preparation of carbon monoxide and as a convenient qualitative test for formic acid (and formates); it also illustrates the relationship between carbon monoxide and formic acid. When formic acid is heated in closed tubes at 160°, carbon dioxide and hydrogen are produced:

$$H.COOH \rightarrow H_2 + CO_2$$

and the same reaction takes place at ordinary temperatures in the presence of finely divided metals of the platinum group.

Being an aldehyde, formic acid is a reducing agent and is readily oxidised to carbon dioxide and water:

$$H{-}C(O{-}H)(=O) + O \rightarrow \left[(H{-}O)_2C{=}O \right] \rightarrow CO_2 + H_2O$$

The estimation of formic acid in warm aqueous solution by potassium permanganate in the presence of dilute sulphuric acid is based on this reaction:

$$5HCOOH + 2KMnO_4 + 3H_2SO_4 \rightarrow K_2SO_4 + 2MnSO_4 + 5CO_2 + 8H_2O$$

As an aldehyde, formic acid readily reduces solutions of silver salts and mercury salts, the metals being obtained in each case, and the formic acid being oxidised to carbon dioxide and water.

Most of the salts of formic acid are readily soluble in water, but the lead and silver salts are somewhat sparingly soluble. Lead formate crystallises in colourless needles. Silver formate, as one would expect, blackens readily on exposure to light. When the alkali salts are heated at about 250° hydrogen is given off and the alkali oxalates are formed:

$$\begin{matrix} H.COONa \\ H.COONa \end{matrix} \rightarrow H_2 + \begin{matrix} COONa \\ | \\ COONa \end{matrix}$$

The oxalate is identified by dissolving the solid residue in water and adding calcium chloride; a white precipitate, calcium oxalate, insoluble in acetic acid is formed:

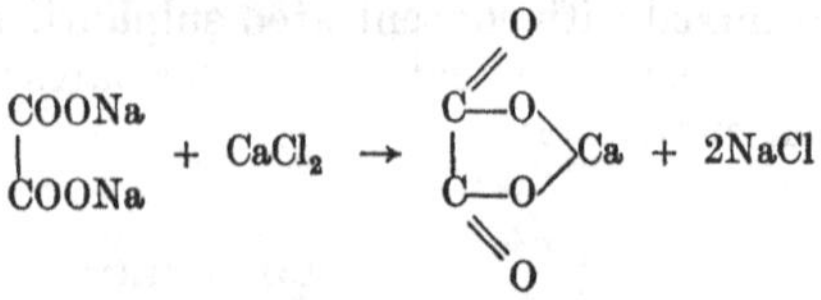

$$\begin{array}{l}COONa\\ |\\ COONa\end{array} + CaCl_2 \rightarrow \begin{array}{l}O{=}C{-}O\\ \quad|\qquad\quad >Ca\\ O{=}C{-}O\end{array} + 2NaCl$$

A convenient method of preparing pure hydrogen is the heating of a mixture of potassium formate and potassium hydroxide:

$$H.\boxed{COOK + KO}H = K_2CO_3 + H_2$$

This reaction is analogous to the action of heat on a mixture of sodium (potassium) hydroxide and the sodium (potassium) salt of acetic acid whereby methane is obtained (p. 203). As may be expected from analogous reactions, formaldehyde is obtained by heating dry calcium formate:

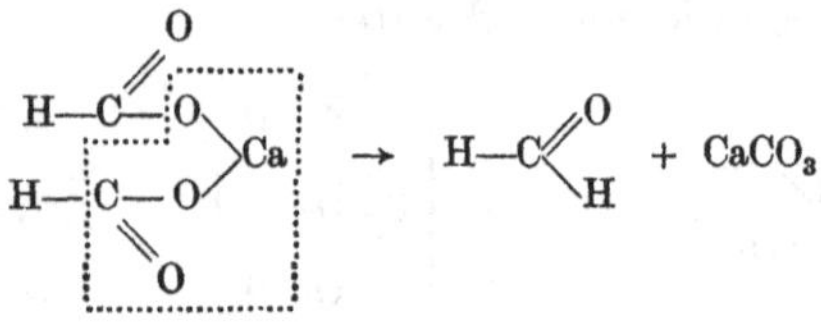

$$\begin{array}{l}H{-}C({=}O){-}O\\ \qquad\qquad\qquad >Ca\\ H{-}C({=}O){-}O\end{array} \rightarrow H{-}C{\begin{array}{l}{=}O\\ {-}H\end{array}} + CaCO_3$$

The following experiment illustrates the action of heat on mercuric formate:

An aqueous solution (50 per cent.) of formic acid is triturated thoroughly in a mortar with mercuric oxide to which some water has been added and the solution filtered. The filtrate contains mercuric formate in solution. If this be heated, carbon dioxide is evolved and a white precipitate (mercurous formate) is produced. On further heating, metallic mercury separates and a further quantity of carbon dioxide is evolved. The reactions taking place are

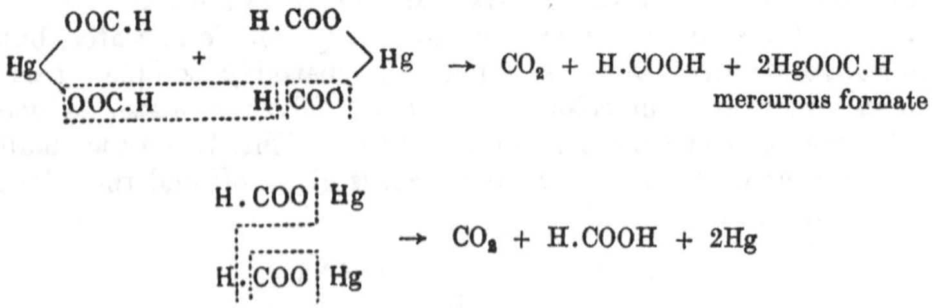

$$Hg\begin{array}{l}\diagup OOC.H\\ \diagdown \boxed{OOC.H}\end{array} + \begin{array}{r}H.COO\diagdown\\ H\boxed{COO}\diagup\end{array}Hg \rightarrow CO_2 + H.COOH + \underset{\text{mercurous formate}}{2HgOOC.H}$$

$$\begin{array}{l}H.COO \mid Hg\\ H.\boxed{COO} \mid Hg\end{array} \rightarrow CO_2 + H.COOH + 2Hg$$

Similarly, the decomposition of silver formate may be represented:

$$\begin{matrix} H.COO \mid Ag \\ H.COO \mid Ag \end{matrix} \rightarrow 2Ag + CO_2 + H.COOH$$

When ammonium formate is heated at 230°, it loses water and a colourless liquid, *formamide** (b.p. 105–106° at 11 mm.) is produced

$$H{-}C(=O){-}O{-}N(H)_4 \rightarrow H{-}C(=O){-}NH_2 + H_2O$$

Some of the esters of formic acid are of importance in certain syntheses of organic compounds and are generally prepared by heating the acid and the alcohol in the presence of hydrochloric acid or sulphuric acid. The general process of *esterification* will be described later (p. 213).

The constitutional formula of formic acid has been given as $H{-}C(=O){-}O{-}H$, which, as already mentioned, is the only possible constitutional formula for a compound possessing the molecular formula, CH_2O_2. This constitutional formula satisfactorily accounts for the following:

(i) of the two atoms of hydrogen in the molecule, only one is ionisable and capable of being replaced by metals or metallic radicals (forming salts) and monovalent hydrocarbon radicals (forming esters);

(ii) the presence of a hydroxyl group capable of being replaced by the amino group (forming formamide);

(iii) the presence of the aldehyde group, accounting for the reducing properties of formic acid and the direct oxidation of formic

* Amides form an important class of organic compounds which can be considered to be derived from the corresponding acids by replacement of hydroxyl by the equivalent $—NH_2$ group. The latter is known as the amino- group. Many amides on further loss of water (usually by heating with phosphorus pentoxide) are converted into the corresponding cyanides (or nitriles), thus:

$$H{-}C(=O){-}NH_2 \rightarrow H{-}C{\equiv}N + H_2O$$

The production of an amide and a cyanide by the dehydration of the ammonium salt of a carboxylic acid is the reversal of the hydrolysis of a cyanide. Compounds containing the $—NH_2$ group are of two classes: (i) those having the $—NH_2$ united to a hydrocarbon radical which are designated amines and (ii) those having the $—NH_2$ united to an acid radical such as 'formyl' ($H—C{=}O$), or 'acetyl' ($CH_3—C{=}O$), known as amides.

acid to carbon dioxide (carbonic acid) containing, like formic acid, one carbon atom in the molecule.

The next homologue is *acetic acid*, $C_2H_4O_2$, which has the constitutional formula,

```
 H       O—H
  \     /
H—C—C
  /     \\
 H       O
```

and which again contains one atom of ionisable hydrogen in the molecule. Obviously, this must be the one attached to oxygen, as this is differently placed from the other three hydrogen atoms which are directly attached to carbon. Acetic acid is isomeric with methyl formate, $H—C\langle^{OCH_3}_{O}$, which is an ester and not an acid and contains no ionisable hydrogen.

Acetic Acid, $CH_3.CO_2H$ (Ethane Acid)

This important compound is actually the earliest described acid,* since it is formed during the 'souring' of alcoholic liquids and the words *vinegar* and *acid* frequently had the same connotation. Wood vinegar, i.e. dilute acetic acid produced in the dry distillation of wood, was known in the middle ages.

When wood and many organic compounds (e.g. sugar, tartaric acid, etc.) are submitted to dry distillation, the very stable acetic acid is produced. Formerly, most of the acetic acid was produced by the dry distillation of wood; the chief volatile products of this process, apart from water, being methyl alcohol (wood spirit), acetone and acetic acid (pyroligneous acid). Methyl alcohol and acetone being more volatile were separated first and the acetic acid was obtained from the highest boiling fraction by neutralising this with lime or calcium hydrate and subsequently distilling the dry calcium acetate with sulphuric acid:

$$(CH_3.COO)_2Ca + H_2SO_4 \rightarrow 2CH_3.COOH + CaSO_4$$

* Proverbs, chap. xxv, verse 20, contains an interesting reference to the chemical reaction between vinegar and nitre (soda, R.V.).

Acetic acid, being unaffected by sulphuric acid, distils over and can be obtained in reasonable purity by this method.

Acetic acid has been synthesised by two methods which have an important bearing on the recognised constitution of the compound. (i) Methyl cyanide on hydrolysis with dilute mineral acids or alkalis is converted into acetic acid or a salt of acetic acid, hence the alternative name of *acetonitrile* for methyl cyanide, the constitution of which has been definitely proved. Starting with methyl alcohol, the synthesis of acetic acid and proof of its constitutional formula is indicated:

$$6CH_3OH + 2P + 3I_2 \rightarrow 6CH_3I + 2H_3PO_3$$

$$CH_3I + KCN \rightarrow CH_3CN + KI$$

$$CH_3.C\equiv N + \begin{matrix} H\ O\ H \\ H\ O\ H \end{matrix} + HCl \rightarrow CH_3.C\begin{matrix} \diagup O{-}H \\ \diagdown\!\!\diagdown O \end{matrix} + NH_4Cl$$

or

$$CH_3.C\equiv N + \begin{matrix} HOH \\ HOH \end{matrix} + KOH \rightarrow CH_3.C\begin{matrix} \diagup O{-}K \\ \diagdown\!\!\diagdown O \end{matrix} + NH_3 + H_2O$$

(ii) When sodium methylate is heated with carbon monoxide under similar conditions for the synthesis of potassium formate from potassium hydroxide and carbon monoxide, sodium acetate is formed:

$$CH_3{-}ONa + CO \rightarrow CH_3.C\begin{matrix} \diagup O{-}Na \\ \diagdown\!\!\diagdown O \end{matrix}$$

One modern development of this reaction is in the claim that methyl alcohol can be quantitatively converted into acetic acid by heating with carbon monoxide under a pressure of from 100 to 200 atmospheres at a temperature of 300–500° in the presence of carbon which has been treated with phosphoric acid; methyl acetate (p. 216) may be formed at the same time:

$$CH_3{-}OH + CO \rightarrow CH_3{-}C\begin{matrix} \diagup\!\!\diagup O \\ \diagdown OH \end{matrix}$$

$$CH_3{-}C\begin{matrix} \diagup\!\!\diagup O \\ \diagdown OH \end{matrix} + HO{-}CH_3 \rightleftharpoons \underset{\text{methyl acetate}}{CH_3{-}C\begin{matrix} \diagup\!\!\diagup O \\ \diagdown OCH_3 \end{matrix}} + H_2O$$

The direct oxidation of ethyl alcohol to acetic acid can be effected by passing the vapour of the alcohol mixed with air over a suitable metallic catalyst (platinum black). The intermediate oxidation product, acetaldehyde, is formed first and some may escape further oxidation:

$$\mathrm{H_3C{-}C(H)(H){-}O{-}H} + \mathrm{O{=}O} \rightarrow \mathrm{H_3C{-}C(=O){-}O{-}H} + \mathrm{H_2O}$$

The direct oxidation of ethyl alcohol to acetic acid is the simplest of all fermentation processes and all vinegar for domestic purposes is manufactured by what is known as the Schützenbach process (1823). The alcoholic liquid (wine of inferior quality) is allowed to trickle through vats provided with perforated shelves on which pine-wood shavings are placed. The oxidation is brought about by the action of enzymes produced during the life process of various bacteria such as *Mycoderma aceti*, *Micrococcus aceti* and *Bacterium aceti*. Such organisms are frequently present in the atmosphere, and pine-wood shavings provide a suitable medium for them. The temperature is kept at 25–30° and the vinegar is collected in suitable vessels at the base of the vats.

Ethyl alcohol can be oxidised directly to acetic acid by refluxing it with an aqueous solution of potassium permanganate in the presence of sulphuric acid and separated by distillation (*v.* below).

Berthelot's synthesis (1870) of acetic acid by allowing air and potassium hydroxide to act upon acetylene in diffused daylight can be regarded as an indication of the modern process for the manufacture of acetic acid:

$$\mathrm{H{-}C{\equiv}C{-}H} + \mathrm{H{-}O{-}H} + \mathrm{O} \rightarrow \mathrm{H_3C{-}C(=O){-}OH}$$

$$CH_3.COOH + HOK \rightarrow CH_3.COOK + H_2O$$

In the modern process, acetylene is converted into acetaldehyde as already indicated (p. 50), and the acetaldehyde is oxidised to acetic acid by passing the vapour with air or oxygen over vanadium pentoxide. Briefly outlined the stages are

$$HC \equiv CH + H_2O \xrightarrow{\text{Hg salts}} CH_3 \cdot CHO$$

$$CH_3 \cdot CHO + O \xrightarrow{V_2O_5} CH_3 \cdot COOH$$

On account of the industrial importance of ethyl alcohol itself, it is not economical to use it for the commercial production of acetic acid.

Pure acetic acid, generally known as a colourless liquid possessing the sharp characteristic 'vinegar' odour, has m.p. 16·7° and b.p. 118·7°. It is an extremely stable substance and is unaffected by such powerful reagents as concentrated sulphuric acid and potassium permanganate. It is usual to purify crude acetic acid by refluxing with potassium permanganate, whereby the impurities are oxidised, and subsequent distillation after addition, if necessary, of a little sulphuric acid. The distillate is cooled until the major part solidifies and the still liquid portion separated. By repeating this process, acetic acid having the above constants is eventually obtained. The recovery of acetic acid from its salts is always carried out by distillation with sulphuric acid.

Glacial acetic acid, that is acetic acid which is solid at a temperature of about 15°, is often regarded as pure acetic acid. This, however, is not necessarily true, since a small quantity of admixed water (and acetic acid is highly hygroscopic) will depress the melting point of 16·7° considerably. It is miscible with water in all proportions, contraction in volume taking place. The maximum density of an aqueous solution of acetic acid is 1·0748; this solution contains 80 per cent. of acetic acid and consists of equimolecular proportions of acetic acid and water. Physical determinations have shown that pure acetic acid at the ordinary temperature is an associated liquid and its composition is more correctly represented as $(CH_3 . CO_2H)_2$. On account of its convenient melting point and its high solvent power, it finds extensive use for the determination of molecular weights of organic compounds by the cryoscopic method; its constant for the molecular lowering of the freezing point is 39 per 100 grams of acetic acid. In using pure acetic acid for physico-chemical determinations, care should be taken to arrange that access of moisture is prevented.

Acetic acid is miscible with all the usual organic solvents and is an important medium for the purification by crystallisation of organic compounds. Its salts are generally highly crystalline and soluble in water. Correlated with its being an associated compound at the ordinary temperature, it is interesting that there are two potassium salts, the normal one having the composition $CH_3.CO_2K$* and the other having the composition $CH_3.CO_2K+CH_3.CO_2H$.

When an aqueous solution of the normal potassium acetate is electrolysed, hydrogen (due to the action of liberated potassium on water) is evolved at the cathode and a mixture of ethane and carbon dioxide is evolved at the anode. The electrolytic action may be represented:

$$\begin{matrix} CH_3 \vdots CO_2 \vdots K \\ CH_3 \vdots CO_2 \vdots K \end{matrix} + 2H_2O \rightarrow \begin{matrix} CH_3 \\ | \\ CH_3 \end{matrix} + 2CO_2 + H_2 + 2KOH$$

It will be realised that this electrolysis has a bearing on the constitution of acetic acid. If the apparatus be so arranged that the gases evolved at the anode are collected over potassium hydroxide in order to absorb the carbon dioxide, a pure specimen of ethane may be prepared.

* By distilling a mixture of potassium acetate and arsenious oxide L. C. Cadet (1760) obtained a distillate having an extremely offensive odour. This is known as Cadet's fuming liquid and, when the reaction is carried out under suitable conditions, constitutes a test for acetic acid or an acetate. Cadet's fuming liquid is spontaneously inflammable and subsequent investigators have shown that the chief reaction taking place may be represented:

$$\begin{matrix} K\,OOC.\,CH_3 \\ K\,OOC.\,CH_3 \\ K\,OOC.\,CH_3 \\ K\,OOC.\,CH_3 \end{matrix} + \begin{matrix} As{=}O \\ \quad >O \\ As{=}O \end{matrix} \rightarrow \begin{matrix} CH_3 \\ CH_3 \end{matrix}\!\!>As{-}O{-}As<\!\!\begin{matrix} CH_3 \\ CH_3 \end{matrix} + 2K_2CO_3 + 2CO_2$$

cacodyl oxide (b.p. 120°)

This, however, does not represent the complete reaction as *cacodyl*,

$$\begin{matrix} CH_3 \\ CH_3 \end{matrix}\!\!>As{-}As<\!\!\begin{matrix} CH_3 \\ CH_3 \end{matrix}$$

b.p. 170°, is also formed. This latter compound is spontaneously inflammable and confers spontaneous inflammability on Cadet's fuming liquid. The name cacodyl is derived from the Greek, κακώδης (stinking). Cacodyl played an important part in the development of the theory of chemistry. It was regarded as a typical organic radical since it gives rise directly to a number of derivatives such as the *oxide* (by direct oxidation), *cacodyl chloride*, $\begin{matrix} CH_3 \\ CH_3 \end{matrix}\!\!>As{-}Cl$, *cacodyl bromide*, *cacodyl iodide* (by the action of halogens) and *cacodyl sulphide*, $\begin{matrix} CH_3 \\ CH_3 \end{matrix}\!\!>As{-}S{-}As<\!\!\begin{matrix} CH_3 \\ CH_3 \end{matrix}$ (by the action of sulphur). These cacodyl compounds form an important series of aliphatic organic arsenicals.

When ammonium acetate is heated and actually distilled, *acetamide* (*v.* below) is obtained by the loss of water:

$$CH_3.C(=O)\text{—}O\text{—}N(H)_4 \rightarrow CH_3.C(=O)NH_2 + H_2O$$

acetamide

A satisfactory method for preparing methane is by heating a mixture of dry sodium acetate and sodium hydroxide (or soda lime) and collecting the gas over water:

$$CH_3.COONa + NaO.H \rightarrow CH_4 + Na_2CO_3$$

The reaction recalls that of the preparation of pure hydrogen from potassium formate and potassium hydroxide.

Calcium acetate is a colourless crystalline salt having a high solubility in water. When the dry salt is heated, acetone distils and calcium carbonate remains:

$$(CH_3.CO.O)_2Ca \rightarrow CaCO_3 + CH_3.CO.CH_3$$

Hence in the old wood distillation process, acetone was not only obtained in the actual distillation but also from the dry calcium acetate produced by the neutralisation of the acetic acid by lime. If a mixture of calcium acetate and calcium formate be heated, acetone, formed as indicated above, and acetaldehyde are produced. The formation of the latter compound is explained:

$$(CH_3.CO.O)_2Ca + Ca(O.CO.H)_2 \rightarrow 2CaCO_3 + 2CH_3.C(=O)H$$

Analogous reactions take place when the corresponding barium salts are employed and they are general throughout this series of acids. If one of the salts in the mixture is a formate, the product is a mixture of an aldehyde and a ketone. If the calcium (barium) salt of a higher fatty acid ($R.CO_2H$) is heated with calcium (barium) acetate, the distillate will contain acetone (formed as indicated above), a mixed ketone ($R.CO.CH_3$) and a higher simple ketone ($R.CO.R$) formed as follows:

$$(R.CO.O)_2Ca + Ca(O.CO.CH_3)_2 \rightarrow 2CaCO_3 + 2R.CO.CH_3$$

$$(R.CO.O)_2Ca \rightarrow CaCO_3 + R.CO.R$$

Ferrous acetate readily undergoes oxidation in aqueous solution and the insoluble basic ferric acetate is precipitated. Ferric and aluminium acetates are not crystalline. These salts are important technically, being employed as mordants in the dyeing or printing of cotton fabrics. The salts unite with or are absorbed into the fibre, and the basic acetates produced on heating retain the dye, probably by chemical reaction, the colour of the dyed fabric depending not only on the colour of the dye itself but also on the nature of the mordant.

Normal lead acetate, $Pb(O_2C.CH_3)_2+3H_2O$ (sugar of lead), is readily obtained by dissolving litharge in 60 per cent. acetic acid and crystallises in colourless glistening prisms which effloresce on exposure to air. By shaking litharge with an aqueous solution of lead acetate two basic acetates of lead can be obtained. These have been shown to have the composition of $Pb(OH)(O_2C.CH_3)$ and $2Pb(OH)_2+Pb(O_2C.CH_3)_2$ respectively, and the particular basic acetate produced depends on the relative amounts of the constituents used. The aqueous solutions of these basic acetates are alkaline in reaction and from them carbon dioxide precipitates basic lead carbonates, the mixture of which constitutes 'white lead'. When red lead (Pb_3O_4) is dissolved in hot glacial acetic, the filtrate, on cooling, deposits colourless prisms of *lead tetraacetate*, $Pb(O_2C.CH_3)_4$, m.p. 175°. This interesting compound, which is easily soluble in hot acetic acid and in chloroform, has been used as an oxidising agent in certain reactions.

Normal cupric acetate, $Cu(O_2C.CH_3)_2+H_2O$, is readily soluble in water and has the characteristic colour of cupric salts. The corresponding basic salt is known as *verdigris*, while the so-called *Schweinfurt Green* is a double cupric arsenite and acetate,

$$3Cu(AsO_2).Cu(O_2C.CH_3)_2$$

Silver acetate, $AgO_2C.CH_3$, is somewhat sparingly soluble in water and crystallises in colourless glistening plates. Like most silver salts, it darkens on exposure to light.

The action of chlorine on acetic acid is similar to that on methane and the paraffin hydrocarbons generally. Direct chlorination in sunlight or in the presence of certain catalysts (iodine, sulphur, phosphorus, etc.) leads to the production of the three *chloroacetic acids*, hydrogen chloride being evolved at the same time:

$$CH_3.CO_2H + Cl_2 \rightarrow \underset{\text{monochloroacetic acid}}{CH_2Cl.CO_2H} + HCl$$

$$CH_2Cl.CO_2H + Cl_2 \rightarrow \underset{\text{dichloroacetic acid}}{CHCl_2.CO_2H} + HCl$$

$$CHCl_2.CO_2H + Cl_2 \rightarrow \underset{\text{trichloroacetic acid}}{CCl_3.CO_2H} + HCl$$

These three compounds are prepared individually by special methods, but the reactions indicated above investigated by Dumas (1839) showed that compounds of a similar type may be produced even when an electro-positive element like hydrogen is replaced by an electro-negative element like chlorine. This led to the formulation of the 'type theory' of organic chemistry and the limitation of Berzelius' (1779–1848) 'dualistic theory' of chemical reaction and the formation of chemical compounds.

Monochloroacetic acid occurs in colourless dimorphic crystals, m.p. 62°, b.p. 189°. It is a convenient starting material for the preparation of *hydroxyacetic acid* or *glycollic acid* (p. 353) and *amino-acetic acid* or *glycine* (p. 318). The former substance is produced when monochloroacetic acid is heated with water or with dilute alkalis:

$$CH_2Cl.CO_2H + H_2O \rightarrow \underset{\substack{\text{glycollic acid} \\ \text{(primary alcohol and monobasic acid)}}}{CH_2OH.CO_2H} + HCl$$

$$CH_2OH.CO_2H + KOH \rightarrow CH_2OH.CO_2K + H_2O$$

Aminoacetic acid is produced by the action of ammonia on mono-chloroacetic acid in aqueous solution; actually, ammonium chloride and the ammonium salt of the 'amino-acid' are formed:

$$ClCH_2.CO_2H + NH_3 \rightarrow \underset{\substack{\text{glycine} \\ \text{(a primary amine or base and monobasic acid} \\ \text{—an amino acid—a neutral substance)}}}{NH_2CH_2.CO_2H} + HCl$$

$$NH_2CH_2.COOH + HCl + 2NH_3 \rightarrow NH_2CH_2.COONH_4 + NH_4Cl$$

Dichloroacetic acid is a colourless liquid, b.p. 194°. When its silver salt is warmed with water *glyoxylic acid* (p. 355) is produced:

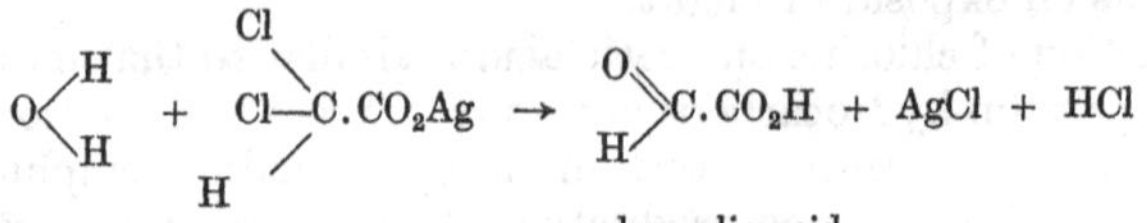

$$H_2O + CHCl_2.CO_2Ag \rightarrow OHC.CO_2H + AgCl + HCl$$

glyoxylic acid
(an aldehyde and monobasic acid, compare formic acid)

Trichloroacetic acid, which has a highly corrosive action on the skin, forms colourless crystals, m.p. 57°, b.p. 197°. It is also prepared by Kolbe's (1845) method, the oxidation of chloral with concentrated nitric acid:

$$CCl_3.CHO + O = CCl_3.CO_2H$$

When boiled with water for some time trichloroacetic acid is hydrolysed, chloroform and carbon dioxide being produced:

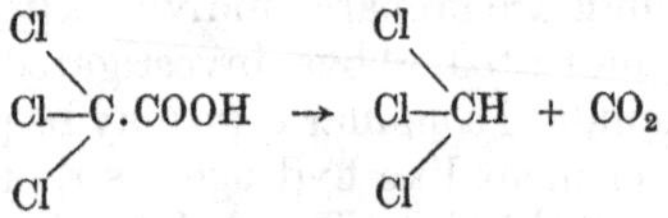

$$CCl_3.COOH \rightarrow CHCl_3 + CO_2$$

If aqueous solutions of alkalis are used for the hydrolysis, the chloroform is hydrolysed, a formate resulting, and alkali chloride and alkali carbonate are produced at the same time.

These three chloroacetic acids are monobasic and all stronger than acetic acid, the strength increasing regularly with increasing chlorine content. When reduced with nascent hydrogen, they are all reconverted into acetic acid. The corresponding bromo- and iodo- acetic acids are also known and all these acids yield derivatives analogous to those of acetic acid.

The typical organic acids all contain the hydroxyl (—OH) group and, like inorganic acids containing this group, yield *acid chlorides* and *anhydrides*. In the case of the inorganic acids, the acid chlorides may be formed directly or indirectly. For example, sulphuryl chloride is the acid chloride of sulphuric acid and is formed by the following reactions:

(a) (i) $SO_2(OH)_2 + PCl_5 = SO_2(OH)Cl + POCl_3 + HCl$
chlorosulphonic acid

(ii) $SO_2(OH)Cl + PCl_5 = SO_2Cl_2 + POCl_3 + HCl$
sulphuryl chloride

(b) $SO_2 + Cl_2 = SO_2Cl_2$

Similarly, phosphorus oxychloride, $POCl_3$, is the acid chloride of orthophosphoric acid; it and sulphuryl chloride yield the corresponding acids on treatment with water:

$$POCl_3 + 3H_2O = H_3PO_4 + 3HCl$$
$$SO_2Cl_2 + 2H_2O = H_2SO_4 + 2HCl$$

The anhydrides of hydroxy-inorganic acids are generally non-metallic oxides which regenerate the corresponding acids by the action of water, thus:

$$N_2O_5 + H_2O = 2HNO_3$$
$$SO_3 + H_2O = H_2SO_4$$
$$P_2O_5 + 3H_2O = 2H_3PO_4$$

Such acid anhydrides, however they may be conveniently prepared, may be regarded as derived from the acid by loss of water. The anhydride of a typical monobasic hydroxy acid is theoretically derived from two molecules of the acid by loss of a molecule of water, that of a dibasic acid from one molecule of the acid by loss of a molecule of water and that of a tribasic acid from two molecules of the acid by loss of three molecules of water.

The importance of the chlorides and anhydrides of certain organic acids is due not so much to the compounds themselves as to the derivatives to which they give rise and which serve for characterisation of certain types of organic compounds. The acid chlorides most frequently used are those derived from acetic, benzoic and benzene sulphonic acids and the anhydrides derived from acetic and benzoic acids are the commonest of this class. These compounds are

$CH_3.C{\scriptstyle\diagup\!\!\!\!\!\!=}O$, $\diagdown OH$ ($CH_3.C(=O)OH$),	$CH_3.C(=O)Cl$,	$CH_3—C(=O)—O—C(=O)—CH_3$
	acetyl* chloride	acetic anhydride
$C_6H_5.C(=O)OH$,	$C_6H_5.C(=O)Cl$,	$C_6H_5.C(=O)—O—C(=O).C_6H_5$
	benzoyl chloride	benzoic anhydride
$C_6H_5.S(=O)(=O)OH$,	$C_6H_5.S(=O)(=O)Cl$	
	phenylsulphonyl chloride (benzene sulphonyl chloride)	

* 'Acetyl' is the name applied to the monovalent radical, $CH_3.CO—$. Acetic anhydride can be designated 'acetyl oxide'. 'Acetyl' and other radicals derived from monobasic acids by the loss of a hydroxyl group are usually referred to as 'acid' or 'acyl' radicals.

Whilst the above acid chlorides and anhydrides are the most common and frequently used for the characterisation of certain types of organic compounds, others are also employed. The characterisation of a particular compound often depends on the production of a crystalline derivative, and if the compound to be characterised is of a type which will react with an acid chloride or anhydride, that one is chosen which will give a crystalline derivative the most easily.

The chlorides of organic acids may generally be prepared by the action of phosphorus pentachloride or trichloride on the acid:

$$R.C\begin{matrix}\nearrow O \\ \searrow OH\end{matrix} + PCl_5 \rightarrow R.C\begin{matrix}\nearrow O \\ \searrow Cl\end{matrix} + POCl_3 + HCl$$

$$3R.C\begin{matrix}\nearrow O \\ \searrow OH\end{matrix} + PCl_3 \rightarrow 3R.C\begin{matrix}\nearrow O \\ \searrow Cl\end{matrix} + H_3PO_3$$

Frequently they are prepared by the action of thionyl chloride on the acid or on its sodium salt:

$$R—C\begin{matrix}\nearrow O \\ \searrow O—H(Na)\end{matrix} + SOCl_2 \rightarrow R—C\begin{matrix}\nearrow O \\ \searrow Cl\end{matrix} + SO_2 + H(Na)Cl$$

This last reaction is particularly convenient when the acid is used, as the sulphur dioxide and hydrogen chloride being gaseous are easily removed from the acid chloride.

Acetyl chloride is usually prepared by the action of phosphorus trichloride on acetic acid, the reaction being:

$$3CH_3.C\begin{matrix}\nearrow O \\ \searrow O—H\end{matrix} + PCl_3 \rightarrow 3CH_3.C\begin{matrix}\nearrow O \\ \searrow Cl\end{matrix} + H_3PO_3$$

The volatile product of this reaction, acetyl chloride, is a colourless liquid, b.p. 55°, which fumes in moist air, owing to the action of water, producing acetic acid and hydrochloric acid:

$$CH_3.C\begin{matrix}\nearrow O \\ \searrow Cl\end{matrix} + H_2O \rightarrow CH_3.C\begin{matrix}\nearrow O \\ \searrow OH\end{matrix} + HCl$$

The action of liquid water is very energetic and that of aqueous solutions of alkali hydroxides proceeds almost explosively; if, for example, sodium hydroxide solution be employed, the products are sodium acetate and sodium chloride.

Acetyl chloride is an important reagent for preparing 'acetyl' derivatives of alcohols or phenols. The reaction can be regarded as analogous to the reaction with water:

$$CH_3.C\begin{matrix}\nearrow O \\ \searrow Cl\end{matrix} + HOR \rightarrow CH_3.C\begin{matrix}\nearrow O \\ \searrow OR\end{matrix} + HCl$$

and the compounds so obtained are esters of acetic acid. For example, *ethyl acetate*, $CH_3.C\langle^{O}_{OC_2H_5}$ (p. 214), and *phenyl acetate*, $CH_3.C\langle^{O}_{OC_6H_5}$ (colourless liquid, b.p. 195°), result from the reaction with ethyl alcohol and phenol respectively. The production of these 'acetyl' derivatives or esters of acetic acid affords a convenient method for the detection of the presence of the hydroxyl group in compounds suspected to be alcohols or phenols and, as will be seen later, for the estimation of the number of hydroxyl groups in the molecules of such compounds. Moreover, the physical constants (especially the melting point and/or the boiling point) of the 'acetyl' derivative of an alcohol or phenol frequently serve for the identification of the latter compound.

Acetyl chloride reacts vigorously with dry sodium acetate, producing *acetic anhydride* ('acetyl oxide'):

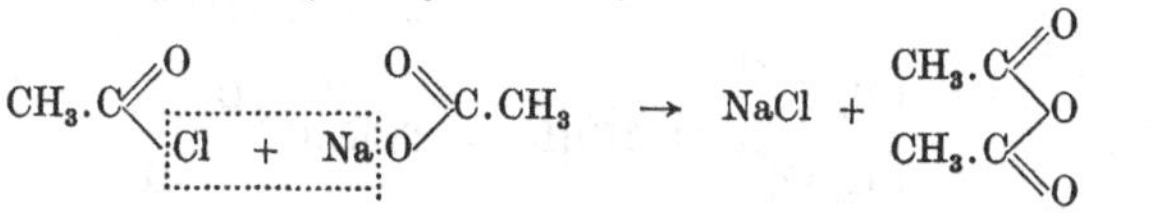

$$CH_3.C\langle^{O}_{Cl} + NaO\rangle C.CH_3 \rightarrow NaCl + \begin{matrix} CH_3.C(=O) \\ \rangle O \\ CH_3.C(=O) \end{matrix}$$

acetic anhydride

This reaction is the basis of the usual method of preparation of acetic anhydride, by collecting the volatile product obtained on heating together four molecular proportions of sodium acetate and one of phosphorus oxychloride, the reaction being expressed:

$$4CH_3.CO.ONa + POCl_3 = 2(CH_3.CO)_2O + NaPO_3 + 3NaCl$$

This reaction may be considered as taking place in two stages:

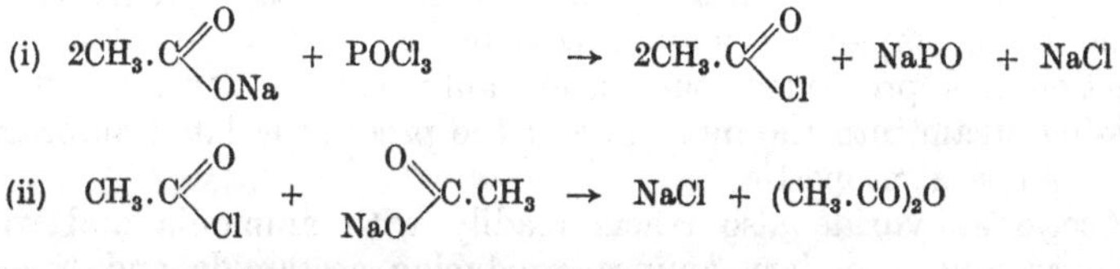

$$\text{(i)}\quad 2CH_3.C\langle^{O}_{ONa} + POCl_3 \rightarrow 2CH_3.C\langle^{O}_{Cl} + NaPO + NaCl$$

$$\text{(ii)}\quad CH_3.C\langle^{O}_{Cl} + NaO\rangle C(=O).CH_3 \rightarrow NaCl + (CH_3.CO)_2O$$

The reaction between acetyl chloride and ammonia is also very energetic and analogous to the reaction with water; the product of the reaction is *acetamide* (p. 211):

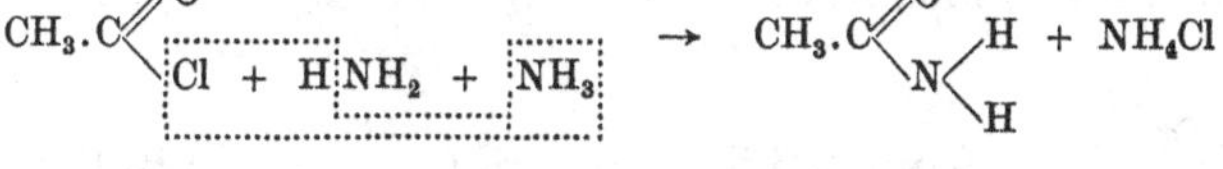

$$CH_3.C\langle^{O}_{Cl} + H\,NH_2 + NH_3 \rightarrow CH_3.C\langle^{O}_{N\langle^{H}_{H}} + NH_4Cl$$

Acetamide is the 'acetyl' derivative of ammonia, and analogous 'acetyl' derivatives (N-substituted acetamides) are formed by the action of acetyl chloride on certain organic derivatives of ammonia, known as primary and secondary amines ($R.NH_2$ and $\begin{matrix} R \\ R_1 \end{matrix}\!\!>\!N{-}H$, R and R_1 representing hydrocarbon or substituted hydrocarbon radicals which may or may not be identical with each other, p. 245). These 'acetyl' derivatives or N-substituted acetamides often provide important evidence in the identification of the substituted derivatives of ammonia of the above types (compare the identification of alcohols and phenols by means of their 'acetyl' derivatives or esters of acetic acid).

Acetic anhydride is a colourless liquid, b.p. 140°, having a characteristic odour and slight lachrymatory power. It has a small solubility in water, with which it reacts slowly at the ordinary temperature and more readily at elevated temperatures, producing acetic acid:

$$\begin{matrix} CH_3.C\!\!\lessgtr^{O} \\ \quad\quad >O \\ CH_3.C\!\!\lessgtr_{O} \end{matrix} + HOH \rightarrow 2CH_3.C\!\!\lessgtr^{O}_{O-H}$$

The reaction with aqueous solutions of the alkalies is more energetic, alkali salts of acetic acid being then the product.

Acetic anhydride reacts with alcohols and phenols, producing acetyl derivatives of the latter (or esters of acetic acid):

$$(CH_3.CO)_2O + HOR \rightarrow CH_3.COOH + CH_3.C\!\!\lessgtr^{O}_{OR}$$

It is generally only necessary to heat the reactants together for some time, employing some excess of the anhydride. After the reaction has proceeded, the excess anhydride is decomposed by passing steam into the mixture and the product isolated according to its physical properties.

Acetic anhydride also reacts readily with ammonia and with primary and secondary amines, producing acetamide and N-substituted acetamides respectively. The reaction may take place on merely mixing the reactants or on heating them together (a drop of concentrated sulphuric acid frequently assists the reaction). The reaction with ammonia is

$$\begin{matrix} CH_3.C\!\!\lessgtr^{O} \\ \quad\quad >O \\ CH_3.C\!\!\lessgtr_{O} \end{matrix} + H\,|\,NH_2 + NH_3 \rightarrow CH_3.C\!\!\lessgtr^{O}_{NH_2} + CH_3.C\!\!\lessgtr^{O}_{ONH_4}$$

For the preparation of these 'acetyl' derivatives, acetic anhydride can thus be used alternatively with acetyl chloride. It is frequently preferred to the latter, because it is not so rapidly affected by moisture and is therefore more easily manipulated.

Acetamide, the monoacetyl derivative of ammonia, obtained by the action of ammonia on acetyl chloride and on acetic anhydride, is also the product of the partial hydrolysis of methyl cyanide (acetonitrile, p. 199) in the presence of mineral acids:

$$CH_3.C{\equiv}N + H_2O \rightarrow CH_3.C\begin{matrix}\nearrow O \\ \searrow NH_2\end{matrix}$$

It is also obtained, but not so conveniently, by the action of ammonia on ethyl acetate (p. 214). It is conveniently prepared by the slow distillation of ammonium acetate:

$$CH_3.C\begin{matrix}\nearrow O \\ \searrow O{-}N{\begin{matrix}H \\ H \\ H \\ H\end{matrix}}\end{matrix} \rightarrow CH_3.C\begin{matrix}\nearrow O \\ \searrow NH_2\end{matrix} + H_2O$$

It will be realised that acetamide is the intermediate product in the complete dehydration of ammonium acetate to methyl cyanide.

Acetamide crystallises in long colourless needles, m.p. 82°, b.p. 222°. It is readily soluble in water and in alcohol and usually possesses a characteristic 'mouse-like' odour which is hardly perceptible in a pure specimen. When heated with phosphorus pentoxide, acetamide is converted by loss of a molecule of water into methyl cyanide; this is a reversal of the partial hydrolysis of methyl cyanide to acetamide:

$$CH_3.C\begin{matrix}\nearrow O \\ \searrow NH_2\end{matrix} \rightarrow CH_3.C{\equiv}N + H_2O$$

On being heated with water, acetamide is reconverted into ammonium acetate:

$$CH_3.C\begin{matrix}\nearrow O \\ \searrow NH_2\end{matrix} + H_2O \rightarrow CH_3.C\begin{matrix}\nearrow O \\ \searrow ONH_4\end{matrix}$$

This hydrolysis of acetamide takes place much more readily in the presence of alkalies and mineral acids. In the former case, an acetate and ammonia are produced, and, in the latter, acetic acid and an ammonium salt are obtained:

$$CH_3.C\begin{matrix}\nearrow O \\ \searrow NH_2\end{matrix} + NaOH \rightarrow CH_3.C\begin{matrix}\nearrow O \\ \searrow ONa\end{matrix} + NH_3$$

$$2CH_3.C\begin{matrix}\nearrow O \\ \searrow NH_2\end{matrix} + H_2SO_4 + 2H_2O \rightarrow 2CH_3.C\begin{matrix}\nearrow O \\ \searrow OH\end{matrix} + (NH_4)_2SO_4$$

As the acetyl derivative of ammonia, acetamide belongs like formamide to a class of substances known as 'acid amides'. These compounds may be regarded as containing a group of acid character (in this case, the acetyl group) and a basic group (in this case, the amino or $—NH_2$ group). As far as the compound itself is concerned, these two groups neutralise each other, but the compound can, however, act as a base or an acid according to conditions and possesses thus an amphoteric character. For example, it forms readily hydrolysable salts with mineral acids such as nitric and hydrochloric acids, and the compounds, $CH_3.CONH_2+HNO_3$ and $(CH_3.CONH_2)_2+HCl$, have been isolated. The possibility of the acid character becoming dominant is shown by the isolation of such a compound as mercuric acetamide, $(CH_3.CONH)_2Hg$, in which a hydrogen atom from each of two molecules of acetamide has been replaced by divalent mercury.

Like all compounds containing the amino group, acetamide reacts readily with nitrous acid (produced *in situ* from sodium nitrite and hydrochloric acid), nitrogen being evolved and the amino group being replaced by the equivalent hydroxyl group so that acetic acid results from this reaction:

$$CH_3.C\begin{smallmatrix}\nearrow O\\ \searrow NH_2\end{smallmatrix} + H—O—N{=}O \rightarrow N_2 + H_2O + CH_3.C\begin{smallmatrix}\nearrow O\\ \searrow O—H\end{smallmatrix}$$

Theoretically, by the loss of carbon monoxide acetamide is converted into the primary amine, monomethylamine:

$$CH_3—C\begin{smallmatrix}\nearrow O\\ \searrow NH_2\end{smallmatrix} \rightarrow CO + CH_3—N\begin{smallmatrix}\nearrow H\\ \searrow H\end{smallmatrix}$$

monomethylamine

The method by which the primary amine is obtained from the corresponding acid amide is one of the important reactions discovered by A. W. von Hofmann (1875). This reaction consists in the formation of N-bromoacetamide by means of bromine in the presence of potassium hydroxide:

$$CH_3.C\begin{smallmatrix}\nearrow O\\ \searrow N\begin{smallmatrix}\nearrow H\\ \searrow H\end{smallmatrix}\end{smallmatrix} + Br_2 + KOH \rightarrow CH_3.C\begin{smallmatrix}\nearrow O\\ \searrow N\begin{smallmatrix}\nearrow H\\ \searrow Br\end{smallmatrix}\end{smallmatrix} + KBr + H_2O$$

and the subsequent hydrolysis of this bromoacetamide by a further quantity of potassium hydroxide. This part of the reaction probably goes in two stages, *methyl isocyanate* (p. 251) being an intermediate product:

$$\text{(i)}\quad CH_3.C\begin{matrix}\diagup O \\ \diagdown N\begin{matrix}\diagup H \\ \diagdown Br\end{matrix}\end{matrix} + KOH \rightarrow \underset{\text{methyl socyanate}}{CH_3{-}N{=}C{=}O} + KBr + H_2O$$

$$\text{(ii)}\quad CH_3{-}N{=}C{=}O + \begin{matrix} KO\,H \\ KO\,H \end{matrix} \rightarrow K_2CO_3 + CH_3{-}N\begin{matrix}\diagup H \\ \diagdown H\end{matrix}$$

The practical details for carrying out the reaction will be given later. The same reaction occurs when acetamide is distilled with bleaching powder made into a paste with water. The Hofmann reaction is applicable to a number of amines of the same type as acetamide and is a satisfactory method for the preparation of primary aliphatic amines. From a theoretical standpoint it is a method for the production of a compound containing one carbon atom less in the molecule than the compound used in the reaction.

It has been shown that the acetyl derivatives of alcohols and phenols are esters of acetic acid which can be conveniently prepared by the action of acetic anhydride (or acetyl chloride) on these hydroxy compounds. These esters may be given the general formula, $CH_3.C\begin{matrix}\diagup O \\ \diagdown O{-}R\end{matrix}$, where R represents the monovalent hydrocarbon (or substituted hydrocarbon) radical originally united to the hydroxy group in the alcohol or phenol. That is, these esters are derivatives of acetic acid in which the ionisable hydrogen of the carboxy group has been replaced by the radical, R, producing compounds which are not electrolytes. It has also been indicated that esters may be formed by the direct interaction of the alcohol and the acid, and this reaction is general and of wide applicability. The direct formation of esters from alcohols and acids has been formulated thus:

$$\begin{array}{ccccccc} \text{acid} & + & \text{alcohol} & \rightleftarrows & \text{ester} & + & \text{water} \\ XH & + & HOR & \rightleftarrows & XR & + & H_2O \end{array}$$

as compared with the normal action of a monobasic acid and a monoacidic hydroxide:

$$\begin{array}{ccccccc} \text{acid} & + & \text{alkali} & = & \text{salt} & + & \text{water} \\ XH & + & HOM & = & XM & + & H_2O \end{array}$$

The most common and typical ester is *ethyl acetate*, which can be obtained, as indicated above, by the action of acetic anhydride or acetyl chloride on ethyl alcohol, but is always prepared directly from ethyl alcohol and acetic acid. When a mixture of these two substances is allowed to stand ethyl acetate is produced, the reaction taking place being represented:

$$CH_3.CO.OH + HOC_2H_5 \rightleftarrows CH_3.CO.OC_2H_5 + H_2O$$

This implies (i) that the reaction is incomplete and as soon as the acid and alcohol are mixed the four substances represented above are present, (ii) that while the acid and alcohol interact to give ester and water, the latter two substances also react to give acid and alcohol, and (iii) that when the two reactions proceed with equal velocities, an equilibrium is established between the four substances. At the ordinary temperature, equilibrium is only established after a very considerable time, but its attainment is hastened by raising the temperature. In the interaction of equimolecular quantities of ethyl alcohol and acetic acid, equilibrium is reached when two-thirds of the acid and of the alcohol have been esterified. The course of such a reaction follows the Law of Mass Action (Guldberg and Waage (1867), van't Hoff (1877)).

Starting with equimolecular quantities of alcohol and acid, the velocity of ester formation (v) at equilibrium is given by:

$$v = \frac{k(1-\frac{2}{3})(1-\frac{2}{3})}{V^2},$$

where V = volume of the reacting masses (= the volume of the solution) and k = the velocity constant. Similarly, the velocity of the hydrolysis of the ester (v') at equilibrium is given by

$$v' = \frac{k'(\frac{2}{3})(\frac{2}{3})}{V^2},$$

where k' is the velocity constant of this reaction. At equilibrium, $v = v'$, and

$$\frac{k(1-\frac{2}{3})^2}{V^2} = \frac{k'(\frac{2}{3})^2}{V^2};$$

$$\therefore \quad \frac{k'}{k} = \frac{1}{4} = K,$$

where K is the equilibrium constant, the value of which is the same whatever the proportions of the reacting substances and is the ratio of the velocity constants of the two opposing reactions. It can be used to determine the extent of the ester formation, using any proportions of ethyl alcohol and acetic acid. If, for example, four molecular proportions of ethyl alcohol and one molecular proportion of acetic acid be employed, then if x = the molecular proportion of acetic acid converted into ethyl acetate when equilibrium is established,

$$\frac{(1-x)(4-x)}{x^2} = \frac{1}{4}.$$

Since x cannot be greater than 1,

$$x = 0{\cdot}94 \text{ (approx.)},$$

i.e. approximately 94 per cent. of the acetic acid will be converted into ethyl acetate when one molecular proportion of acetic acid is allowed to react with four molecular proportions of ethyl alcohol.

From the above general expression of ester formation, it is clear that esterification is assisted by removal of one of the products (ester or water) from the reaction mixture, thus preventing the reverse (right to left) reaction or hydrolysis (saponification) of the ester. This can be accomplished, if the ester be sufficiently volatile, by distillation during esterification or the use of suitable dehydrating agents for the removal of the water. The velocity of the reaction, i.e. the speed in reaching equilibrium, can also be greatly accelerated by the presence of mineral acids (hydrochloric and sulphuric acids are usually employed) and other strong acids which act catalytically both in this and other reactions. Sulphuric acid in sufficient quantity, provided that its use in the particular reaction is not contra-indicated for chemical reasons, will also act as a dehydrating agent. In the usual method of preparation of ethyl acetate, not only is sulphuric acid used as a catalyst and possibly as a dehydrating agent, but advantage is taken of the volatility of ethyl acetate, whereby the latter is continuously removed from the reaction mixture resulting in an increase in the relative amount of ester produced.

Ethyl acetate may also be obtained by heating together in a suitable medium (diethyl ether) silver acetate and ethyl iodide. This reaction goes to completion on account of the insolubility of silver iodide:

$$CH_3.CO_2Ag + IC_2H_5 = AgI + CH_3.CO.OC_2H_5$$

An interesting method for the production of ethyl acetate is the polymerisation which acetaldehyde is described as undergoing in the presence of aluminium ethylate, $Al(OC_2H_5)_3$ (p. 158):

$$CH_3.C\begin{matrix}\diagup H \\ \diagdown\!\!\diagdown O\end{matrix} + \begin{matrix}H\diagdown \\ O\diagup\!\!\diagup\end{matrix}C.CH_3 \rightarrow CH_3.C\begin{matrix}\diagup\!\!\diagup O \\ \diagdown OC_2H_5\end{matrix}$$

The general methods for the preparation of carboxylic esters are well illustrated from what has been indicated in the case of ethyl acetate. These methods may be summarised: (i) action of acid chlorides or anhydrides on alcohols, (ii) action of silver salts of the acids on the necessary iodide, (iii) the direct interaction of acids and alcohols in the presence of a suitable catalyst, generally hydrochloric or sulphuric acid. Dimethyl and diethyl sulphate may often also be employed in the preparation of methyl and ethyl esters.

Ethyl acetate is a colourless liquid, b.p. 77°. It possesses a characteristic and pleasant 'ester-like' odour. It is miscible with all organic solvents in all proportions and partially miscible with water. Ethyl acetate which has been exposed to air generally contains free acetic acid and ethyl alcohol on account of the hydrolysis of some small quantity of the ester:

$$CH_3.C\begin{matrix}\diagup\!\!\diagup O \\ \diagdown OC_2H_5\end{matrix} + H_2O \rightleftarrows CH_3.C\begin{matrix}\diagup\!\!\diagup O \\ \diagdown O{-}H\end{matrix} + C_2H_5OH$$

The reaction leading to the production of acetamide by the action of ammonia is slow:

$$CH_3.C\begin{matrix}\nearrow O\\ \searrow OC_2H_5\end{matrix} + HNH_2 \rightarrow CH_3.C\begin{matrix}\nearrow O\\ \searrow NH_2\end{matrix} + C_2H_5OH$$

When ethyl acetate is heated with hydriodic acid, ethyl iodide and acetic acid are produced:

$$CH_3.C\begin{matrix}\nearrow O\\ \searrow OC_2H_5\end{matrix} + HI \rightarrow CH_3.C\begin{matrix}\nearrow O\\ \searrow O—H\end{matrix} + C_2H_5I$$

Many esters of acetic acid are known. As might be expected, *methyl acetate*, b.p. 57·5°, occurs in wood spirit. The acetic esters of the higher alcohols have industrial applications in perfumery on account of their possessing odours closely resembling those of many fruits. Iso*amyl acetate* (b.p. 140°), for example, has an odour almost indistinguishable from that of pears.

Acetic acid being a fundamentally important organic compound, it is desirable to consider the evidence on which its constitutional formula is based, and this will exemplify the manner of demonstrating the constitutional formula of other fundamental compounds. Elementary analysis shows that its empirical formula is CH_2O (this, incidentally, is the molecular formula of formaldehyde and the empirical formula of a number of sugars), but determinations of its molecular weight by various methods prove that the molecular formula is $C_2H_4O_2$. Now, acetic acid is a monobasic acid, having only one atom of ionisable hydrogen in the molecule, which indicates that the other three hydrogen atoms must be differently combined from that one which is ionisable in the presence of water and replaceable by metals and by hydrocarbon radicals or substituted hydrocarbon radicals. Again, of the four hydrogen atoms in the molecule of acetic acid only three can be replaced directly by chlorine, the final product of this substitution being trichloroacetic acid, $C_2HO_2Cl_3$, which is also a monobasic acid. From trichloroacetic acid, chloroform and carbon dioxide can be obtained by a simple decomposition and therefore the chlorine atoms in the acid must be combined as in chloroform, that is, directly to a carbon atom. Hence, the hydrogen atoms which they have displaced in acetic acid must also be directly attached to a carbon atom. Therefore the formula of acetic acid must be $CH_3.CHO_2$. The fourth or ionisable hydrogen, which cannot be directly united to carbon, must be united to oxygen, and this would indicate the constitutional formula of acetic acid to be:

$$\begin{matrix} & & H & & O & & & \\ & & | & & \| & & & \\ H & — & C & — & C & — & O & — & H \\ & & | & & & & & \\ & & H & & & & & \end{matrix}$$

which is in keeping with the normal valencies of carbon, hydrogen and oxygen. This formula implies the presence of the hydroxyl group, of which we have definite evidence from the reaction of the acid with phosphorus pentachloride when acetyl chloride, hydrogen chloride and phosphorus oxychloride are produced. The formula also indicates the dissimilarity between the hydrogen atoms in the molecule, viz. those which are directly joined to carbon and which are replaceable by chlorine and the one which is united to oxygen and which is ionisable.

The decomposition of potassium acetate in aqueous solution by electrolysis (p. 202), the products of which are ethane, carbon dioxide and hydrogen, is in keeping with the above formula, and, finally, the production of acetic acid from methyl cyanide—the constitution of which is adequately demonstrated by synthesis—by hydrolysis is also in keeping with the formula:

$$H_3C.C\equiv N + \begin{matrix} H.O H \\ H.O H \end{matrix} \rightarrow H_3C.C\begin{matrix} \diagup O-H \\ \diagdown\!\!\!\diagdown O \end{matrix} + NH_3$$

CHAPTER VIII

HOMOLOGUES OF ACETIC ACID AND UNSATURATED ACIDS

The next homologue in the $C_nH_{2n+1}.COOH$ series is *propionic acid*, the direct final oxidation product of *n*-propyl alcohol:

$$CH_3.CH_2.CH_2OH \rightarrow \underset{\text{propionaldehyde}}{CH_3.CH_2.C\begin{matrix}\diagup H\\ \diagdown\!\!\diagdown O\end{matrix}} \rightarrow \underset{\text{propionic acid}}{CH_3.CH_2.C\begin{matrix}\diagup O{-}H\\ \diagdown\!\!\diagdown O\end{matrix}}$$

An alternative name for propionic acid is methylacetic acid,* which indicates its theoretical relationship with acetic acid. The relationship of propionic acid with *n*-propyl alcohol and of the latter with propane accounts for the systematic name of 'propane acid'. It may also be regarded as a carboxy derivative of ethane and described as 'carboxyethane'; since the hydrogen atoms in ethane are all similarly placed in the molecule, there can only be one 'carboxyethane' or propionic acid.

Starting with ethyl alcohol, the synthesis of propionic acid is carried out by reactions entirely analogous to those employed for the synthesis of acetic acid and are briefly:

$$C_2H_5OH \xrightarrow[\text{red phosphorus}]{\text{I and}} C_2H_5I \xrightarrow{\text{KCN}} \underset{\substack{\text{ethyl cyanide}\\ \text{(propionitrile)}}}{C_2H_5.C{\equiv}N} \xrightarrow{\text{hydrolysis}} \underset{\text{propionic acid}}{C_2H_5.C\begin{matrix}\diagup\!\!\diagup O\\ \diagdown O{-}H\end{matrix}}$$

Propionic acid is a colourless liquid, m.p. −22°, b.p. 141°. It possesses a somewhat rancid odour and is miscible with water, and organic liquids generally, in all proportions. It is a monobasic acid and its dissociation constant indicates that it is a weaker acid than acetic acid. As a general rule, the strengths of acids in a homologous series diminish with increasing complexity of the hydrocarbon radical. As far as the reactions due to the carboxyl group are concerned, propionic acid forms salts, esters, propionyl chloride, propionic anhydride, propionamide, etc., derivatives entirely analogous to those of acetic acid. Propionyl chloride, propionic anhydride and

* Propionic acid is isomeric with methyl acetate, $CH_3.C\begin{matrix}\diagup O{-}CH_3\\ \diagdown\!\!\diagdown O\end{matrix}$.

propionamide all undergo reactions analogous to those of the corresponding derivatives of acetic acid. The direct action of halogens on propionic acid is, however, more complicated than in the case of acetic acid.

The constitutional formula of propionic acid indicates, as far as the chlorine substitution products are concerned, that theoretically there are two monochloropropionic acids, three dichloropropionic acids, three trichloropropionic acids, two tetrachloropropionic acids and one pentachloropropionic acid capable of existence. The mono-halogenated propionic acids are the most important, and two mono-chloro-, two monobromo- and two monoiodo- propionic acids are known. These have the constitutional formulae:

$$\underset{\beta}{CH_3}.\underset{\alpha}{CHX}.CO_2H \quad \text{and} \quad \underset{\beta}{CH_2X}.\underset{\alpha}{CH_2}.CO_2H$$

where $X = $ Cl, Br and I. The chloro- compounds, for example, are known as α-chloropropionic acid and β-chloropropionic acid respectively; the Greek letters α and β being used to designate the position of the carbon atom to which the halogen is attached relative to the carboxy or 'key' group of the compound. Direct halogenation of propionic acid by chlorine or bromine leads to the production of the α-halogeno- compound.

α-*Chloropropionic acid* (b.p. 186°) and α-*bromopropionic acid* (m.p. 25·7°, b.p. 205°) are analogous to monochloro- and monobromo-acetic acids respectively. By similar reactions to those which the latter compounds undergo, they both yield hydroxy- and amino-derivatives of propionic acid which are known as α-*hydroxypropionic acid* and α-*aminopropionic acid* and have the following constitutional formulae respectively:

$$\begin{array}{c} CH_3 \\ | \\ H—C—OH \\ | \\ COOH \end{array} \quad \text{and} \quad \begin{array}{c} CH_3 \\ | \\ H—C—NH_2 \\ | \\ COOH \end{array}$$

α-Hydroxypropionic acid is usually referred to under the name of *lactic acid*, a compound of fundamental chemical and biochemical importance; the corresponding derivative of acetic acid is hydroxy-acetic acid or glycollic acid (p. 205). α-*Aminopropionic acid* corresponds to glycine or aminoacetic acid (p. 205) and is also of fundamental biochemical importance. It is generally referred to as *alanine* (p. 320). All these α-derivatives of propionic acid may exhibit optical activity (p. 113).

The most important higher homologues of the $C_nH_{2n+1}.COOH$

series are those acids which occur as *glyceryl* esters (esters of the trihydroxy alcohol, glycerol*) in animal and vegetable fats.

The homologues having the formula $C_3H_7.CO_2H$ are the final oxidation products of the two isomeric primary alcohols, C_4H_9OH, viz.

$$CH_3.CH_2.CH_2.CH_2OH, \qquad H-\overset{\displaystyle CH_3}{\underset{\displaystyle CH_3}{C}}-CH_2OH$$

n-butyl alcohol — *i*-propyl carbinol

and the two acids known as n-*butyric acid* and i-*butyric acid* have the respective constitutional formulae:

$$CH_3.CH_2.CH_2.COOH, \qquad H-\overset{\displaystyle CH_3}{\underset{\displaystyle CH_3}{C}}-COOH$$

* *Glycerol* will be discussed in detail later (p. 385). For the present, it is sufficient to state that glycerol is trihydroxypropane, its constitutional formula being:

$$\begin{array}{l} CH_2OH \\ | \\ CHOH \\ | \\ CH_2OH \end{array}$$

which indicates that it is a trihydroxy alcohol. Just as a monohydroxy alcohol (R.OH) can yield esters, so glycerol can yield esters through each and all of the three alcoholic hydroxyl groups, thus:

$$\begin{array}{lclcll} CH_2OH & & HOOC.C_nH_{2n+1} & & CH_2.O.OC.C_nH_{2n+1} & \\ | & & & & | & \\ CHOH & + & HOOC.C_nH_{2n+1} & \rightleftarrows & CH.O.OC.C_nH_{2n+1} & + \ 3H_2O \\ | & & & & | & \\ CH_2OH & & HOOC.C_nH_{2n+1} & & CH_2.O.OC.C_nH_{2n+1} & \end{array}$$

The glyceryl ester of the general formula represented above is by definition a fat (although the fats are generally complicated mixtures of glyceryl esters).

Compare:

$$C_2H_5OH + HOOC.CH_3 \rightleftarrows C_2H_5.O.OC.CH_3 + H_2O$$

ethyl acetate

$$\begin{array}{lclcll} CH_2OH & & HOOC.CH_3 & & CH_2.O.OC.CH_3 & \\ | & & & & | & \\ CHOH & + & HOOC.CH_3 & \rightleftarrows & CH.O.OC.CH_3 & + \ 3H_2O \\ | & & & & | & \\ CH_2OH & & HOOC.CH_3 & & CH_2.O.OC.CH_3 & \end{array}$$

glyceryl triacetate

Glyceryl triacetate (generally known as triacetin) being a glyceryl ester of a fatty acid, in this case acetic acid, is a fat according to definition. The term 'fat' is, however, generally specifically applied to those 'fats' which are products of animal and vegetable metabolism.

They may be considered as the two isomeric monocarboxy propanes (or butane acids), just as the two propyl alcohols are the two monohydroxy propanes. *n*-Butyric acid is ethylacetic acid and *i*-butyric acid is dimethylacetic acid. They may be synthesised in an analogous manner to propionic acid from the *n*- and *i*- propyl alcohols through the iodides, cyanides and hydrolysis of the latter compounds.

Of the two compounds, *n*-butyric acid is the more important. The glyceryl ester, $C_3H_5(O.OC.C_3H_7)_3$, occurs to the extent of approximately 5 per cent. in butter made from cow's milk. The acid occurs free in perspiration and is also found in rancid butter. It is produced by the butyric acid fermentation of sugar (glucose), starch and lactic acid. The fermentation of the sugar by the particular bacilli (present in rancid cheese) is carried out in the presence of calcium carbonate, which keeps the fermenting solution neutral, otherwise the fermentation ceases. As the acid forms, it is at once neutralised by the calcium carbonate, producing calcium butyrate. The butyric acid is obtained from this somewhat sparingly soluble salt by treatment with the calculated quantity of sulphuric acid filtering from the calcium sulphate and evaporating the aqueous filtrate until free from water. The butyric acid fermentation of glucose may take place via lactic acid, the decomposition of this into formic acid and acetaldehyde, the polymerisation of the latter to aldol and the conversion of the aldol to *n*-butyric acid by intramolecular change:

$$\underset{\text{glucose}}{C_6H_{12}O_6} \rightarrow \underset{\text{lactic acid}}{2C_3H_6O_3}$$

$$2\underset{\text{lactic acid}}{\begin{array}{c}CH_3.CHOH\\ |\\ COOH\end{array}} \rightarrow \underset{\text{formic acid}}{\left[2\begin{array}{c}H\\ |\\ COOH\end{array}\right]} + \underset{\text{acetaldehyde}}{\begin{array}{c}CH_3.CHO\\ \\ OHC.CH_3\end{array}} \rightarrow \underset{\text{aldol}}{\begin{array}{c}CH_3.CHOH\\ |\\ OHC.CH_2\end{array}} \rightarrow \underset{n\text{-butyric acid}}{\begin{array}{c}CH_3.CH_2\\ |\\ HOOC.CH_2\end{array}}$$

n-Butyric acid is a thick liquid with a rancid odour. It has b.p. 163°. It is readily soluble in water and in alcohol, and yields derivatives analogous to those of acetic and propionic acids.

i-Butyric acid generally has properties similar to those of the *normal* acid, but it is less soluble in water. It has b.p. 155°.

Corresponding to the four isomeric primary alcohols having the formula $C_5H_{11}OH$ or $C_4H_9.CH_2OH$, there are four isomeric acids (pentane acids—direct and final oxidation products of the alcohols) having the formula $C_4H_9.COOH$. These acids are all known and are collectively described as valeric acids, since the two commonest have the odour of valerian and actually occur in valerian root. The constitutional formulae and systematic names of these four acids are:

$CH_3.CH_2.CH_2.CH_2.CO_2H$ n-*propylacetic acid*, n-*valeric acid*, b.p. 186°*

$$H-\underset{\displaystyle CH_3}{\overset{\displaystyle CH_3}{\underset{|}{\overset{|}{C}}}}-CH_2.CO_2H$$

i-*propylacetic acid*, i-*valeric acid*, b.p. 175°

$$H-\underset{\displaystyle C_2H_5}{\overset{\displaystyle CH_3}{\underset{|}{\overset{|}{C}}}}-CO_2H$$

α-methyl-n-*butyric acid*, b.p. 177° (which can occur in optically active forms, p. 113)*

$$CH_3-\underset{\displaystyle CH_3}{\overset{\displaystyle CH_3}{\underset{|}{\overset{|}{C}}}}-CO_2H$$

trimethylacetic acid, b.p. 164°

Of the other acids of this series, the most important are *palmitic acid*, $C_{15}H_{31}.CO_2H$, and *stearic acid*, $C_{17}H_{35}.CO_2H$, which together with *oleic acid* (p. 224) occur as glycerides in animal fats.† The formulae of the glycerides of palmitic acid (often referred to as *tripalmitin*) and of stearic acid (*tristearin*) may be briefly represented thus:

$$\begin{array}{l} CH_2.O.OC.C_{15}H_{31} \\ | \\ CH.O.OC.C_{15}H_{31} \\ | \\ CH_2.O.OC.C_{15}H_{31} \end{array} \qquad \begin{array}{l} CH_2.O.OC.C_{17}H_{35} \\ | \\ CH.O.OC.C_{17}H_{35} \\ | \\ CH_2.O.OC.C_{17}H_{35} \end{array}$$

tripalmitin tristearin

These and *triolein*, while being the chief, are not the only constituents of animal fats, which apart from containing small quantities of other substances of an entirely different nature may contain so-called mixed glycerides, that is, glycerides whose acid radicals are not identical with each other. The acids may be obtained from the glycerides by hydrolysis (i) by superheated steam under pressure:

$$\begin{array}{l} CH_2.O.OC.R \\ | \\ CH\ .O.OC.R \\ | \\ CH_2.O.OC.R \end{array} + \begin{array}{l} HOH \\ \\ HOH \\ \\ HOH \end{array} \rightleftarrows 3R.CO_2H + \begin{array}{l} CH_2OH \\ | \\ CHOH \\ | \\ CH_2OH \end{array}$$

glyceride or ester water acid glycerol or alcohol

* If these two acids be regarded as carboxy derivatives of *n*-butane, they may be described as 1-carboxy-*n*-butane and 2-carboxy-*n*-butane respectively. The other two acids are similarly carboxy derivatives of *i*-butane.

† It is worthy of note that these acids occurring as their glyceryl esters in animal fats all have an even number of carbon atoms in the molecule. The acids having an odd number of carbon atoms in the molecule and which are members of this series are entirely synthetic products and do not occur as products of animal metabolism.

when the free acid or acids are obtained, (ii) by boiling with sodium or potassium hydroxide in aqueous solution, when the sodium or potassium salt of the acid or acids are obtained:

$$\begin{array}{l} CH_2.O.\boxed{OC.R \quad + \quad NaO}H \\ \vert \\ CH.O.\;OC.R \quad + \quad NaO\;H \rightarrow 3R.CO_2Na + \begin{array}{l}CH_2OH\\ \vert \\ CHOH \\ \vert \\ CH_2OH\end{array} \\ \vert \\ CH_2.O.\;OC.R \qquad\quad NaO\;H \end{array}$$

The sodium and potassium salts of these acids are *soaps*. When sodium hydroxide is used as the hydrolysing or saponifying agent, the sodium salt goes into solution. On addition of sodium chloride this salt separates, being less soluble in an aqueous solution of sodium chloride than it is in water, and rises to the surface, forming a cake which solidifies on cooling. This may be collected and pressed so as to free it to some extent from aqueous material and then constitutes a *hard soap*. Potassium hydroxide produces what is known as *soft soap*. The free acids can be obtained from their alkali salts by treatment of their aqueous solutions with mineral acid. Palmitic and stearic acids are thus precipitated, being insoluble in water and still less soluble in aqueous solutions containing alkali salts and free mineral acids.

Palmitic acid is a colourless and odourless solid, m.p. 64°. Although 'waxy' in appearance it is crystalline and may be recrystallised from ethyl alcohol, in which as in ether it is somewhat readily soluble; it is sparingly soluble in water. It has been proved that the carbon atoms in the molecules of palmitic and stearic acids are 'in a straight chain' and consequently the constitution of palmitic acid may be briefly represented as $CH_3.[CH_2]_{14}.CO_2H$.

Stearic acid is very similar to palmitic acid in general properties. It has m.p. 69·2° and its constitution may be represented as $CH_3.[CH_2]_{16}.CO_2H$.*

* The constitutions of palmitic and stearic acids have been proved by converting them into the barium salts, heating these with barium acetate and separating the mixed ketone. The latter is oxidised and the next lower acid in the series obtained. In this way the acid is broken down in stages until finally an acid of known constitution is obtained. The process may be represented thus:

$$\begin{array}{l} C_{17}H_{35}.CO_2 \\ C_{17}H_{35}.CO_2 \end{array}\!\!>\!Ba \xrightarrow{(CH_3.CO_2)_2Ba} \begin{array}{l} C_{17}H_{35} \\ CH_3 \end{array}\!\!>\!CO \xrightarrow{\text{oxidation}} C_{16}H_{33}.CO_2H$$

barium stearate — methylheptadecyl ketone — margaric acid

$$\begin{array}{l} C_{16}H_{33}.CO_2 \\ C_{16}H_{33}.CO_2 \end{array}\!\!>\!Ba \xrightarrow{(CH_3.CO_2)_2Ba} \begin{array}{l} C_{16}H_{33} \\ CH_3 \end{array}\!\!>\!CO \xrightarrow{\text{oxidation}} C_{15}H_{31}.CO_2H$$

barium margarate — methylhexadecyl ketone — palmitic acid

etc. down to the acid, $C_6H_{13}.CO_2H$, known as *œnanthic acid*, which has been shown

While the alkali salts of these acids are soluble in water, the aqueous solutions react alkaline and are always cloudy owing to hydrolysis of these salts of weak acids. The cloudiness of the solutions is due to incipient precipitation of free acid. The calcium and magnesium salts are sparingly soluble in water and form a scum when the sodium salt or potassium salt is used with 'hard' water.

Triolein, which also occurs along with tripalmitin and tristearin in animal fats, is the chief constituent of *olive oil.* Unlike tripalmitin and tristearin, which are solids, it is a liquid at the ordinary temperature and has a constitution which may be briefly represented:

$$\begin{array}{l} CH_2.O.OC.C_{17}H_{33} \\ | \\ CH.O.OC.C_{17}H_{33} \\ | \\ CH_2.O.OC.C_{17}H_{33} \end{array}$$

Oleic acid, $C_{17}H_{33}.CO_2H$, can be obtained from triolein by the usual methods of hydrolysis and is a liquid which, when pure, has m.p. 14°, b.p. 223° at 14 mm. If the animal fat contains the three glycerides and especially much triolein, the mixture of acids obtained from it by hydrolysis or saponification is generally liquid. When pure, oleic

to have the constitution $CH_3.[CH_2]_5.CO_2H$ by reactions which may be briefly outlined:

$$\underset{\text{alcohol}}{CH_3OH} \longrightarrow \underset{\text{iodide}}{CH_3I} \xrightarrow{\text{KCN}} \underset{\text{cyanide}}{CH_3CN} \xrightarrow{\text{hydrolysis}} \underset{\text{acid}}{CH_3.CO_2H}$$

$$\xrightarrow[\text{with Ca formate}]{\text{heating Ca salt}} \underset{\text{aldehyde}}{CH_3.CHO} \xrightarrow{\text{reduction}} \underset{\text{alcohol}}{CH_3.CH_2OH}$$

$\rightarrow$ $CH_3.CH_2I$ $\rightarrow$ $CH_3.CH_2.CN$ $\rightarrow$ $CH_3.CH_2.COOH$ $\rightarrow$ $CH_3.CH_2.CHO$
$\rightarrow$ $CH_3.CH_2.CH_2OH$ $\rightarrow$ $CH_3.CH_2.CH_2I$ $\rightarrow$ $CH_3.CH_2.CH_2.CN$
$\rightarrow$ $CH_3.CH_2.CH_2.CO_2H$ $\rightarrow$ $CH_3.CH_2.CH_2.CHO$ $\rightarrow$ $CH_3.CH_2.CH_2.CH_2OH$
$\rightarrow$ $CH_3.CH_2.CH_2.CH_2I$ $\rightarrow$ $CH_3.CH_2.CH_2.CH_2.CN$
$\rightarrow$ $CH_3.CH_2.CH_2.CH_2.COOH$ $\rightarrow$ $CH_3.CH_2.CH_2.CH_2.CHO$
$\rightarrow$ $CH_3.CH_2.CH_2.CH_2.CH_2OH$ $\rightarrow$ $CH_3.CH_2.CH_2.CH_2.CH_2I$
$\rightarrow$ $CH_3.CH_2.CH_2.CH_2.CH_2.CN$ $\rightarrow$ $CH_3.[CH_2]_4.COOH$ $\rightarrow$ $CH_3.[CH_2]_4.CHO$
$\rightarrow$ $CH_3.[CH_2]_4.CH_2OH$ $\rightarrow$ $CH_3.[CH_2]_5I$ $\rightarrow$ $CH_3.[CH_2]_5.CN$
$\rightarrow$ $CH_3.[CH_2]_5.COOH$.

The final substance obtained in this latter series of reactions is oenanthic acid, identical with the product finally obtained by breaking down stearic acid stage by stage by continuing the first series of reactions.

These two series of reactions are also theoretically interesting; they indicate (i) the preparation from one acid of the next lower one in the series, (ii) the preparation from one acid of the next higher one in the series, and (iii) the preparation from one alcohol of the next higher one in the series.

acid has little or no odour and does not affect the colour of litmus. It is, of course, a monobasic acid and its alkali salts are soluble in water. It may be purified and separated from other acids by taking advantage of the fact that lead oleate, $Pb(O_2C.C_{17}H_{33})_2$, is soluble in ether. Oleic acid is closely related to stearic acid and belongs to the series $C_nH_{2n-1}.CO_2H$. Its constitution is discussed below.

The determination of the *saponification value* of esters and particularly of glyceryl esters and their mixtures (as in fats) is used (in addition to the *iodine value* when the substances are unsaturated, p. 233) for the characterisation and estimation of these substances. The saponification value is defined as the 'number of milligrams' of potassium hydroxide (usually in alcoholic solution) required to hydrolyse or saponify 1 gram of the material. The saponification values of some typical esters are:

	Saponification Value
ethyl acetate	636
glyceryl acetate (triacetin)	770
glyceryl butyrate (tributyrin)	556
glyceryl palmitate (tripalmitin)	208
glyceryl stearate (tristearin)	189
glyceryl oleate (triolein)	190

The amount of adulteration in commercial olive oil may be readily determined through the saponification value of the material.

Apart from connective tissue, animal fats contain what is described as 'unsaponifiable matter' associated with the fats and dissolved along with the latter when the fats are extracted by fat solvents. Among the constituents of the unsaponifiable matter are substances known as *sterols*, among which are *cholesterol* and (in small quantity) *ergosterol*.

Cholesterol is the chief constituent of gall stones. It occurs in brain tissue and, to a small extent, in blood. It can be readily extracted from gall stones by means of alcohol from which it crystallises in characteristic colourless plates, m.p. 148·5°, which are *laevo*rotatory (p. 113) in ethereal solution. Cholesterol has the molecular formula $C_{27}H_{46}O$ and is an alcohol of a type similar to *cyclo*hexanol (p. 123) and its constitution may be represented briefly:

$$C_{27}H_{44}\langle\begin{matrix}H\\OH\end{matrix}$$

It is an unsaturated compound (the molecule contains one double bond) belonging to the *iso*cyclic series. Cholesterol is identified by its acetyl derivative (*cholesteryl acetate*), m.p. 114–115°, its benzoyl derivative (*cholesteryl benzoate*), m.p. 150–151°, and by various colour reactions.

Ergosterol, which is related to and isomeric with *calciferol* (until recently identified as *Vitamin D*), has the molecular formula $C_{28}H_{44}O$. It is also an alcohol of a similar type to cholesterol and its constitution may be briefly expressed:

$$C_{28}H_{42}\langle\begin{matrix}H\\OH\end{matrix}$$

Ergosterol crystallises from alcohol in anhydrous crystals, m.p. 163°, and its acetyl derivative (*ergosteryl acetate*) has m.p. 175–176°. It is *dextro*rotatory (compare cholesterol).

ACIDS OF THE $C_nH_{2n-1}.CO_2H$ SERIES

The difference of two hydrogen atoms between stearic acid belonging to the $C_nH_{2n+1}.CO_2H$ series and oleic acid is the same as the difference between the corresponding members of the paraffin and ethylene hydrocarbon series and, apart from being acids, the members of this series of compounds have, like the ethylene hydrocarbons, properties associated with unsaturated compounds.

The first member of this series is *acrylic acid* which has the constitutional formula:

```
H—C—H
  ||
  C—H
  |
  COOH
```

being theoretically derived from propionic acid by the loss of two hydrogen atoms. It may also be regarded as being derived from ethylene by the substitution of a carboxyl group for a hydrogen atom, and, just as propionic acid may be described as 'propane acid', so acrylic acid may be designated as 'propylene acid' or 'propene acid'.

Acrylic acid, obtained by the action of heat on hydracrylic acid (p. 366), is generally prepared by the oxidation of the corresponding aldehyde, *acrolein*,* or acrylic aldehyde, which has the constitutional formula $CH_2:CH.CHO$. This aldehyde has typical aldehydic properties and at the same time is, like ethylene, an unsaturated

* Acrolein is an unsaturated aldehyde belonging to the series $C_nH_{2n-1}.CHO$. It is prepared by the dehydration of glycerol, which is accomplished by distilling that substance with potassium bisulphate, boric acid, phosphoric acid or anhydrous magnesium sulphate, or by passing water-free glycerol over heated aluminium oxide. The dehydration of glycerol can be represented as taking place in stages:

```
     OH
     |
H—C—[H]           H—C—OH          H—C=O             H—C=O
  |        →        ||      →       |         →       |
H—C—[OH]            CH            H—C—[H]             C—H
  |                 |               |                 ||
  CH2OH             CH2OH         H—C—[OH]          H—C—H
                                    |
                                    H
```

Acrolein is a colourless liquid, b.p. 52°. It has an intolerable odour and exhibits characteristic aldehydic properties. It possesses the property of resinification to a remarkable degree and Moureu (1920) has shown that this can be retarded or hastened by certain catalysts. The acrolein, undergoing polymerisation, is first converted into a thick transparent liquid and then suddenly into a gel which rapidly hardens. The solidified gel has insulating properties similar to those of amber, a substance which it simulates to a remarkable degree. Artificial resins made from acrolein would appear to have important industrial applications. The unsaturated character of acrolein is shown by its combining directly with chlorine and bromine

compound. It is readily oxidised to the corresponding acid, acrylic acid, and this oxidation takes place on mere exposure to air:

$$\begin{array}{c} CH_2 \\ \| \\ CH \\ | \\ C\!\!\begin{smallmatrix}\diagup H \\ \diagdown\!\!\diagdown O\end{smallmatrix} \end{array} + O \rightarrow \begin{array}{c} CH_2 \\ \| \\ CH \\ | \\ C\!\!\begin{smallmatrix}\diagup OH \\ \diagdown\!\!\diagdown O\end{smallmatrix} \end{array}$$

Acrylic acid is a colourless liquid, m.p. 13°, b.p. 141°, which has an odour resembling that of acetic acid. It is a monobasic acid and forms salts and esters. Its unsaturated nature is shown by its combining directly with chlorine and with bromine to give *αβ-dichloro-* and *αβ-dibromo-propionic acids* respectively:

CH_2Cl $\vert$ $CHCl$ $\vert$ CO_2H	CH_2Br $\vert$ $CHBr$ $\vert$ CO_2H	CH_2Cl $\vert$ CH_2 $\vert$ CO_2H	CH_2Br $\vert$ CH_2 $\vert$ CO_2H
αβ-dichloro-propionic acid m.p. 50°	αβ-dibromo-propionic acid* Stable form m.p. 64° Less stable form m.p. 51°	β-chloro-propionic acid m.p. 61°	β-bromo-propionic acid m.p. 62·5°

in a manner similar to ethylene, to give αβ-dihalogenopropionaldehydes, which are directly oxidised to the corresponding dihalogenopropionic acids:

$$\begin{array}{c} CH_2 \\ \| \\ CH \\ | \\ C\!\!\begin{smallmatrix}\diagup H \\ \diagdown\!\!\diagdown O\end{smallmatrix} \end{array} + \underset{\text{(chlorine or bromine)}}{X_2} \rightarrow \begin{array}{c} CH_2X \\ | \\ CHX \\ | \\ C\!\!\begin{smallmatrix}\diagup H \\ \diagdown\!\!\diagdown O\end{smallmatrix} \end{array} \xrightarrow{\text{oxidation}} \begin{array}{c} CH_2X \\ | \\ CHX \\ | \\ C\!\!\begin{smallmatrix}\diagup O\text{—}H \\ \diagdown\!\!\diagdown O\end{smallmatrix} \end{array}$$

The reduction products of acrolein are indicated below:

$$\begin{array}{c} CH_2 \\ \| \\ CH \\ | \\ CHO \end{array} \begin{array}{l} \nearrow \begin{array}{c} CH_2 \\ \| \\ CH \\ | \\ CH_2OH \end{array} \\ \searrow \begin{array}{c} CH_3 \\ | \\ CH_2 \\ | \\ C\!\!\begin{smallmatrix}\diagup H \\ \diagdown\!\!\diagdown O\end{smallmatrix} \end{array} \rightarrow \begin{array}{c} CH_3 \\ | \\ CH_2 \\ | \\ CH_2OH \end{array} \end{array}$$

allyl alcohol (p. 116) also obtained by dehydration of glycerol

propionaldehyde *n*-propyl alcohol

* αβ-Dibromopropionic acid is dimorphic, a stable form and a less stable form, the latter having the higher solubility and lower melting point, having been isolated.

It also combines directly with hydrogen chloride and hydrogen bromide yielding *β-chloropropionic acid* and *β-bromopropionic acid* respectively.* From these two latter acids, by the usual reactions, can be obtained the corresponding *β-hydroxypropionic acid* (hydracrylic acid) and *β-aminopropionic acid*, which, unlike the isomeric α-hydroxypropionic acid (lactic acid) and α-aminopropionic acid (alanine), are incapable of exhibiting optical activity:

$$\begin{array}{l} CH_2OH \\ | \\ CH_2 \\ | \\ CO_2H \end{array} \qquad\qquad \begin{array}{l} CH_2NH_2 \\ | \\ CH_2 \\ | \\ CO_2H \end{array}$$

β-hydroxypropionic acid β-aminopropionic acid

Hydracrylic acid is readily distinguished from the isomeric lactic acid, because, unlike lactic acid, it loses water on being heated, yielding acrylic acid:

$$\begin{array}{c} H \\ | \\ H-C-\boxed{O-H} \\ | \\ H-C-\boxed{H} \\ | \\ CO_2H \end{array} \rightarrow H_2O + \begin{array}{c} H-C-H \\ \| \\ CH \\ | \\ CO_2H \end{array}$$

The next homologue to acrylic acid must have the molecular formula $C_3H_5.CO_2H$, and four isomeric acids possess this formula. These are all known and possess the following constitutional formulae:

$$\begin{array}{c} H-C-CH_3 \\ \| \\ HO_2C-C-H \end{array} \quad \begin{array}{c} H-C-CH_3 \\ \| \\ H-C-CO_2H \end{array} \quad \begin{array}{c} H-C-H \\ \| \\ CH_3-C-CO_2H \end{array} \quad \begin{array}{c} H-C-H \\ \| \\ H-C-CH_2.CO_2H \end{array}$$

crotonic acid, *iso*crotonic acid (β-methylacrylic acids); α-methylacrylic acid m.p. 16°, b.p. 160·5°; vinylacetic acid b.p. 168°

Of these acids, the more important are the β-methylacrylic acids or *crotonic acids*, which are stereoisomerides; they belong to the substituted ethylene types:

$$\begin{array}{c} X-C-Y \\ \| \\ Z-C-X \end{array} \qquad\qquad \begin{array}{c} X-C-Y \\ \| \\ X-C-Z \end{array}$$

(p. 41). The determination as to whether the higher melting crotonic acid (m.p. 72°) possesses the *trans-* formula and, in that case, the lower melting *iso*crotonic acid (m.p. 15°) has the *cis-* configuration as indicated above, has been a problem of some difficulty. It is probable

* These two acids are isomeric with α-chloro- and α-bromo-propionic acids respectively, which are the chief products of the direct halogenation of propionic acid (p. 219).

that the above constitutional formulae for these acids are correct. They are based on a considerable accumulation of chemical and physical evidence and comparison of the reactions and properties of these acids with those of other *cis-* and *trans-* isomerides.

Crotonic acid is made by the oxidation of crotonaldehyde (p. 158). If the *trans-* formula of crotonic acid be correct, it would appear that crotonaldehyde also possesses the *trans-* configuration: in this case, the production of crotonic acid may be represented:

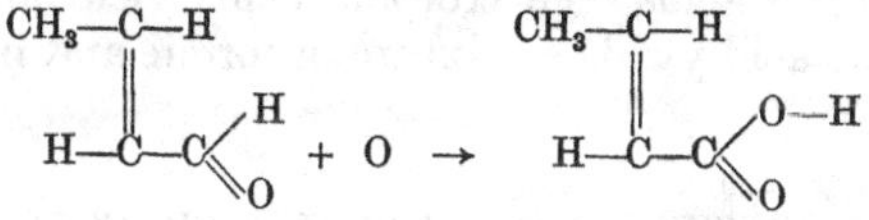

$$\begin{array}{c} CH_3{-}C{-}H \\ \| \\ H{-}C{-}C{\lt}^{H}_{O} \end{array} + O \rightarrow \begin{array}{c} CH_3{-}C{-}H \\ \| \\ H{-}C{-}C{\lt}^{O-H}_{O} \end{array}$$

crotonaldehyde crotonic acid

Crotonic acid is colourless and crystalline. Apart from being an acid it possesses the properties of a typical unsaturated compound. Acid and unsaturation properties are, of course, common to acrylic acid and the above four isomeric acids.

Iso*crotonic acid* cannot be prepared so directly as crotonic acid. It is best prepared by heating $\alpha\beta$-dichlorobutyric acid in an alcoholic solution of potassium hydroxide, when, as usual in these conditions, hydrogen chloride is eliminated and the unsaturated β-chloro*iso*-crotonic acid is formed. From the latter compound *iso*crotonic acid is obtained by gentle reduction (sodium amalgam). Assuming the *cis-* configuration for *iso*crotonic acid, its production by the above reaction may be expressed:

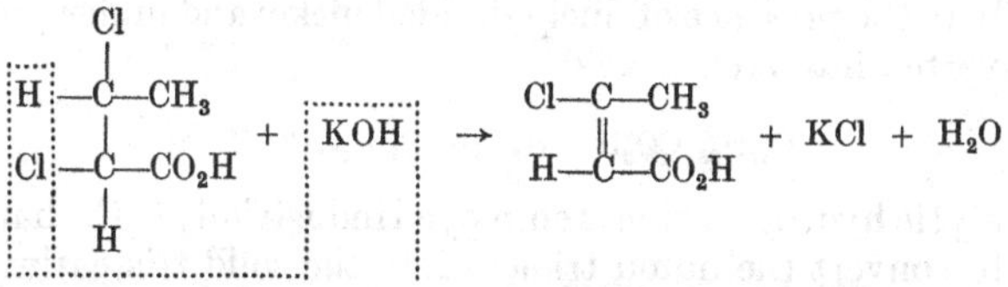

$$\begin{array}{c} Cl \\ | \\ \boxed{H}{-}C{-}CH_3 \\ | \\ \boxed{Cl}{-}C{-}CO_2H \\ | \\ H \end{array} + \boxed{KOH} \rightarrow \begin{array}{c} Cl{-}C{-}CH_3 \\ \| \\ H{-}C{-}CO_2H \end{array} + KCl + H_2O$$

$\alpha\beta$-dichlorobutyric acid β-chloro*iso*crotonic acid

$$\begin{array}{c} Cl{-}C{-}CH_3 \\ \| \\ H{-}C{-}CO_2H \end{array} + H_2 \rightarrow \begin{array}{c} H{-}C{-}CH_3 \\ \| \\ H{-}C{-}CO_2H \end{array} + HCl$$

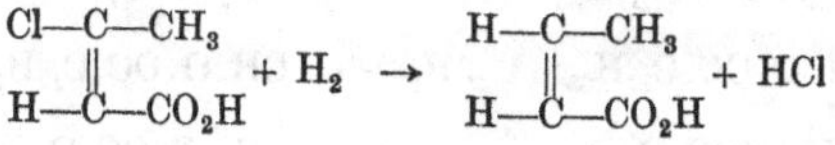

*iso*crotonic acid

The analogous derivatives of crotonic and *iso*crotonic acids are, like the parent acids, generally stereoisomerides. On reduction with hydrogen (best effected by a catalytic method) both acids, however, yield the same saturated *n*-butyric acid, $CH_3.CH_2.CH_2.CO_2H$. *Iso*crotonic acid can be easily converted into the stereoisomeric and higher melting crotonic acid by heating in a sealed tube at 170–180°, or by the action of sunlight on its aqueous or carbon disulphide solution, especially, in the latter case, if a small quantity of bromine be present.

The behaviour of acids of the $C_nH_{2n-1}.CO_2H$ series on fusion with potassium hydroxide is characteristic and of fundamental importance as affording evidence of the constitution of these unsaturated acids. In these conditions, the molecule of the particular acid breaks at the double bond, a mixture of two saturated acids (in the form of their potassium salts) resulting. Thus, acrylic acid yields a mixture of formic and acetic acids, the crotonic acids yield acetic acid, and α-methylacrylic acid yields a mixture of formic and propionic acids:

$$CH_2{=}CH{-}CO_2H \rightarrow H.CO_2H + CH_3.COOH$$

$$\left.\begin{array}{l} H{-}C{-}CH_3 \\ \quad \| \\ HO_2C{-}C{-}H \\ \\ H{-}C{-}CH_3 \\ \quad \| \\ H{-}C{-}CO_2H \end{array}\right\} \rightarrow CH_3.CO_2H + CH_3.CO_2H$$

$$\begin{array}{l} H{-}C{-}H \\ \quad \| \\ CH_3{-}C{-}CO_2H \end{array} \rightarrow H.CO_2H + CH_3.CH_2.CO_2H$$

Oleic acid is related to stearic acid in the same way as are the crotonic acids to *n*-butyric acid. When it is reduced with hydrogen (generally in the presence of finely divided nickel and under pressure), it is converted into stearic acid:

$$C_{17}H_{33}.CO_2H + H_2 \rightarrow C_{17}H_{35}.CO_2H$$

This catalytic hydrogenation is employed industrially in the hardening of fats, to convert the liquid triolein into the solid tristearin:

$$\begin{array}{l} CH_2.O.OC.C_{17}H_{33} \\ | \\ CH.O.OC.C_{17}H_{33} \\ | \\ CH_2.O.OC.C_{17}H_{33} \end{array} + 3H_2 \rightarrow \begin{array}{l} CH_2.O.OC.C_{17}H_{35} \\ | \\ CH.O.OC.C_{17}H_{35} \\ | \\ CH_2.O.OC.C_{17}H_{35} \end{array}$$

Similarly oleic acid combines directly with bromine and is converted into *dibromostearic acid* (liquid):

$$C_{17}H_{33}.CO_2H + Br_2 \rightarrow C_{17}H_{33}Br_2.CO_2H$$

Since the carbon atoms in stearic acid are 'in a straight chain' and oleic acid is converted by direct hydrogenation into stearic acid, it must be assumed that the carbon atoms in oleic acid are also 'in a straight chain'. The position of the double bond in the molecule of oleic acid has been established in the following way:

(i) Oleic acid combines with bromine yielding dibromostearic acid.

(ii) Dibromostearic acid when heated with an alcoholic solution of potassium hydroxide yields first *monobromoleic acid* and then *stearolic acid* by the loss of one and two molecules of hydrogen bromide respectively:*

$$C_{17}H_{33}Br_2.CO_2H + KOH \rightarrow \underset{\text{monobromoleic acid}}{C_{17}H_{32}Br.CO_2H} + KBr + H_2O$$

$$C_{17}H_{32}Br.CO_2H + KOH \rightarrow \underset{\text{stearolic acid}}{C_{17}H_{31}.CO_2H} + KBr + H_2O$$

(iii) When stearolic acid is oxidised with potassium permanganate, the molecule is, as in many analogous cases, disrupted at the triple bond and equimolecular quantities of two acids, known as pelargonic and azelaic acids respectively are produced.

Now *pelargonic acid* (m.p. 12°, b.p. 254°) is an acid of the $C_nH_{2n+1}.CO_2H$ series and actually has the formula $C_8H_{17}.CO_2H$, having all the carbon atoms 'in a straight chain'. Its constitution may be briefly written $CH_3.[CH_2]_7.CO_2H$.

Further *azelaic acid* (m.p. 106·5°) has been shown to be a dibasic acid of the oxalic or $C_nH_{2n}(CO_2H)_2$ series and to have the formula $C_7H_{14}(CO_2H)_2$, and, since the carbon atoms in the molecule of the acid have been proved to be 'in a straight chain', its constitutional formula may be represented as $CO_2H.[CH_2]_7.CO_2H$.

Obviously, these two acids have been derived by disrupting the molecule of stearolic acid at the triple bond, the position of which must be the same as that of the double bond in the molecule of oleic acid: the constitutional formula of oleic acid is thus established:

$$\underset{\text{pelargonic acid}}{CH_3.[CH_2]_7.CO_2H} + \underset{\text{azelaic acid}}{CO_2H.[CH_2]_7.CO_2H} \leftarrow \underset{\text{stearolic acid}}{CH_3.[CH_2]_7.C{\equiv}C.[CH_2]_7.CO_2H}$$

$$\nearrow$$

$$\underset{\text{oleic acid}}{CH_3.[CH_2]_7.CH{=}CH.[CH_2]_7.COOH}$$

Oleic acid is an ethylene derivative of the type $\begin{matrix} X\text{—}C\text{—}Y \\ \| \\ Z\text{—}C\text{—}Y \end{matrix}$, and a stereoisomer of oleic acid is therefore possible. On mixing oleic acid

* Compare the production of acetylene starting with ethylene bromide (p. 46). *Stearolic acid*, m.p. 48°, belongs to the $C_nH_{2n-3}.CO_2H$ or 'acetylene' series of monocarboxylic acids. The lowest member of this series is *propiolic acid*, having the constitutional formula $CH{\equiv}C.CO_2H$, and may be considered to be derived from acetylene by the substitution of a carboxyl group for a hydrogen atom. Propiolic acid bears the same relationships to acrylic and propionic acids as stearolic acid does to oleic and stearic acids respectively.

with a small quantity of sodium nitrite in aqueous solution and then adding a little dilute sulphuric acid, the liquid oleic acid becomes transformed into a solid compound when the mixture is allowed to stand. This remarkable action of nitrous acid on oleic acid has not been adequately explained, but the solid compound, known as *elaïdic acid*, produced under these conditions is the stereoisomeride of oleic acid. Elaïdic acid has m.p. 61° and its relationship to oleic acid is similar to that existing between crotonic and *iso*crotonic acids:

$$\begin{array}{c} CH_3.[CH_2]_7\text{—}C\text{—}H \\ \| \\ HO_2C.[CH_2]_7\text{—}C\text{—}H \\ \text{oleic acid} \end{array} \qquad \begin{array}{c} CH_3.[CH_2]_7\text{—}C\text{—}H \\ \| \\ H\text{—}C\text{—}[CH_2]_7.COOH \\ \text{elaïdic acid} \end{array}$$

While the corresponding derivatives of both these acids are distinct and isomeric with each other, both acids yield stearic acid on reduction. It is interesting that when oleic acid is heated at 220° with potassium hydroxide, the potassium salt of palmitic acid is produced. This is explained by the double bond in oleic acid changing its place in the molecule to the α-β position, thus:

$$CH_3.[CH_2]_7.CH{=}CH.[CH_2]_7.COOH \rightarrow CH_3.[CH_2]_{14}.CH{=}CH.COOH$$

After this change has been effected, oxidation takes place at the double bond with the production of the potassium salts of palmitic and acetic acids, thus:

$$\begin{array}{l} CH_3.[CH_2]_{14}.CH{=}\vdots CH.COOH \\ \qquad\qquad\qquad\quad \vdots H \\ \qquad\qquad\quad O\langle \vdots \\ \qquad\qquad\qquad\quad \vdots H \\ \qquad\qquad\quad O \;\vdots \end{array} \rightarrow \begin{array}{c} CH_3.[CH_2]_{14}.COOH \\ \text{palmitic acid} \end{array} + \begin{array}{c} CH_3.COOH \\ \text{acetic acid} \end{array}$$

The presence of the glyceride of an unsaturated acid like oleic acid in a fat or oil is indicated by the tendency of such fats or oils to oxidise on exposure to air: pure oleic acid becomes rancid under such conditions. Vegetable oils, such as linseed oil, which oxidise readily on exposure to air are known as 'drying' oils and are extensively used in the paint industry. These drying oils consist of glycerides of acids more highly unsaturated than oleic acid, and the drying is due to oxidation at the positions of unsaturation in the compound. Qualitatively, the presence of glycerides of such unsaturated acids can be rapidly determined by their decolorising bromine. The amount of unsaturation can be determined by Wijs' method or some suitable modification of it. This method depends on determining the amount of iodine monochloride (reckoned as iodine) which 100 grams of the fat or oil will combine with. The weighed quantity of the fat or oil dissolved in pure carbon tetrachloride containing a known excess of a solution of iodine monochloride in pure acetic acid is allowed to stand in the dark for one to three hours. The more highly unsaturated the fat or oil, the longer should the mixture stand. In the meantime, the iodine monochloride solution is standardised by adding an excess of potassium iodide and titrating the free iodine against standard sodium thiosulphate. This gives the strength of the iodine monochloride in terms of iodine. When the combination of iodine monochloride by the fat or oil is completed, the excess iodine monochloride (reckoned as iodine) is determined in the same way. The difference between the two amounts of iodine gives the amount of iodine which has theoretically combined with the weight of fat or oil taken. The amount in grams of iodine which combines theoretically with 100 grams of the fat or oil is the *iodine value.* It is usual to carry out a 'blank' determination, using only carbon tetrachloride under the same conditions and at the same time as the actual determination of the iodine value of the fat or oil.

Since a molecule of pure olein requires theoretically six atoms of iodine for complete saturation, its iodine value is approximately 86; good commercial olive oil has an iodine value varying from 77 to 88 and human fat has an iodine value of about 60. A good drying oil, such as linseed oil, may have an iodine value as high as 190.

CHAPTER IX

AROMATIC ACIDS AND THEIR DERIVATIVES

THE monocarboxylic acids already described may be considered to be derived from the various types of aliphatic hydrocarbons by the substitution of a carboxy group for an atom of hydrogen. Corresponding to these aliphatic acids, there is a very large class of monobasic aromatic acids which can also be considered to be derived from aromatic hydrocarbons by the substitution of a hydrogen atom by the carboxy group. These carboxylic acids are of two types: (*a*) those in which the carboxy group is directly attached to the aromatic nucleus, and (*b*) those in which the carboxy group is in the side chain. Thus, the monocarboxylic acids which can be derived theoretically from the two simplest aromatic hydrocarbons are:

Benzene, C_6H_6 → Benzoic acid, $C_6H_5.CO_2H$ or [ring: CO_2H; H, H, H, H, H]

Three isomeric toluic acids, $C_6H_4\begin{matrix}CH_3\\CO_2H\end{matrix}$,

whose constitutions are conveniently represented:

Toluene, $C_6H_5.CH_3$

[ring: CH_3, CO_2H, H, H, H, H] *o*-toluic acid

[ring: CH_3, H, CO_2H, H, H, H] *m*-toluic acid

[ring: CH_3, H, H, H, H, CO_2H] *p*-toluic acid

Phenylacetic acid, $C_6H_5.CH_2.CO_2H$, or [ring: $CH_2.CO_2H$; H, H, H, H, H]

Phenylacetic acid and all acids belonging to this series in which the carboxy group is in the side chain need not be specially considered since, as far as the acid properties of such compounds are concerned, their behaviour is similar to that of the carboxylic acids of the aliphatic series.*

* Corresponding to the carboxylic acids like benzoic and toluic acids, in which the carboxy group is directly attached to the nucleus, are the aromatic sulphonic acids, which are readily produced by the direct action of concentrated

Benzoic acid, $C_6H_5.CO_2H$, the simplest aromatic acid, occurs both free and combined in certain natural products. The resin known as 'gum benzoin' contains the free acid which can be obtained from it by sublimation. The benzyl ester of benzoic acid, $C_6H_5.CO.O.CH_2C_6H_5$, occurs in 'balsam of Tolu'. Hippuric acid, which occurs in the urine of herbivora, is an important derivative of glycine or aminoacetic acid,* and benzoic acid is readily obtained from it by boiling for

sulphuric (sulphonation) on the hydrocarbon. The more important of these acids are:

Benzene sulphonic acid, $C_6H_5.SO_2OH$

o-Toluene sulphonic acid, $C_6H_4\langle{}^{CH_3\ (1)}_{SO_2OH\ (2)}$ or [benzene ring: CH_3 at top, SO_2OH adjacent, H at the other positions]

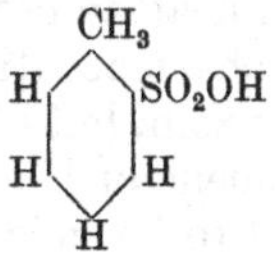

p-Toluene sulphonic acid, $C_6H_4\langle{}^{CH_3\ (1)}_{SO_2OH\ (4)}$ or [benzene ring: CH_3 at top, SO_2OH at bottom, H at the other positions]

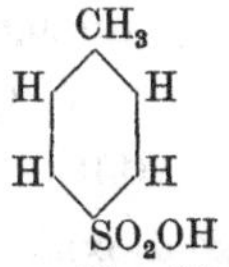

α-Naphthalene sulphonic acid, $C_{10}H_7.SO_2OH$ or [naphthalene ring with SO_2OH at an α-position, H at the other positions]

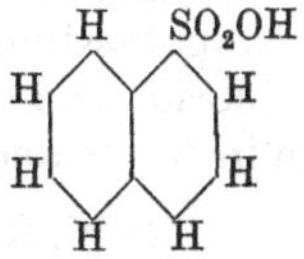

β-Naphthalene sulphonic acid, $C_{10}H_7.SO_2OH$ or [naphthalene ring with SO_2OH at a β-position, H at the other positions]

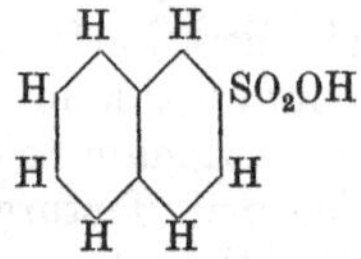

The general method of preparing and isolating these acids has been described in the case of benzene sulphonic acid (p. 59). When naphthalene is sulphonated at 80°, the chief product is the α- acid; when the sulphonation is carried out at a higher temperature (160°) and with an excess of sulphuric acid, the chief product is the β- acid. The most important derivatives of these sulphonic acids are the acid chlorides, in which the hydroxy group of the sulphonic or $—SO_2OH$ group is replaced by chlorine. Typical 'sulphonic chlorides' will be referred to in comparison with the acid chlorides derived from the typical aromatic carboxylic acids.

* The constitutional formula of hippuric acid is

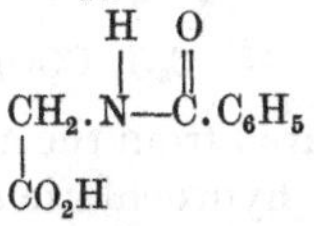

$$\begin{array}{l} \quad\quad\ \ H\ \ O \\ \quad\quad\ \ |\ \ \ \| \\ CH_2.N—C.C_6H_5 \\ | \\ CO_2H \end{array}$$

It is glycine or amino acetic acid in which one of the hydrogen atoms of the amino or $—NH_2$ group has been replaced by the 'benzoyl' or $C_6H_5.CO—$ group. On boiling with concentrated hydrochloric acid the compound is hydrolysed, thus:

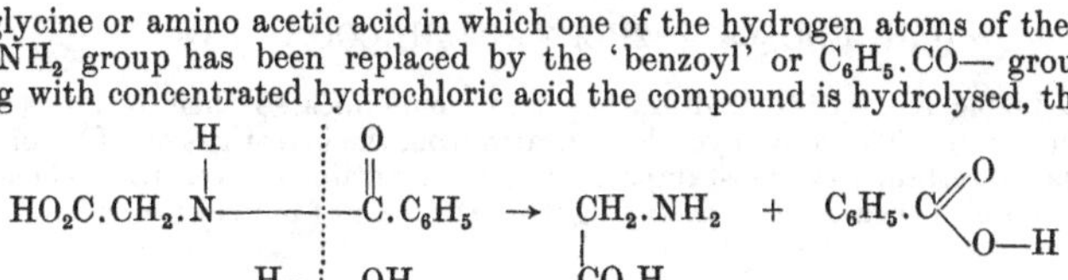

$$\begin{array}{l} HO_2C.CH_2.\overset{H}{\overset{|}{N}}———\overset{O}{\overset{\|}{C}}.C_6H_5 \\ \quad\quad\quad\quad\quad H—\ \ —OH \end{array} \rightarrow \begin{array}{l} CH_2.NH_2 \\ | \\ CO_2H \end{array} + C_6H_5.C\langle{}^{O}_{O—H}$$

about twenty minutes with somewhat concentrated hydrochloric acid. Part of the benzoic acid sublimes during the hydrolysis.

On account of the ease with which it can be obtained from natural products, benzoic acid was one of the first organic compounds to be systematically examined. Scheele investigated the acid in 1775 and its composition was determined by Liebig and Wöhler in 1832 at an early stage in their classical *Researches on the Radical of Benzoic Acid*. 'They showed that oil of bitter almonds, benzoic acid and a number of compounds derived from them group themselves round a "compound element", which they called Benzoyl, C_7H_5O.'* This was the first 'radical' in organic chemistry to be described.

For pharmaceutical purposes benzoic acid is still prepared from gum benzoin, but otherwise it is prepared almost exclusively starting with toluene or benzyl chloride. Toluene and benzyl chloride can be oxidised to benzoic acid by long boiling with an aqueous solution of potassium permanganate:

$$C_6H_5.CH_3 + 3O \rightarrow C_6H_5.COOH + H_2O$$

$$C_6H_5.CH_2Cl \rightarrow C_6H_5.CH_2OH \rightarrow C_6H_5.COOH$$

Manganese dioxide is precipitated and the aqueous filtrate from this contains potassium benzoate in solution. On acidification with hydrochloric acid, benzoic acid is precipitated, being sparingly soluble in water and even less so in dilute hydrochloric acid:

$$C_6H_5.CO_2K + HCl \rightarrow C_6H_5.CO_2H + KCl$$

The filtration from the manganese dioxide may be avoided by passing sulphur dioxide into the solution which is rendered acid, soluble manganese sulphate being formed, and the benzoic acid is precipitated.

The manufacture of benzoic acid is usually carried out by converting the toluene into *benzotrichloride*, by passing chlorine into the boiling hydrocarbon:

$$C_6H_5.CH_3 + 3Cl_2 \rightarrow C_6H_5.CCl_3 + 3HCl$$

This benzotrichloride (b.p. 213°), which may be described as phenylchloroform, undergoes hydrolysis in a similar manner to chloroform when boiled with alkalis, and milk of lime is employed on the manufacturing scale:

$$2C_6H_5.CCl_3 + 4Ca(OH)_2 \rightarrow (C_6H_5.COO)_2Ca + 3CaCl_2 + 4H_2O$$

The benzoic acid is precipitated from the aqueous solution of calcium benzoate by the addition of hydrochloric acid:

$$(C_6H_5.COO)_2Ca + 2HCl \rightarrow 2C_6H_5.COOH + CaCl_2$$

* To Liebig and Wöhler, after they had communicated their results to him, Berzelius wrote: 'The results you have drawn from the investigation of oil of bitter almonds are certainly the most important hitherto attained in vegetable chemistry; they promise to shed an unexpected light on this part of our science.'

Benzoic acid may be obtained by other methods, among which may be mentioned (*a*) the oxidation of benzyl alcohol and benzaldehyde:

$$C_6H_5.C(H)(H)(OH) + O \rightarrow C_6H_5.C(H){=}O + H_2O$$

$$C_6H_5.C(H){=}O + O \rightarrow C_6H_5.C(O{-}H){=}O$$

(*b*) the fusion of sodium benzene sulphonate with sodium cyanide, whereby *phenyl cyanide* or *benzonitrile* (b.p. 191°) is produced, and the hydrolysis of the latter by boiling with aqueous alkalies or better with mineral acids (especially sulphuric acid) to benzoic acid:

$$C_6H_5.SO_3Na + NaCN \rightarrow C_6H_5:C\vdots N + Na_2SO_3$$

$$C_6H_5.CN + 2H_2O \rightarrow C_6H_5.CO_2H + NH_3$$

and (*c*) from benzene, which is first converted into mononitrobenzene, then into aniline from which phenyl cyanide is obtained by reactions described later (p. 277) and then by hydrolysis of the cyanide as mentioned above. The last method of obtaining benzoic acid may be represented schematically:

$$C_6H_6 \rightarrow C_6H_5.NO_2 \rightarrow C_6H_5.NH_2 \rightarrow C_6H_5.CN^* \rightarrow C_6H_5.CO_2H$$

Since benzene can be prepared from acetylene and this gas can be synthesised from carbon and hydrogen, we have two methods of obtaining benzoic acid starting with these two elements.

Benzoic acid is conveniently recrystallised from hot water and is readily soluble in alcohol and ether. It crystallises in colourless glistening plates, m.p. 121·5°, and is easily volatile in steam. It has a dissociation constant (k) of 0·0060 and is therefore a much stronger acid than acetic acid (k=0·0018). It is, of course, a monobasic acid and forms characteristic salts, many of which are readily soluble in water. From the aqueous solutions of the salts the acid is readily precipitated by the addition of the stronger mineral acids. The ferric salt, $(C_6H_5.COO)_3Fe$, forms a characteristic buff-coloured precipitate when a soluble ferric salt is added to an aqueous solution of a soluble salt of benzoic acid. The ethyl ester, *ethyl benzoate*, a colourless liquid, b.p. 213°, having a characteristic pleasant aromatic odour, is readily formed when benzoic acid is warmed with an excess of ethyl alcohol containing a little sulphuric acid. When the product is poured

* Phenyl cyanide can also be obtained from benzamide (the amide of benzoic acid) by the action of phosphorus pentoxide (p. 242).

into water, the ester separates as an oil and can be isolated in the usual manner.

Mitscherlich discovered in 1834 that when benzoic acid is heated with lime benzene is produced:

$$C_6H_5.CO_2H + CaO \rightarrow CaCO_3 + C_6H_6$$

This reaction forms a convenient test for benzoic acid and provides a method of converting toluene into benzene, thus:

$$C_6H_5.CH_3 \xrightarrow{\text{oxidation}} C_6H_5.CO_2H \xrightarrow{CaO} C_6H_6$$

The product of nitration of benzoic acid with a mixture of concentrated nitric and sulphuric acids is chiefly m-*nitrobenzoic acid* (m.p. 141°): also the chief product of the direct chlorination of benzoic acid is m-*chlorobenzoic acid* (m.p. 158°):

CO_2H CO_2H

NO_2 Cl

the carboxyl group being essentially *meta*-directing with regard to the position taken up by another group directly entering the nucleus.

Benzoic acid is used to a small extent in medicine as an antiseptic and a diuretic and also sometimes as an antipyretic. Subject to certain limitations as to the quantity which may be used, benzoic acid finds applications as an important food preservative.

Benzoyl chloride is one of the most important derivatives of benzoic acid and is prepared by the action of phosphorus pentachloride on the acid:

$$C_6H_5.C\begin{matrix}\diagup O \\ \diagdown O-H\end{matrix} + PCl_5 \rightarrow C_6H_5.C\begin{matrix}\diagup O \\ \diagdown Cl\end{matrix} + POCl_3 + HCl$$

Benzoyl chloride is a colourless liquid, m.p. −1°, b.p. 198°, which fumes in moist air and possesses a characteristic odour and has slight lachrymatory properties. It reacts slowly with water forming benzoic acid and hydrochloric acid, which explains its fuming in moist air:

$$C_6H_5.COCl + H_2O \rightarrow C_6H_5.COOH + HCl$$

Its reaction with aqueous alkali solutions is much more rapid, salts of benzoic and hydrochloric acid being produced:

$$C_6H_5.COCl + 2NaOH \rightarrow C_6H_5.COONa + NaCl + H_2O$$

Benzoyl chloride is an important reagent for the introduction of the benzoyl group ($C_6H_5.CO.$) in place of a hydrogen atom which is attached to oxygen in alcohols and phenols and to nitrogen in primary and secondary amines. This process is generally described as *benzoylation*, and is effected by the Schotten-Baumann reaction. The normal method for carrying out the reaction is to mix the compound gradually with rather more than the calculated quantity of benzoyl chloride, adding also alkali, usually an aqueous solution of sodium hydroxide, from time to time to neutralise the hydrogen chloride produced. Thus, for the preparation of *phenyl benzoate*, m.p. 70°, from phenol and benzoyl chloride, the reaction may be expressed:

$$C_6H_5.OH + \begin{matrix} O \\ Cl \end{matrix}\!\!>\!C.C_6H_5 + NaOH \rightarrow C_6H_5{-}O{-}\overset{\overset{O}{\|}}{C}.C_6H_5 + NaCl + H_2O$$

In this case the phenol may be dissolved in water containing the necessary quantity of sodium hydroxide and the benzoyl chloride added gradually. Heat is generally developed during the reaction and the mixture may be warmed towards the end. The conditions of the reaction can be varied considerably. It may be necessary or desirable to exclude water during the actual reaction and, in that case, the aqueous solution of sodium hydroxide can be replaced by *pyridine* (p. 297), C_5H_5N, which is a liquid organic base, a tertiary amine (p. 246), capable of combining with hydrochloric acid and unaffected by benzoyl chloride. If pyridine is used, it is only necessary to mix pyridine solutions of the reacting substances, when the benzoyl derivative is formed either immediately or on warming. The benzoyl derivative, usually insoluble in water, is freed from the excess pyridine and pyridine hydrochloride by pouring the mixture into water and isolating in any convenient manner.

Hippuric acid (p. 235) can be readily prepared by the Schotten-Baumann reaction by treating an aqueous solution of the sodium salt of aminoacetic acid with benzoyl chloride and a solution of sodium hydroxide as indicated above. On acidification of the final alkaline solution, hippuric acid mixed with some benzoic acid is precipitated. The benzoic acid is eliminated from this mixture by extracting with ether, and the hippuric acid is then recrystallised from hot water. The formation of hippuric acid may be represented:

$$HOOC.CH_2.N\!<\!\begin{matrix} H \\ H \end{matrix} + Cl.\overset{\overset{O}{\|}}{C}.C_6H_5 + NaOH \rightarrow \underset{\text{hippuric acid}}{HOOC.CH_2.N\!<\!\begin{matrix} H \\ CO.C_6H_5 \end{matrix}} + NaCl + H_2O$$

The Schotten-Baumann reaction has been very considerably extended and almost any aromatic acid chloride may be employed in the place of benzoyl chloride. Aliphatic acid chlorides do not generally require the presence of alkali to assist the reaction, and the Schotten-

Baumann reaction implies the action of an aromatic acid chloride in the presence of the corresponding amount of alkali on an alcohol, phenol, primary and secondary amine.*

* Corresponding to benzoyl chloride is the analogous compound derived from benzenesulphonic acid (or its sodium salt) by the action of phosphorus pentachloride:

$$C_6H_5.SO_2.OH + PCl_5 \rightarrow C_6H_5.SO_2Cl + POCl_3 + HCl$$

Benzenesulphonyl chloride, a colourless liquid, m.p. 14·5°, b.p. 120° at 10 mm., is also acted upon by water (but less rapidly than is benzoyl chloride) and by aqueous alkalis:

$$C_6H_5.SO_2Cl + H_2O \rightarrow C_6H_5.SO_2OH + HCl$$

$$C_6H_5.SO_2Cl + 2NaOH \rightarrow C_6H_5.SO_2ONa + NaCl + H_2O$$

It is frequently used for the preparation of benzenesulphonyl derivatives similarly to benzoyl chloride for the preparation of benzoyl derivatives.

o-*Toluenesulphonyl chloride*, a colourless liquid, and *p*-toluenesulphonyl chloride, m.p. 69°,

$$C_6H_4\begin{cases}CH_3\ (1)\\SO_2Cl\ (2)\end{cases} \qquad C_6H_4\begin{cases}CH_3\ (1)\\SO_2Cl\ (4)\end{cases}$$

obtained in an analogous manner from the corresponding sulphonic acids, are more frequently used than benzenesulphonyl chloride.

o-*Toluenesulphonyl chloride* is interesting as being an intermediate product in the manufacture of *saccharin*, a crystalline substance, estimated as being 500 times sweeter than cane sugar. The preparation of saccharin may be outlined:

$$C_6H_4\begin{cases}CH_3\ (1)\\SO_2Cl\ (2)\end{cases} + NH_3 + NH_3 \rightarrow C_6H_4\begin{cases}CH_3\ (1)\\SO_2NH_2\ (2)\end{cases} + NH_4Cl$$

o-toluenesulphonamide

$$C_6H_4\begin{cases}CH_3\ (1)\\SO_2NH_2\ (2)\end{cases} + 3O \rightarrow C_6H_4\begin{cases}COOH\ (1)\\SO_2NH_2\ (2)\end{cases} + H_2O$$

benzoic acid *o*-sulphonamide

$$C_6H_4\begin{cases}C{\overset{\displaystyle O}{\diagup}}\ \boxed{OH}\\ \underset{O_2}{S}-N\begin{cases}\boxed{H}\\H\end{cases}\end{cases} \rightarrow C_6H_4\begin{cases}CO\\SO_2\end{cases}NH + H_2O$$

saccharin (Remsen 1879)
(imide of benzoic acid *o*-sulphonic acid)

Saccharin, m.p. 220°, is colourless and sparingly soluble in water. Compounds containing the imide group (: NH united to two groups such as : CO or : SO_2) form salts by replacement of the hydrogen of the : NH group by metals. The sodium derivative of saccharin is much more soluble in water than saccharin and is estimated as being 400 times sweeter than cane sugar. Saccharin is also known as *glucide*.

p-Toluenesulphonyl chloride is employed for preparing *p*-toluenesulphonyl derivatives as alternative to benzoyl derivatives. The constitution of the *phenyl ester of* p-*toluene sulphonic acid* (prepared by the action of *p*-toluenesulphonyl chloride on phenol in the presence of alkali) may be represented as:

$$C_6H_5O.\overset{(4)}{SO_2}.C_6H_4.\overset{(1)}{CH_3} \quad \text{or} \quad CH_3\langle\bigcirc\rangle SO_2.O.C_6H_5$$

The constitution of *p*-toluenesulphonyl glycine (prepared from *p*-toluenesulphonyl

Benzoyl chloride bears the same relationship to benzoic acid that acetyl chloride does to acetic acid, and from benzoyl chloride other compounds are derived analogous to those derived from acetyl chloride. Thus, the anhydride of benzoic acid, *benzoic anhydride*,

$$\begin{matrix} C_6H_5.C\langle^{O}_{} \\ \quad\rangle O \\ C_6H_5.C\langle_{O} \end{matrix}$$

is easily prepared by the action of sodium benzoate and benzoyl chloride:

$$C_6H_5.C\langle^{O}_{Cl} + {}^{O}_{NaO}\rangle C.C_6H_5 \rightarrow NaCl + \begin{matrix} C_6H_5.C\langle^{O} \\ \rangle O \\ C_6H_5.C\langle_{O} \end{matrix}$$

Benzoic anhydride is a colourless crystalline compound, m.p. 42°, b.p. 360°. It is occasionally used instead of benzoyl chloride for benzoylation and thus corresponds in this respect to the use of acetic anhydride for acetylation; acetic anhydride is, however, used for acetylation far more frequently than acetyl chloride.

Benzamide, $C_6H_5.C\langle^{O}_{NH_2}$, is prepared by reactions similar to some of those employed for the preparation of acetamide and particularly by the action of ammonia (usually in the form of ammonium carbonate) on benzoyl chloride, the reaction taking place being represented:

$$C_6H_5.C\langle^{O}_{Cl} + (NH_4)_2CO_3 \rightarrow C_6H_5.C\langle^{O}_{NH_2} + NH_4Cl + H_2O + CO_2$$

The benzamide is recrystallised from boiling water and obtained in colourless crystals, m.p. 130°. It behaves like other amides when heated with alkalis, ammonia being evolved:

$$C_6H_5.C\langle^{O}_{NH_2} + NaOH \rightarrow C_6H_5.C\langle^{O}_{ONa} + NH_3$$

Benzamide may be regarded as the intermediate dehydration product between ammonium benzoate and phenyl cyanide, just as acetamide is intermediate between ammonium acetate and methyl

chloride and glycine or aminoacetic acid in the presence of alkali) may be represented as:

$$HO_2C.CH_2.\overset{H}{\overset{|}{N}}—\overset{(4)}{S}O_2.C_6H_4.\overset{(1)}{C}H_3 \quad \text{or} \quad HO_2C.CH_2.\overset{H}{\overset{|}{N}}—SO_2—\langle\_\_\rangle—CH_3$$

cyanide (acetonitrile). When benzamide is heated with phosphorus pentoxide, *phenyl cyanide* distils over as a colourless oil:

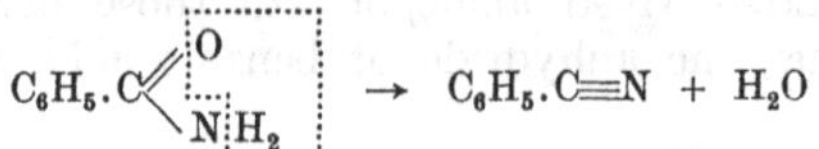

$$C_6H_5.C\langle^{O}_{NH_2} \rightarrow C_6H_5.C{\equiv}N + H_2O$$

phenyl cyanide

Phenyl cyanide, like the aliphatic cyanides, is hydrolysed to the corresponding acid (benzoic acid) when boiled with aqueous solutions of alkalis and mineral acids, especially sulphuric acid:

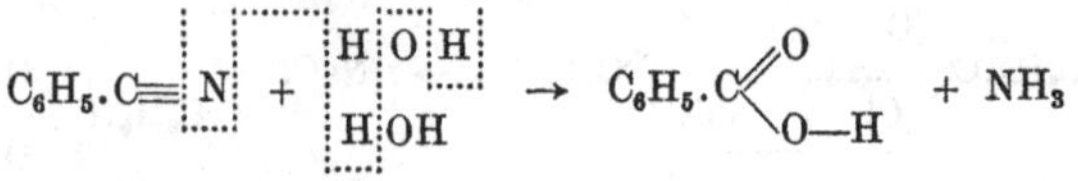

$$C_6H_5.C{\equiv}N + \begin{matrix} H\,O\,H \\ H\,OH \end{matrix} \rightarrow C_6H_5.C\langle^{O}_{O-H} + NH_3$$

Phenyl cyanide may therefore be designated *benzonitrile.* Phenyl cyanide cannot be produced by the action of potassium cyanide on monochloro- or monobromo- benzene. It is prepared by the two methods already mentioned (p. 237).

Unsaturated Monobasic Aromatic Acids

The simplest unsaturated monobasic acid is acrylic acid,

$$CH_2{=}CH{-}COOH$$

and the simplest monobasic acid corresponding to this aliphatic acid in the aromatic series is *phenylacrylic acid*, theoretically derived from the acrylic group by substituting a phenyl group for a hydrogen atom in the methylene group. This acid is known more generally as *cinnamic acid*, $C_6H_5.CH{=}CH.CO_2H$, and has already been referred to as a derivative of benzaldehyde, from which it is easily prepared by Perkin's reaction (p. 173).

Cinnamic acid is monobasic and exists in stereoisomeric forms, of which the one presumed to have the *trans-* configuration is the more stable and has been the more completely investigated. The *amide* is a colourless crystalline substance, m.p. 141·5°, and the *ethyl ester* (prepared by heating for 5 hours a mixture of the acid with excess of ethyl alcohol and a small quantity of sulphuric acid) is also colourless and has m.p. 12° and b.p. 271°.

On nitration with concentrated nitric acid, cinnamic acid yields a mixture of *ortho-* and *para-* nitrocinnamic acids:

$$o\text{-}NO_2C_6H_4{-}CH{=}CH{-}CO_2H \qquad p\text{-}NO_2C_6H_4{-}CH{=}CH{-}CO_2H$$

m.p. 240° m.p. 286°

These two acids, along with the isomeric m-*nitrocinnamic acid* (m.p. 197°), can be prepared from the corresponding nitrobenzaldehydes by the Perkin reaction.

Cinnamic acid and all its derivatives in which the ethylenic linkage or double bond is still present have the usual properties of unsaturated compounds and react towards bromine, alkaline potassium permanganate, etc., similarly to ethylene. From cinnamic acid is obtained, for example, *cinnamic acid dibromide* (β-phenyl-$\alpha\beta$-dibromopropionic acid), m.p. 195°:

$$\begin{array}{c} C_6H_5-C-H \\ \| \\ H-C-CO_2H \end{array} + Br_2 \rightarrow \begin{array}{c} H \\ | \\ C_6H_5-C-Br \\ | \\ H-C-Br \\ | \\ CO_2H \end{array}$$

which is capable of exhibiting optical activity (p. 113). When cinnamic acid is reduced with nascent hydrogen, the corresponding saturated acid, *β-phenylpropionic acid* or *hydrocinnamic acid*, m.p. 49°, b.p. 280°, is obtained. This acid is the next homologue to phenylacetic acid, belonging to the same series as benzoic acid (or phenyl formic acid, as it may be regarded in this connexion):

$$\begin{array}{c} C_6H_5.CH \\ \| \\ CH \\ | \\ CO_2H \end{array} + H_2 \rightarrow \begin{array}{c} C_6H_5.CH_2 \\ | \\ CH_2 \\ | \\ CO_2H \end{array} \quad \text{cf.} \quad \begin{array}{c} \beta CH_3 \\ | \\ \alpha CH_2 \\ | \\ CO_2H \end{array}$$

$$\begin{array}{c} C_6H_5.CH_2 \\ | \\ COOH \end{array} \quad \text{cf.} \quad \begin{array}{c} CH_3 \\ | \\ COOH \end{array}; \quad \begin{array}{c} C_6H_5 \\ | \\ COOH \end{array} \quad \text{cf.} \quad \begin{array}{c} H \\ | \\ COOH \end{array}$$

The action of alkaline potassium permanganate on cinnamic acid follows the usual course, the dihydroxy saturated acid, *β-phenylglyceric acid* or *β-phenyl-$\alpha\beta$-dihydroxypropionic acid*, m.p. 141°, being produced. The molecular configuration of this acid is similar to that of the above dibromo- acid and, like the latter, it is capable of exhibiting optical activity.

The cinnamic acid dibromides react similarly to ethylene dibromide towards an alcoholic solution of potassium hydroxide. In the case of the cinnamic acid dibromides, the reactions taking place may be formulated:

(i) $C_6H_5.CHBr.CHBr.CO_2H + KOH \rightarrow C_6H_5.CH:CBr.CO_2H + KBr + H_2O$
cinnamic acid dibromides — α-monobromocinnamic acids

(ii) $C_6H_5.CH:CBr.CO_2H + KOH \rightarrow C_6H_5.C:C.CO_2H + KBr + H_2O$
phenylpropiolic acid

Two stereoisomeric (*cis*- and *trans*-) α-monobromocinnamic acids are known and both yield *phenylpropiolic acid* as indicated. This acetylenic (because of the triple bond or acetylenic linkage) monobasic acid is colourless and has m.p. 136°. On reduction it is converted into cinnamic and hydrocinnamic acids. By addition of hydrogen chloride and hydrogen bromide it is converted into *β-monochloro-* and *β-monobromo-* cinnamic acids respectively, $C_6H_5.CCl:CH.CO_2H$ and $C_6H_5.CBr:CH.CO_2H$. Both of these acids, like the α-monohalogeno- cinnamic acids, are capable of existing in stereoisomeric (*cis-* and *trans-*) forms.

By a similar series of reactions, starting with the *o*-nitrocinnamic acid dibromides, *o-nitrophenylpropiolic acid* may be obtained. This acid was first obtained by A. von Baeyer in 1882. It decomposes at 156° and when heated with water yields *o-nitrophenylacetylene*, m.p. 81°, and carbon dioxide:

$$NO_2.C_6H_4.C\vdots C.CO_2H \rightarrow CO_2 + NO_2.C_6H_4.C\vdots CH$$

o-Nitrophenylacetylene, since it has a hydrogen atom attached to a triply linked carbon like acetylene, yields a cuprous-copper derivative. When this copper derivative is oxidised by a gentle oxidising agent, particularly potassium ferricyanide,* *oo'-dinitrodiphenyldiacetylene* (m.p. 212° with decomposition) is produced:

$$C_6H_4\begin{cases}C{\equiv}C.H \\ NO_2\end{cases} \quad O \quad \begin{cases}H.C{\equiv}C \\ NO_2\end{cases}C_6H_4 \rightarrow H_2O + C_6H_4\begin{cases}C{\equiv}C{-}C{\equiv}C \\ NO_2 \quad\quad NO_2\end{cases}C_6H_4$$

oo'-dinitrodiphenyldiacetylene

*o-Nitro*phenylacetylene and *oo'*-dinitrodiphenyldiacetylene are of historical interest. These two compounds were the starting materials for the first systematic syntheses of the important vat dye-stuff, indigo blue or indigotin (p. 334) by A. von Baeyer (1880—1882).

* In the presence of alkali, potassium ferricyanide acts as oxidising agent, being converted at the same time into potassium ferrocyanide:

$$2K_3Fe(CN)_6 + 2KOH \rightarrow 2K_4Fe(CN)_6 + H_2O + O$$

CHAPTER X

ORGANIC BASES

It is frequently useful, and the historical development of organic chemistry justifies it, to compare the simpler organic compounds with analogous simple inorganic compounds. From water, alcohols and ethers—essentially neutral compounds—may be considered to be derived by the replacement of one and both hydrogen atoms respectively in the molecule by monovalent alkyl groups (CH_3—, C_2H_5—, C_3H_7—, etc. $=R$).* Similarly, acids and acid anhydrides may be referred to the 'water' type by the replacement of one and both hydrogen atoms respectively in the molecule by monovalent acyl groups [acetyl ($CH_3.CO$—), benzoyl ($C_6H_5.CO$—), etc. $=R.CO$—]:

H—O—H

R—O—H alcohols   *R*.CO—O—H acids

R—O—*R* ethers   *R*.CO—O—OC.*R* anhydrides

The amides of the acids may be considered to be derived from ammonia by the replacement of one hydrogen atom in the molecule by the acyl (*R*.CO—) group, thus:

$$\mathrm{N}\begin{cases}\mathrm{H}\\ \mathrm{H}\\ \mathrm{H}\end{cases} \rightarrow \mathrm{N}\begin{cases}\mathrm{H}\\ \mathrm{H}\\ \mathrm{OC}.R\end{cases}$$

the amides being thus referred to the 'ammonia' type.

The organic bases may be considered similarly to be derived from ammonia (and in some cases, ammonia is the starting material in their preparation) by the replacement of the hydrogen atoms by monovalent alkyl and aryl groups (*R*—). There are, therefore, three types of these compounds, substituted ammonias or *amines* as they are designated, in which one, two and three of the hydrogen atoms in the molecule of ammonia are replaced, and these three types

* If $R=$ monovalent aromatic (aryl) group (C_6H_5—, $CH_3.C_6H_4$—, etc.), the compound *R*—O—H is a phenol which is slightly acidic in character, due to the phenyl or substituted phenyl group being potentially more electro-negative in character than a monovalent aliphatic (alkyl) group.

are known as primary, secondary and tertiary amines respectively, thus:

H / N—H \ H	R / N—H \ H	R / N—R \ H	R / N—R \ R
	primary amine	secondary amine	tertiary amine

If R is an alkyl (CH_3—, C_2H_5—, etc.) group, then the aqueous solution of the amine, presuming its solubility in water, is more strongly basic than that of ammonia. These compounds* are known as aliphatic amines. If R is an aryl group (C_6H_5—, $CH_3.C_6H_4$—, etc.), the amines belong to the aromatic series and are either very sparingly soluble or insoluble in water; experimental evidence indicates that the aromatic amines are less strongly basic than ammonia, due to the potentially electro-negative or acidic character of the aromatic group or groups.

The groups substituting the hydrogen atoms attached to the nitrogen in secondary or tertiary amines may be the same or different from each other. In the former case the compounds are frequently described as simple amines and in the latter case as mixed amines: a mixed amine may also contain both aliphatic and aromatic groups.

The alkalinity of aqueous solutions of primary, secondary and tertiary amines arises from these compounds behaving with water in a manner similar to ammonia: neither the substituted ammonium radicals, $[NRH_3]$—, $[NR_2H_2]$—, $[NR_3H]$—, nor the substituted ammonium hydroxides, $[NRH_3]OH$, $[NR_2H_2]OH$, $[NR_3H]OH$, have been isolated. The corresponding salts are, however, well known and, in most cases, stable; they may be formed, just as ammonium salts are formed, by the direct combination of the amines and acids, thus:

$$NRH_2 + HCl \rightarrow [NRH_3]Cl \qquad \text{cf. } [NH_4]Cl$$

$$2NR_2H + H_2SO_4 \rightarrow [NR_2H_2]_2SO_4 \qquad \text{cf. } [NH_4]_2SO_4$$

$$NR_3 + HBr \rightarrow [NR_3H]Br \qquad \text{cf. } [NH_4]Br$$

These salts behave exactly like the corresponding ammonium salts

* Ammonia and the amines are usually described as basic substances or bases. When pure none of these substances, however, have an alkaline reaction towards indicators. In the case of ammonia, the alkaline reaction of its aqueous solution is due to the presence of an excess of hydroxyl ions in the solution, which contains the following constituents in equilibrium:

$$NH_3 + H_2O \rightleftarrows NH_4OH \rightleftarrows [NH_4]^+ + [OH]^-$$

The alkalinity of the aqueous solutions of the amines is similarly explained.

Ammonia and the amines—the latter whether soluble in water or not—all combine directly with acids, forming ammonium and substituted ammonium salts respectively.

in almost all respects, and are decomposed by alkalis similarly to ammonium salts, with liberation of the amine:

$$[NRH_3]Cl + NaOH \rightarrow NRH_2 + NaCl + H_2O$$

$$[NR_2H_2]_2SO_4 + 2NaOH \rightarrow 2NR_2H + Na_2SO_4 + 2H_2O$$

The salts with hydrochloroauric and hydrochloroplatinic acids are generally highly crystalline and are important derivatives. The general formulae for these compounds are

$$[NRH_3]AuCl_4 \qquad [NRH_3]_2PtCl_6$$

$$[NR_2H_2]AuCl_4 \qquad [NR_2H_2]_2PtCl_6$$

$$[NR_3H]AuCl_4 \qquad [NR_3H]_2PtCl_6$$

derived from the acids having the formulae $HAuCl_4$ and H_2PtCl_6, and the primary, secondary and tertiary amines respectively. These chloroaurates and chloroplatinates are conveniently employed not only for distinguishing and identifying amines but also for the determination of their molecular weights. When these salts are heated they decompose completely, leaving a residue of the metal.

In the case of a chloroaurate, the amount of gold remaining after decomposing by heating a known amount of the compound can be determined. From this, the molecular weight in grams of the chloroaurate can be calculated as the amount which would contain the atomic weight in grams of gold. By subtracting from this molecular weight (in grams) of the salt the molecular weight (in grams) of hydrochloroauric acid ($HAuCl_4$), the molecular weight (in grams) of the amine is obtained. By decomposing the chloroplatinate in a similar manner, the molecular weight of the salt can be determined. On subtracting from this the molecular weight of hydrochloroplatinic acid (H_2PtCl_6), twice the molecular weight of the amine is obtained (p. 11).

The tertiary amines generally combine readily, frequently on simple mixing in a suitable nonhydroxylic solvent, with alkyl halides to form *quaternary ammonium compounds*. These usually highly crystalline salts have the general formula $[NR_4]X$, where X = halogen from the alkyl halide (RX). For example, tetramethylammonium iodide which superficially resembles ammonium iodide is formed:

$$N(CH_3)_3 + CH_3I \rightarrow [N(CH_3)_4]I$$

Such a quaternary ammonium compound is dissociated in aqueous solution thus:

$$[N(CH_3)_4]I \rightarrow [N(CH_3)_4]^+ + [I]^-$$

When heated it decomposes thus:

$$[N(CH_3)_4]I \rightarrow N(CH_3)_3 + CH_3I$$

These quaternary ammonium halides are salts of the quaternary ammonium hydroxides, which can be obtained from them by mixing (and warming, if necessary) the aqueous solution of the halide with silver hydroxide:

$$[N(R_4)]X + AgOH \rightarrow AgX + [N(R_4)]OH$$

On evaporation at ordinary temperatures of the aqueous filtrate from the silver halide and excess of silver hydroxide, the quaternary ammonium hydroxide remains. These hydroxides are usually difficult to obtain crystalline, being highly hygroscopic. The aliphatic quaternary ammonium hydroxides are readily soluble in water and are strong bases, approximating in strength to sodium and potassium hydroxides. Tetramethylammonium hydroxide, the simplest of the quaternary ammonium hydroxides, on being distilled decomposes into methyl alcohol and trimethylamine:

$$[N(CH_3)_4]OH \rightarrow CH_3OH + N(CH_3)_3$$

In other cases the decomposition of the hydroxide may be more complicated; for example, tetraethylammonium hydroxide on being heated decomposes thus:

$$[N(C_2H_5)_4]OH \rightarrow \underset{\text{triethylamine}}{N(C_2H_5)_3} + C_2H_4 + H_2O$$

The salts of these bases are dissociated in aqueous solution to an extent almost equal to that of the corresponding salts of the alkali metals. The formulae of some typical salts are:

Chlorides $[NR_4]Cl$ Sulphates $[NR_4]_2SO_4$

Chloroaurates $[NR_4]AuCl_4$ Chloroplatinates $[NR_4]_2PtCl_6$

When all four radicals in a quaternary ammonium salt or hydroxide are different from each other, $[NR^{i}.R^{ii}.R^{iii}.R^{iv}]X$, the compounds are capable of exhibiting optical activity (p. 348).

The isomerism which may be exhibited by primary, secondary and tertiary amines to a very considerable extent may be illustrated by the following three classes of isomeric compounds:

Molecular formula		
C_2H_7N	$N(C_2H_5)(H)(H)$ — N bonded to C_2H_5, H, H	$N(CH_3)(CH_3)(H)$ — N bonded to CH_3, CH_3, H
	monoethylamine	dimethylamine

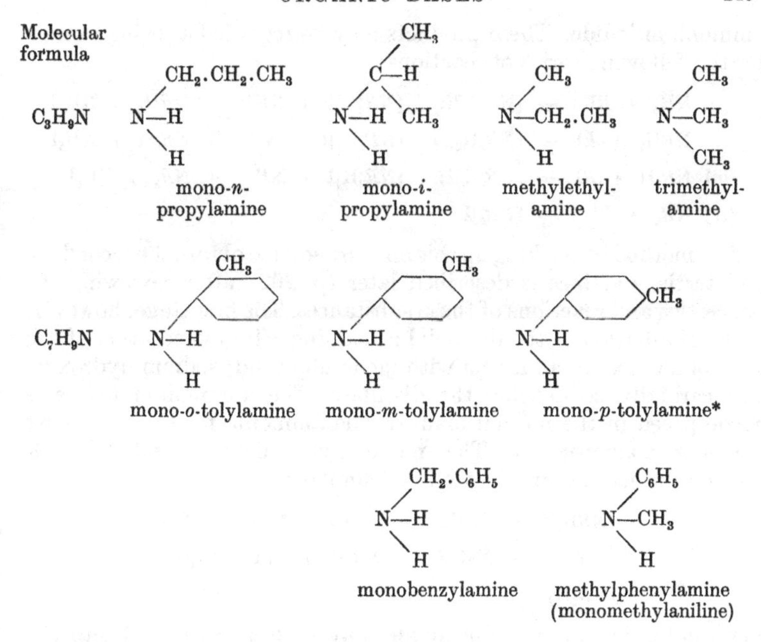

The characteristic reactions of primary and secondary amines are due to their containing hydrogen atoms directly attached to nitrogen; since the tertiary amines do not contain hydrogen atoms directly attached to nitrogen they are characterised by being chemically less reactive.

The close relationship between ammonia and ammonium salts and the simple aliphatic amines and substituted ammonium salts (including quaternary ammonium salts) is illustrated by the fact that these organic compounds can be prepared directly from ammonia. The reaction, described as the sealed tube reaction, was investigated by Hofmann in 1849. It is carried out by heating an alkyl iodide (RI)—the bromide or chloride may be used—with an alcoholic solution of ammonia in a sealed glass tube at about 100°. If an alkyl iodide be used, the reaction product consists of a mixture of the monoalkyl-, dialkyl-, trialkyl- and tetraalkyl-ammonium iodides together with

* The primary tolylamines, usually known as toluidines, may have their constitutional formulae written more briefly:

$$C_6H_4\begin{cases}NH_2\ (1)\\CH_3\ (2)\end{cases},\quad C_6H_4\begin{cases}NH_2\ (1)\\CH_3\ (3)\end{cases},\quad C_6H_4\begin{cases}NH_2\ (1)\\CH_3\ (4)\end{cases}$$

ammonium iodide. These products may be regarded as being formed by the following series of reactions:

(i) $NH_3 + RI \rightarrow [NRH_3]I$; $[NRH_3]I + NH_3 \rightarrow NRH_2 + NH_4I$

(ii) $NRH_2 + RI \rightarrow [NR_2H_2]I$; $[NR_2H_2]I + NH_3 \rightarrow NR_2H + NH_4I$

(iii) $NR_2H + RI \rightarrow [NR_3H]I$; $[NR_3H]I + NH_3 \rightarrow NR_3 + NH_4I$

(iv) $NR_3 + RI \rightarrow [NR_4]I$

The method of working up this mixture so as to obtain the secondary and tertiary amines is described later (p. 262), after reviewing the properties and reactions of these substances. The first stage, however, is the distillation of the dry solid remaining after evaporation of the alcohol and excess ammonia with (generally solid) sodium hydroxide and carefully condensing the distillate. The ammonium iodide is decomposed in the normal manner, the ammonia for the most part escaping condensation. The monoalkyl-, dialkyl- and trialkyl-ammonium iodides are decomposed similarly:

$$[NRH_3]I + NaOH \rightarrow NRH_2 + NaI + H_2O$$

$$[NR_2H_2]I + NaOH \rightarrow NR_2H + NaI + H_2O$$

$$[NR_3H]I + NaOH \rightarrow NR_3 + NaI + H_2O$$

The quaternary ammonium iodide may be first converted into the corresponding hydroxide, thus:

$$[NR_4]I + NaOH \rightarrow [NR_4]OH + NaI$$

which is then decomposed into the tertiary amine, NR_3, and the alcohol, ROH (or the olefine and water), as already indicated. The distillate is essentially therefore a mixture of the primary, secondary and tertiary amines.

PRIMARY ALIPHATIC AMINES*

Although these compounds are obtained in the Hofmann sealed tube reaction they are not usually isolated from the product. They are prepared from the amides of the fatty acids by Hofmann's reaction, which has already been described (p. 213). From acetamide, for example, monomethylamine is obtained through a series of reactions schematically represented:

$$\underset{}{CH_3.CONH_2} \xrightarrow{Br_2} \underset{\text{monobromoacetamide}}{CH_3.CONHBr} \xrightarrow{KOH} \underset{\text{methyl}\textit{iso}\text{cyanate}}{CH_3.N:C:O} \xrightarrow{KOH} \underset{\text{monomethylamine}}{CH_3.NH_2}$$

* These primary amines may be considered as amino derivatives of the aliphatic hydrocarbons; thus, monomethylamine and monoethylamine may be systematically described as aminomethane and aminoethane respectively.

This is probably the simplest method for preparing the lower primary aliphatic amines in a state of purity.

They can also be obtained by reduction of the aliphatic cyanides with nascent hydrogen. For example, when methyl cyanide is reduced with sodium and ethyl alcohol, monoethylamine is obtained (Mendius' reaction, 1862):

$$CH_3.C\vdots N + 2H_2 \rightarrow CH_3.CH_2.NH_2$$

It has been indicated that the alkyl *iso*cyanates (R.N : C : O) are probably intermediate products in the conversion of amides into amines by Hofmann's reaction, and it has been shown that the alkyl *iso*cyanates yield primary amines on hydrolysis with alkalis, thus:

$$CH_3\text{—}N\text{=}C\text{=}O + \begin{matrix} KO\vdots H \\ KO\vdots H \end{matrix} \rightarrow K_2CO_3 + CH_3.NH_2$$

The alkyl *iso*cyanides (RNC),* isomeric with the alkyl cyanides, are also converted into primary amines by hydrolysis with alkalis, a formate being produced at the same time:

$$R\text{—}N\text{=}C + \begin{matrix} KO\vdots H \\ HO\vdots H \end{matrix} \rightarrow RNH_2 + H.COOK\dagger$$

Monomethylamine, $CH_3.NH_2$, is a colourless gas, b.p. $-6°$, and *monoethylamine* is a colourless liquid, b.p. 16·5°. They are more inflammable than ammonia, of which their odour is strongly reminiscent. In their chemical properties these primary aliphatic amines are typical; they are freely soluble in water, giving

* The constitutional formula of the *iso*cyanides at one time was written R—N≡C, in which the carbon being regarded as quadrivalent necessitates the nitrogen being quinquevalent. For stereochemical and other reasons, the constitution is better represented as R—N=C (compare carbon monoxide), in keeping with their properties as unsaturated compounds. The alkyl *iso*cyanides are produced with the cyanides when alkyl iodides are heated with potassium cyanide and constitute the main product of the action of silver cyanide and alkyl iodides. They can be prepared from primary amines (p. 254).

† Other methods for obtaining primary aliphatic amines may be briefly mentioned. The *nitroparaffins* (obtained by the action of silver nitrite on alkyl iodides:

$$RI + AgNO_2 \rightarrow RNO_2 + AgI)$$

when reduced with nascent hydrogen (tin and hydrochloric acid) are converted finally into primary amines, the corresponding alkylhydroxylamines being intermediate products:

$$R\text{—}NO_2 + 2H_2 \rightarrow R\text{—}N\begin{matrix} H \\ OH \end{matrix} + H_2O;\quad R\text{—}N\begin{matrix} H \\ OH \end{matrix} + H_2 \rightarrow R\text{—}N\begin{matrix} H \\ H \end{matrix} + H_2O$$

Another method, often actually employed in preparing pure primary aliphatic amines, involves the use of monoalkylanilines as starting material. This is described later (p. 286).

alkaline solutions due to the formation and ionisation of a substituted ammonium hydroxide:

$$NRH_2 + H_2O \rightleftarrows [NRH_3]OH \rightleftarrows [NRH_3]^+ + [OH]^-$$

Under ordinary conditions, like ammonia, they combine directly with hydrogen chloride, producing colourless substituted ammonium chlorides:

$$NRH_2 + HCl \rightarrow [NRH_3]Cl$$

The chlorides and other salts of these substituted ammonium bases, the formulae of which have already been indicated, are, as a rule, somewhat less soluble in water than the corresponding ammonium salts.

When a solution of a primary aliphatic amine in excess of a dilute solution of mineral acid is treated with a cold aqueous solution of sodium nitrite, the nitrous acid produced reacts at once with the amine, nitrogen being evolved and an alcohol produced by the replacement of the $—NH_2$ group by hydroxyl:

$$R—NH_2 + HO—N{=}O \rightarrow R—OH + N_2 + H_2O^*$$

This reaction of nitrous acid with primary aliphatic amines is markedly different from that with primary aromatic amines. By this reaction, also, primary aliphatic amines are distinguished from secondary and tertiary aliphatic amines. This reaction has additional importance in that it provides a method for obtaining the next lower compound in a homologous series of primary alcohols and the corresponding acids. For example, from ethyl alcohol and acetic acid, methyl alcohol and formic acid can be obtained by a series of reactions represented schematically:

$$CH_3.CH_2OH \xrightarrow{\text{oxidation}} CH_3.CO_2H \rightarrow CH_3.CO_2NH_4 \xrightarrow{-H_2O} CH_3.CONH_2 \xrightarrow{\text{Hofmann's reaction}}$$

$$CH_3.NH_2 \xrightarrow{HNO_2} CH_3OH \xrightarrow{\text{oxidation}} H.CO_2H$$

Theoretically, starting with a primary alcohol containing n carbon atoms and repeating this series of reactions $n-1$ times, it is possible to descend the homologous series completely, ending with the lowest member, methyl alcohol.

The reverse process, the ascending of a homologous series step by step, can be accomplished by a series of reactions each one of which has already been described and which may be represented:

$$CH_3OH \xrightarrow{I_2 + \text{phosphorus}} CH_3I \xrightarrow{KCN} CH_3.CN \xrightarrow{\text{reduction}} CH_3.CH_2NH_2 \xrightarrow{HNO_2} CH_3.CH_2OH$$

* This reaction recalls the preparation of nitrogen by the action of heat on ammonium nitrite:

$$NH_4—O—N{=}O \rightarrow 2H_2O + N_2$$

Theoretically, starting with the lowest primary alcohol, methyl alcohol, any primary alcohol of the same series containing n carbon atoms can be obtained by repeating this series of reactions $n-1$ times.

Primary amines react with acid chlorides, anhydrides and esters yielding *N*-monosubstituted amides. For example, N-*methylacetamide*, m.p. 28°, b.p. 206°, is easily obtained as indicated:

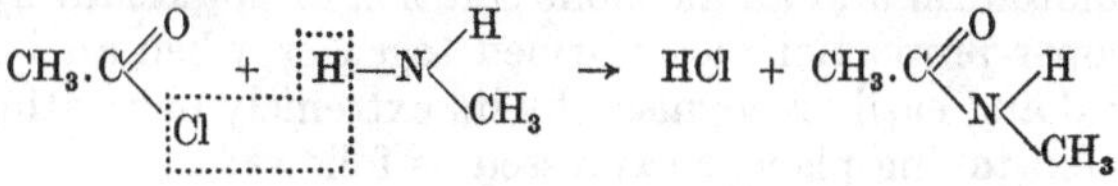

$$CH_3.C\begin{matrix}\nearrow O \\ \searrow Cl\end{matrix} + H{-}N\begin{matrix}\nearrow H \\ \searrow CH_3\end{matrix} \rightarrow HCl + CH_3.C\begin{matrix}\nearrow O \\ \searrow N\begin{matrix}\nearrow H \\ \searrow CH_3\end{matrix}\end{matrix}$$

N-*Methylbenzamide*, $C_6H_5.C\begin{matrix}\nearrow O \\ \searrow N\begin{matrix}\nearrow H \\ \searrow CH_3\end{matrix}\end{matrix}$, m.p. 78°, is similarly formed using benzoyl chloride. The production of the corresponding di-*N*-alkyl amide from the diethyl ester of oxalic acid (p. 263) is used in the separation of the products of the Hofmann sealed tube reaction. The reaction takes place on simply mixing the two reactants:

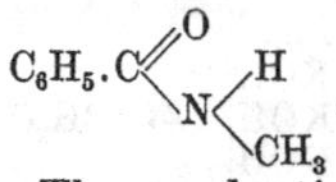

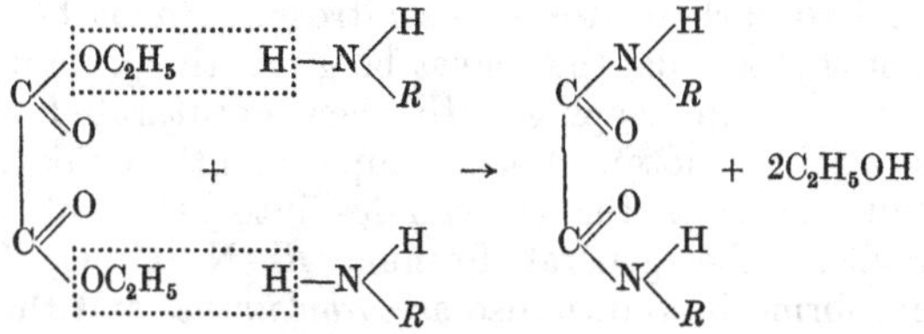

$$\begin{matrix} C\begin{matrix}\nearrow OC_2H_5 \\ \searrow O\end{matrix} \\ | \\ C\begin{matrix}\nearrow O \\ \searrow OC_2H_5\end{matrix}\end{matrix} + \begin{matrix} H{-}N\begin{matrix}\nearrow H \\ \searrow R\end{matrix} \\ \\ H{-}N\begin{matrix}\nearrow H \\ \searrow R\end{matrix}\end{matrix} \rightarrow \begin{matrix} C\begin{matrix}\nearrow N\begin{matrix}\nearrow H \\ \searrow R\end{matrix} \\ \searrow O\end{matrix} \\ | \\ C\begin{matrix}\nearrow O \\ \searrow N\begin{matrix}\nearrow H \\ \searrow R\end{matrix}\end{matrix}\end{matrix} + 2C_2H_5OH$$

Dimethyloxamide, $(CO.NHMe)_2$, m.p. 210°, and *diethyloxamide*, m.p. 179°, are both colourless substances and soluble in water.

The final product of the action of an alkyl iodide on a primary, secondary and a tertiary amine under similar conditions to those employed for the Hofmann sealed tube reaction must in each case be a quaternary ammonium iodide. A molecule of a primary amine would thus react with three molecules of an alkyl iodide, a secondary amine with two and a tertiary amine with one molecule of an alkyl iodide. The alkyl iodide used being known, analysis of the quaternary iodide produced may afford information as to the nature of the amine. For example, it has been pointed out that there are four isomeric amines having the molecular formula C_3H_9N, viz. (*a*) *mono*-n-*propylamine* (b.p. 50°), (*b*) *mono*-i-*propylamine* (b.p. 33·5°), (*c*) *methylethylamine* (b.p. 35°) and (*d*) *trimethylamine* (b.p. 3·5°). If these be treated with excess of methyl iodide (*a*) would yield trimethyl-*n*-propylammonium iodide, (*b*) would yield trimethyl-*i*-propylammonium iodide, (*c*) would yield trimethylethylammonium iodide and (*d*) would yield tetramethylammonium iodide. Analysis would distinguish between the two last quaternary compounds and would distinguish between the last two and the first two quaternary compounds, but would not distinguish between the first two. From this it would be possible to say whether the amine used was primary, secondary or tertiary, i.e. whether the original amine was either of the two monopropylamines, methylethylamine or trimethylamine.

The characteristic reaction of primary amines with chloroform in the presence of alkalis was discovered by Hofmann in 1868. The amine, chloroform and an alcoholic solution of potassium hydroxide in the proper proportions are warmed together, when an *iso*cyanide is produced and easily recognised by its extremely nauseating odour. The reaction taking place is expressed as follows:

$$R\text{—N}\begin{matrix}\text{H}\\ \text{H}\end{matrix} + \text{H—C}\begin{matrix}\text{Cl}\\ \text{Cl}\\ \text{Cl}\end{matrix} + \begin{matrix}\text{KOH}\\ \text{KOH}\\ \text{KOH}\end{matrix} \rightarrow 3H_2O + 3KCl + R\text{—N=C}$$

The *iso*cyanides are obtained unmixed with the isomeric cyanides (or nitriles) by this reaction, and it is clear from the above mode of formation that the nitrogen atom in these compounds must be connected to two carbon atoms. As nitrogen cannot be quinquevalent in such compounds, they must be given the general formula R—N=C, which is in keeping with their unsaturated character (compare p. 251 footnote). For example, gentle oxidation with mercuric oxide converts the *iso*cyanides into esters of *iso*cyanic acid which have the general formula R—N=C=O. The *iso*cyanides were formerly known also as *carbylamines* and their odour is still frequently referred to as the carbylamine odour. *Methyl*iso*cyanide*, CH_3—N=C, prepared from monomethylamine has b.p. 59°; *ethyl*iso*cyanide*, C_2H_5—N=C, has b.p. 79° and when heated in a sealed tube at about 200° is converted into ethyl cyanide or propionitrile:

$$C_2H_5\text{—N=C} \rightarrow C_2H_5\text{—C}\equiv\text{N}$$

Secondary Aliphatic Amines

These compounds having the general formula NR_2H are separated from the products of the Hofmann sealed tube reaction.

They may be conveniently obtained in a state of purity from mixed tertiary aliphatic-aromatic amines of the type $\text{N}\begin{matrix}Ar\\ R\\ R\end{matrix}$, where Ar represents a phenyl or suitably substituted phenyl group and R represents an aliphatic (alkyl) group. Such compounds as dimethylaniline, $C_6H_5.N(CH_3)_2$, and diethylaniline, $C_6H_5.N(C_2H_5)_2$, are prepared starting with aniline (or phenylamine, $C_6H_5.NH_2$)—the typical aromatic primary amine—and are described later (p. 286). The preparation of either dimethylamine, $N(CH_3)_2H$, or diethylamine,

$N(C_2H_5)_2H$, from dimethylaniline or diethylaniline respectively is carried out as indicated below; it is only necessary to describe the preparation of dimethylamine, that of diethylamine being analogous.

Dimethylaniline is dissolved in an aqueous solution containing rather more than two equivalents of hydrochloric acid, and this solution is carefully mixed with an aqueous solution of one equivalent of sodium nitrite at the ordinary temperature, when the *hydrochloride of* p-*nitrosodimethylaniline* (m.p. 177°) separates in yellow needles, sparingly soluble in water:

$$(CH_3)_2N{-}C_6H_4{-}\boxed{H + HO}NO + HCl$$

$$\rightarrow O{:}N{-}C_6H_4{-}N(CH_3)_2.HCl + H_2O$$

This compound may be filtered off and when suspended in water and treated with the requisite quantity of alkali the p-*nitrosodimethylaniline*, bright green leaflets, m.p. 85°, insoluble in water and easily soluble in ether, is formed.

$$O{:}N{-}C_6H_4{-}N(CH_3)_2.HCl + NaOH$$

$$\rightarrow O{:}N{-}C_6H_4{-}N(CH_3)_2 + NaCl + H_2O$$

When this *p*-nitroso- compound is heated with alkali—it is only necessary to submit it to steam distillation in the presence of alkali—it undergoes hydrolysis, *dimethylamine* distilling over, the residue being a solution of the sodium salt of *p*-nitrosophenol:

$$O{:}N{-}C_6H_4{-}\vdots N(CH_3)_2 \;\; (Na{-}O{-}\vdots H) \rightarrow O{:}N{-}C_6H_4{-}O{-}Na + H{-}N(CH_3)_2$$

The strongly cooled distillate consists of an aqueous solution of the secondary aliphatic amine, which can be isolated by cautiously making strongly alkaline with a large excess of sodium or potassium hydroxide. The secondary amine forms an oily layer and can be separated, dried still further with solid potassium hydroxide and redistilled.

Dimethylamine, $N(CH_3)_2H$, b.p. 7·2°, and *diethylamine*, $N(C_2H_5)_2H$, b.p. 56°, are colourless and freely soluble in water, forming alkaline solutions more strongly basic than the corresponding solutions of ammonia. Their odour, while strongly ammoniacal, is distinctly fish-like and this becomes more accentuated in the case of the tertiary aliphatic amines. They combine directly with hydrogen chloride under ordinary conditions and generally form substituted ammonium salts of the types already indicated.

By their reaction with nitrous acid, the secondary aliphatic amines are distinguished from the primary and tertiary compounds. When a solution of a secondary aliphatic amine in excess of dilute hydrochloric acid is treated at ordinary temperatures or in the cold with sodium nitrite, a yellow oil is produced; this is a nitrosoamine which is insoluble in, and lighter than, water and which is formed as indicated:

$$(CH_3)_2N\text{—}H + H\text{—}O\text{—}N{=}O \rightarrow (CH_3)_2N\text{—}N{=}O + H_2O$$

and can be readily separated from the reaction mixture. *Dimethylnitrosoamine*, $(CH_3)_2N.NO$, having b.p. 148°, and *diethylnitrosoamine*, $(C_2H_5)_2N.NO$, having b.p. 177°, are representatives of this class of compounds, which are decomposed by boiling with concentrated hydrochloric acid, forming the hydrochloride of the secondary amine and nitrous acid which is, of course, decomposed during the reaction

$$(CH_3)_2N\text{—}N{=}O + HCl \rightarrow \left[(CH_3)_2N\langle{}^{H}_{H}\right]Cl + H\text{—}O\text{—}N{=}O$$

When the oxides of nitrogen have been driven off, the aqueous residue may be made alkaline and the secondary amine isolated in the usual manner. This procedure may be adopted, as indicated below, for separating secondary aliphatic amines from the mixture of products of the Hofmann sealed tube reaction.

Secondary aliphatic amines do not react with chloroform (compare primary amines). They react with acid chlorides giving dialkylamides; for example, from dimethylamine and acetyl chloride (or acetic anhydride) N-*dimethylacetamide*, b.p. 165·5°, is obtained:

$$CH_3\text{—}C(=O)Cl + HN(CH_3)_2 \rightarrow CH_3\text{—}C(=O)N(CH_3)_2 + HCl$$

The product of the reaction of dimethylamine and benzoyl chloride is N-*dimethylbenzamide*, m.p. 41°:

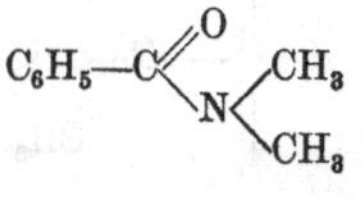

$$C_6H_5{-}C(=O){-}N(CH_3)_2$$

These compounds are also formed from esters of monocarboxylic acids. The product, however, of interaction between a secondary aliphatic amine and ethyl oxalate is an ester-amide. The reaction in the case of dimethylamine is:

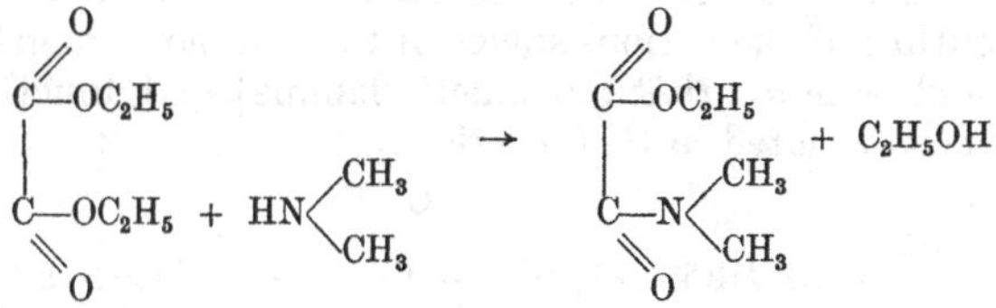

$$\begin{array}{l} O{=}C{-}OC_2H_5 \\ \vert \\ O{=}C{-}OC_2H_5 \end{array} + HN(CH_3)_2 \rightarrow \begin{array}{l} O{=}C{-}OC_2H_5 \\ \vert \\ O{=}C{-}N(CH_3)_2 \end{array} + C_2H_5OH$$

The *ethyl ester of* N-*dimethyloxamic* acid, b.p. 242–245°, is insoluble in water, and this property is made use of in separating the products of the Hofmann sealed tube reaction by means of ethyl oxalate. By distilling the compound with potassium hydroxide solution, the secondary amine can be obtained:

$$\begin{array}{l} COOC_2H_5 \\ \vert \\ CON(CH_3)_2 \end{array} + 2KOH \rightarrow \begin{array}{l} COOK \\ \vert \\ COOK \end{array} + C_2H_5OH + N(CH_3)_2H$$

TERTIARY ALIPHATIC AMINES

Tertiary aliphatic amines are the most easily isolated product of the Hofmann sealed tube reaction. They are less soluble in water than the corresponding primary and secondary amines, and their aqueous solutions are again more strongly alkaline than those of ammonia.

Trimethylamine, $N(CH_3)_3$, b.p. 3·5°, and *triethylamine*, $N(C_2H_5)_3$, b.p. 89°, are typical tertiary aliphatic amines. They are colourless and possess a strong ammoniacal and fish-like odour; they combine directly with acids generally, forming substituted ammonium salts of the types such as those mentioned above.

Trimethylamine can be produced in considerable quantity by the distillation of herring brine and beet molasses. Its production from these substances and its occurrence in many plants are probably due to the fact that it there exists in the form of a derivative of aminoacetic acid or glycine (p. 205). This compound is known as *betaïne* and has the constitution

$$\begin{array}{c} \text{H} \quad\quad \text{CH}_3{}^{*} \\ \text{H—}\overset{|}{\text{C}}\text{——}\overset{+}{\text{N}}\text{—CH}_3 \\ \quad\quad\quad \text{CH}_3 \\ \text{O=}\underset{|}{\text{C}}\text{——}\bar{\text{O}} \end{array}$$

from which the trimethylamine may be produced by bacterial decomposition. The characteristic odour of herring brine is due to trimethylamine.

Trimethylamine (in the form of trimethylammonium chloride) is the final product of the action of formaldehyde on ammonium chloride. Only the outline of the various stages of this reaction are given here, and it should be noted that monomethylamine† and dimethylamine have also been isolated in this reaction:

(i) $H{-}C{\overset{H}{\underset{O}{\lessdot}}} + NH_3(HCl) \rightarrow H{-}C(OH)(H){-}NH_2 \rightarrow H_2C{=}NH.HCl + H_2O$

(ii) $H_2C{=}NH.HCl + H_2O + H{-}C{\overset{H}{\underset{O}{\lessdot}}} \rightarrow \underline{(CH_3)NH_2.HCl} + H.CO_2H$

(iii) $H{-}C{\overset{H}{\underset{O}{\lessdot}}} + H_2N{-}CH_3.HCl \rightarrow H_2C{=}N(CH_3).HCl + H_2O$

(iv) $H_2C{=}N(CH_3).HCl + H_2O + H{-}C{\overset{H}{\underset{O}{\lessdot}}} \rightarrow \underline{(CH_3)_2NH.HCl} + H.CO_2H$

(v) $H{-}C{\overset{H}{\underset{O}{\lessdot}}} + 2HN(CH_3)_2.HCl \rightarrow H_2C[N(CH_3)_2.HCl]_2 + H_2O$

(vi) $H_2C[N(CH_3)_2.HCl]_2 \rightarrow \underline{N(CH_3)_3.HCl} + H_2C{=}N(CH_3).HCl$

* Compare this formula of betaïne with that of a typical ammonium salt:

$$CH_3{-}\overset{+}{N}(CH_3)_2{-}CH_2{-}\underset{\|}{C}{-}\bar{O} \quad (C{=}O) \qquad \left[CH_3{-}\overset{+}{N}(CH_3)_2{-}CH_3\right]\bar{X}$$

The constitution of betaïne was formerly written:

$$\begin{array}{c} CH_2{-}N{\equiv}(CH_3)_3 \\ | \quad\quad | \\ O{=}C{-}{-}O \end{array}$$

where the nitrogen is represented as being ordinarily quinquevalent.

† Hence, this is frequently a hydrolysis product of hexamethylenetetramine (p. 148).

This series of reactions indicates that formaldehyde is capable of acting as a methylating agent.

Apart from their ability to form salts with acids of the types already described, the aliphatic tertiary amines are characterised by chemical inactivity in comparison with the corresponding primary and secondary amines. Having no hydrogen directly attached to nitrogen, they do not react with chloroform, nitrous acid, acid chlorides (and/or acid anhydrides) or esters, the four reagents—apart from formaldehyde—which have been specifically mentioned in reactions of primary and secondary amines of the aliphatic series. On the other hand, they react energetically with alkyl halides, yielding quaternary ammonium salts:

$$NR_3 + RX \rightarrow [NR_4]X$$

the reaction usually being carried out by mixing the reactants in a suitable solvent (usually ether), cooling being frequently necessary.

The quaternary ammonium halides are the best known of the quaternary ammonium salts, since they are formed so easily in the manner indicated. A quaternary salt like dimethyldiethylammonium iodide can be formed in at least four ways:

$$\left.\begin{array}{ll}\text{(i)} & N(CH_3)_2H + 2C_2H_5I \rightarrow [N(CH_3)_2(C_2H_5)_2]I + HI \\ \text{(ii)} & N(C_2H_5)_2H + 2CH_3I \rightarrow [N(C_2H_5)_2(CH_3)_2]I + HI\end{array}\right\} \text{(p. 250)}$$

$$\text{(iii)} \quad N(CH_3)_2(C_2H_5) + C_2H_5I \rightarrow [N(CH_3)_2(C_2H_5)_2]I$$

$$\text{(iv)} \quad N(C_2H_5)_2(CH_3) + CH_3I \rightarrow [N(C_2H_5)_2(CH_3)_2]I$$

and the quaternary ammonium iodide produced by the four reactions is the same compound. The constitution of these quaternary ammonium compounds should therefore be expressed in such a way as to indicate no difference, as far as its relation to the nitrogen atom is concerned, between any of the alkyl groups. These salts in aqueous solution are dissociated similarly to the corresponding salts of the alkali metals and to ordinary ammonium salts:

$$NaX \rightarrow [Na]^+ + [X]^-$$
$$NH_4X \rightarrow [NH_4]^+ + [X]^-$$

and their constitution is expressed according to the electronic theory of valency as:

$$\left[\begin{array}{c} R \\ \cdot\cdot \\ R : N : R \\ \cdot\cdot \\ R \end{array}\right] X \rightarrow \left[\begin{array}{c} R \\ \cdot\cdot \\ R : N : R \\ \cdot\cdot \\ R \end{array}\right]^+ + [X]^-$$

compare

$$\left[\begin{array}{c} H \\ \cdot\cdot \\ H : N : H \\ \cdot\cdot \\ H \end{array}\right] X \rightarrow \left[\begin{array}{c} H \\ \cdot\cdot \\ H : N : H \\ \cdot\cdot \\ H \end{array}\right]^+ + [X]^-$$

While the nitrogen atom has been regarded as trivalent in ammonia and the amines becoming quinquevalent in ammonium and quaternary ammonium compounds, the five valencies in the latter are not all of the same kind. One of the five valencies is an electrovalency and the nitrogen in forming the stable ammonium complex becomes quadri-covalent like the carbon atom in methane or a substituted methane. It is shown later (p. 347) that when the four groups attached to the nitrogen atom by covalencies in an ammonium compound—no matter what the atom or group attached by an electrovalency may be—are different from each other, such a compound of the type [N*wxyz*]*X* is capable of exhibiting optical activity just like a carbon compound of the type C*wxyz*, e.g. lactic acid; and the four atoms or groups attached by covalencies to the nitrogen atom in an ammonium compound must have a tetrahedral configuration in space as is the case of the four atoms or groups attached to a carbon atom in methane or a substituted methane.

The general properties of quaternary salts and hydroxides have already been described (p. 247).

An interesting quaternary ammonium compound which can be isolated from *lecithin** is *trimethyl-β-hydroxyethylammonium hydroxide* or *choline*, the constitution of which is:

$$\left[\begin{matrix}CH_3 & & CH_3 \\ & N & \\ CH_3 & & CH_2.CH_2OH\end{matrix}\right] OH$$

* Lecithin belongs to a class of organic compounds known as *lipins* and particularly to the sub-class *monoaminophospholipins* in which the nitrogen and phosphorus are present in the atomic ratio of 1 : 1. Lecithin is isolated from the *phosphorus containing fats* chiefly of the tissues of the brain, heart and liver and it is usually obtained from egg yolk. When such tissues are extracted with ether and acetone is added to the ethereal extract, the phospholipins are precipitated along with other substances. Of this precipitate, part is soluble in cold ethyl alcohol and that portion of this cold-alcohol-soluble material which is soluble in ether consists chiefly of lecithin.

The hydrolytic products of lecithin are fatty acid, frequently the unsaturated oleic acid, choline and glycerophosphoric acid, and part of the glycerophosphoric acid may be further hydrolysed to glycerol and phosphoric acid:

Lecithin $\rightarrow$ fatty acid + oleic acid + choline + glycerophosphoric acid.

The glycerophosphoric acid obtained from lecithin may be optically active (p. 113), and in this case it has the constitution indicated:

$$\begin{matrix}CH_2OH \\ | \\ CHOH \\ | \\ CH_2{-}O{-}\overset{}{P}(O)(OH)_2\end{matrix} + H_2O \xrightarrow{\text{hydrolysis}} \begin{matrix}CH_2OH \\ | \\ CHOH \\ | \\ CH_2OH\end{matrix} + \begin{matrix}H{-}O \\ H{-}O{-}PO \\ H{-}O\end{matrix}$$

The fatty acid and oleic acid residues are present in lecithin as in an ordinary fat,

It is usually prepared from lecithin obtained from egg yolk by hydrolysis with baryta and isolated in the form of its orange-coloured chloroplatinate, $[(CH_3)_3N(CH_2.CH_2OH)]_2PtCl_6$, which is insoluble in alcohol.

Choline was synthesised by Wurtz (1868) by the action of ethylene oxide on a concentrated aqueous solution of trimethylamine at the ordinary temperature:

$$(CH_3)_3N + \begin{matrix} CH_2 \\ | \\ CH_2 \end{matrix}\!\!>O + H_2O \rightarrow \left[\begin{matrix} CH_3 \\ CH_3 \end{matrix}\!\!>N\!\!<\!\begin{matrix} CH_3 \\ CH_2.CH_2OH \end{matrix}\right]OH$$

The corresponding chloride was synthesised by heating a mixture of trimethylamine and glycol chlorhydrin (β-chloroethyl alcohol) in a closed tube in a boiling water-bath:

$$(CH_3)_3N + \begin{matrix} CH_2Cl \\ | \\ CH_2OH \end{matrix} \rightarrow \left[\begin{matrix} CH_3 \\ CH_3 \end{matrix}\!\!>N\!\!<\!\begin{matrix} CH_3 \\ CH_2.CH_2OH \end{matrix}\right]Cl$$

Choline is a colourless, crystalline, highly hygroscopic substance which has a strongly alkaline reaction. Being an alcohol, it yields an acetyl derivative, *acetyl choline*, $[(CH_3)_3N(CH_2.CH_2.O.OC.CH_3)]OH$, which has a pronounced physiological action. When choline is boiled

while the choline is present as an ester of the phosphoric acid residue through the alcoholic group of the quaternary ammonium hydroxide:

$$\begin{matrix} CH_2.O.OC.R \\ | \\ \mathbf{C}H.O.OC.R' \\ | \\ CH_2.O{-}P(O)(OH)_2 \end{matrix} \qquad \begin{matrix} CH_2.O.OC.R \\ | \\ \mathbf{C}H.O.OC.R' \\ | \\ CH_2.O{-}P(O)(OH){-}O.CH_2.CH_2.N(CH_3)_3\,OH \end{matrix}$$

The groups R and R' may be the same as or different from each other.

There are several lecithins known, and these differ according to the nature and position of the fatty and other acid residues present. The most common lecithin and the one usually known as *lecithin* yields stearic acid and oleic acid on hydrolysis and the following formulae indicate the relationships between triolein, tristearin and lecithin:

$$\underset{\text{triolein}}{\begin{matrix} CH_2.O.OC.C_{17}H_{33} \\ | \\ CH.O.OC.C_{17}H_{33} \\ | \\ CH_2.O.OC.C_{17}H_{33} \end{matrix}} \qquad \underset{\text{tristearin}}{\begin{matrix} CH_2.O.OC.C_{17}H_{35} \\ | \\ CH.O.OC.C_{17}H_{35} \\ | \\ CH_2.O.OC.C_{17}H_{35} \end{matrix}} \qquad \underset{\text{lecithin}}{\begin{matrix} CH_2.O.OC.C_{17}H_{35} \\ | \\ \mathbf{C}H.O.OC.C_{17}H_{33} \\ | \\ CH_2.O.P(O)(OH){-}O.CH_2.CH_2.N(CH_3)_3\,OH \end{matrix}}$$

Like the glycerophosphoric acid obtained from it, lecithin is optically active. In both compounds the carbon atom (in clarendon type) is attached to four different groups, rendering the molecular structure of these compounds asymmetric.

for some time in concentrated aqueous solution, it is decomposed into glycol (β-hydroxyethyl alcohol) and trimethylamine:

$$\left[\begin{matrix}CH_3 \\ CH_3\end{matrix}\!\!>\!N\!<\!\!\begin{matrix}CH_3 \\ CH_2.CH_2OH\end{matrix}\right]OH \rightarrow \begin{matrix}CH_3 \\ CH_3\end{matrix}\!\!>\!N\!-\!CH_3 + HO.CH_2.CH_2OH$$

and when it is oxidised under controlled conditions it is converted into betaïne (p. 258):

$$\left[\begin{matrix}CH_2.N\!\!\begin{matrix}\diagup CH_3 \\ -CH_3 \\ \diagdown CH_3\end{matrix} \\ | \\ CH_2OH\end{matrix}\right]OH + O_2 \rightarrow \begin{matrix}CH_2\!-\!\overset{+}{N}\!\!\begin{matrix}\diagup CH_3 \\ -CH_3 \\ \diagdown CH_3\end{matrix} \\ | \qquad \overset{-}{} \\ C\!-\!\!-\!O \\ \diagdown\!\!\diagdown O\end{matrix} + 2H_2O$$

betaïne

Choline is converted into the corresponding iodo-iodide, *trimethyl-β-iodoethylammonium iodide*, when heated with concentrated hydriodic acid and red phosphorus in a sealed tube at 120°:

$$\left[\begin{matrix}CH_3 \\ CH_3\end{matrix}\!\!>\!N\!<\!\!\begin{matrix}CH_3 \\ CH_2.CH_2OH\end{matrix}\right]OH + 2HI \rightarrow 2H_2O + \left[\begin{matrix}CH_3 \\ CH_3\end{matrix}\!\!>\!N\!<\!\!\begin{matrix}CH_3 \\ CH_2.CH_2I\end{matrix}\right]I$$

and this compound when heated with silver hydroxide (moist silver oxide) is converted into *neurine*, which, unlike choline, is highly poisonous and is a product (probably from choline) of putrefaction:

$$\left[\begin{matrix}CH_3 \\ CH_3\end{matrix}\!\!>\!N\!<\!\!\begin{matrix}CH_3 \\ CH_2.CH_2I\end{matrix}\right]I + 2AgOH \rightarrow \left[\begin{matrix}CH_3 \\ CH_3\end{matrix}\!\!>\!N\!<\!\!\begin{matrix}CH_3 \\ CH:CH_2\end{matrix}\right]OH + 2AgI + H_2O$$

Neurine is *trimethylvinylammonium hydroxide.* It has been synthesised by heating the product of condensation of trimethylamine and ethylene bromide (trimethyl-β-bromoethylammonium bromide) with silver hydroxide:

$$\left[\begin{matrix}CH_3 \\ CH_3\end{matrix}\!\!>\!N\!<\!\!\begin{matrix}CH_3 \\ CH_2.CH_2Br\end{matrix}\right]Br + 2AgOH \rightarrow \left[\begin{matrix}CH_3 \\ CH_3\end{matrix}\!\!>\!N\!<\!\!\begin{matrix}CH_3 \\ CH:CH_2\end{matrix}\right]OH + 2AgBr + H_2O$$

Separation of Amines from the Product of the Hofmann Sealed Tube Reaction

For the purpose of illustration, the reaction may be supposed to have been carried out using ammonia and ethyl iodide in the presence of ethyl alcohol. In such a case, the organic reaction product will be a mixture of monoethylammonium iodide $[N(C_2H_5)H_3]I$, diethylammonium iodide $[N(C_2H_5)_2H_2]I$, triethylammonium iodide $[N(C_2H_5)_3H]I$ and tetraethylammonium iodide $[N(C_2H_5)_4]I$ (p. 250).

The reaction mixture is evaporated to dryness, and the residue made strongly alkaline, by mixing with excess of potassium hydroxide, and distilled. The first three ammonium salts are decomposed as shown:

$$[N(C_2H_5)H_3]I + KOH \rightarrow KI + H_2O + N(C_2H_5)H_2$$

$$[N(C_2H_5)_2H_2]I + KOH \rightarrow KI + H_2O + N(C_2H_5)_2H$$

$$[N(C_2H_5)_3H]I + KOH \rightarrow KI + H_2O + N(C_2H_5)_3$$

The tetraethylammonium iodide may be converted into tetraethylammonium hydroxide and this is decomposed into triethylamine, ethylene and water during the distillation:

$$[N(C_2H_5)_4]I + KOH \rightarrow KI + H_2O + [N(C_2H_5)_4]OH \rightarrow N(C_2H_5)_3 + C_2H_4 + H_2O$$

The distillate, therefore, will consist of an aqueous solution of the three bases, which mixture can be freed from water by adding cautiously solid potassium hydroxide and separating the lower layer of strong aqueous solution of potassium hydroxide as long as it is formed.

(i) In order to obtain the three individual amines, the dried mixture of bases is treated with ethyl oxalate. The primary amine, monoethylamine, is converted into *diethyloxamide*, a colourless crystalline substance, soluble in hot water:

$$\begin{array}{l} O{=}C{-}OC_2H_5 \\ \quad | \\ O{=}C{-}OC_2H_5 \end{array} + \begin{array}{l} H.NH(C_2H_5) \\ H.NH(C_2H_5) \end{array} \rightarrow \begin{array}{l} O{=}C{-}N(H){-}C_2H_5 \\ \quad | \\ O{=}C{-}N(H){-}C_2H_5 \end{array} + 2C_2H_5OH$$

(*soluble in hot water*)

The secondary amine, diethylamine, is converted into the *ethyl ester of diethyloxamic acid*, a colourless substance insoluble in hot water:

$$\begin{array}{l} O{=}C{-}OC_2H_5 \\ \quad | \\ \quad COOC_2H_5 \end{array} + HN(C_2H_5)_2 \rightarrow \begin{array}{l} O{=}C{-}N(C_2H_5)_2 \\ \quad | \\ \quad COOC_2H_5 \end{array} + C_2H_5OH$$

(*insoluble in hot water*)

The tertiary amine, triethylamine, is unaffected. On distilling the reaction mixture the triethylamine distils and can be collected. The residue remaining is extracted with hot water, when the diethyloxamide dissolves and the aqueous solution can be filtered from the insoluble ethyl ester of diethyloxamic acid. The aqueous solution can either be evaporated to dryness and then distilled in the presence of excess potassium hydroxide or it can be mixed with excess alkali and distilled at once, when potassium oxalate is formed and the primary amine, monoethylamine, distils over:

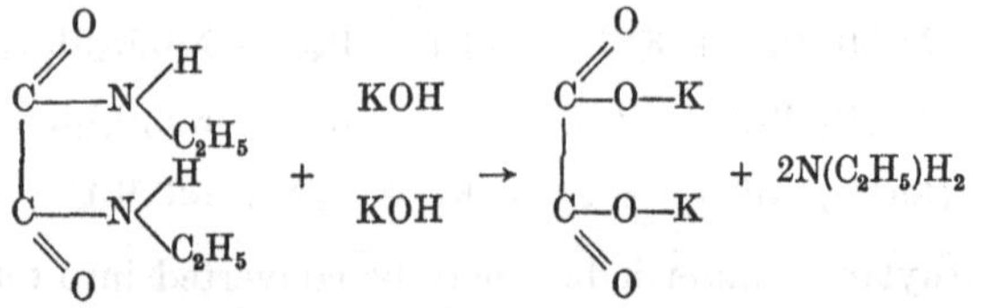

$$\begin{array}{l} O{=}C{-}N(H)(C_2H_5) \\ | \\ O{=}C{-}N(H)(C_2H_5) \end{array} + \begin{array}{c} KOH \\ KOH \end{array} \rightarrow \begin{array}{l} O{=}C{-}O{-}K \\ | \\ O{=}C{-}O{-}K \end{array} + 2N(C_2H_5)H_2$$

The solid ethyl ester of diethyloxamic acid is distilled separately also with potassium hydroxide, when the secondary amine, diethylamine, distils together with ethyl alcohol:

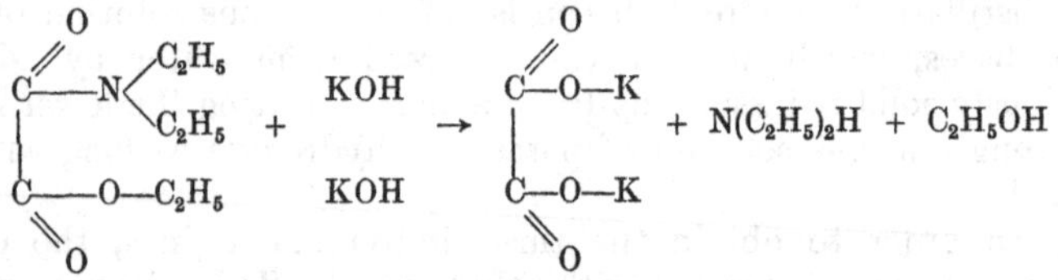

$$\begin{array}{l} O{=}C{-}N(C_2H_5)_2 \\ | \\ O{=}C{-}O{-}C_2H_5 \end{array} + \begin{array}{c} KOH \\ KOH \end{array} \rightarrow \begin{array}{l} O{=}C{-}O{-}K \\ | \\ O{=}C{-}O{-}K \end{array} + N(C_2H_5)_2H + C_2H_5OH$$

The aqueous solutions of monoethylamine and of triethylamine and the alcoholic solution of diethylamine can be freed from water and alcohol respectively by treatment with potassium hydroxide in the manner indicated above.

(ii) The other method for working up the mixture of products of the Hofmann sealed tube reaction involves the use of nitrous acid and leads only to the isolation of the secondary and tertiary amines. The mixture of the three bases obtained as indicated above is dissolved in an excess of dilute hydrochloric acid and to this cooled solution is slowly added an aqueous solution of sodium nitrite until the presence of free nitrous acid is just indicated in the solution by the potassium iodide-starch paper test. The primary amine reacts with the nitrous acid, nitrogen being evolved and the corresponding alcohol produced:

$$C_2H_5.NH_2 + O{:}N.OH \rightarrow C_2H_5OH + N_2 + H_2O$$

The secondary amine is converted into the nitrosoamine which separates as a yellow oil above the aqueous solution:

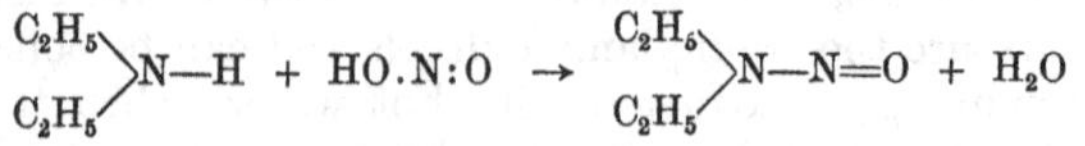

$$(C_2H_5)_2{>}N{-}H + HO.N{:}O \rightarrow (C_2H_5)_2{>}N{-}N{=}O + H_2O$$

while the tertiary amine is unaffected and remains dissolved in the dilute hydrochloric acid. When the reaction is completed the mixture is transferred to a separating funnel, and the nitrosoamine separated from the aqueous layer and washed with water. The nitrosoamine is decomposed by boiling with concentrated hydrochloric acid:

$$(C_2H_5)_2N{-}N{=}O + H_2O + HCl \rightarrow [N(C_2H_5)_2H_2]Cl + \underset{\text{(decomposes)}}{HNO_2}$$

and the secondary amine isolated from the resulting acid solution in the manner already described. The tertiary amine is isolated in a similar manner from the aqueous solution remaining after the separation of the nitrosoamine.

Primary Aromatic Amines

The primary amines of the aromatic series are, in all cases, most easily and conveniently produced by the reduction of the corresponding nitro-compounds. The simplest and most typical primary aromatic amine is *aniline*,* phenylamine, or aminobenzene, $C_6H_5.NH_2$.

Aniline is always produced by the reduction of nitrobenzene with nascent hydrogen, the reaction being generally represented:

$$C_6H_5.NO_2 + 3H_2 = C_6H_5{-}N\begin{matrix}H\\H\end{matrix} + 2H_2O$$

Tin and hydrochloric acid may be conveniently used as the reducing agent in the laboratory preparation, the reaction taking place being represented:

$$2C_6H_5NO_2 + 3Sn + 14HCl \rightarrow [(C_6H_5)NH_3]_2SnCl_6 + 2SnCl_4 + 4H_2O$$

In this case, concentrated hydrochloric acid in small portions is poured down an air condenser attached to a suitable flask containing the nitrobenzene and granulated tin. A brisk reaction takes place, and after this has subsided the mixture is gently boiled until all odour of the nitrobenzene has disappeared. The salt, *phenylammonium chlorostannate*, formed from aniline hydrochloride (phenylammonium chloride) and stannic chloride can be isolated by crystallisation, but it is usual to isolate the aniline at once by distilling the reaction mixture in steam after making it strongly alkaline with sodium hydroxide. The aniline passes over with the steam and is isolated from the cooled distillate by extraction of the latter with ether, drying the ethereal solution with solid potassium hydroxide, distilling off the ether on the water-bath, and finally distilling the residue in the ordinary way using an air condenser.

Other reducing agents, such as iron and hydrochloric acid and iron and acetic acid, may be used for the reduction of nitrobenzene and

* The name aniline was given to the compound by Fritzsche, who isolated it in 1841 by distilling indigo (Spanish, *anil*) with strong potassium hydroxide solution. Actually, Unverdorben obtained the same product by distillation of indigo by itself in 1826. Hofmann in 1843 proved that the product from indigo, the basic substance isolated in small quantity from coal tar oil by Runge in 1834 and named by him *cyanol*, and the substance obtained by Zinin in 1842 by the reduction of nitrobenzene by means of ammonium sulphide and named by him *benzidam*, are identical.

other aromatic nitro-compounds. On the technical scale, iron and hydrochloric acid are employed for the reduction of nitrobenzene to aniline; in this case it has been found that only about one-fortieth of the amount of hydrochloric acid required by the following equation is actually necessary:

$$C_6H_5NO_2 + 2Fe + 6HCl \rightarrow C_6H_5NH_2 + 2FeCl_3 + 2H_2O$$

This is explained by the fact that ferrous chloride, an intermediate product, is also a reducing agent capable of converting nitrobenzene into aniline.* Similarly, when acetic acid is used, the amount of the acid actually necessary is much smaller than that required theoretically according to an equation analogous to the above, owing to ferrous acetate being capable of reducing nitrobenzene to aniline.

Nitrobenzene may be reduced electrolytically to aniline, and this reduction may also be effected catalytically as shown by Senderens by passing water gas (carbon monoxide and hydrogen) and nitrobenzene vapour over such finely divided metals as copper and nickel. Another method of reduction which has been successfully employed is heating nitrobenzene with sodium disulphide:

$$C_6H_5NO_2 + Na_2S_2 + H_2O \rightarrow C_6H_5NH_2 + Na_2S_2O_3$$

Although it is not usually employed for the reduction of nitrobenzene, hydrogen sulphide in ethyl alcohol-ammonia solution can be employed; this method is often of convenience in reducing one of two or more nitro groups present in an aromatic compound. For example, *m*-dinitrobenzene dissolved in ethyl alcohol containing ammonia is converted into *m*-nitraniline when hydrogen sulphide is passed into the solution:

$$C_6H_4\langle\begin{matrix} NO_2\,(1) \\ NO_2\,(3) \end{matrix} + 3H_2S = C_6H_4\langle\begin{matrix} NH_2\,(1) \\ NO_2\,(3) \end{matrix} + 2H_2O + 3S\dagger$$

* The following equations have been suggested as representing the course of the reduction of nitrobenzene to aniline by means of iron and hydrochloric acid:

(i) $C_6H_5.NO_2 + 3Fe + 6HCl \rightarrow C_6H_5.NH_2 + 3FeCl_2 + 2H_2O$

(ii) $C_6H_5.NO_2 + 6FeCl_2 + H_2O \rightarrow C_6H_5.NH_2 + 3Fe_2Cl_4O$ (basic ferric chloride)

(iii) $4Fe_2Cl_4O + 3Fe \rightarrow Fe_3O_4 + 8FeCl_2$

† The methods indicated above by no means exhaust those available for the reduction of nitrobenzene and other aromatic nitro-compounds, and the particular reduction product isolated depends on the material and conditions employed. It has been shown, for example, that when nitrobenzene is reduced in acid or neutral solution the reaction proceeds in the following stages:

$$C_6H_5.NO_2 \rightarrow \underset{\text{nitrosobenzene}}{C_6H_5{-}N{=}O} \rightarrow \underset{\beta\text{-phenylhydroxylamine}}{\begin{matrix} C_6H_5 \\ H \end{matrix}\rangle N{-}O{-}H} \rightarrow C_6H_5.NH_2$$

Aniline has also been oxidised through β-phenylhydroxylamine and nitrosobenzene

Aniline is a colourless liquid, m.p. 6·2°, b.p. 181°. It is soluble in all neutral organic solvents and sparingly soluble in water; at ordinary temperatures about 35 c.c. of aniline dissolve in 1000 c.c. of water and about 52 c.c. of water dissolve in 1000 c.c. of aniline. Aniline is a

to nitrobenzene. *Nitrosobenzene* crystallises in colourless needles, m.p. 68°, and is readily oxidised to nitrobenzene or reduced to aniline. *β-Phenylhydroxylamine* is also a crystalline compound, m.p. 81°. On oxidation, it is rapidly converted first into *azoxybenzene* (*v.* below) and then into nitrosobenzene. β-Phenylhydroxylamine forms salts with acids and when warmed with mineral acids undergoes an interesting intramolecular change into p-*aminophenol*, m.p. 184°,

$$H(C_6H_4)-N(H)(O-H) \rightarrow H-O-(C_6H_4)-N(H)(H), \text{ or}$$

$$C_6H_5.NH(OH) \rightarrow C_6H_4\begin{cases}NH_2\,(1)\\OH\,(4)\end{cases}$$

an example of the mobility of the hydrogen in the *para-* position to a suitable substituting group in the benzene nucleus.

When the reduction of nitrobenzene is carried out in alkaline solution, other intermediate compounds are produced:

$$\begin{matrix}C_6H_5.NO_2\\C_6H_5.NO_2\end{matrix} \rightarrow \begin{matrix}C_6H_5-N\\ \|\\ C_6H_5-N{:}O\end{matrix} \rightarrow \begin{matrix}C_6H_5-N\\ \|\\ C_6H_5-N\end{matrix} \rightarrow \begin{matrix}C_6H_5-N-H\\ |\\ C_6H_5-N-H\end{matrix} \rightarrow \begin{matrix}C_6H_5.NH_2\\C_6H_5.NH_2\end{matrix}$$

	azoxybenzene	*azobenzene*	*hydrazobenzene*
	yellow crystals m.p. 36°	orange red crystals m.p. 68°	(*sym-diphenyl-hydrazine*) colourless crystals m.p. 131°

Azoxybenzene may also be produced by reaction between nitrosobenzene and β-phenylhydroxylamine:

$$\begin{matrix}C_6H_5-N{=}O\\ C_6H_5-N\begin{cases}O-H\\H\end{cases}\end{matrix} \rightarrow \begin{matrix}C_6H_5-N\\ \|\\ C_6H_5-N{:}O\end{matrix} + H_2O$$

These various products have been isolated and the conditions of their formation studied chiefly by the electrolytic (cathodic) reduction of nitrobenzene, during which the conditions, current density, etc., can easily be altered quantitatively.

Azoxybenzene is no longer given the constitution

$$C_6H_5-N\overset{}{—}N-C_6H_5 \quad (\text{N—N bridged by } O)$$

owing to the fact that azoxy compounds of this type $R'.N_2O.R$ occur in isomeric forms. In the above formula for azoxybenzene, the bond between one of the nitrogen atoms and the oxygen atom is described as a *co-ordinate link* (or semipolar double bond) formed by the 'lone pair' of electrons of a tervalent nitrogen being donated to the oxygen, thus completing the electronic octet of the latter. The co-ordinate link may conveniently be represented thus, $\equiv N \rightarrow O$. The electronic theory of valency similarly indicates the constitution of the *nitro* group as $-N\begin{cases}{=}O\\ \rightarrow O\end{cases}$ which is not only in keeping with the two oxygen atoms being dissimilarly joined

poison to which some individuals are particularly susceptible; in these cases the aniline causes marked cyanosis.

The aqueous solution of aniline is neutral, unlike that of a primary aliphatic amine. The acid nature of the phenyl group in reducing the basic nature of ammonia (in aqueous solution) is also shown by the properties of the salts of aniline; e.g. the hydrochloride, $C_6H_5NH_2.HCl$ or $[C_6H_5NH_3]Cl$, and the normal sulphate, $(C_6H_5NH_2)_2.H_2SO_4$ or $[C_6H_5NH_3]_2SO_4$, which in the dry state are colourless and crystalline, are extensively hydrolysed in aqueous solutions, which consequently have an acid reaction (compare the properties of phenol with those of a typical aliphatic alcohol such as methyl or ethyl alcohol).

Aniline is readily oxidised and the products of its oxidation are dependent on the conditions employed. They may be simple substances, such as those already mentioned (β-phenylhydroxylamine, nitrosobenzene, nitrobenzene, etc.), or much more complicated compounds or mixtures, of which *aniline black* is an example. The production of this material constitutes a delicate test for aniline and is carried out by adding an aqueous solution of potassium dichromate to a solution of the base in sulphuric acid. Another delicate test for aniline is the deep violet colour produced when its aqueous solution is mixed with an aqueous solution of bleaching powder.

The action of halogens on aniline affords another illustration of the readiness with which an already substituted benzene compound yields further substitution products (compare the action of halogens on benzene and on phenol). When an aqueous solution of chlorine or bromine is added to an aqueous solution of aniline or of an aniline salt, trichloro- or tribromo-aniline is precipitated immediately.

to the nitrogen atom but is also in keeping with the physical properties of nitro compounds.

When the electrolytic reduction of nitrobenzene is carried out in strongly acid solutions, β-phenylhydroxylamine is not isolated, but, instead, one obtains *p*-aminophenol on account of the above intramolecular change.

Hydrazobenzene is another substance which undergoes intramolecular change rapidly in the presence of mineral acids. In these conditions it is converted into *pp'-diaminodiphenyl* or *benzidine*. Analogous *benzidine transformations* also take place with certain homologues of hydrazobenzene and its substitution products:

$$C_6H_5-\underset{H}{N}-\underset{H}{N}-C_6H_5 \xrightarrow[\text{acids}]{\text{mineral}} H_2N-C_6H_4-C_6H_4-NH_2$$

Benzidine crystallises in colourless leaflets, m.p. 122°, and is employed in the manufacture of many important azo dyestuffs. It is also used in the presence of hydrogen peroxide for testing for blood (characteristic blue colour). Benzidine is usually prepared by reducing nitrobenzene to hydrazobenzene by means of zinc dust in the presence of an aqueous solution of sodium hydroxide, and then treating the product with acid. The compound may be conveniently isolated as the sparingly soluble sulphate.

Tribromoaniline, having the bromine atoms in the 2, 4 and 6 positions, is the more important. Its constitution may be represented:

NH_2
Br Br
Br

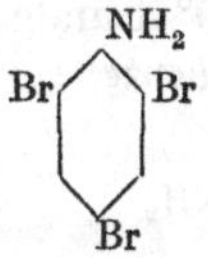

being analogous to that of tribromophenol. It has m.p. 119°, and owing to its low solubility in water may be used as the basis of a method for the quantitative estimation of aniline.

When aniline is heated with fuming sulphuric acid (concentrated sulphuric acid containing 8–10 per cent. excess sulphur trioxide), the aniline sulphate first formed undergoes intramolecular change and *sulphanilic acid* or p-*aminobenzenesulphonic acid* is formed:

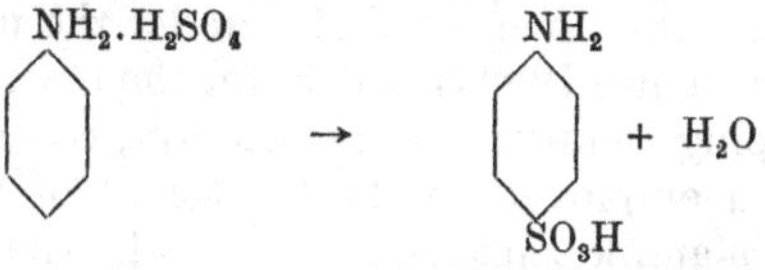

$NH_2.H_2SO_4$ → NH_2, SO_3H + H_2O

Sulphanilic acid is technically important for the production of certain dye-stuffs. It is a colourless crystalline compound, sparingly soluble in water. The crystals have the composition $C_6H_4(NH_2)\,(SO_3H)_{+}2H_2O$ and effloresce on exposure to air. When treated with bromine water it is converted into 2 : 4 : 6-tribromoaniline, the sulphonic group being replaced by bromine.

Sulphanilic acid forms salts with acids owing to the presence of the amino group, and also with bases owing to the presence of the sulphonic acid group. It belongs to the class of amino acids and is by itself a neutral substance. To signify the amphoteric nature of the compound its constitution may be represented:

$\overset{+}{N}$ (H, H, H)
$S\overset{-}{O_3}$

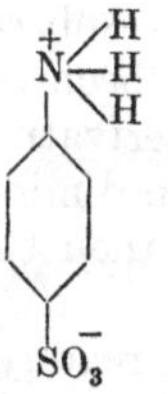

indicating the neutrality of the compound of which the hydrochloride and sodium salt have the following formulae respectively:

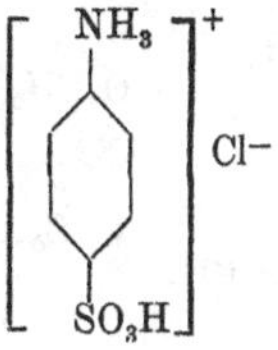

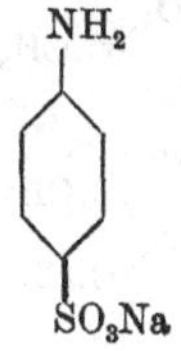

$[NH_3 \ldots SO_3H]^+ Cl^-$ $NH_2 \ldots SO_3Na$

Aniline also forms a salt with arsenic acid, and this *aniline arsenate* may be represented by the formula $C_6H_5NH_2.H_3AsO_4$. When aniline arsenate is heated at about 250° it undergoes a change similar to that of aniline sulphate and p-*aminobenzene arsonic acid* (Béchamp, Ehrlich) is formed:

$$\underset{\text{(benzene ring)}}{NH_2.H_3AsO_4} \rightarrow \underset{AsO(OH)_2}{\overset{NH_2}{\text{(benzene ring)}}} + H_2O \qquad \underset{\substack{AsO(OH)(ONa)\\ \text{atoxyl}}}{\overset{NH_2}{\text{(benzene ring)}}}$$

This colourless crystalline compound, like sulphanilic acid, is sparingly soluble in water and forms salts with acids and bases. The mono-sodium salt (*v.* above) as prepared always contains water of crystallisation and is known generally under the name of *atoxyl*; it has been employed since 1902 (Ehrlich) for the treatment of trypanosomiasis or sleeping sickness. More recently, certain derivatives of atoxyl have been employed in the treatment of parasitic disease. Starting from *p*-aminobenzenearsonic acid, the preparation of 3:3′-*diamino*-4:4′-*dihydroxyarsenobenzene*, the dihydrochloride of which is known as *arsphenamine* or *salvarsan* (compound "606" of Ehrlich's series, 1912), constituted Ehrlich's greatest achievement in this field. The compound has the constitution:

$$\begin{array}{cc} OH & OH \\ ClH.H_2N\text{(ring)} & \text{(ring)}NH_2.HCl \\ As & =\!=\!= \; As \end{array}$$

and is employed successfully in the treatment of syphilis and other protozoal diseases. As a result of detailed investigation both atoxyl and arsphenamine have been replaced in medical treatment by relatively easily obtained derivatives of these compounds. The object of such investigation is to diminish the toxicity of the parent substances without diminishing their therapeutic properties.

As a primary amine, aniline readily yields an acetyl derivative known as *acetanilide* or N-*phenylacetamide* when acted upon by acetyl chloride or acetic anhydride:

$$C_6H_5.N\begin{matrix}H\\H\end{matrix} + \begin{matrix}O\\Cl\end{matrix}\!\!>C.CH_3 \rightarrow HCl + C_6H_5.N\begin{matrix}H\\CO.CH_3\end{matrix}$$

$$2C_6H_5.N\begin{matrix}H\\H\end{matrix} + O\begin{matrix}\langle^{O}_{}\!C.CH_3\\ \langle_{O}\!C.CH_3\end{matrix} \rightarrow C_6H_5.N\begin{matrix}H\\CO.CH_3\end{matrix} + \underset{\text{aniline acetate}}{C_6H_5NH_2.HOOC.CH_3}$$

The reagents are carefully mixed together and the reaction completed by heating the mixture for two hours on the water-bath. Glacial acetic acid* can be used instead of the above reagents; in this case a considerable excess of the acid is employed and the mixture is gently boiled during about 8 hours. If acetic anhydride or acetyl chloride is used, the excess of the acetylating agent is destroyed by steam distillation or addition of water respectively. If glacial acetic acid is employed, most of the excess acid should be neutralised with alkali. The acetanilide which separates is recrystallised from hot water and obtained in colourless plates, m.p. 114°, b.p. 305°. Under the name *antifebrin* it has been used as a febrifuge or antipyretic. It is hydrolysed by sodium and potassium hydroxides and more readily by gently boiling with hydrochloric acid, aniline and acetic acid being obtained:

$$C_6H_5.N{<}^{H}_{C(=O)CH_3} + H_2O \rightarrow C_6H_5.NH_2 + CH_3.CO_2H$$

When acetyl chloride is used in the preparation of acetanilide some *diacetanilide*, colourless crystals, m.p. 37°, is always produced, and this compound can be obtained by the action of acetyl chloride on acetanilide or by the action of a large excess of acetic anhydride on aniline:

$$C_6H_5.N{<}^{H}_{C(=O)CH_3} + Cl.CO.CH_3 \rightarrow HCl + C_6H_5.N(COCH_3)_2$$

The use of acetanilide in medicine has been replaced by the acetyl derivative of the ethyl ether of *p*-aminophenol. This ethyl ether is known as *p*-phenetidine and the acetyl derivative is known as *phenacetin*; its systematic name is p-*acetylaminophenetole*:†

$$NH.CO.CH_3 \text{ (para to) } OC_2H_5 \text{ on benzene ring}$$

* When aniline is heated with formic acid, *formanilide* or N-*phenylformamide* is produced:

$$C_6H_5.N{<}^{H}_{H} + HO{-}C{<}^{O}_{H} \rightarrow C_6H_5.N{<}^{H}_{C(=O)H} + H_2O$$

Formanilide, isomeric with benzamide, has m.p. 47°, b.p. 284°. When it is heated with hydrochloric acid or with alkali it is hydrolysed to aniline and formic acid, the reverse reaction to the above taking place.

† *Phenetole*, b.p. 172°, is phenylethyl ether, $C_6H_5.O.C_2H_5$. The corresponding methyl ether, b.p. 154°, $C_6H_5.O.CH_3$, is known as anisole (p. 130).

Phenacetin has m.p. 135° and is described as the safest of the antipyretics.

Corresponding to acetanilide is *benzanilide*, *benzoylaniline* or N-*phenylbenzamide*, made by the action of benzoyl chloride on aniline in the presence of an equivalent quantity of sodium or potassium hydroxide in aqueous solution (Schotten-Baumann reaction):

$$C_6H_5.NH_2 + Cl.CO.C_6H_5 + NaOH \rightarrow C_6H_5.N\begin{cases}H \\ CO.C_6H_5\end{cases} + NaCl + H_2O$$

Benzanilide forms colourless crystals, m.p. 161°. The analogous *p*-toluenesulphonyl derivative of aniline, p-*toluenesulphonyl aniline* or *N-phenyl*-p-*toluenesulphonamide*, $CH_3.C_6H_4.SO_2.NH.C_6H_5$, made by the above reaction using *p*-toluenesulphonyl chloride, is a colourless crystalline substance, m.p. 103°.

These acyl derivatives of aniline and analogous primary aromatic amines are, in many cases, soluble in alkali and therefore behave as weak acids. The sodium derivatives have, in some cases, been isolated and these are believed to be derived from a tautomeric form as indicated:

$$C_6H_5.NH.CO.C_6H_5 + NaOH \rightarrow C_6H_5.N:C(ONa).C_6H_5 + H_2O$$

Another derivative of aniline is made by gently warming equimolecular quantities of the base and benzaldehyde in alcoholic solution. The reaction which takes place is:

$$C_6H_5.NH_2 + O:CH.C_6H_5 \rightarrow H_2O + C_6H_5.N:CH.C_6H_5$$

The product, *benzylideneaniline* or *benzalaniline* crystallises very readily from water and has m.p. 54°. It is a member of a class of compounds known as Schiff's bases.

Derivatives of aniline in which hydrogen atoms in the nucleus are replaced by nitro groups are known as *nitroanilines* or *nitranilines*; they cannot be prepared by the direct action of nitric acid on aniline. Nitric acid reacts vigorously with aniline, giving resinous products from which it is difficult to isolate any one individual product. If the aniline be dissolved in a large excess of sulphuric acid and this solution be nitrated with concentrated nitric acid, o-*nitraniline* and p-*nitraniline* are produced in the greater quantity together with a small quantity of m-*nitraniline*:

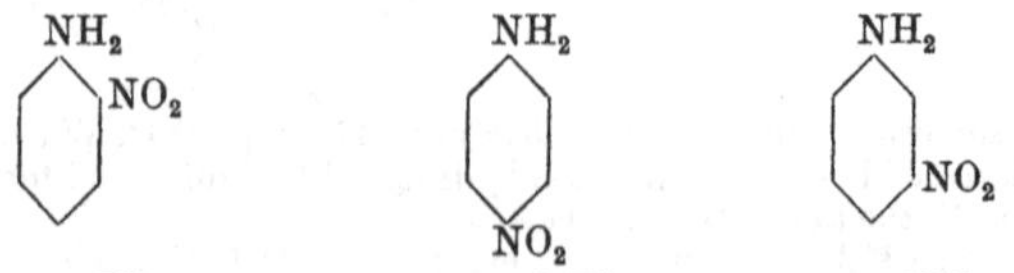

NH_2 / NO_2 (ortho) — m.p. 71°

NH_2 / NO_2 (para) — m.p. 147°

NH_2 / NO_2 (meta) — m.p. 114°

The sulphuric acid can be regarded as protecting the amino group and, actually, in nitrating aniline in the presence of an excess of sulphuric acid it is aniline sulphate rather than aniline which is being nitrated. More generally, in preparing nitro derivatives of aromatic primary amines, it is usual to protect the amino group by employing the acetyl derivative. For example, the direct nitration of acetanilide leads to a mixture of o-*nitroacetanilide* and p-*nitroacetanilide*:

NH.CO.CH_3 (benzene ring, NO_2 ortho) — m.p. 92–93°

NH.CO.CH_3 (benzene ring, NO_2 para) — m.p. 213°

and from these two compounds the corresponding nitranilines can be readily obtained by hydrolysis by means of hydrochloric acid. *m*-Nitraniline is prepared by the 'half' reduction of *m*-dinitrobenzene with ammonium sulphide or sodium disulphide. In the laboratory, this is usually carried out by passing hydrogen sulphide into a solution of *m*-dinitrobenzene in an alcoholic solution of ammonia:

$$C_6H_4(NO_2)_2 + 3H_2S \rightarrow C_6H_4(NH_2)(NO_2) + 3S + 2H_2O$$

The *m*-nitraniline is easily separated from its admixture with the precipitated sulphur by recrystallisation from alcohol. m-*Nitroacetanilide*, prepared by the acetylation of *m*-nitraniline with acetic anhydride, has m.p. 155°.

The nitranilines are much less basic than aniline, the acidic influence of the nitro group being more pronounced in the *ortho*- and *para*- positions than in the *meta*- position.

The diazo reaction

The most characteristic reaction of aniline and generally of all aromatic primary amines is that of these bases in acid solution, i.e. of their salts, with nitrous acid. This reaction is known as diazotisation and the diazo reaction was discovered by Griess in 1858.

Under any conditions of temperature, when a solution of an aliphatic primary amine, containing at least two equivalents of acid (sulphuric or hydrochloric), is treated with an aqueous solution of one equivalent of sodium nitrite, nitrogen is evolved and an alcohol is formed; and the result of the reaction may be summarised as the replacement of the amino group by hydroxyl:

$$R.NH_2 + O{:}N.OH \rightarrow R.OH + N_2 + H_2O$$

The course of the reaction between nitrous acid and aromatic primary amines depends entirely on the temperature at which the reaction is allowed to take place. At the ordinary and at more elevated temperatures it appears superficially to take the same course as in the case of the aliphatic primary amines, the amino group being replaced by the hydroxyl, a phenol resulting and nitrogen being evolved at the same time. This formation of the phenols is, however, only a secondary reaction resulting from the decomposition of a less stable intermediate compound, which in some cases has been isolated.

The diazo reaction (or diazotisation) is normally carried out by slowly adding an aqueous solution containing the calculated quantity of sodium nitrite to an aqueous solution of the aromatic primary amine (kept in most cases below $+5°$) which contains at least two equivalents of acid, one equivalent of the acid being necessary to form the salt of the amine and the other to react with the sodium nitrite in aqueous solution to form nitrous acid. The slow addition of the sodium nitrite is continued until the presence of free nitrous acid is just manifest—liberation of iodine from potassium iodide, using potassium iodide-starch paper as outside indicator. Using aniline and hydrochloric acid, the reaction may be represented in its simplest form:

$$C_6H_5.NH_2.HCl + O{:}N.OH \rightarrow C_6H_5.N_2.Cl + 2H_2O$$

The compound produced in this case is *phenyldiazonium chloride* or *diazobenzene chloride*. This and analogous compounds produced from other aromatic primary amines are usually soluble in water. As a general rule, they are not isolated, their reaction products being more important than the diazonium compounds themselves, but they can be isolated if necessary by modifying the conditions of the reaction in various ways.

The constitution of the diazonium compounds may be considered to follow from their mode of formation. For example, the above equation is more completely expressed as follows:

$$\left[\begin{matrix} C_6H_5 \searrow \\ \quad\quad N \\ H_3 \nearrow\!\!\!\!/\!\!/ \end{matrix}\right] Cl + HNO_2 \rightarrow \left[\begin{matrix} C_6H_5 \searrow \\ \quad\quad N \\ N \nearrow\!\!\!\!/\!\!/ \end{matrix}\right] Cl + 2H_2O$$

aniline hydrochloride — phenyl (or benzene) diazonium chloride

This constitution of the diazonium compounds is in keeping with their resemblance to quaternary ammonium salts; like the latter, they yield chloroplatinates and chloroaurates:

$$[C_6H_5.N_2]_2PtCl_6, \qquad [C_6H_5.N_2]AuCl_4$$

On the other hand, a compound of the type $C_6H_5.N_2.X$, in which it is known that the nitrogen atoms are connected together, might well have the constitution $C_6H_5—N=N—X$. This structure is known as the diazo structure and diazo compounds can be obtained from the diazonium compounds. When a diazonium salt is treated with alkali, it is converted into an alkali salt of a diazo hydroxide which functions as an acid (cf. other compounds containing the $=N—OH$ group):

$$[C_6H_5.N_2]Cl + 2KOH \rightarrow C_6H_5—N=N—OK + KCl + H_2O$$

Such diazotates are theoretically capable of existing in stereoisomeric forms as follows:

$$\begin{array}{c} C_6H_5—N \\ \| \\ KO—N \end{array} \qquad\qquad \begin{array}{c} C_6H_5—N \\ \| \\ N—OK \end{array}$$

syn-diazotate *anti*-diazotate

According to this, the *syn*- and *anti*- diazotates represented above are the two possible oxime-like products of nitrosobenzene ($C_6H_5—N=O$), and both of these have been produced by the action of hydroxylamine on nitrosobenzene in the presence of potassium hydroxide:

$$C_6H_5—N=O + H_2=N—OH + KOH \rightarrow C_6H_5—N=N—OK + 2H_2O$$

The diazotates are reconverted into diazonium salts in the presence of acids.

The simple phenyl or benzene diazonium salts and diazotates are colourless substances, easily soluble in water. It must be assumed that when the product of the reaction between the salt of an aromatic primary amine and nitrous acid reacts in acid solution it does so as a diazonium salt, and in alkaline solution it reacts as a diazotate.

The products of the diazotisation of aniline and of other aromatic primary amines are highly important on account of the reactions which they undergo. These reactions generally consist in the replacement of the $—N_2X$ complex by other univalent atoms or groups; and for these reactions it is rarely necessary to isolate the immediate product of the diazotisation and, generally, the 'diazotised' solution is used directly.

The following are the chief types of diazo reactions:

(i) *Production of phenols:* i.e. the replacement of the $—N_2X$ group, and indirectly the $—NH_2$ group of the amine, by hydroxyl. When the diazotised solution is heated, nitrogen is evolved and a phenol is produced:

$$C_6H_5.N_2.Cl + H_2O \rightarrow C_6H_5.OH + N_2 + HCl$$

If the phenol is volatile in steam, as is the case with ordinary phenol, it may be freed from by-products of the reaction by steam distillation.

This reaction indicates that the formation of the 'diazo' compound in the case of primary aromatic amines is the intermediate product of the reaction of nitrous acid on the salt of the amine and constitutes an alternative and frequently convenient method for the preparation of phenols when they cannot be prepared by the fusion of the sulphonic acid with sodium and potassium hydroxides (compare p. 252).

(ii) *Replacement of the* —N_2X *group (and indirectly the* —NH_2 *group of the amine) by hydrogen.* Usually this is done by heating the diazo solution with ethyl alcohol,* the latter being oxidised to aldehyde, and from benzene diazonium chloride, benzene is produced:

$$C_6H_5.N_2.Cl + CH_3.CH_2OH \rightarrow C_6H_6 + N_2 + HCl + CH_3.CHO$$

In this case phenetole would also be produced, but if there are potentially electro-negative substituents in the benzene nucleus the replacement of the —N_2X group by hydrogen is the main reaction; thus, diazotised 2 : 4 : 6-tribromoaniline is almost quantitatively converted into 1 : 3 : 5-tribromobenzene:

$$C_6H_2Br_3.N_2Cl \ (\text{1-}N_2Cl;\ 2,4,6\text{-}Br;\ 3,5\text{-}H) + CH_3.CH_2OH \rightarrow C_6H_3Br_3 \ (\text{1,3,5-}Br;\ 2,4,6\text{-}H) + HCl + N_2 + CH_3.CHO$$

To minimise the formation of by-products (chiefly the ether) in this reaction, a solution of stannous chloride in excess of alkali (an aqueous solution of sodium stannite) may be employed instead of ethyl alcohol as the reducing agent. In this case, the diazotised solution, which should not be too strongly acid, is cooled to 0° and made just alkaline, avoiding any heating of the solution as far as possible, before addition of the sodium stannite solution. Nitrogen is steadily evolved as soon as the mixture reaches the ordinary temperature, and when this ceases the product is isolated in any convenient manner. This method of replacement of the —N_2X group by hydrogen (reduction) is generally much more satisfactory than when alcohol is employed. By this method, aniline has been converted almost quantitatively into benzene, sulphanilic acid into benzene-

* If an alcohol be used as the medium of the reaction some of the corresponding ether is produced; but the reaction is complicated in the case of a primary alcohol by the reduction of the alcohol to the corresponding aldehyde (*v.* below):

$$C_6H_5.N_2.Cl + C_2H_5OH \rightarrow \underset{\text{phenetole}}{C_6H_5.O.C_2H_5} + N_2 + HCl$$

sulphonic acid and α-naphthylamine into naphthalene through their respective diazo compounds:

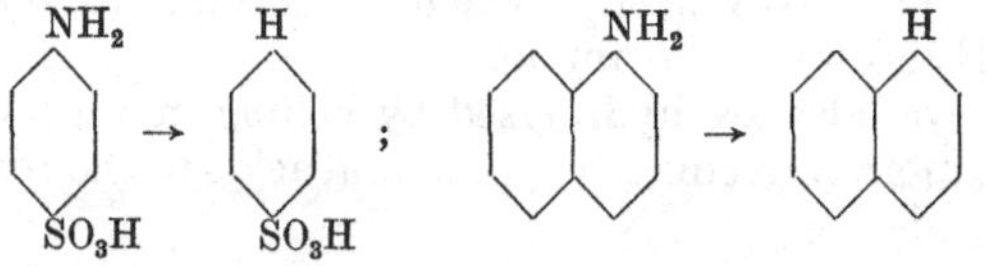

$$\text{(benzene ring with } NH_2 \text{ and } SO_3H\text{)} \rightarrow \text{(benzene ring with H and } SO_3H\text{)}\ ; \quad \text{(naphthalene with } NH_2\text{)} \rightarrow \text{(naphthalene with H)}$$

In the case of the production of benzene, the reaction may be expressed:

$$C_6H_5.N_2.OK + K_2SnO_2 + H_2O \rightarrow C_6H_6 + N_2 + K_2SnO_3 + KOH$$

(iii) *Production of aromatic iodo- compounds:* replacement of the —N_2X group (and indirectly the —NH_2 group of the amine) by iodine. This is carried out by the simple addition of an aqueous solution of potassium iodide to the diazotised solution and subsequently warming the mixture to complete the reaction. This is the only method for introducing iodine into the aromatic nucleus and, for example, *iodobenzene*, b.p. 188°, is obtained by the following reaction:

$$C_6H_5.N_2.Cl + KI \rightarrow C_6H_5I + KCl + N_2$$

(iv) *Production of (a) chloro-, (b) bromo- and (c) cyano- compounds:* replacement of the —N_2X group (and indirectly the —NH_2 group of the amine) by (*a*) chlorine, (*b*) bromine and (*c*) cyanogen. When the diazo solution is treated with cuprous salts (chloride, bromide, cyanide), nitrogen is evolved and the —N_2X group is replaced by chlorine, bromine and cyanogen respectively. In the case of benzene diazonium chloride the reactions taking place may be expressed:

$$2C_6H_5.N_2.Cl + Cu_2Cl_2 \rightarrow 2C_6H_5Cl^{*} + N_2 + Cu_2Cl_2$$

$$2C_6H_5.N_2.Cl + Cu_2Br_2 \rightarrow 2C_6H_5Br^{\dagger} + N_2 + Cu_2Cl_2$$

$$2C_6H_5.N_2.Cl + Cu_2(CN)_2 \rightarrow 2C_6H_5CN^{\ddagger} + N_2 + Cu_2Cl_2$$

This reaction, known as Sandmeyer's reaction, was discovered in 1884, and Gattermann showed in 1890 that benzene diazonium chloride is converted into monochlorobenzene when finely divided copper is added to the aqueous solution:

$$C_6H_5.N_2.Cl \rightarrow C_6H_5Cl + N_2$$

Sandmeyer's reaction is of great importance; it is capable of application to many classes of primary aromatic amines which yield diazo compounds. It is *the* method for preparing *chloro-*, *bromo-* and

* *Chlorobenzene*, b.p. 132°.

† *Bromobenzene*, b.p. 155°.

‡ Phenyl cyanide (benzonitrile), also obtained by the fusion of the potassium salt of benzenesulphonic acid with potassium cyanide (p. 237).

cyano- derivatives of aromatic compounds of known constitution, since the entering group takes the position in the nucleus previously occupied by the $—N_2X$ group, which is the same as that occupied by the $—NH_2$ group in the amine.

Since the cyanides are hydrolysed by boiling in aqueous solutions of alkalis or, more conveniently, of mineral acids to the corresponding carboxylic acids:

$$R—C{\equiv}N + 2H_2O \rightarrow R—C\begin{matrix}{/\!\!/}O \\ \diagdown OH\end{matrix} + NH_3$$

Sandmeyer's reaction affords a convenient method for the preparation of aromatic acids from amino compounds of known constitution, the position of the carboxyl group being that previously occupied by the amino group (compare p. 236).

(v) *Production of aromatic arsonic acids:* replacement of the $—N_2X$ group (and indirectly the $—NH_2$ group of the amine) by the arsonic acid ($—AsO(OH)_2$) group. This reaction was discovered by Bart in 1912. The simplest example is the preparation of phenylarsonic acid, $C_6H_5.AsO(OH)_2$, by the action of an aqueous solution of sodium arsenite on the solution of the diazonium compound:

$$C_6H_5.N_2.Cl + Na_3AsO_3 \rightarrow C_6H_5.AsO(ONa)_2 + NaCl + N_2$$

The Bart reaction now constitutes the most important method for the preparation of aromatic arsenical compounds of various types. These arsenical compounds are of the highest importance in the treatment of protozoal diseases (*chemotherapy*). The reaction has been applied to many types of primary aromatic amines and has been extended to the preparation of aromatic stibonic acids (e.g. phenylstibonic acid, $C_6H_5.SbO(OH)_2$), using an aqueous solution of sodium antimonite in place of sodium arsenite.

(vi) *Production of aromatic hydrazines:* replacement of the $—N_2X$ group (and indirectly of the $—NH_2$ group of the amine) by the $—NH.NH_2$ group. When the diazonium chlorides in aqueous solution are reduced with a solution of stannous chloride in hydrochloric acid the corresponding hydrazines are produced. From benzene diazonium chloride *phenylhydrazine* is obtained by a reaction which may be represented:

$$C_6H_5.N_2.Cl + 2SnCl_2 + 4HCl \rightarrow \begin{matrix}C_6H_5\diagdown \\ H\diagup\end{matrix}N—N\begin{matrix}\diagup H \\ \diagdown H\end{matrix}.HCl + 2SnCl_4$$

The white solid compound which separates is a complex compound of phenylhydrazine hydrochloride and stannic chloride,

$$(C_6H_5NH.NH_2.HCl)_2SnCl_4 \text{ or } \left[\begin{matrix}C_6H_5\diagdown \\ H\diagup\end{matrix}N—N\begin{matrix}\diagup H \\ —H \\ \diagdown H\end{matrix}\right]_2 SnCl_6$$

which can be filtered off, decomposed with an excess of alkali and the phenylhydrazine extracted with ether and isolated in the usual manner.

Phenylhydrazine, $\begin{matrix} C_6H_5 \\ H \end{matrix}\!\!>\!N{-}N\!<\!\!\begin{matrix} H \\ H \end{matrix}$, crystallises in colourless tabular crystals, m.p. 19·6°. When distilled at ordinary pressure (b.p. 241–242°) it undergoes slight decomposition and is therefore distilled for purification under reduced pressure (b.p. 120° at 12 mm.). The compound soon becomes darker and almost black in colour on standing, owing to the formation of small quantities of complex oxidation products by contact with oxygen of the air. Like aniline, it combines with hydrogen chloride and the salt, *phenylhydrazine hydrochloride*,

$$\begin{matrix} C_6H_5 \\ H \end{matrix}\!\!>\!N{-}N\!<\!\!\begin{matrix} H \\ H \end{matrix}.HCl \quad \text{or} \quad \left[\begin{matrix} C_6H_5 \\ H \end{matrix}\!\!>\!N{-}N\!\!\begin{matrix} H \\ \leftarrow H \\ H \end{matrix}\right]Cl$$

forms glistening colourless crystals.*

Phenylhydrazine is an important reagent; it and the corresponding compounds containing substituted groups in the benzene nucleus (substituted phenylhydrazines) are largely employed for producing derivatives—phenylhydrazones—by which aldehydes and ketones are conveniently identified and characterised.

Phenylhydrazine was discovered by Emil Fischer in 1875; he used it for characterising aldehydes and ketones and particularly for preparing important derivatives, *osazones*, of reducing sugars, whereby our knowledge of this important group of compounds known as carbohydrates was very greatly extended. Phenylhydrazine is also used for preparing other types of compounds, a derivative of one of these is the well-known medicinal *antipyrine* (p. 371).

Phenylhydrazine is a reducing agent; it will precipitate cuprous oxide from an alkaline solution of cupric axide (Fehling's solution). In this reaction the phenylhydrazine is decomposed as follows:

$$\begin{matrix} C_6H_5 \\ H \end{matrix}\!\!>\!N{-}N\!<\!\!\begin{matrix} H \\ H \end{matrix} \rightarrow C_6H_6 + N_2 + H_2$$

* If this salt be heated with hydrochloric acid in a sealed tube at about 200°, it is converted into p-*phenylenediamine hydrochloride*. The change is of a similar type to other intramolecular changes which have been previously mentioned and depends on the mobility of the hydrogen atom in the *para*- position of the substituted benzene nucleus:

$$\begin{matrix} C_6H_5 \\ H \end{matrix}\!\!>\!N{-}N\!<\!\!\begin{matrix} H \\ H \end{matrix} \rightarrow H_2N{-}C_6H_4{-}NH_2$$

p-phenylenediamine
(1:4-diaminobenzene)

p-*Phenylenediamine*, m.p. 147°, is a diacid base having two primary amino groups. It is isomeric with the corresponding *ortho*- and *meta*- phenylenediamines.

In the presence of certain reducing agents phenylhydrazine is reduced to aniline and ammonia and then functions as an oxidising agent:

$$\begin{matrix} C_6H_5 \\ H \end{matrix}\!\!>N-N\!<\!\!\begin{matrix} H \\ H \end{matrix} + H_2 \rightarrow C_6H_5.NH_2 + NH_3$$

This reaction is one of the stages in the formation of osazones of reducing sugars by means of phenylhydrazine (p. 409).

(vii) *Production of* azo-dyes. These substances, of great technical and scientific importance, are produced by interaction of diazonium compounds and aromatic amines and phenols; these reactions are often referred to technically as 'couplings'.

(*a*) When diazonium salts in solution are mixed with equimolecular quantities of primary and secondary aromatic amines, coupling takes place and coloured diazoamino-compounds are formed, for example:

$$C_6H_5.N_2.Cl + H_2N.C_6H_5 \rightarrow C_6H_5.N:N.N\!<\!\!\begin{matrix} H \\ C_6H_5 \end{matrix} + HCl$$

diazoaminobenzene

Diazoaminobenzene crystallises in golden yellow leaflets or prisms, m.p. 98°, exploding when heated to a higher temperature. It is insoluble in water and sparingly soluble in cold alcohol. Like other diazoamino-compounds it undergoes intramolecular change when allowed to stand in alcohol, and this change is hastened by the presence of aniline hydrochloride; the diazoaminobenzene is transformed into p-*aminoazobenzene*, m.p. 127°, thus:

$$C_6H_5.N:N.NH-C_6H_5 \rightarrow C_6H_5.N:N-C_6H_4-NH_2$$

p-aminoazobenzene

The intramolecular change takes place readily when the *para*- position to the :NH group is free, i.e. only occupied by hydrogen. If the *para*-position is not free, the change takes place less readily and the amino group then enters the *ortho*- position to the *azo* (—N=N—) group. For example, diazobenzene-*p*-toluidide, produced by coupling benzene-diazonium chloride with *p*-toluidine:

$$C_6H_5.N_2.Cl + H_2N-C_6H_4-CH_3 \rightarrow C_6H_5.N_2.NH-C_6H_4-CH_3 + HCl$$

undergoes intramolecular change as above described and *benzene-azo-4-methyl-2-aminobenzene* (*benzene-azo-2-amino-*p-*toluene*) is formed:

$$C_6H_5.N_2.NH-C_6H_4-CH_3 \rightarrow C_6H_5.N_2-C_6H_3(NH_2)-CH_3$$

(*b*) With tertiary mixed aromatic-aliphatic amines, diazonium salts couple directly, forming aminoazo-compounds. For example, benzene diazonium chloride and dimethylaniline (p. 286) react together to give p-*dimethylaminoazobenzene*:

$$C_6H_5.N_2.Cl + H-C_6H_4-N(CH_3)_2 \rightarrow C_6H_5-N{=}N-C_6H_4-N(CH_3)_2 + HCl$$

p-*Dimethylaminoazobenzene* crystallises in golden yellow plates, m.p. 116°. It is a monoacid base and its *hydrochloride* crystallises in purple red needles. That the coupling takes place and the compound has the constitution indicated, are proved by the fact that the product of the reaction, when reduced with tin and hydrochloric acid, is disrupted at the double bond of the azo group, aniline and N-*dimethyl*-p-*phenylenediamine* being formed, thus:

$$C_6H_5-N{=}N-C_6H_4-N(CH_3)_2 + 2H_2 \rightarrow C_6H_5NH_2 + H_2N-C_6H_4-N(CH_3)_2$$

N-dimethyl-*p*-phenylenediamine

If the *para*- position in the tertiary amine is occupied by a group other than hydrogen, then coupling with the diazonium salt takes place in the *ortho*- position to the azo group. If only the *meta*- position is free from a substituting group, coupling does not take place.

(*c*) Coupling takes place between phenols and alkaline diazo solutions; in ordinary cases coupling takes place in the *para*- position to the hydroxyl group in the phenol. If the *para*- position is occupied by a group other than hydrogen coupling takes place in the *ortho*-position, while coupling does not take place if only the *meta*- position remains free from a substituting group. p-*Hydroxyazobenzene*, yellow crystals, m.p. 152°, is formed when a solution of phenol in excess of sodium hydroxide is added to an aqueous solution of benzene diazonium chloride,

$$C_6H_5.N_2.Cl + H-C_6H_4-OH + NaOH$$
$$\rightarrow C_6H_5.N{=}N-C_6H_4-OH + NaCl + H_2O$$

Azo dye-stuffs

p-Aminoazobenzene, *p*-dimethylaminoazobenzene and *p*-hydroxyazobenzene are the simplest examples of what are described as monoazo dyes. The azo group itself confers a yellow colour on a compound containing it. Such a colour-producing group is known as a *chromophore*. A compound containing a chromophore, in this case the azo group, is not necessarily a dye-stuff; it is only when an *auxochrome* group such as $-SO_3H$, $-OH$ or $-NH_2$ which is capable of conferring acid or basic properties is also present that the compound becomes a dye, i.e. capable of fixing itself on various types of material. Wool and silk are dyed directly by solutions of azo dyes; cotton, on the other hand, requires to be *mordanted* with acid materials such as tannic acid or with basic materials such as aluminium, iron or chromium hydroxides (generally produced by steaming the fabric after soaking in an aqueous solution of the acetate), depending on whether a basic or acidic dye-stuff is being employed. Not only do the colours of the monoazo dyes vary according to the particular diazonium salt employed, but also according to the particular aromatic amine or phenol with which it is coupled. If a di-primary amine (e.g. *o*-, *m*- or *p*- phenylenediamine or benzidine) be diazotised, dis-azo dyes can be obtained; similarly tris-azo and tetra-azo dyes are also known, and additions to all these classes of azo dye-stuffs are constantly being made.

One azo dye which, however, rarely finds technical application is *helianthine* or *methyl orange*, of importance as an indicator in volumetric analysis. The starting-point for the preparation of this substance is sulphanilic acid, which is diazotised in the normal way; the product is sparingly soluble in water and can be separated, since it is stable in the moist condition:

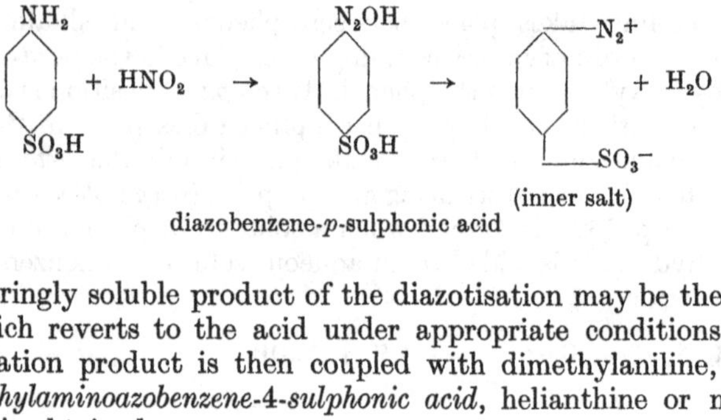

$$NH_2C_6H_4SO_3H + HNO_2 \rightarrow N_2OH{\cdot}C_6H_4{\cdot}SO_3H \rightarrow {}^{+}N_2{\cdot}C_6H_4{\cdot}SO_3^{-} + H_2O$$

(inner salt)

diazobenzene-*p*-sulphonic acid

The sparingly soluble product of the diazotisation may be the inner salt which reverts to the acid under appropriate conditions. The diazotisation product is then coupled with dimethylaniline, when 4′-*dimethylaminoazobenzene*-4-*sulphonic acid*, helianthine or methyl orange, is obtained:

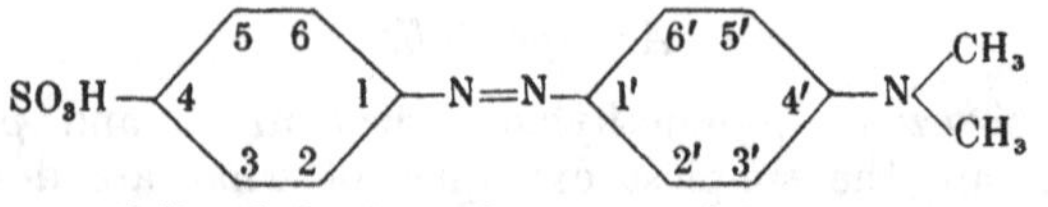

$$SO_3H-C_6H_4-N{=}N-C_6H_4-N(CH_3)_2$$

(ring positions: 1–6 on the sulphonic acid ring, 1′–6′ on the dimethylamino ring)

4′-dimethylaminoazobenzene-4-sulphonic acid

This is the simplest way of expressing the constitution of the substance. From its reactions, it appears to react in a tautomeric form and its constitution may be more fully expressed:

SO_3H ... $N{=}N$ ... $N(CH_3)_2$ $\underset{\text{alkalis}}{\overset{\text{acids}}{\rightleftharpoons}}$ SO_3^- ... NH* ... N ... $\overset{+}{N}(CH_3)_2$

On this view of the constitution of methyl orange, its orange-red colour in the presence of acids is due to its having the *para*-quinonoid-internal salt structure under these conditions; in the presence of alkalis, its colour is yellow, the alkali salts being necessarily derived from the ordinary acid-azo form.

The simplest *homologues of aniline* are the three isomeric *toluidines* or *aminotoluenes*, $C_6H_4{<}^{CH_3}_{NH_2}$. o-*Toluidine* is a liquid, b.p. 199°, m-*toluidine* has b.p. 203° and p-*toluidine* is a colourless crystalline substance, m.p. 45°, b.p. 199°. These substances behave similarly to aniline in all typical reactions and are isomeric with *benzylamine*, $C_6H_5.CH_2.NH_2$, b.p. 187°, which has all the typical reactions of an aliphatic primary amine and which is a stronger base than the isomeric toluidines.

AROMATIC SECONDARY AND TERTIARY AMINES

These compounds are of three types: (i) those which contain only aromatic (aryl) groups united to the nitrogen atom, (ii) those which contain aromatic and aliphatic (alkyl) groups united to the nitrogen atom, and (iii) those which contain at least one nitrogen atom as part of a ring structure. Examples of class (i) are *diphenylamine*, $(C_6H_5)_2NH$, and *triphenylamine*, $(C_6H_5)_3N$. Examples of

* What is known as the *quinonoid* (*para*-quinonoid, [structure], is more frequent than *ortho*-quinonoid, [structure]) grouping is also a chromophore, and the colour of many compounds is believed to be due to the presence of this grouping.

class (ii) are *monomethylaniline*, $(C_6H_5)(CH_3)NH$, and *dimethylaniline*, $C_6H_5N(CH)_2$. A large number of both classes of these bases and particularly of the mixed aromatic-aliphatic amines are known.

Monomethylaniline and dimethylaniline are prepared on a technical scale by heating aniline (or aniline hydrochloride) with methyl alcohol and hydrochloric acid in an autoclave at about 250°. This reaction can be considered to be a modification of Hofmann's sealed tube reaction for preparing aliphatic amines. The resulting mixture of monomethylaniline and dimethylaniline cannot be separated by fractional distillation, their boiling points being almost identical.

Monomethylaniline can be prepared starting with acetanilide. This is dissolved in toluene and to the solution is added gradually the calculated quantity of sodium. Hydrogen is evolved and the colourless sparingly soluble *sodium-acetanilide* is formed:

$$C_6H_5.N\begin{matrix} \diagup CO.CH_3 \\ \diagdown H \end{matrix} + Na \rightarrow \underset{\text{sodium acetanilide}}{C_6H_5.N\begin{matrix} \diagup CO.CH_3 \\ \diagdown Na \end{matrix}} \rightleftarrows C_6H_5.N:C(ONa).CH_3$$

Sodium acetanilide reacts immediately with methyl iodide, forming the acetyl derivative of monomethylaniline (N-methylacetanilide or N-methylphenylacetamide) and sodium iodide:

$$C_6H_5.N\begin{matrix} \diagup CO.CH_3 \\ \diagdown Na \end{matrix} + ICH_3 \rightarrow C_6H_5.N\begin{matrix} \diagup CO.CH_3 \\ \diagdown CH_3 \end{matrix} + NaI$$

N-*Methylacetanilide* is a colourless crystalline substance, m.p. 101°. When it is hydrolysed by heating with hydrochloric acid, monomethylaniline is obtained after making the resulting solution alkaline:

$$C_6H_5.N\begin{matrix} \diagup CO.CH_3 \\ \diagdown CH_3 \end{matrix} + H_2O \rightarrow CH_3.COOH + C_6H_5.N\begin{matrix} \diagup H \\ \diagdown CH_3 \end{matrix}$$

Monomethylaniline has b.p. 196° and is a colourless liquid. It forms salts with acids analogous to those of aniline. When allowed to react with acetic anhydride it forms the above-mentioned acetyl derivative. Since dimethylaniline, being a tertiary amine, does not form an acetyl derivative, it is easily possible to separate the two methylanilines from each other in the technical mixture.

For this purpose, the mixture is allowed to react with some excess of acetic anhydride, as described for the preparation of acetanilide. The resulting product is submitted to steam distillation, when excess of acetic anhydride is decomposed and the unchanged dimethylaniline is also separated, since, like acetic acid, it is volatile in steam. The residue in the distillation apparatus should crystallise on cooling and will consist of N-methylacetanilide. This can be separated, recrystallised if necessary, and then hydrolysed to give pure monomethylaniline.

The formation of an acetyl derivative is one of the proofs that monomethylaniline is a secondary amine. Another acyl derivative of importance is p-*toluenesulphonylmethylaniline* (N-methylphenyl-*p*-toluene sulphonamide), $\begin{matrix} C_6H_5 \\ CH_3 \end{matrix}\!\!>\!N.SO_2.C_6H_4.CH_3$, a colourless crystalline substance, m.p. 94–95°, prepared by the action of *p*-toluenesulphonyl chloride on monomethylaniline in the presence of sodium hydroxide (Schotten-Baumann reaction). This *p*-toluenesulphonyl derivative is frequently employed instead of the acetyl derivative for the separation of monomethylaniline from dimethylaniline.*

The benzoyl derivative of monomethylaniline (N-*methylphenylbenzamide*), $\begin{matrix} C_6H_5 \\ CH_3 \end{matrix}\!\!>\!N.CO.C_6H_5$, prepared from benzoyl chloride and monomethylaniline by the Schotten-Baumann reaction, is a colourless crystalline substance, m.p. 68°.

Like an aliphatic secondary amine, monomethylaniline reacts with nitrous acid, giving a nitrosoamine:

$$\begin{matrix} C_6H_5 \\ CH_3 \end{matrix}\!\!>\!N{-}H \;+\; H{-}O{-}N{=}O \;\rightarrow\; \begin{matrix} C_6H_5 \\ CH_3 \end{matrix}\!\!>\!N{-}N{=}O \;+\; H_2O$$

Methylphenylnitrosoamine at ordinary temperatures is a yellow oil, m.p. 15°. When a nitroso derivative of a secondary amine is gently warmed with phenol and sulphuric acid and the product poured into water and finally made alkaline, a characteristic bright blue-coloured solution is obtained. This is *Liebermann's reaction* (p. 124) and may be used as a characteristic test for phenols and for secondary amines.

An interesting reaction of the nitroso derivatives of the mixed aromatic-aliphatic secondary amines such as the above is the intramolecular change they undergo when warmed with alcoholic-hydro-

* Another method for the preparation of monomethylaniline, more convenient than that described above, is to start with *p*-toluenesulphonylaniline. This compound is soluble in sodium hydroxide solution, and when this solution is shaken with dimethyl sulphate a fairly vigorous reaction takes place which may be expressed:

$$C_6H_5.N\!\!<\!\begin{matrix} SO_2.C_6H_4.CH_3 \\ Na \end{matrix} + (CH_3)_2SO_4 \rightarrow C_6H_5.N\!\!<\!\begin{matrix} SO_2.C_6H_4.CH_3 \\ CH_3 \end{matrix} + (CH_3)NaSO_4$$

When the reaction is completed, any excess of dimethyl sulphate is hydrolysed with alkali to methyl alcohol and sodium sulphate and the p-*toluenesulphonylmethylaniline* separated. This compound is then hydrolysed in the usual way by boiling with hydrochloric acid and the base obtained by making the resultant solution alkaline.

chloric acid. The nitroso (—NO) group enters the *para-* position, when this is possible, of the benzene nucleus, thus:

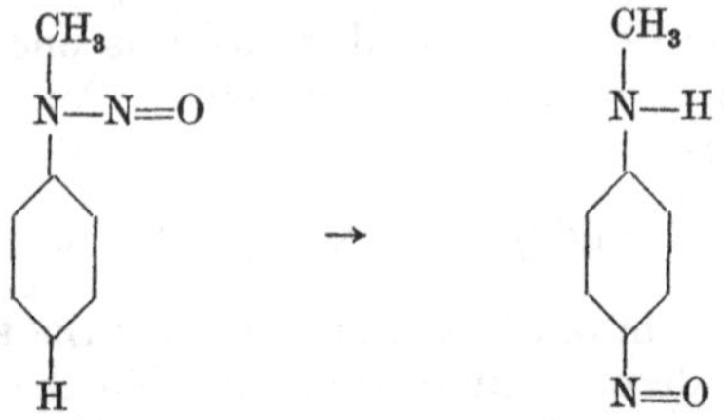

$$\begin{array}{ccc} CH_3 & & CH_3 \\ | & & | \\ N{-}N{=}O & & N{-}H \\ | & & | \\ \bigcirc & \rightarrow & \bigcirc \\ | & & | \\ H & & N{=}O \end{array}$$

The compound formed in this case is p-*nitrosomonomethylaniline*, a green crystalline substance, m.p. 118°, which is soluble in aqueous alkali solutions and precipitated from them by means of carbon dioxide. When it is boiled with alkalis, it is hydrolysed into monomethylamine and *p*-nitrosophenol (p. 124):

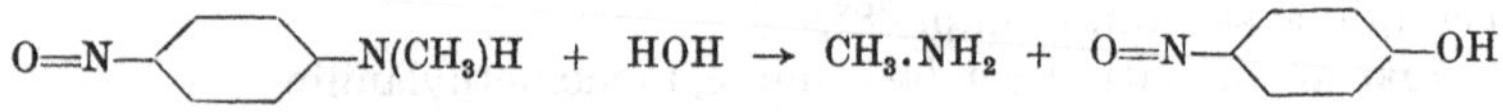

$$O{=}N{-}\langle\ \ \rangle{-}N(CH_3)H \ + \ HOH \ \rightarrow \ CH_3.NH_2 \ + \ O{=}N{-}\langle\ \ \rangle{-}OH$$

The formation of *p*-nitrosomonomethylaniline by the above intramolecular change is another example of the mobility of the hydrogen atom in the *para-* position of a substituted benzene nucleus.

Dimethylaniline, $C_6H_5N(CH_3)_2$, is a colourless oil, m.p. 2°, b.p. 193°. The technical product is purified either with acetic anhydride or *p*-toluenesulphonyl chloride as described above. Dimethylaniline being a tertiary amine does not react with either acid anhydrides or with acid chlorides.

Dimethylaniline forms salts by direct combination with acids; as a general rule, the salts of this amine are not so easily crystallisable as the corresponding salts of aniline and of monomethylaniline. It combines readily with methyl iodide forming the well-crystallised quaternary ammonium compound, *trimethylphenylammonium iodide*, $[N(C_6H_5)(CH_3)_3]I$. This compound on treatment with moist silver oxide is converted into *trimethylphenylammonium hydroxide*, $[N(C_6H_5)(CH_3)_3]OH$, a strong base. A convenient way of obtaining pure dimethylaniline on the small scale is to distil pure trimethylphenylammonium iodide in a current of hydrogen chloride, when methyl iodide is liberated:

$$[N(C_6H_5)(CH_3)_3]I \ \rightarrow \ CH_3I \ + \ N(C_6H_5)(CH_3)_2$$

All mixed aromatic-aliphatic tertiary amines of the type of dimethylaniline react with nitrous acid to form *p*-nitroso- derivatives, thus:

$$(CH_3)_2N-C_6H_4-H + H-O-N=O \rightarrow (CH_3)_2N-C_6H_4-N=O + H_2O$$

The reaction is carried out in the usual manner, using a solution of the hydrochloride of the base in dilute hydrochloric acid and sodium nitrite. p-*Nitrosodimethylaniline* (which, of course, is not a nitroso-amine) can be separated by making the resulting solution alkaline and extracting with ether, in which the compound is soluble. It crystallises in green leaflets, m.p. 85°, and forms a *hydrochloride*, crystallising in small yellow needles and dissolving in water to an intensely yellow solution. On reduction by means of tin or zinc and hydrochloric acid, it is converted into *dimethyl*-p-*phenylenediamine* (1-amino-4-dimethylaminobenzene):

$$(CH_3)_2N-C_6H_4-NH_2$$

When *p*-nitrosodimethylaniline is boiled with sodium or potassium hydroxide in aqueous solution it is hydrolysed and yields dimethylamine and *p*-nitrosophenol (p. 124), the latter remaining dissolved as the potassium derivative in the alkaline solution after steam distillation:

$$(CH_3)_2N-C_6H_4-N=O + KOH \rightarrow (CH_3)_2N-H + KO-C_6H_4-N=O$$

This reaction provides a convenient method for obtaining the secondary aliphatic amines (p. 254) in a state of purity and can be extended, by using the nitroso- derivatives of homologues of dimethylaniline, to the preparation of secondary aliphatic amines generally. Incidentally, this reaction furnishes a proof that the nitroso- group in the nitrosodimethylaniline is in the *para*- position to the dimethylamino group.

When dimethylaniline hydrochloride is heated at 300° it is converted by intramolecular rearrangement into 2 : 4-*dimethylaniline* (or 2 : 4-*xylidine*) *hydrochloride*. Under the same conditions, monomethylaniline hydrochloride yields a mixture of the hydrochlorides of *o*- and *p*- toluidines. 2 : 4-Xylidine is prepared commercially by this method and is an important intermediate for the preparation of certain azo dyes.

$$\left[\begin{array}{c} N(CH_3)_2H \\ H \quad H \\ H \quad H \\ H \end{array}\right]Cl \rightarrow \left[\begin{array}{c} NH_3 \\ H \quad CH_3 \\ H \quad H \\ CH_3 \end{array}\right]Cl \quad \text{or} \quad C_6H_3\begin{cases} NH_2.HCl \\ CH_3\,(2) \\ CH_3\,(4) \end{cases}$$

2 : 4-xylidine hydrochloride
(2 : 4-dimethylphenylammonium chloride)

$$\left[\begin{array}{c} N(CH_3)H_2 \\ H \quad H \\ H \quad H \\ H \end{array}\right]Cl \nearrow \left[\begin{array}{c} NH_3 \\ H \quad CH_3 \\ H \quad H \\ H \end{array}\right]Cl \quad \text{or} \quad C_6H_4\begin{cases} NH_2.HCl \\ CH_3\,(2) \end{cases}$$

o-toluidine hydrochloride
(2-methylphenylammonium chloride)

$$\searrow \left[\begin{array}{c} NH_3 \\ H \quad H \\ H \quad H \\ CH_3 \end{array}\right]Cl \quad \text{or} \quad C_6H_4\begin{cases} NH_2.HCl \\ CH_3\,(4) \end{cases}$$

p-toluidine hydrochloride
(4-methylphenylammonium chloride)

The reactivity of the hydrogen in the *para*- position of the benzene nucleus in dimethylaniline is further indicated by the condensation which the latter undergoes with carbonyl chloride ($COCl_2$, phosgene) in the presence of aluminium chloride:

$$O{=}C\begin{cases} Cl \\ Cl \end{cases} + \begin{array}{c} H\langle C_6H_4\rangle N(CH_3)_2 \\ H\langle C_6H_4\rangle N(CH_3)_2 \end{array} \rightarrow O{=}C\begin{cases} \langle C_6H_4\rangle{-}N(CH_3)_2.HCl \\ \langle C_6H_4\rangle{-}N(CH_3)_2.HCl \end{cases}$$

when the hydrochloride of p-*tetramethyldiaminobenzophenone* (Michler's ketone) is produced. The free base, tetramethyldiaminobenzophenone, is a colourless crystalline substance, m.p. 173°, and is an important intermediate material in the technical production of dye-stuffs belonging to the triphenylmethane class.

Diphenylamine, $(C_6H_5)_2NH$, the typical secondary amine of the wholly aromatic class (i), is another important intermediate compound in the production of certain dye-stuffs. It was originally obtained by Hofmann (1864) by heating a mixture of aniline and aniline hydrochloride under pressure, i.e. in a sealed apparatus, at about 250° for some hours. The reaction taking place may be represented:

$$C_6H_5.NH_2 + C_6H_5.NH_2.HCl \rightarrow (C_6H_5)_2NH + NH_4Cl$$

This reaction is still employed for the technical production of diphenylamine.

A convenient method for synthesising diphenylamine and substituted diphenylamines of known constitution has been devised by Chapman (1929). The reactions leading to the preparation of diphenylamine may be outlined:

$$C_6H_5NH_2 + Cl.CO.C_6H_5 \xrightarrow{NaOH} C_6H_5{-}N\begin{smallmatrix}H\\CO.C_6H_5\end{smallmatrix} + NaCl + H_2O$$

aniline → benzoylaniline, benzanilide or N-phenylbenzamide

$$C_6H_5{-}N\begin{smallmatrix}H\\CO.C_6H_5\end{smallmatrix} \rightleftharpoons C_6H_5{-}N{=}\overset{OH}{\overset{|}{C}}{-}C_6H_5$$

$$C_6H_5{-}N{=}\overset{OH}{\overset{|}{C}}{-}C_6H_5 + PCl_5 \rightarrow C_6H_5{-}N{=}\overset{Cl}{\overset{|}{C}}{-}C_6H_5 + HCl + POCl_3$$

benzanilideiminochloride or *N-phenylbenziminochloride* (b.p. 173° at 15 mm.)

$$C_6H_5{-}N{=}\overset{Cl}{\overset{|}{C}}{-}C_6H_5 + NaOC_6H_5 \rightarrow C_6H_5{-}N{=}\overset{OC_6H_5}{\overset{|}{C}}{-}C_6H_5 + NaCl$$

N-phenylbenziminophenyl ether

$$C_6H_5{-}N{=}\overset{OC_6H_5}{\overset{|}{C}}{-}C_6H_5 \xrightarrow{280-300^\circ} (C_6H_5)_2N.CO.C_6H_5$$

benzoyldiphenylamine (or N-diphenylbenzamide, m.p. 177°)

$$(C_6H_5)_2N.CO.C_6H_5 \xrightarrow{KOH} (C_6H_5)_2NH + C_6H_5.COOK$$

Starting with the above N-phenylbenziminochloride, *phenyl*-m-*tolylamine* may be obtained by the following series of reactions:

$$C_6H_5{-}N{=}\overset{Cl}{\overset{|}{C}}{-}C_6H_5 + NaO{-}C_6H_4{-}CH_3 \rightarrow C_6H_5{-}N{=}\overset{OC_6H_4.CH_3}{\overset{|}{C}}{-}C_6H_5 + NaCl$$

sodium derivative of *m*-cresol → *N-phenylbenzimino*-m-*tolyl ether* (colourless needles, m.p. 65°)

$$C_6H_5{-}N{=}\overset{OC_6H_4.CH_3}{\overset{|}{C}}{-}C_6H_5 \xrightarrow{280-300^\circ} \begin{smallmatrix}C_6H_5\\CH_3.C_6H_4\end{smallmatrix}{>}N{-}\overset{O}{\overset{\|}{C}}.C_6H_5$$

benzoyl derivative of phenyl-m-*tolylamine* (N-phenyl-*m*-tolylbenzamide)—(colourless prisms, m.p. 104–106°)

$$\begin{matrix} C_6H_5 \\ CH_3.C_6H_4 \end{matrix} \Big> N.CO.C_6H_5 \xrightarrow{KOH} \begin{matrix} C_6H_5 \\ CH_3.C_6H_4 \end{matrix} \Big> N{-}H + C_6H_5.COOK$$

*phenyl-*m*-tolylamine*
(colourless crystals, m.p. 27·5°)

The same end-product can be obtained by starting with the benzoyl derivative of *m*-toluidine,

$$CH_3.C_6H_4.N \Big< \begin{matrix} H \\ CO.C_6H_5 \end{matrix}$$

converting this by means of phosphorus pentachloride into N-*m*-tolylbenziminochloride,

$$CH_3.C_6H_4.N{=}C \Big< \begin{matrix} Cl \\ C_6H_5 \end{matrix}$$

allowing this to react with sodium phenate and obtaining N-*m*-tolylbenziminophenyl ether,

$$CH_3.C_6H_4.N{=}C \Big< \begin{matrix} OC_6H_5 \\ C_6H_5 \end{matrix}$$

which is isomeric with the above N-phenylbenzimino-*m*-tolyl ether. *N-*m*-Tolylbenziminophenyl ether*, colourless prisms, m.p. 60°, when heated at 280–300° for two hours is converted into the above benzoyl derivative of phenyl-*m*-tolylamine (N-phenyl-*m*-tolylbenzamide), and this on hydrolysis with potassium hydroxide in alcoholic solution is readily converted into phenyl-*m*-tolylamine and potassium benzoate. Prior to the discovery of the above series of reactions phenyl-*m*-tolylamine had not been obtained in a state of purity.

Diphenylamine is a colourless crystalline substance, m.p. 54°, b.p. 302°. It is insoluble in water and readily soluble in alcohol, ether and organic solvents generally. The introduction of a second phenyl group in place of a hydrogen atom attached to the nitrogen atom in aniline renders the compound less strongly basic than aniline, whose aqueous solution has no alkaline reaction. Diphenylamine forms a hydrochloride which can be isolated in the solid state by shaking a benzene solution of diphenylamine with concentrated hydrochloric acid. This *diphenylamine hydrochloride* (colourless crystals, m.p., with decomposition, at about 180°) is almost completely dissociated when treated with water, the base being left as an insoluble precipitate. The acidic influence of the phenyl radicle is further emphasised by the fact that the remaining hydrogen attached to nitrogen in diphenylamine can be replaced by sodium and potassium.

Like other secondary amines, diphenylamine reacts with nitrous acid yielding *diphenylnitrosoamine*, $(C_6H_5)_2N{-}N{=}O$. At the same time diphenylamine is readily attacked by oxidising agents, yielding a product which gives an intense blue colour with sulphuric acid.

For this reason, diphenylamine is used for detecting nitrous and nitric acid, the test being very delicate.

As a secondary amine, diphenylamine yields a benzoyl derivative $(C_6H_5)_2N.OC.C_6H_5$ (*v.* above) when treated with benzoyl chloride in the presence of sodium hydroxide. When warmed with acetic anhydride, diphenylamine is converted into its acetyl derivative, *acetyl-diphenylamine* or N-*diphenylacetamide*, $(C_6H_5)_2N.OC.CH_3$, colourless crystals, m.p. 103°.

When diphenylamine is heated with sulphur, hydrogen sulphide is evolved and a tricyclic compound, *thiodiphenylamine* (colourless crystals, m.p. 150°), is produced.

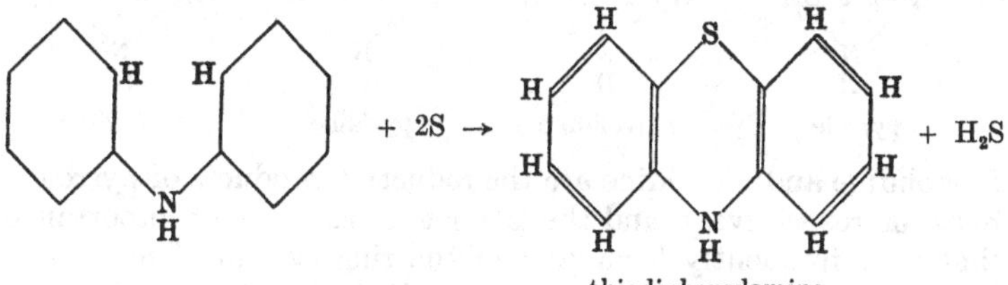

+ 2S → + H_2S

thiodiphenylamine

The derivatives of thiodiphenylamine find important application in the production of certain classes of dye-stuffs. The action of arsenious chloride on diphenylamine is similar to that of sulphur. When the two compounds are heated together, hydrogen chloride is evolved and a yellow highly crystalline compound, m.p. 194°, known as 10-*chloro*-5 : 10-*dihydrophenarsazine*, is produced:

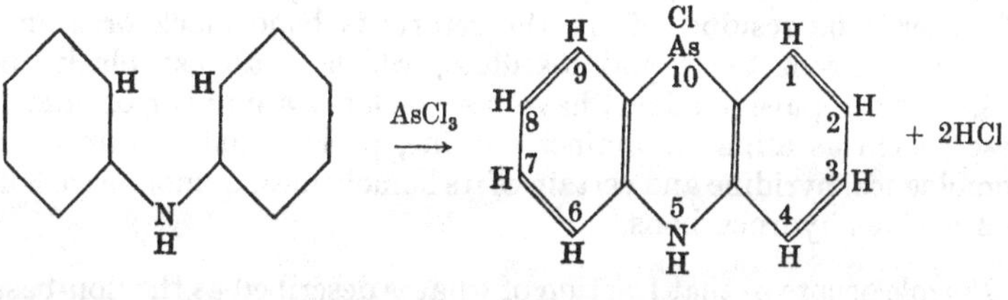

$\xrightarrow{AsCl_3}$ + 2HCl

Diphenylamine is the simplest wholly aromatic secondary amine, and a large number of aromatic amines belonging to the same class having analogous reactions are known.

The simplest aromatic tertiary amine is *triphenylamine*, large colourless crystals, m.p. 127°, which can be obtained by the action of monobromobenzene on the sodium derivative of diphenylamine:

$$(C_6H_5)_2NNa + BrC_6H_5 \rightarrow NaBr + (C_6H_5)_3N$$

No salts of triphenylamine are known and the diminution in basicity from ammonia, through aniline and diphenylamine to triphenylamine, is very striking.

Aromatic amines belonging to class (iii) may belong to the secondary or tertiary series. In these compounds nitrogen atoms form part of the ring system, giving rise to compounds belonging to the *hetero-cyclic* series. Some of the simplest of these bases are

```
                                                  H                H2
                                                  C                C
                                                //γ\              /4\
HC———CH          H2C———CH2                    HCβ  β'CH      H2C3   5CH2
||β  β'||           | 3  4 |                    |     ||          |      |
HCα  α'CH        H2C2   5CH2                  HCα  α'CH      H2C2   6CH2
  \   /              \1 /                       \\   /            \1 /
    N                  N                          N                 N
    H                  H                                            H
 pyrrole          pyrrolidine                  pyridine          piperidine
```

Pyrrolidine and piperidine are the reduction products of pyrrole and pyridine respectively, and the latter two bases are of importance in that they frequently form part of the ring system of many of the naturally occurring vegetable bases or alkaloids. Further, derivatives of pyrrole have been obtained as hydrolytic products of certain proteins; and haemoglobin and chlorophyll are essentially derivatives of pyrrole.

Both pyrrole and pyridine occur in coal-tar, but the chief source of these compounds is bone-oil (known since 1771 as Dippel's oil or animal oil). Bone-oil is obtained by the dry distillation (in closed retorts) of bones which have been roughly dried after removing the greater part of adhering fatty matter by boiling in a large quantity of water. The residue left in the retorts is bone black or animal charcoal. From the liquid distillate, which is almost black and possesses a repulsive odour, have been isolated a number of organic bases, such as aliphatic amines, aniline, pyrrole and certain of its homologues, pyridine and certain of its homologues, quinoline (p. 302) and various hydrocarbons.

Pyrrole occurs in that fraction of what is described as the non-basic portion of the bone-oil distilling at 100–150°. It has been synthesised by passing acetylene and ammonia through a tube heated to redness. This pyrogenic reaction, which may be represented:

```
HC≡CH                  HC=CH
        + NH3  →       |     >NH + H2
HC≡CH                  HC=CH
```

may explain the presence of pyrrole in coal-tar. The simplest synthesis of pyrrole is effected by heating succinimide (p. 379) with zinc dust containing some zinc hydroxide (a powerful reducing agent):

$$\begin{array}{l} CH_2—CO \\ \quad\quad\quad\rangle NH \\ CH_2—CO \end{array} \rightleftharpoons \begin{array}{l} CH{=}C(OH) \\ \quad\quad\quad\rangle NH \\ CH{=}C(OH) \end{array}$$

succinimide — tautomeric form of succinimide

$$\begin{array}{l} CH{=}C(OH) \\ \quad\quad\quad\rangle NH \\ CH{=}C(OH) \end{array} + 2H_2 \rightarrow \begin{array}{l} CH{=}CH \\ \quad\quad\quad\rangle NH \\ CH{=}CH \end{array} + 2H_2O$$

This synthesis is evidence of the correctness of the constitution assigned to pyrrole.

Pyrrole is a colourless liquid, b.p. 130–131°; it rapidly turns brown on standing. It is sparingly soluble in water and miscible with most organic solvents. It has an odour somewhat reminiscent of that of chloroform. A characteristic test for pyrrole is the turning an intense red colour of a pine wood shaving which has been moistened with hydrochloric acid when submitted to the action of pyrrole vapour (πυῤῥός = fire-red). In spite of its having the constitution of a secondary amine, pyrrole is only very feebly basic. It is only partially soluble in dilute mineral acids in the cold and tends to give resinous products when these solutions are heated; the resinous products are probably polymerisation products of pyrrole which have not been fully investigated. Pyrrole is also feebly acidic and like phenol will react with potassium evolving hydrogen. The colourless product, *potassium-pyrrole*, $\begin{array}{l} HC{=}CH \\ \quad\quad\quad\rangle NK \\ HC{=}CH \end{array}$, is insoluble in ether and is decomposed by water, forming pyrrole and potassium hydroxide (compare potassium phenate). Potassium-pyrrole reacts readily with alkyl and aryl iodides (e.g. methyl iodide and iodobenzene), yielding N-alkyl or N-aryl derivatives of pyrrole which are tertiary amines.

When pyrrole vapour and hydrogen are passed over reduced nickel at about 180°, the completely reduced product, *pyrrolidine*, is obtained. Pyrrolidine, b.p. 87°, is also a secondary amine, but much more strongly basic than pyrrole. This compound has also been synthesised by various methods, one of which may be represented in outline, thus:

$$\begin{array}{l} CH_2.Br \\ | \\ CH_2.Br \end{array} \xrightarrow{2KCN} \begin{array}{l} CH_2.C\vdots N \\ | \\ CH_2.C\vdots N \end{array} \xrightarrow{\text{reduction}} \begin{array}{l} CH_2.CH_2.NH_2 \\ | \\ CH_2.CH_2.NH_2 \end{array} \xrightarrow{HCl}$$

ethylene bromide — ethylene cyanide — tetramethylenediamine (diprimary amine)

$$\begin{array}{l} CH_2.CH_2.NH_2.HCl \\ | \\ CH_2.CH_2.NH_2 \end{array} \xrightarrow{\text{heat}} \begin{array}{l} CH_2—CH_2 \\ \quad\quad\quad\rangle NH \\ CH_2—CH_2 \end{array} + NH_4Cl$$

monohydrochloride of tetramethylenediamine

When pyrrole is shaken with aqueous alkali hydroxide and iodine, *tetraiodopyrrole* (known technically as 'iodol') is precipitated. This brown crystalline compound has the constitution

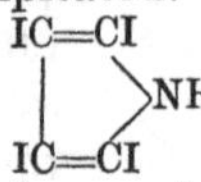

IC=CI
| >NH; it
IC=CI

decomposes at 140° and is used as an antiseptic, being preferred to iodoform on account of its lack of odour. The ease of replacement of hydrogen atoms attached to the carbon atoms in pyrrole by halogens is reminiscent of the behaviour of phenol.

Constitutionally pyrrole can be regarded as being derived from thiophen (p. 57) by replacing the bivalent sulphur by the :NH (imino) group. Substituted derivatives of pyrrole are denoted

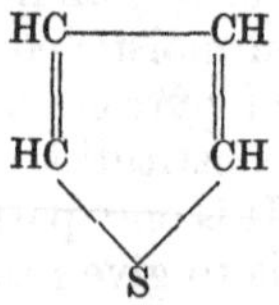

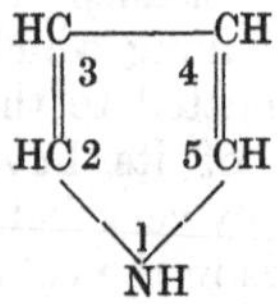

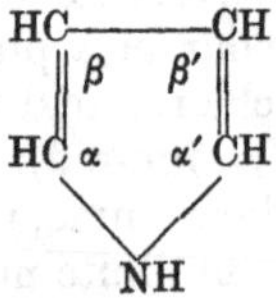

HC——CH　　HC——CH　　HC——CH
‖　　‖　　‖ 3　4 ‖　　‖ β　β′ ‖
HC　　CH　　HC 2　5 CH　　HC α　α′ CH
S　　1 NH　　NH

by the numbers or letters indicated. In view of the symmetry of the molecule, for each substituting group there are two mono-substituted derivatives of pyrrole, e.g. there are two monomethyl-pyrroles known which may be described as 2-methylpyrrole and 3-methylpyrrole or α- and β- methylpyrroles respectively. A dimethylpyrrole which occurs in bone-oil has been shown to be 2:5-dimethylpyrrole or αα′-dimethylpyrrole. Under certain conditions pyrrole and its derivatives (5-atom ring system) give rise to pyridine derivatives (6-atom ring system), which thus appear to have a more stable configuration. A simple example of this change is the conversion of potassium-pyrrole into β-chloropyridine by heating with sodium ethylate and chloroform:

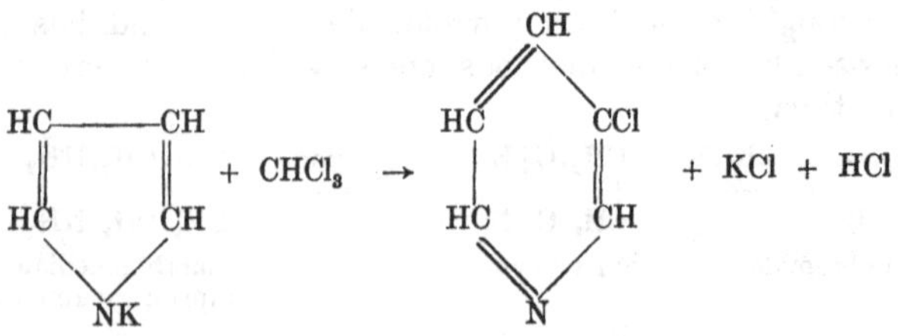

HC——CH　　　　　　CH
‖　　‖　　　　HC　　CCl
HC　　CH + $CHCl_3$ → |　　‖ + KCl + HCl
NK　　　　　　HC　　CH
　　　　　　　　　　N

β-chloropyridine

An important derivative of pyrrole is *indole*, which occurs in coal-tar and which has the constitution:

```
        H                                 H
        C                                 C
      //4 \                             //  \
  HC 5     C———————CH               HC       C———————CH
  |        ‖     3 ‖                |        ‖     β ‖
  HC 6     C     2 CH      or       HC       C     α CH
     \\ 7 / \  1  /                   \\    / \     /
        C     NH                         C     NH
        H                                H
```

It is an example of a condensed ring system, condensation of the benzene and pyrrole rings, the system having the grouping =C=C= common to both rings (compare benzene and naphthalene). Indole can actually be prepared by a remarkable reaction which is effected by allowing a solution of pyrrole in a 10 per cent. aqueous solution of sulphuric acid to stand for some two hours and then submitting the product to steam distillation. The formation of indole may be due to the decomposition of polymerised pyrrole, tripyrrole, as indicated:

```
HC———CH—CH—CH—CH—CH        HC=CH—C———CH        HC———CH
‖     |  |  |  |  ‖   →    |     ‖    ‖    +   ‖    |    + NH3
HC    CH—CH CH—CH CH       HC=CH—C    CH       HC   CH
  \  /     \  /    \ /            \  /           \  /
   NH       NH      NH             NH             NH
                                indole
```

Indole crystallises in colourless plates, m.p. 52°, and ordinarily possesses a strong faecal odour. When carefully purified and suitably diluted it may be added to other perfumes to simulate the odour of natural flowers. Jasmine-flower oil contains some 2½ per cent. of indole. Indole and its 3- or β- methyl derivative, skatole, occur in faeces, their presence being due to the bacterial decomposition of proteins containing the tryptophan* nucleus. Like pyrrole, indole is a secondary amine with very feebly basic properties. It also gives the pyrrole reaction and undergoes resinification (? polymerisation) with acids.

* *Tryptophan* is one of the important α-amino-acids, its constitutional name being β-indole-α-aminopropionic acid (p. 324):

```
 /‾‾\———C—CH2—CH(NH2).COOH
|    |   ‖
 \__/   CH
    \  /
     NH
```

Skatole has the constitutional formula:

```
 /‾‾\———C.CH3
|    |   ‖
 \__/   CH
    \  /
     NH
```

Indole has a close relationship with indigo (p. 332) and can be regarded as the parent of that substance. An intermediate reduction product of indigo is indoxyl (β-hydroxy-indole), and this colourless substance when further reduced is converted into indole:

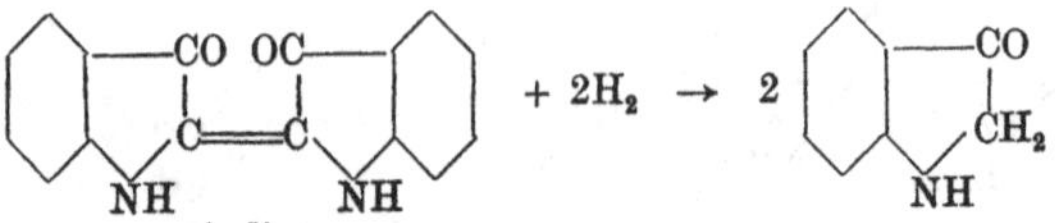

$$\text{indigo (CO OC, C=C, NH NH)} + 2H_2 \rightarrow 2\ \text{(CO, CH}_2\text{, NH)}$$

indigo

$$\text{(CO, CH}_2\text{, NH)} \rightleftarrows \text{(COH, CH, NH)} + H_2 \rightarrow \text{(CH, CH, NH)}$$

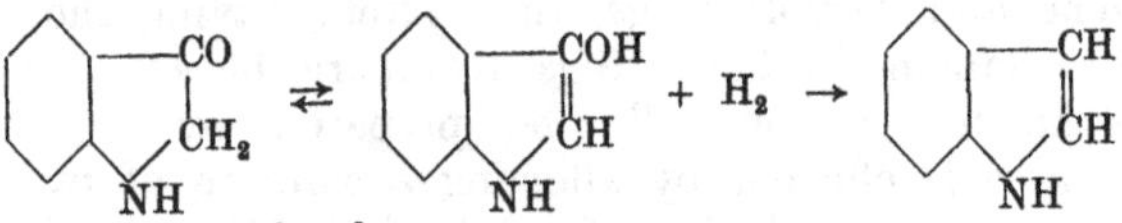

indoxyl

When bacterial decomposition takes place in the alimentary canal, indole from tryptophan is present as 3-hydroxyindole or indoxyl which is finally excreted in the urine as indoxyl sulphate or its potassium salt (*urinary 'indican'*). The presence of indoxyl sulphate in urine can easily be determined by converting it into indigo. To the urine, add an equal volume of strong hydrochloric acid (to hydrolyse the indoxyl sulphate) and then not more than one or two drops of a 2 per cent. aqueous solution of potassium chlorate as oxidising agent. Indigo is produced by changes which can be represented:

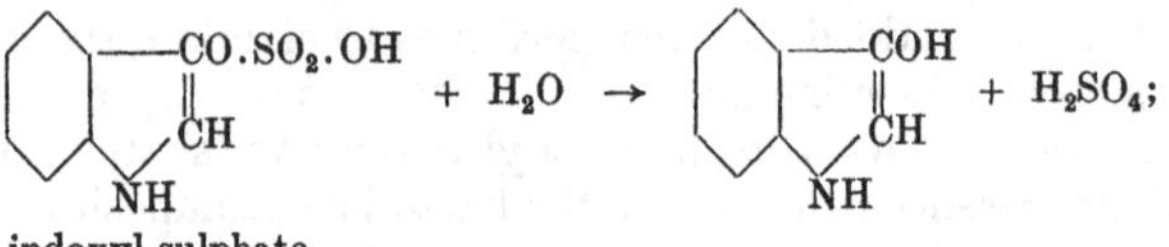

$$\text{(CO.SO}_2\text{.OH, CH, NH)} + H_2O \rightarrow \text{(COH, CH, NH)} + H_2SO_4;$$

indoxyl sulphate

$$\text{(COH, CH, NH)} \rightleftarrows \text{(CO, CH}_2\text{, NH)}$$

$$\text{(CO, CH}_2\text{, NH)} + \text{(OC, H}_2\text{C, NH)} + O_2 \rightarrow \text{(CO OC, C=C, NH NH)} + 2H_2O$$

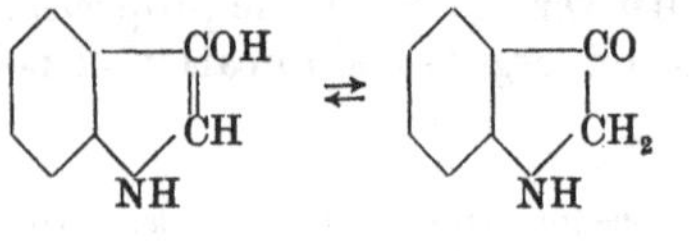

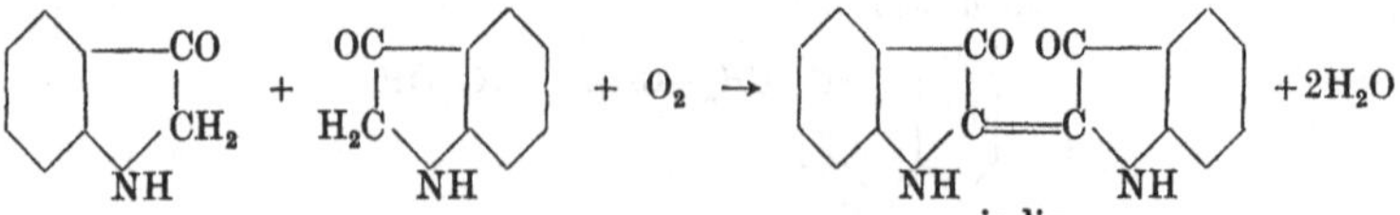

indigo

and the presence of a small quantity of indigo can be more readily observed by shaking the mixture with a little chloroform, which dissolves the blue colouring matter. The conversion of indoxyl into indigo is the reverse of the reduction of indigo.

Pyridine occurs in the basic fraction of bone-oil which distils at above 100°. From these it can be separated by means of its sparingly soluble picrate, $C_5H_5N.C_6H_2(OH)(NO_2)_3$. It is also separated from the distillate, b.p. 80–170°, from coal-tar, the basic constituents being separated from the hydrocarbons by washing with dilute sulphuric acid and treating the acid liquor with excess of slaked lime and water.

Pyridine has been synthesised pyrogenically by passing a mixture of acetylene and hydrogen cyanide through a red-hot tube:

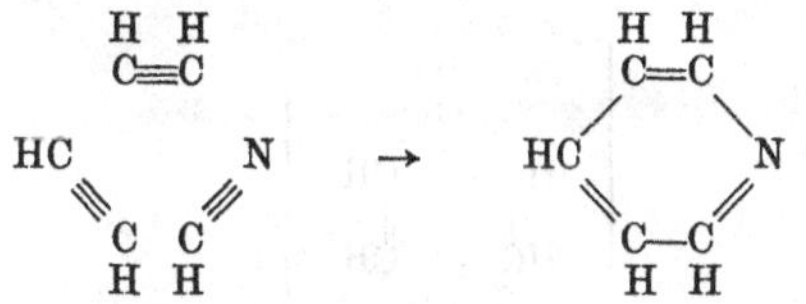

```
      H H                        H H
      C≡C                        C=C
                                /   \
 HC         N      →        HC       N
   \\\     ///                 \\   //
     C   C                      C—C
     H   H                      H H
```

This resembles the pyrogenic production of benzene from acetylene under analogous conditions, and benzene will obviously be produced along with pyridine in the above reaction. The following outlined synthesis affords proof of the correctness of the constitution given to pyridine and piperidine:

```
    CH2Br       KCN                  CH2—C≡N
   /                                /
 CH2        +             →       CH2            + 2KBr
   \                                \
    CH2Br       KCN                  CH2—C≡N
trimethylene bromide              trimethylene cyanide
(1 : 3-dibromopropane)            (1 : 3-dicyanopropane)

    CH2—C≡N      2H2                 CH2—CH2—NH2
   /                                /
 CH2          +           →       CH2
   \                                \
    CH2—C≡N      2H2                 CH2—CH2—NH2
                                  pentamethylene diamine

   CH2.CH2.:NH2.HCl:                          CH2—CH2
  /         :.....:                          /       \
 CH2              :       heated   NH4Cl + H2C        NH
  \              :H       ————→              \       /
   CH2.CH2.N<  :......                        CH2—CH2
             \H
                                              piperidine

                                                   H H
    CH2—CH2                                        C=C
   /       \             oxidation                /   \
 CH2        NH  +  3O   ——————————→  3H2O  +  HC       N
   \       /                                    \\   //
    CH2—CH2                                       C—C
                                                  H H
                                                pyridine
```

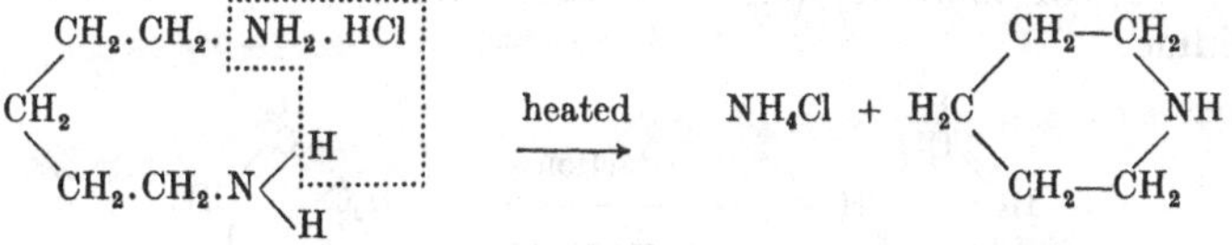

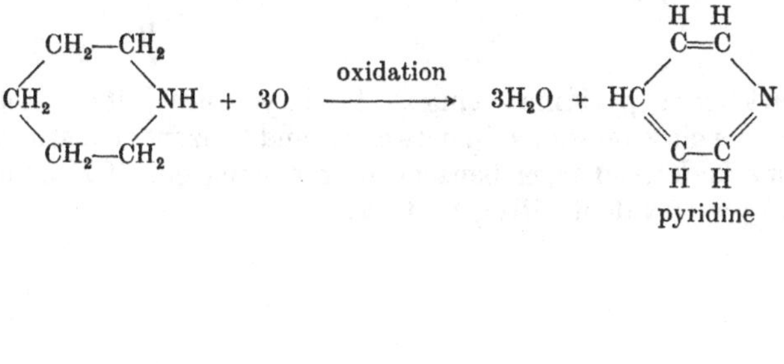

Pyridine is a colourless liquid, b.p. 115°, possessing a characteristic pungent odour. It is soluble in water and this solution is feebly alkaline. It forms salts by direct combination with acids, e.g. the hydrochloride, $C_5H_5N.HCl$ or $[C_5H_5NH]Cl$, and chloroplatinate, $(C_5H_5N)_2.H_2PtCl_6$ or $[(C_5H_5N)_2H_2]PtCl_6$. Being a tertiary amine it combines directly with methyl iodide, forming the colourless *pyridine methiodide* which is a *quaternary* compound and is also designated *methylpyridinium iodide* having the constitution:

```
 ┌        H        ┐
 │        C        │
 │    HC     CH    │
 │    HC     CH    │  I
 │        N        │
 │       CH3       │
 └                 ┘
```

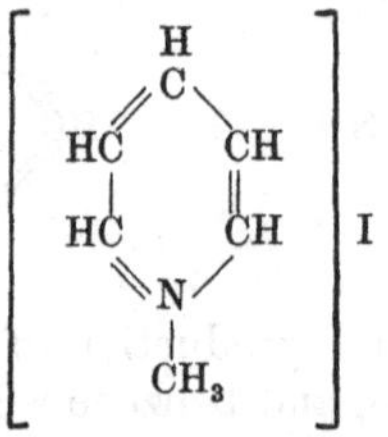

This compound has m.p. 117° and, when heated to 300°, it is converted into the hydriodide of *α-methylpyridine* or *α-picoline* (*v.* below) (α-methylpyridinium iodide) having the constitution:

```
        H                     ┌        H         ┐
        C                     │        C         │
    HC     CH                 │    HC     CH     │
    HC     C.CH3              │    HC     C.CH3  │  I
        N                     │        N         │
                              │        H         │
                              └                  ┘
```

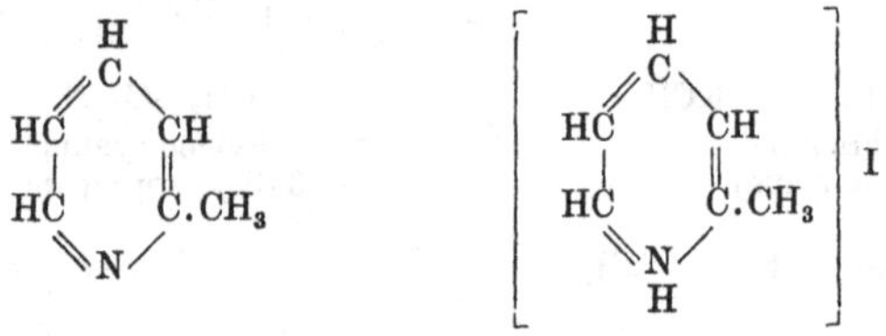

When pyridine in alcoholic solution reacts with sodium it is reduced to piperidine (hexahydropyridine), which again can be oxidised by a variety of methods (e.g. hot concentrated sulphuric acid) to pyridine:

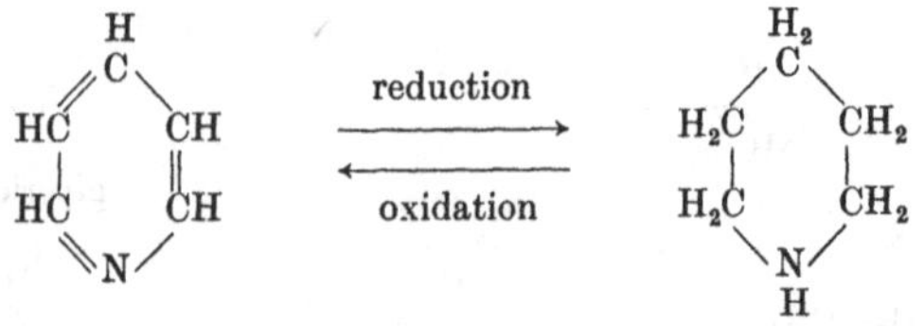

```
        H                                  H2
        C            reduction             C
    HC     CH      ------------->     H2C     CH2
    HC     CH      <-------------     H2C     CH2
        N            oxidation             N
                                           H
```

Apart from pyridine being a tertiary amine, its constitution indicates a close relationship between it and benzene; constitutionally pyridine is derived from benzene by replacing one ⫶CH or methine group by a tervalent nitrogen atom:

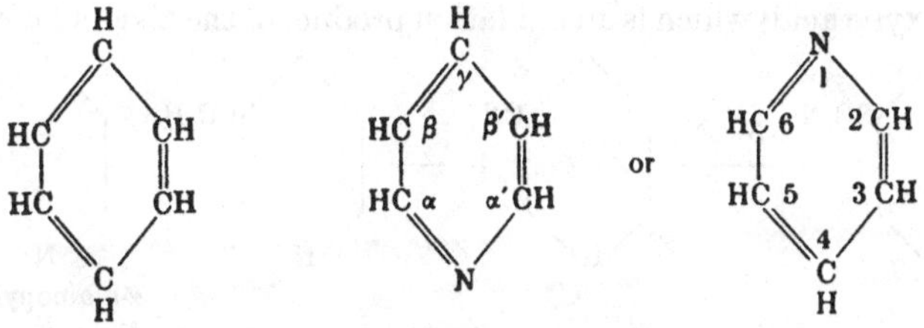

The presence of the nitrogen atom renders the isomerism greater among pyridine derivatives than among those of benzene. It is usual to differentiate the carbon atoms in pyridine as indicated. On account of their different positions relative to the nitrogen atom, by substitution of each of the three hydrogen atoms in the α-, β- and γ-positions three isomeric monosubstituted pyridines are possible for each substituting group. There are, for example, three monomethyl pyridines known:

α-methylpyridine or *α-picoline* (2-methylpyridine), b.p. 130°.

β-methylpyridine or *β-picoline* (3-methylpyridine), b.p. 143°.

γ-methylpyridine or *γ-picoline* (4-methylpyridine), b.p. 144°.

α- and β-methylpyridines occur in bone-oil and γ-methylpyridine is formed in small quantity along with α-methylpyridine during the heating of methylpyridinium iodide.* Just as methylbenzene or toluene is oxidised to benzoic acid (benzene carboxylic acid), these three methylpyridines are converted into the corresponding pyridine carboxylic acids by oxidation:

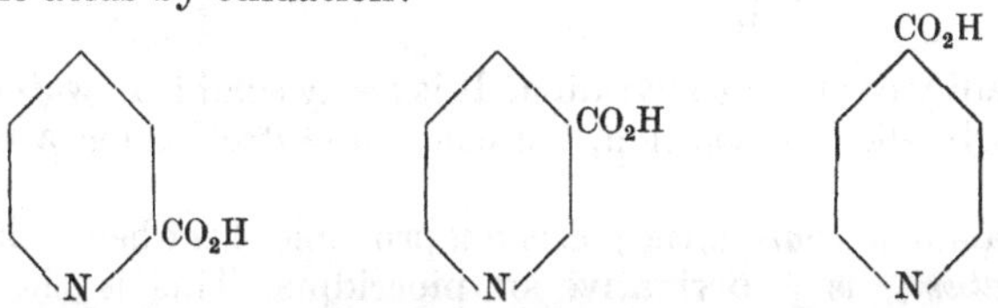

picolinic acid nicotinic acid *iso*nicotinic acid

Pyridine is more resistant to attack by reagents than benzene. It can be nitrated but with greater difficulty than benzene, and requires to be heated to a high temperature with concentrated sulphuric acid for sulphonation to take place. Under these conditions *pyridine-β-sulphonic acid* is formed. This compound reacts similarly to benzenesulphonic acid; when its sodium salt is fused with sodium cyanide, *β-cyanopyridine* is formed and the latter on hydrolysis in the usual manner is converted into *nicotinic acid* (pyridine-β-

* β-Methylpyridine is obtained during the distillation of the important alkaloid *strychnine*.

carboxylic acid) which is an oxidation product of the alkaloid *nicotine*:

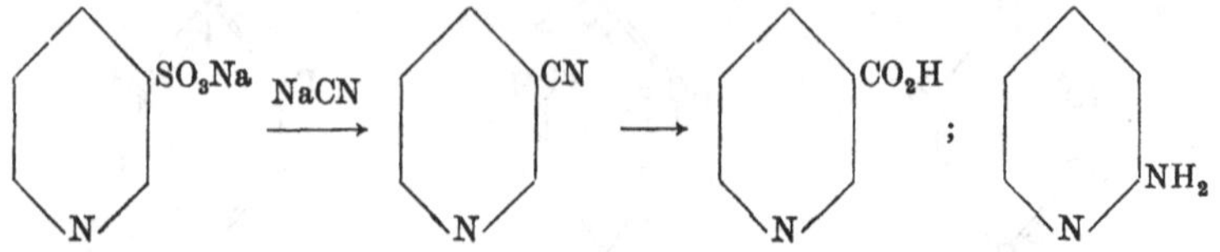

SO_3Na N $\xrightarrow{NaCN}$ CN N $\longrightarrow$ CO_2H N ; NH_2 N

α-aminopyridine

A convenient method for preparing an aminopyridine analogous to aniline consists in allowing sodamide to react with pyridine at 100–120° for some hours. The chief product of this reaction is *α-amino-pyridine* or 2-aminopyridine, which forms colourless crystals, m.p. 56°, b.p. 204°, which besides being a pyridine derivative has all the usual reactions of a primary amine.

Piperidine is a colourless liquid, b.p. 105·7°, which has a characteristic basic odour somewhat resembling that of pepper. It is the reduction product of pyridine and it is also obtained by heating *piperine* (the alkaloid from pepper) with soda lime or alcoholic potash. By synthesis it has been shown to be hexahydropyridine and to have the constitution:

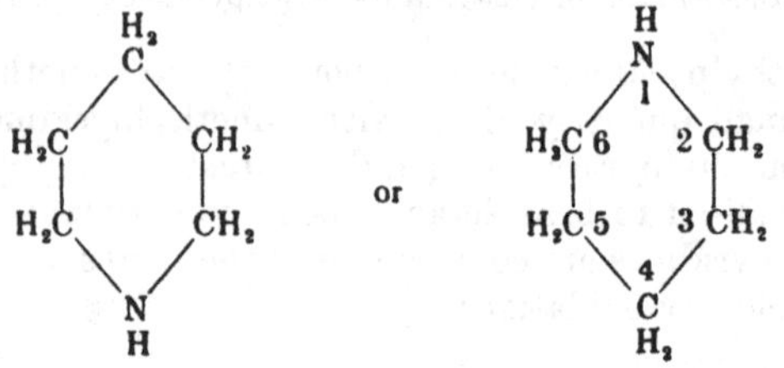

H_2 C; H_2C CH_2; H_2C CH_2; N H

or

H N 1; H_2C 6 2 CH_2; H_2C 5 3 CH_2; 4 C H_2

It is readily oxidised to pyridine. It is freely soluble in water yielding a strongly alkaline solution, the compound itself being a secondary amine.

The alkaloid *coniine*, the poisonous principle of the hemlock (*Conium masculatum*), is a derivative of piperidine. This is the simplest alkaloid* known and was the first alkaloid to be synthesised (Ladenburg, 1889). Coniine is *α-n-propylpiperidine* or 2-*n*-propylpiperidine:

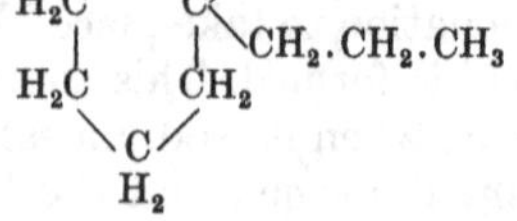

H N; H_2C C H; $CH_2.CH_2.CH_3$; H_2C CH_2; C H_2

* Most vegetable alkaloids contain carbon, hydrogen, oxygen and nitrogen. Coniine and nicotine are among the very few which contain only carbon, hydrogen and nitrogen. The constitutions of these alkaloids are much simpler than of those containing oxygen.

and the natural alkaloid (extracted generally from the seeds of the hemlock) is optically active and *dextro*rotatory; the α- or 2- carbon atom is asymmetric (p. 113). The synthesis of coniine was effected as follows: α- or 2- methylpyridine was condensed with aldehyde by heating it with paraldehyde in the presence of zinc chloride and the α- or 2- allylpyridine was reduced to the piperidine derivative by means of sodium in alcoholic solution:

```
                  H                                  
     N           /        H              N
  HC    C—C=[H2   +   O].C.CH3        HC    C—CH:CH.CH3
   |    ||                       →     |    ||              + H2O
  HC    CH                            HC    CH
     C                                   C
     H                                   H
                                      α-allylpyridine

            H  H                         H
     N      |  |                         N       H
  HC    C—C=C—CH3                     H2C    C—CH2.CH2.CH3
   |    ||          + 4H2    →         |     |
  HC    CH                            H2C    CH2
     C                                   C
     H                                   H2
                                      α-n-propylpiperidine
```

α-*n*-Propylpiperidine thus obtained is not identical with the natural coniine. Like all synthetic products, it is optically inactive. It is usually described as externally compensated, or *dl*-α-*n*-propylpiperidine.

The synthetic or externally compensated compound is resolved by the crystallisation of its salt with *d*-tartaric acid. A mixture of the two salts, *d*-α-*n*-propylpiperidine *d*-tartrate and *l*-α-*n*-propylpiperidine *d*-tartrate, is obtained, the former salt being the less soluble (see p. 346). The base isolated from the less soluble salt is *dextro*-α-*n*-propylpiperidine and identical with the natural coniine. This is a colourless, highly poisonous liquid, b.p. 167°; its specific rotatory power, $[\alpha]_D$, is +15·7°.

Like piperidine it is a secondary amine and its aqueous solution reacts strongly alkaline.

Occurring also among the basic constituents of coal-tar and bone-oil is another heterocyclic tertiary amine, *quinoline*, which has also been obtained from various alkaloids by heating them with sodium hydroxide. Its constitution is represented:

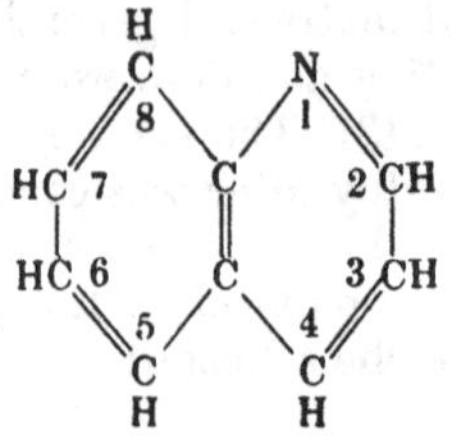

H C 8 N 1
HC 7 C 2 CH
HC 6 C 3 CH
5 C H 4 C H

being an αβ- or 2 : 3- substituted derivative of pyridine, and it may be designated αβ- or 2 : 3- *benzopyridine.* Its relationship to pyridine corresponds with that of naphthalene to benzene, and it may be considered to be derived constitutionally from naphthalene by the substitution of a trivalent nitrogen atom for one ⋮CH or methine group in the position indicated.

The isomerism of the derivatives of quinoline is more complicated than that of naphthalene derivatives; the unsymmetrical nature of the molecule necessitates that no two hydrogen atoms are similarly placed with respect to the nitrogen atom, and consequently for each substituting atom or group there are seven monosubstituted derivatives of quinoline theoretically capable of existence.

Quinoline has been synthesised from aniline by Skraup's reaction (1896), which has also been applied to other aromatic primary amines. Aniline, glycerol and sulphuric acid are heated together at 140° with an oxidising agent such as nitrobenzene or arsenic acid. The reaction taking place may be summarised by the equation:

$$C_6H_5.NH_2 + C_3H_8O_3 \rightarrow C_9H_7N + 3H_2O + H_2$$

The simplest explanation of the course of the reaction is to consider that acroleïn, formed by the dehydration of glycerol:

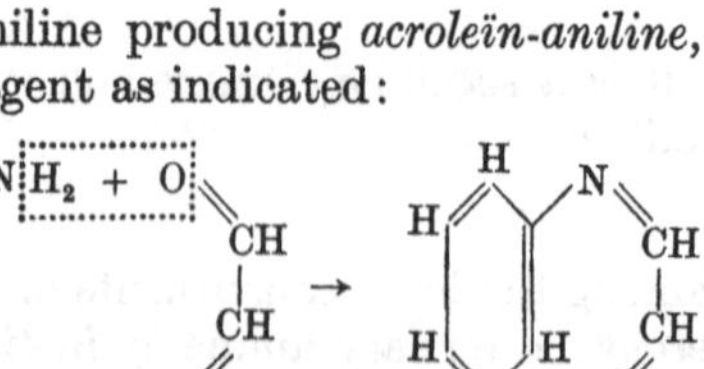

$$H_2C{-}OH,\ HC{-}OH,\ H_2C(H)(OH) \rightarrow CH_2{=}CH{-}C(H){=}O + 2H_2O$$

reacts with the aniline producing *acroleïn-aniline,* which is oxidised by the oxidising agent as indicated:

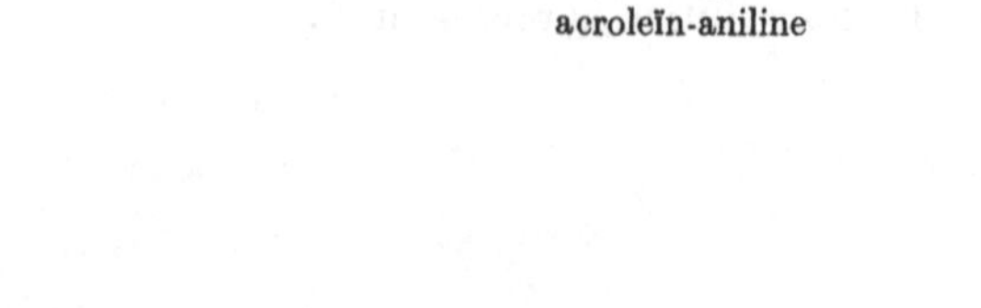

$$C_6H_5{-}NH_2 + O{=}CH{-}CH{=}CH_2 \rightarrow C_6H_5{-}N{=}CH{-}CH{=}CH_2 + H_2O$$

acroleïn-aniline

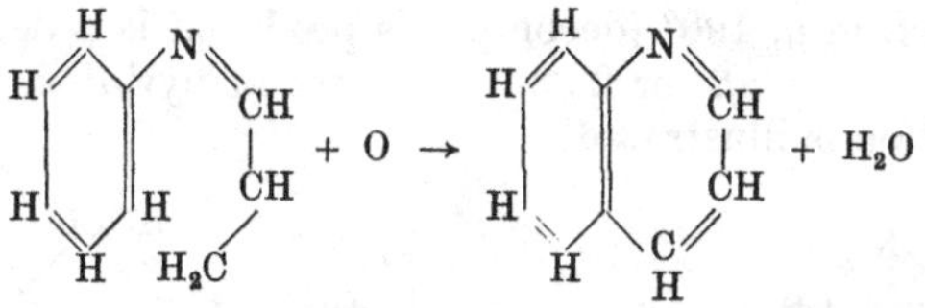

(dihydroquinoline structure: H, H, H, H, N, CH, CH, H, H_2C) + O → (quinoline structure: H, H, H, H, N, CH, CH, C, H) + H_2O

Quinoline is a colourless liquid, m.p. −19·5°, b.p. 240°, which is miscible with all neutral organic solvents, almost insoluble in water and, like pyridine, is a tertiary amine. It forms stable salts and combines directly with alkyl iodides, *methylquinolinium iodide*, m.p. 133°, having the constitution:

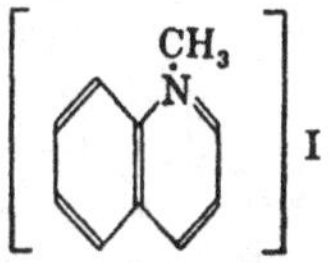

$$\left[C_9H_7N\text{-}CH_3 \right] I$$

being formed when methyl iodide is used.

When quinoline is reduced with sodium in alcoholic solution, *tetrahydroquinoline*, m.p. 20°, b.p. 251°, is produced, only the pyridine ring being affected:

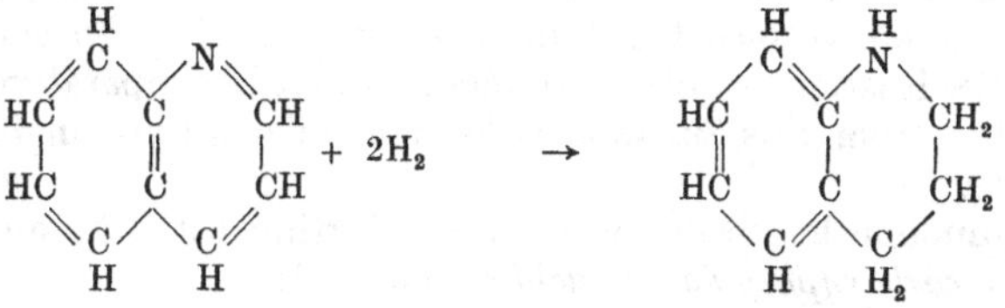

(quinoline: H, C, N, HC, C, CH, HC, C, CH, C, C, H, H) + $2H_2$ → (H, H, C, N, HC, C, CH_2, HC, C, CH_2, C, C, H, H_2)

tetrahydroquinoline

This compound is a secondary amine. The true piperidine derivative corresponding to quinoline is, however, the fully reduced compound, *decahydroquinoline*, b.p. 206°, produced by the catalytic reduction of quinoline. Its properties are very similar to those of pyridine and it resembles an aliphatic secondary amine in yielding stable salts and a strongly alkaline aqueous solution. Its constitution is represented:

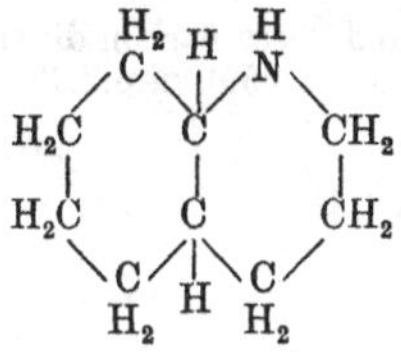

(H_2, H, H, C, N, H_2C, C, CH_2, H_2C, C, CH_2, C, H, C, H_2, H_2)

The behaviour of quinoline on oxidation with potassium permanganate whereby oxalic acid and αβ- or 2 : 3- *dicarboxypyridine* or *quinolinic acid*, m.p. 190° (decomp.), is produced is a definite proof that quinoline is an $\alpha\beta$- or 2 : 3- derivative of pyridine. The course of this oxidation is illustrated:

```
       H                                                HOOC
      C    N                                               \ /N\
   HC    C    CH                      HOOC                  C    CH
   |     ||   |    + 8O    →          |          +          ||   |
   HC    C    CH                      HOOC                  C    CH
      C    C                                             /    C
      H    H                                        HOOC      H
                                                      quinolinic acid*
```

The one possible isomeride of quinoline is iso*quinoline*, which has the constitution:

```
       H    H
      C    C
   HC    C    N
   |     ||   |
   HC    C    CH
      C    C
      H    H
```

This compound does not occur in bone-oil but occurs along with quinoline in coal-tar. It is a colourless liquid, b.p. 241°, generally remaining supercooled below its melting point, 25°. In many of its properties it is very similar to quinoline and it is comparatively easily separated from this substance through its more sparingly soluble sulphate.

*Iso*quinoline has been synthesised, starting with the ammonium salt of o-*carboxyphenylacetic acid* as outlined:

```
             O
       H     ||                                  H     O
      C      C—ONH4                             C     C
   HC    C        ONH4     heat            HC    C       NH
   |     ||       |        ——→             |     ||      |      + NH3 + 2H2O
   HC    C        C=O                      HC    C       CO
      C     C                                 C     C
      H     H2                                H     H2
                                               homophthalimide
```

* Quinolinic acid, when heated, loses carbon dioxide and is converted into β- or 3- *carboxypyridine*, nicotinic acid (p. 299), m.p. 229°:

```
       N                                  N
   HC     C—COOH                      HC     CH
   ||     |            ——→            ||     |             + CO2
   HC     C—COOH                      HC     C—COOH
       C                                  C
       H                                  H
                                       nicotinic acid
```

```
   H    O                H    OH                          H    Cl
   C    C                C    C                           C    C
HC   C    NH   ⇄     HC    C    N      POCl3         HC    C    N
|    ‖    |          |     ‖    |      ----->        |     ‖    |    + HPO3 + HCl
HC   C    CO         HC    C    C(OH)   heat         HC    C    CCl
   C    C                C    C                           C    C
   H    H2               H    H                           H    H
                                                     dichloroisoquinoline
```

```
     H    Cl                            H    H
     C    C                             C    C
HC     C    N          4H          HC     C    N
|      ‖    |       --------->     |      ‖    |     +   2HCl
HC     C    CCl     reduction      HC     C    CH
     C    C                             C    C
     H    H                             H    H
```

That *iso*quinoline is a $\beta\gamma$- or 3 : 4- derivative of pyridine, having the alternative name of $\beta\gamma$- or 3 : 4- benzopyridine, is proved by the course of the oxidation and the nature of the products formed when *iso*quinoline is oxidised with potassium permanganate (compare quinoline). In this case, dicarboxylic acids derived from benzene and from pyridine are obtained:

```
                                H
                                C
                          HC       C.COOH
     H    H               |        ‖            phthalic acid (o-dicarboxy-
     C    C        ↗      HC       C.COOH         benzene, p. 74)
HC     C    N                   C
|      ‖    |                   H
HC     C    CH
     C    C        ↘            N
     H    H               HC       CH           cinchomeronic acid* or βγ-
                          |        ‖             or 3 : 4- dicarboxypyridine,
                     HOOC.C        CH            m.p. 266° (decomp.)
                                C
                               COOH
```

* Cinchomeronic acid on being heated loses carbon dioxide and is converted into iso*nicotinic acid* (γ- or 4- carboxypyridine), m.p. 304°; compare action of heat on quinolinic acid:

```
        N                          N
   HC       CH               HC       CH
   ‖        |        →       ‖        |     +   CO2
   HC       C.COOH           HC       CH
        C                          C
       COOH                       COOH
```

A dibenzopyridine, related to pyridine as benzene is to anthracene, is *acridine*, which occurs in coal-tar. Its constitution may be represented in two ways:

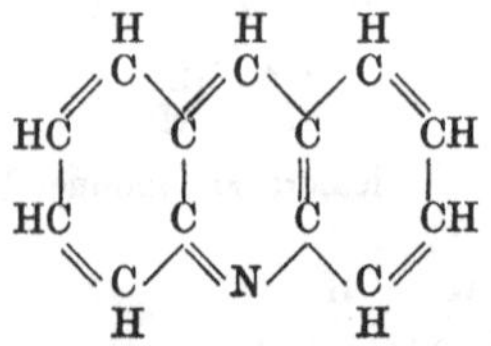

or

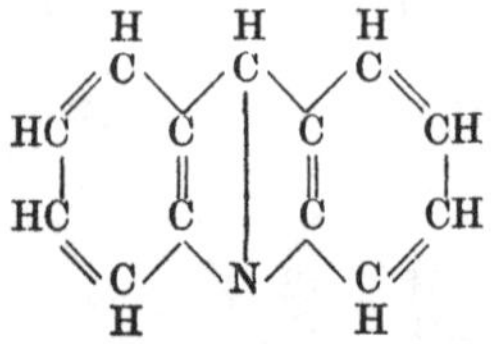

It has been synthesised and its constitutional formula verified by heating the formyl derivative of diphenylamine with zinc chloride:

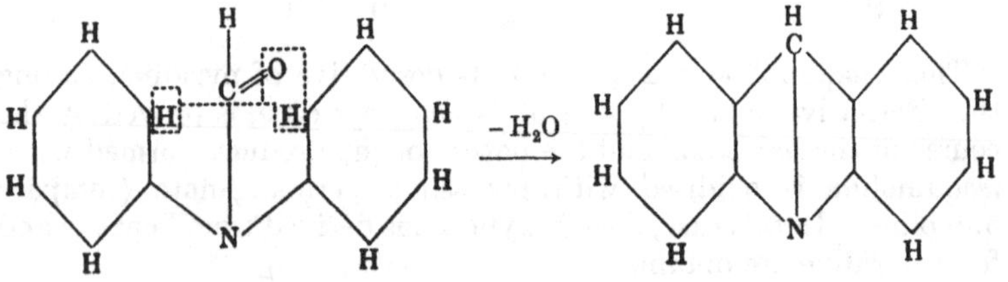

$-H_2O$

formyldiphenylamine

Acridine crystallises in colourless needles, m.p. 110°. It is a very weak base, characterised by great stability and by yielding strongly fluorescent solutions. It is the parent member of an important group of dye-stuffs.

The constitution of some more complex cyclic amines may be briefly indicated. Pyridine has been shown to be derived constitutionally from benzene by replacing a ⋮CH or methine group in the latter by a tervalent nitrogen atom:

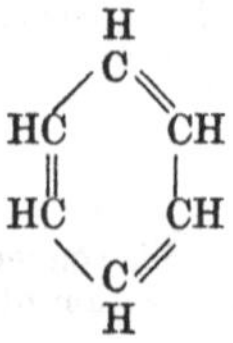

benzene 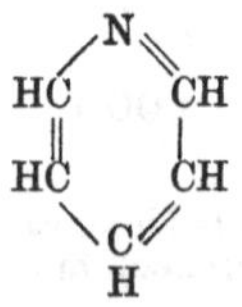 pyridine 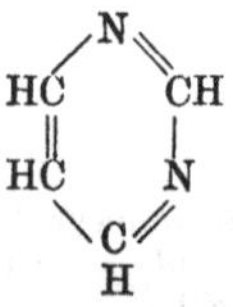 pyrimidine

By substituting another nitrogen atom for another ⋮CH group in the alternate or *β*- position in pyridine, the constitutional formula of pyrimidine is obtained. *Pyrimidine*, m.p. 21°, b.p. 124°, has a peculiar narcotic odour and is readily soluble in water. It belongs to a group of basic compounds known as *metadiazines*, and analogous *ortho-* and *para-* diazines are known.

In a similar way, starting with the constitutional formula of pyrrole the constitutional formula of *glyoxaline* or *iminazole* may be derived:

```
      H                    H                          
      N                    N                          N
    /   \                /   \                      /   \\
  HC     CH           HC 5  1  2 CH      or      HC 5  1  2 CH
  ||     ||           ||          ||              ||         |
  ||     ||           ||  4    3  ||              ||  4   3  |
  HC —— CH            HC ——————— N               HC ——————— NH
   pyrrole             glyoxaline        or        iminazole
```

It is impossible to distinguish between positions 4 and 5 in the glyoxaline nucleus, and glyoxaline may be represented by the alternative formula indicated. Glyoxaline derives its name from the fact of its being obtained by the action of ammonia on glyoxal (p. 354). In this reaction it is assumed that a part of the glyoxal is first hydrolysed into formaldehyde and formic acid, and that glyoxal, formaldehyde and ammonia condense as indicated:

```
H—C=O                          O                 O
  |     + H2O  →  H—C  //   +   H—C  //
H—C=O                          \ H               \ O—H
```

```
       H                                   H
       N                                   N
HC¦O   H2¦  ¦O¦CH                        /   \
  |         ¦ H¦                       HC     CH
  |         ¦  ¦         →             ||     ||      + 3H2O
  |         ¦ H¦                       HC ——— N
HC¦O        H2¦N
```

Glyoxaline is a weak base and crystallises in colourless prisms, m.p. 89°, b.p. 255°. The important amino acid *histidine* (p. 325) and the base *histamine* are derivatives of glyoxaline.

Just as the constitutional formula of naphthalene can be derived by the fusion of two benzene rings (having the =C=C= grouping common to both) and the constitutional formulae of quinoline and *iso*quinoline similarly derived from a benzene and pyridine ring:

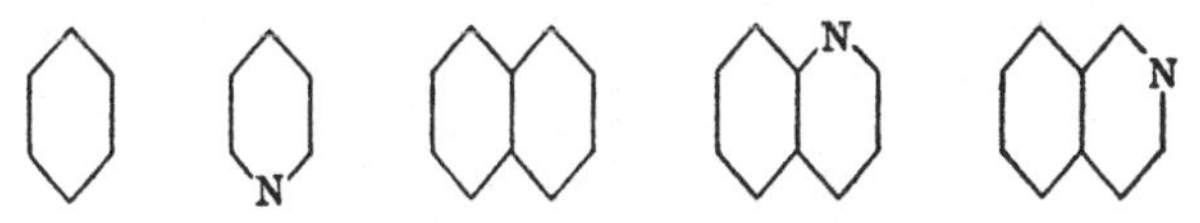

so, by the fusion of a pyrimidine and glyoxaline ring, the constitutional formula of *purine* is obtained:

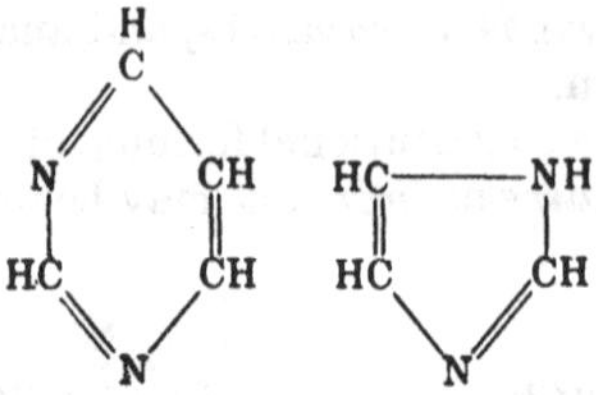

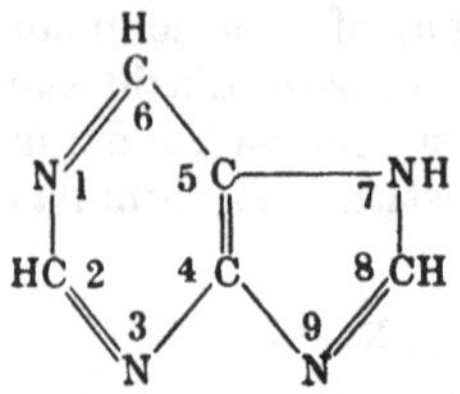

purine

Purine has been synthesised; it is a colourless crystalline substance, m.p. 217°, readily soluble in water and possesses both acidic and basic properties. Its importance lies in the fact that it is the parent member of a group of compounds, the *purines*, which contains the physiologically important *uric acid* and such alkaloids as *theobromine* and *caffeine*, and also other compounds such as *guanine* and *adenine* which are fundamental constituents of *nucleic acids*, the essential constituents of the living cell.

CHAPTER XI

AMINO ACIDS

MANY organic compounds may be described as compounds of a single function due to the presence in their molecules of one typical group such as —OH (alcohols and phenols), —CHO (aldehydes), =CO (ketones), —CO_2H (carboxylic acids), —NH_2 (primary amines), etc., attached to a carbon atom. Organic compounds may contain two or more identical functional groups, and these, like those referred to above, are also described as compounds of only one function. On the other hand, many compounds have already been referred to which contain at least two different functional groups, including the above and others in the same molecule. Examples of such compounds are salicylic acid (at the same time a phenol and a carboxylic acid) and choline (a quaternary ammonium base and a primary alcohol). It is a significant fact that substances of biochemical importance are frequently compounds of more than one function. For example, the sugars react as aldehyde- or ketone- alcohols; lactic acid, tartaric acid and citric acid are alcohol-acids, and the amino acids are, as their name implies, at the same time bases and acids and are typical amphoteric substances.*

The typical amino acids of biochemical importance are carboxylic acids in which one or more hydrogen atoms of the hydrocarbon radicle have been replaced by the amino (—NH_2) group, and they belong to the aliphatic, aromatic and heterocyclic series. In the case of those biochemically important amino acids belonging to the aromatic and heterocyclic series they all have the amino and carboxyl groups in the same side chain attached to the aromatic or heterocyclic nucleus.

The largest number and the most important of the amino acids are those which result as the final hydrolytic products of the *proteins* (p. 328); they are also the products of hydrolysis of proteins by ferments in the digestive tract. The simplest amino acids obtained in this way are substituted fatty acids higher in the series than formic acid.

Bearing in mind the properties of the monobasic amino (—NH_2) and of the monoacidic carboxy (—CO_2H) groups, if one of each of

* The typical amides (possessing the —$CONH_2$ group) have both acid and basic properties, but this is due to the nature of the group itself and not to two distinct functional groups in the particular compound.

these groups be present in the molecule, the amino acid will be neutral in character; if two amino groups and one carboxy group be present, the amino acid will be basic in character; and if two carboxy groups and one amino group be present in the molecule at the same time, the amino acid will be acidic in character. Thus, the general properties of the amino acid depend on the preponderance or otherwise of the basic amino groups over the acidic carboxy groups in the molecule. Many amino acids of importance contain only one amino and one carboxy group.

Starting with acetic acid, the monoamino acids derived from the lower normal members of the fatty acid series indicate the isomerism in this series of compounds:

$$\begin{array}{l} CH_3 \\ | \\ CO_2H \end{array} \qquad \begin{array}{l} CH_2.NH_2 \\ | \\ CO_2H \end{array} \quad \textit{aminoacetic acid} \text{ or } \textit{glycine}$$

acetic acid

$$\begin{array}{l} \beta CH_3 \\ | \\ \alpha CH_2 \\ | \\ CO_2H \end{array} \qquad \begin{array}{l} CH_3 \\ | \\ CH.NH_2 \\ | \\ CO_2H \end{array} \quad \begin{array}{l} \alpha\textit{-aminopropionic} \\ \textit{acid} \text{ or } \textit{alanine}^{*} \end{array} \quad : \quad \begin{array}{l} CH_2.NH_2 \\ | \\ CH_2 \\ | \\ CO_2H \end{array} \quad \begin{array}{l} \beta\textit{-aminopropionic} \\ \textit{acid}^{\dagger} \end{array}$$

propionic acid

$$\begin{array}{l} \gamma CH_3 \\ | \\ \beta CH_2 \\ | \\ \alpha CH_2 \\ | \\ CO_2H \end{array} \qquad \begin{array}{l} CH_3 \\ | \\ CH_2 \\ | \\ CH.NH_2 \\ | \\ CO_2H \end{array} \quad \begin{array}{l} \alpha\textit{-amino-} \\ \text{n-}\textit{butyric} \\ \textit{acid} \end{array} \qquad \begin{array}{l} CH_3 \\ | \\ CH.NH_2 \\ | \\ CH_2 \\ | \\ CO_2H \end{array} \quad \begin{array}{l} \beta\textit{-amino-} \\ \text{n-}\textit{butyric} \\ \textit{acid} \end{array} \qquad \begin{array}{l} CH_2.NH_2 \\ | \\ CH_2 \\ | \\ CH_2 \\ | \\ CO_2H \end{array} \quad \begin{array}{l} \gamma\textit{-amino-} \\ \text{n-}\textit{butyric} \\ \textit{acid} \end{array}$$

n-butyric acid

Similarly, corresponding to *n*-valeric acid, $CH_3.(CH_2)_3.CO_2H$, four isomeric monoamino acids having the amino groups respectively attached to the α-, β-, γ- and δ- carbon atoms are known. While it is hardly necessary to consider the specific properties of the β-, γ-, δ-, etc. monoamino acids on account of the preponderating importance of the α-amino acids, it may be pointed out that they can be distinguished by the nature of the product they yield on being heated. *α-Amino acids* readily lose two molecules of water from two molecules of the amino acid yielding cyclic 'anhydrides', known as 2 : 5-diketopiperazines,‡ which have properties analogous to those of N-substituted amides:

* Sometimes referred to as α-alanine.

† Sometimes, but infrequently, referred to as β-alanine.

‡ The same compounds are obtained by the ready loss of two molecules of ethyl

$$\begin{array}{l} H H \\ H_2C-N-H \quad HO-CO \quad \rightarrow \quad H_2C-N-CO \\ \quad + \qquad\qquad\qquad\qquad\qquad\qquad 6\ \ 1\ \ 2 \qquad +2H_2O \\ OC-O-H \quad H\,N-CH_2 \qquad\quad 5\ \ 4\ \ 3 \\ H \qquad\qquad\quad OC-N-CH_2 \\ H \end{array}$$

glycine 'anhydride' or
2 : 5-diketopiperazine

β-Amino acids lose ammonia at their melting points, yielding unsaturated acids:

$$\begin{array}{l} H_2C-NH_2 \qquad\qquad\qquad H_2C \\ | \qquad\qquad\qquad\qquad\quad \ \| \\ HC-H \qquad \rightarrow \qquad HC \qquad + NH_3 \\ | \qquad\qquad\qquad\qquad\quad\ | \\ CO_2H \qquad\qquad\qquad\quad CO_2H \end{array}$$

β-aminopropionic acid   acrylic acid

γ- and δ-*Amino acids* on being heated lose a molecule of water from one molecule of the amino acid, yielding inner 'anhydrides' of a type known as *lactams*. The lactams from the γ-amino acids are pyrrolidones or ketopyrrolidines (p. 292) derived by substitution of an oxygen atom for two hydrogens on the α- or 2- carbon atom:

$$\begin{array}{l} CH_2-CH_2 \qquad\qquad\qquad\qquad CH_2-CH_2 \\ | \qquad\qquad \backslash \qquad\qquad\qquad\quad | \qquad\quad \backslash \\ | \qquad\qquad NH\ H \quad \rightarrow \quad | \qquad\qquad NH \ + H_2O \\ | \qquad\qquad\qquad\qquad\qquad\quad | \qquad\quad / \\ CH_2-C-OH \qquad\qquad\qquad CH_2-CO \\ O \end{array}$$

γ-amino-*n*-butyric acid   γ-butyrolactam or α-pyrrolidone

In the case of δ-amino acids, loss of water leads to the production of δ-lactams which are α-piperidones. From δ-amino-*n*-valeric acid δ-*valerolactam* or α-*piperidone* (i.e. α-ketopiperidine) is obtained:

$$\begin{array}{l} CH_2-CH_2 \qquad\qquad\qquad\qquad\quad H_2\ \ H_2 \\ \quad / \qquad\qquad \backslash \qquad\qquad\qquad\qquad\ \ C-C \\ H_2C \qquad\qquad NH \quad \rightarrow \quad H_2C \qquad\qquad NH \ + H_2O \\ \quad \backslash \qquad\qquad\ \ H \qquad\qquad\qquad \backslash \qquad\quad / \\ CH_2-C-OH \qquad\qquad\qquad\quad C-C \\ O \qquad\qquad\qquad\qquad\quad H_2\ \ O \end{array}$$

alcohol from two molecules of the ethyl ester of the α-amino acids. This loss of alcohol takes place when the ethyl ester is kept for some time:

$$\begin{array}{l} H\quad H \qquad\qquad\qquad\quad O \qquad\qquad\qquad H\ \ H \\ CH_3.C-N\,H \quad C_2H_5O-C \qquad\qquad CH_3.C-N-CO \\ | \qquad\quad + \qquad\qquad\ \ | \quad \rightarrow \qquad\quad | \qquad\quad | \qquad + 2C_2H_5OH \\ OC-OC_2H_5 \qquad H\,N-C.CH_3 \qquad OC-N-C.CH_3 \\ H\quad H \qquad\qquad\quad\ H\ \ H \end{array}$$

3 : 6-dimethyl-2 : 5-diketopiperazine
(alanine 'anhydride')

Emil Fischer—on whose work in this field, which began in about 1900, our knowledge of the proteins and their hydrolytic products is based—showed that almost exclusively α-amino acids are the final hydrolytic products of the proteins. He employed as hydrolytic agents dilute sulphuric acid and aqueous solutions of barium hydroxide, and separated the amino acids, after esterifying the hydrolysis product, by fractional distillation of the mixture of the carboxylic esters under highly reduced pressure. From the (ethyl) esters, the amino acids were easily obtained by the action of water and, although there was always a considerable loss of material in the process, it is clear that proteins must be built up of units of α-amino acids in some way and that the proteins differ essentially according to their constituent amino-acid units.

Fischer showed that all α-amino acids with the exception of glycine and sarcosine (N-methylglycine) obtained from proteins are optically active. Since even the simplest proteins contain different amino-acid units, all the amino-acid units with the above exceptions must be present in the protein molecule in the optically active form. All the α-amino acids derived from fatty acids (higher in the series than acetic acid) and substituted fatty acids can be represented by the general formula:

$$\begin{array}{c} R \\ | \\ \mathrm{H—C^*—NH_2} \\ | \\ \mathrm{CO_2H} \end{array}$$

where R represents an alkyl (CH_3—, etc.) group or a substituted alkyl group. In such compounds the carbon atom distinguished by an asterisk (*) has four different groups attached to it, and consequently the molecule has an asymmetric configuration and the compound is capable of existing in optically active forms. Almost all the α-amino acids which have been obtained from the proteins have been prepared synthetically. These synthetic compounds being optically inactive, a large number of investigators, basing their work on that of Emil Fischer, have investigated methods for resolving them into their optically active components so as to be able to use the optically active forms of the α-amino acids in attempts to synthesise a protein or protein-like substance.

The α-amino acids which are immediate derivatives of the fatty acids may be synthesised:

(i) By the action of ammonia on α-halogeno fatty acids. The action of ammonia, for example, on monochloroacetic acid and the method of working up the product so as to obtain glycine may be outlined:

$$\begin{matrix} CH_2Cl \\ | \\ CO_2H \end{matrix} + HNH_2 + 2NH_3 \rightarrow NH_4Cl + \begin{matrix} CH_2.NH_2 \\ | \\ COONH_4 \end{matrix}$$

$$2\begin{matrix} CH_2.NH_2 \\ | \\ COONH_4 \end{matrix} + CuSO_4 \rightarrow (NH_4)_2SO_4 + \left[\begin{matrix} CH_2.NH_2 \\ | \\ COO- \end{matrix}\right]_2 Cu$$

or

$$\begin{matrix} H_2C\text{---}NH_2 & & O\text{---}CO^* \\ | & \searrow \quad \swarrow & | \\ | & Cu & | \\ | & \swarrow \quad \nwarrow & | \\ OC\text{---}O & & H_2N\text{---}CH_2 \end{matrix}$$

$$\left[\begin{matrix} CH_2.NH_2 \\ | \\ C\text{—}O\text{—} \\ \| \\ O \end{matrix}\right]_2 Cu + H_2S \rightarrow 2\begin{matrix} CH_2.NH_2 \\ | \\ C(=O)OH \end{matrix} + CuS$$

One of the difficulties in carrying out this method of preparation is due to the formation of the by-products:

$$N\begin{cases} CH_2.COOH \\ CH_2.COOH \\ H \end{cases} \qquad \text{and} \qquad N\begin{cases} CH_2.COOH \\ CH_2.COOH \\ CH_2.COOH \end{cases}$$

iminodiacetic acid   *nitrilotriacetic acid*

which may be described as secondary and tertiary amino acids and from which the α-monoamino acid must be separated.

(ii) By the action of potassium phthalimide (p. 332) on the esters of the α-halogeno fatty acids the corresponding phthalyl derivatives of the amino acids are obtained. When these are hydrolysed by means of hydrochloric acid, the α-amino acids in the form of their hydrochlorides are produced, and from these the α-amino acids can be isolated by suitable methods. These reactions are outlined:

$$\begin{matrix} CH_2Cl \\ | \\ CO_2C_2H_5 \end{matrix} + KN\begin{matrix} CO \\ \diagup \quad \diagdown \\ \diagdown \quad \diagup \\ CO \end{matrix}C_6H_4 \rightarrow \begin{matrix} CH_2\text{—}N\begin{matrix} CO \\ \diagup \quad \diagdown \\ \diagdown \quad \diagup \\ CO \end{matrix}C_6H_4 \\ | \\ CO_2C_2H_5 \end{matrix} + KCl;$$

potassium phthalimide   *ethyl ester of phthalylglycine*

$$\begin{matrix} CH_2\text{—}N\begin{matrix} CO \\ \diagup \quad \diagdown \\ \diagdown \quad \diagup \\ CO \end{matrix}C_6H_4 \\ | \\ CO_2C_2H_5 \end{matrix} + 3H_2O + HCl \rightarrow \begin{matrix} CH_2.NH_2.HCl \\ | \\ CO_2H \end{matrix} + C_6H_4\begin{matrix} CO_2H \\ CO_2H \end{matrix} + C_2H_5OH$$

$$2\begin{matrix} CH_2.NH_2.HCl \\ | \\ CO_2H \end{matrix} + Ba(OH)_2 \rightarrow 2\begin{matrix} CH_2.NH_2 \\ | \\ CO_2H \end{matrix} + BaCl_2 + 2H_2O$$

* See p. 315.

(iii) By Strecker's (1850) synthesis or its suitable modifications the α-amino acids are obtained from the corresponding cyanohydrins of the aldehydes or ketones. The preparation of alanine, for example, may be carried out by the interaction of equimolecular proportions of acetaldehyde, potassium cyanide and ammonium chloride in alcoholic or aqueous solution. The reactions in this case may be outlined:

$$CH_3{-}C{\genfrac{}{}{0pt}{}{\diagup H}{\diagdown\!\!\diagdown O}} + HCN \rightarrow CH_3{-}\overset{\displaystyle H}{\underset{\displaystyle OH}{C}}{-}CN$$

acetaldehyde cyanohydrin (lactonitrile)

$$CH_3{-}\overset{\displaystyle H}{\underset{\displaystyle OH}{C}}{-}CN + NH_3 \rightarrow CH_3{-}\overset{\displaystyle H}{\underset{\displaystyle NH_2}{C}}{-}CN + H_2O$$

nitrile of α-aminopropionic acid

$$CH_3{-}\overset{\displaystyle H}{\underset{\displaystyle NH_2}{C}}{-}CN + 2H_2O \rightarrow CH_3{-}\overset{\displaystyle H}{\underset{\displaystyle NH_2}{C}}{-}CO_2H + NH_3$$

α-aminopropionic acid or alanine

The hydrocyanic acid is produced by the hydrolysis of the potassium cyanide:

$$KCN + H_2O \rightleftarrows KOH + HCN,$$

and the ammonia is produced by the action of potassium hydroxide on the ammonium chloride:

$$NH_4Cl + KOH \rightarrow KCl + H_2O + NH_3.$$

The final product of the reaction is the ammonium salt of alanine (α-aminopropionic acid) from which the alanine is obtained through the copper salt (or copper derivative) and the decomposition of the latter as indicated above in the case of glycine.

Such simple α-amino acids as glycine and alanine are colourless crystalline substances often possessing a sweetish taste. They are readily soluble in water, very sparingly soluble in ethyl alcohol and insoluble in ether. Aminoacetic acid or glycine is still often known under the name of glycocoll (German, *Glykokoll*), owing to its possessing a sweet taste and being obtained from colloidal substances such as glue and gelatin (proteins). Possessing both the acidic carboxy group and the basic amino group, they are at the same time acids and bases and form easily hydrolysable salts with bases and acids. Thus, in the case of glycine, typical salts are:

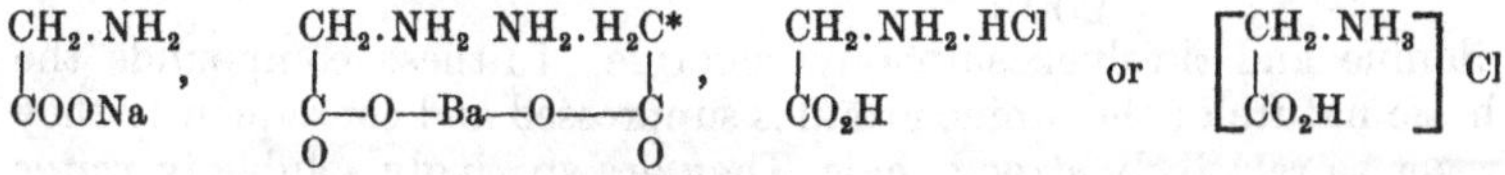

$$\begin{array}{l} CH_2.NH_2 \\ | \\ COONa \end{array}, \quad \begin{array}{l} CH_2.NH_2 \quad NH_2.H_2C^{*} \\ | \qquad\qquad\qquad\quad | \\ C{-}O{-}Ba{-}O{-}C \\ O \qquad\qquad\qquad\quad O \end{array}, \quad \begin{array}{l} CH_2.NH_2.HCl \\ | \\ CO_2H \end{array} \text{ or } \left[\begin{array}{l} CH_2.NH_3 \\ | \\ CO_2H \end{array}\right]Cl$$

They also form esters (in which the carboxy group no longer functions) which behave as relatively strong bases like typical primary amines and which form crystalline salts of the usual types, for example:

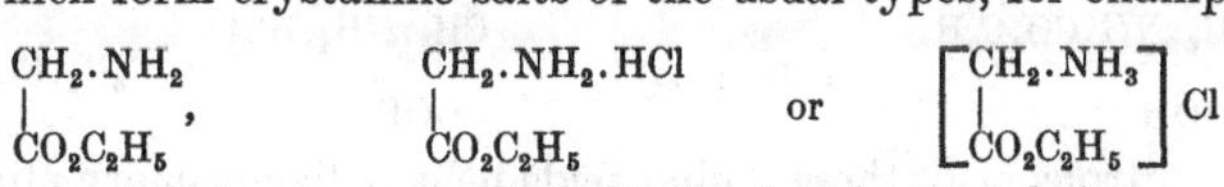

$$\begin{array}{l} CH_2.NH_2 \\ | \\ CO_2C_2H_5 \end{array}, \qquad \begin{array}{l} CH_2.NH_2.HCl \\ | \\ CO_2C_2H_5 \end{array} \text{ or } \left[\begin{array}{l} CH_2.NH_3 \\ | \\ CO_2C_2H_5 \end{array}\right]Cl$$

ethyl ester of glycine *hydrochloride of ethyl ester of glycine*

By the usual methods, they can be converted into acid chlorides, containing the $-C\begin{array}{l}{\diagup}O \\ {\diagdown}Cl\end{array}$ group, and these compounds are important synthetically.

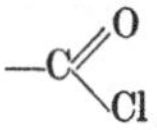

On account of the presence of the amino group in the molecule of these α-amino acids, they will undergo many of the reactions of primary amines. By methylation, for example, glycine can be converted into the following compounds:

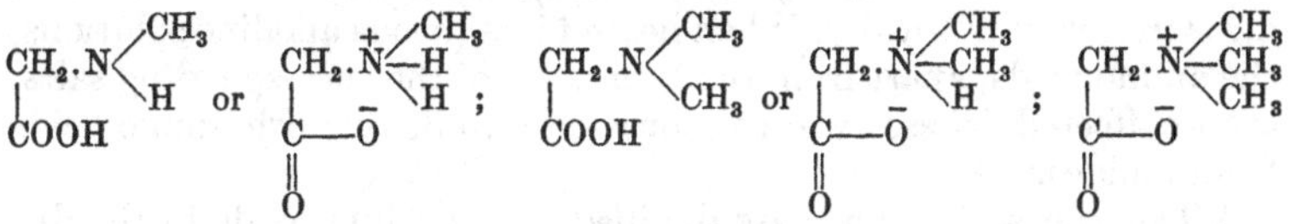

$$\begin{array}{l} CH_2.N\begin{array}{l}{\diagup}CH_3 \\ {\diagdown}H\end{array} \\ | \\ COOH \end{array} \text{ or } \begin{array}{l} CH_2.\overset{+}{N}\begin{array}{l}{\diagup}CH_3 \\ {-}H \\ {\diagdown}H\end{array} \\ | \\ C{-}\bar{O} \\ \| \\ O \end{array}; \quad \begin{array}{l} CH_2.N\begin{array}{l}{\diagup}CH_3 \\ {\diagdown}CH_3\end{array} \\ | \\ COOH \end{array} \text{ or } \begin{array}{l} CH_2.\overset{+}{N}\begin{array}{l}{\diagup}CH_3 \\ {-}CH_3 \\ {\diagdown}H\end{array} \\ | \\ C{-}\bar{O} \\ \| \\ O \end{array}; \quad \begin{array}{l} CH_2.\overset{+}{N}\begin{array}{l}{\diagup}CH_3 \\ {-}CH_3 \\ {\diagdown}CH_3\end{array} \\ | \\ C{-}\bar{O} \\ \| \\ O \end{array}$$

N-methylglycine or sarcosine N-dimethylglycine trimethylglycine or betaïne

the fully methylated compound being betaïne, which was formerly given the constitution

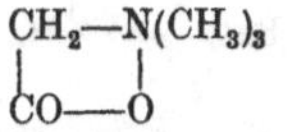

$$\begin{array}{l} CH_2{-}N(CH_3)_3 \\ | \qquad\quad | \\ CO{-}O \end{array}$$

They will also react with acid chlorides as do primary amines under analogous conditions. Such typical acyl derivatives are *benzoylglycine*, $\begin{array}{l} CH_2.NH.CO.C_6H_5 \\ | \\ COOH \end{array}$, prepared from glycine and benzoyl

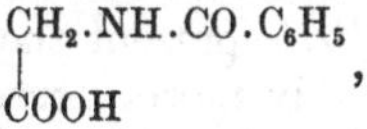

* The copper derivatives of α-amino acids have a particularly deep violet colour both in the crystalline condition and in aqueous solution. This colour in aqueous solution is quite different from that due to the cupric ion. This and other evidence indicate that the copper derivative may be a 'co-ordinated' compound having the constitution shown by the alternative formula on p. 313. Neutral aqueous solutions of copper derivatives of α-amino acids do not show the normal reactions of cupric ions; they show these reactions after acidification with acetic acid.

chloride in the presence of two equivalents of alkali, and p-*toluenesulphonylalanine*, $\underset{\text{COOH}}{\overset{\text{C(CH}_3\text{)H.NH.SO}_2\text{.C}_6\text{H}_4\text{.CH}_3}{|}}$, prepared similarly from alanine and *p*-toluenesulphonyl chloride. In these compounds the basic nature of the amino group is suppressed and consequently they react as relatively strong acids. They are sparingly soluble in water and highly crystalline, and when hydrolysed by boiling with concentrated hydrochloric acid are converted into the amino acid (as hydrochloride) and the aromatic acid:

$$\begin{array}{l}CH_2.NH.CO.C_6H_5\\|\\COOH\end{array} + H_2O + HCl \rightarrow \begin{array}{l}CH_2.NH_2.HCl\\|\\COOH\end{array} + C_6H_5.COOH$$

With nitrous acid, these amino acids behave like primary aliphatic amines, nitrogen is evolved and the amino group is replaced by hydroxyl, the corresponding *hydroxy acids* being obtained. The reaction taking place may be represented:

$$H-\underset{CO_2H}{\overset{R}{C}}-NH_2 + O{:}N.OH \rightarrow H-\underset{CO_2H}{\overset{R}{C}}-OH + N_2 + H_2O$$

This reaction is employed for determining the 'amino nitrogen' in protein substances (van Slyke's method), the nitrogen evolved being measured volumetrically. When heated in aqueous alkaline solutions, the amino acids, apart from the formation of the corresponding salts, are unaffected. These two reactions serve to distinguish amino acids from amides.

When the amino acids are distilled with barium oxide in the dry condition, the corresponding primary amines and barium carbonate are produced. Thus, from alanine, monoethylamine is obtained by decarboxylation:

$$H-\underset{CO_2H}{\overset{CH_3}{C}}-NH_2 + BaO \rightarrow \overset{CH_3}{CH_2}.NH_2 + BaCO_3$$

In aqueous solution these amino acids are neutral in reaction. This is due to the mutual neutralisation of the acid and basic groups, and although it is frequently convenient to represent the α-amino acids by the general formula I, it is probably more correct to represent them by the so-called betaïne formula, which in its modern form is indicated by II:

$$\underset{\text{I}}{H-\underset{CO_2H}{\overset{R}{C}}-NH_2} \qquad \underset{\text{II}}{H-\underset{C\overset{\displaystyle -\bar{O}}{\searrow O}}{\overset{R}{C}}-\overset{+}{N}H_3}$$

Since under appropriate conditions the α-amino acids can react as either acids or bases they are classified as *amphoteric compounds*, being both very weak acids and very weak bases.* While they are practically neutral in aqueous solution, they may be titrated as acids in alcoholic suspension using phenolphthalein as indicator. This, however, is not a convenient method for the rapid estimation of amino acids and the only satisfactory method apart from the action of nitrous acid for the rapid estimation of these compounds is to convert them into acids which can be titrated in the normal way (their conversion into titratable bases, for example, the carboxylic esters, is too difficult to be employed as part of an analytical process). For this purpose, it is necessary to suppress the basic nature of the amino group and, as already mentioned, this can be done by forming the N-acyl derivative, which can then be titrated as a typical organic acid in the usual manner. This method implies the isolation of the N-acyl derivative quantitatively from the amino acid and is therefore not suitable for the estimation of the amino acid as such. A highly convenient method for the estimation of amino acids in aqueous solution is to treat that solution with an excess of a neutral solution of formaldehyde. Under these conditions, Sörensen has shown that amino acids form methylene derivatives which can be titrated with standard alkali solutions as ordinary organic acids. This is the basis of Sörensen's formaldehyde method for the estimation of amino acids:

$$\underset{\text{neutral}}{H{-}\overset{R}{\underset{CO_2H}{C}}{-}N\boxed{H_2 + O}{=}C\begin{matrix}H\\H\end{matrix}} \rightarrow \underset{\text{acid}}{H{-}\overset{R}{\underset{CO_2H}{C}}{-}N{=}C\begin{matrix}H\\H\end{matrix}} + H_2O$$

Amino acids will also react with other aldehydes, such as benzaldehyde, forming analogous derivatives; but the formaldehyde reaction takes place rapidly and at the ordinary temperature. In carrying out Sörensen's method the methylene derivative is not isolated.

* The following is a comparison of the dissociation constants (K) in approximate figures of the α-amino acids, glycine and alanine, with those of the corresponding fatty acids (acetic and propionic) and the corresponding primary amines (monomethylamine and monoethylamine) under analogous conditions:

	Substance	K at 25°
Acids	Acetic acid	$1{\cdot}8 \times 10^{-5}$
	Aminoacetic acid (glycine)	$3{\cdot}4 \times 10^{-10}$
	Propionic acid	$1{\cdot}4 \times 10^{-5}$
	α-Aminopropionic acid (alanine)	$9{\cdot}0 \times 10^{-10}$
Bases	Monomethylamine	$5{\cdot}0 \times 10^{-4}$
	Aminoacetic acid (glycine)	$2{\cdot}7 \times 10^{-12}$
	Monoethylamine	$5{\cdot}6 \times 10^{-4}$
	α-Aminopropionic acid (alanine)	$5{\cdot}1 \times 10^{-12}$

Although for all practical purposes these amino acids are neutral in aqueous solution, the above values for the dissociation constants indicate that they are slightly stronger as acids than as bases.

It has been remarked that all α-amino acids, with the exception of glycine and sarcosine, isolated from the proteins by hydrolysis are optically active, and it is clearly necessary, where it is possible, to obtain α-amino acids in their optically active forms in order to be able to build up proteins or protein-like substances which are themselves optically active. Before applying the ordinary methods for the resolution of externally compensated substances into their optically active components to the resolution of α-amino acids, it is necessary to convert these practically neutral substances into sufficiently strong acids or bases, and it is more convenient to convert them into acidic rather than basic compounds. For this purpose, the externally compensated N-acyl derivatives are employed, and a number of these have been resolved into their optically active components by Fischer, and later workers. Having resolved the externally compensated N-acyl derivatives by the usual methods applicable to acids (p. 346), the *dextro-* and *laevo-* forms of the α-amino acids can be obtained from the optically active N-acyl derivatives by hydrolysis, generally by boiling the compound with concentrated hydrochloric acid (p. 316). The hydrolysis of these N-acyl derivatives takes place at very different rates according to the nature of the acyl group, and the amino acid can be isolated from the solution neutralised after the hydrolysis through the copper salt or derivative or by some other convenient method.

Glycine, aminoacetic acid,

$$\begin{array}{l} CH_2.NH_2 \\ | \\ COOH \end{array} \quad \text{or} \quad \begin{array}{l} CH_2.\overset{+}{N}H_3 \\ | \\ CO.\bar{O} \end{array}$$

is conveniently prepared from monochloroacetic acid and ammonia or from the ethyl ester of monochloroacetic acid and potassium phthalimide. It was first synthesised by Perkin and Duppa in 1858 by the action of ammonia on monobromoacetic acid and first obtained by the hydrolysis of glue or gelatin; its original name of glycocoll (γλυκύς = sweet, κόλλα = glue) signifies its sweet taste and the material from which it was first isolated. Its benzoyl derivative, N-benzoylglycine, occurs in the urine of herbivora, hence its usual name, ***hippuric acid***; when it is boiled with concentrated hydrochloric acid it is hydrolysed to glycine (in the form of its hydrochloride) and benzoic acid:

$$\begin{array}{l} CH_2.N\begin{array}{l} \diagup H \\ \diagdown CO.C_6H_5 \end{array} \\ | \\ CO_2H \end{array} + H_2O + HCl \rightarrow \begin{array}{l} CH_2.NH_2.HCl \\ | \\ CO_2H \end{array} + C_6H_5.CO_2H$$

Glycine is a colourless crystalline compound, m.p. 232–236°

(decomp.), which is readily soluble in water and sparingly insoluble in alcohol. On being heated with barium oxide, glycine yields monomethylamine:

$$\begin{array}{l} CH_2.NH_2 \\ | \\ CO_2H \end{array} + BaO \rightarrow CH_3.NH_2 + BaCO_3$$

It is converted into *hydroxyacetic acid* or *glycollic acid* when treated with nitrous acid:

$$\begin{array}{l} CH_2.NH_2 \\ | \\ CO_2H \end{array} + O:N.OH \rightarrow \begin{array}{l} CH_2OH \\ | \\ CO_2H \end{array} + N_2 + H_2O$$

It yields a characteristic deep violet-blue crystalline copper derivative and a hydrochloride. The ethyl ester has b.p. 149°, and this forms a colourless crystalline hydrochloride, m.p. 144°. When treated with benzoyl chloride in the presence of alkali, glycine is converted into benzoyl glycine or hippuric acid:

$$\begin{array}{l} CH_2.NH_2 \\ | \\ CO_2H \end{array} + ClCO.C_6H_5 + NaOH \rightarrow \begin{array}{l} CH_2.NH(CO.C_6H_5) \\ | \\ CO_2H \end{array} + NaCl + H_2O$$

Hippuric acid is a colourless crystalline substance, m.p. 187·5°, very sparingly soluble in water, sparingly soluble in cold alcohol and practically insoluble in ether.

The N-monomethyl derivative of glycine is known as *sarcosine*. It was first obtained by von Liebig (1847) by the hydrolysis of meat extract containing *creatine* (p. 485) and named by him (σαρκός = of meat); it is prepared by the action of monomethylamine on monochloroacetic acid (compare synthesis of glycine):

$$\begin{array}{l} CH_2Cl \\ | \\ CO_2H \end{array} + HN\begin{array}{l} H \\ CH_3 \end{array} + 2NH_2(CH_3) \rightarrow \begin{array}{l} CH_2.N\begin{array}{l} H \\ CH_3 \end{array} \\ | \\ CO.ONH_3(CH_3) \end{array} + [NH_3(CH_3)]Cl$$

monomethylammonium salt of sarcosine; monomethyl-ammonium chloride

The sarcosine, $\begin{array}{l} CH_2.N\begin{array}{l} H \\ CH_3 \end{array} \\ | \\ CO_2H \end{array}$, is isolated from its monomethylammonium salt through the copper salt and decomposition of the latter by means of hydrogen sulphide. Sarcosine is very similar to glycine in appearance, m.p. 210° (decomp.); it is readily soluble in water and, on being heated, decomposes into carbon dioxide and dimethylamine:

$$\underset{\displaystyle CO_2H}{\underset{|}{CH_2}}.N\begin{matrix}H\\CH_3\end{matrix} \rightarrow CH_3.N\begin{matrix}H\\CH_3\end{matrix} + CO_2$$

The typical α-amino acids like glycine, alanine, etc., in so far as they are bases, are primary amines. Sarcosine, being a secondary amine, is thus not a typical amino acid, and is easily distinguished from glycine.

Alanine, α-aminopropionic acid,

$$\begin{matrix}CH_3\\|\\CH.NH_2\\|\\CO_2H\end{matrix} \quad \text{or} \quad \begin{matrix}(CH_3)CH.\overset{+}{N}H_3\\|\\CO.\bar{O}\end{matrix}$$

has been synthetically prepared by the action of ammonia on α-chloro- and α-bromo- propionic acids. Its convenient preparation by Strecker's synthesis has been outlined (p. 314). The alanine prepared by these methods is the externally compensated form, *dl*-alanine. *d*-Alanine is obtained by the hydrolysis of natural raw silk, using hydrochloric acid. Natural silk on hydrolysis also yields glycine. The products of the hydrolysis are obtained in the form of their ethyl esters, which are separated by fractional distillation under reduced pressure, the ethyl ester of *d*-alanine boiling at a higher temperature (at the same pressure) than the ethyl ester of glycine. The ethyl ester of *d*-alanine is boiled in aqueous solution, and on evaporation of the resulting solution *d*-alanine having $[\alpha]_D^{20^\circ} = +2{\cdot}7^\circ$ is obtained.

dl-Alanine, m.p. 195°, is very similar to glycine in appearance and general physical and chemical properties. Its copper derivative has a deeper violet colour than that of glycine. Its N-benzoyl derivative has m.p. 160° and this has been resolved into *d*-benzoylalanine and *l*-benzoylalanine, both of which have m.p. 148°. From these optically active benzoylalanines the *d*- and *l*- alanines have been obtained by hydrolysis with hydrochloric acid, and the *d*-alanine, m.p. 293°, obtained in this way is identical in all respects with that obtained by the hydrolysis of natural raw silk. When *dl*-alanine is allowed to react with nitrous acid, it is converted into dl-α-*hydroxypropionic acid* or dl-*lactic acid* (p. 364):

$$\begin{matrix} & CH_3 & \\ & | & \\ H— & C & —NH_2 \\ & | & \\ & CO_2H & \end{matrix} + O{:}N.OH \rightarrow \begin{matrix} & CH_3 & \\ & | & \\ H— & C & —OH \\ & | & \\ & CO_2H & \end{matrix} + N_2 + H_2O$$

When *d*-alanine is treated in the same way it is interesting that *l*-lactic acid is obtained, and therefore unless a configurational change has taken place during this reaction these two optically active compounds having rotatory powers of opposite sign have the same stereochemical configurations.

Other α-amino acids which may all be considered to be substitution products of alanine are:

Serine, α-amino-β-hydroxypropionic acid:

$$\begin{array}{c} CH_2.OH \\ | \\ H-C-NH_2 \\ | \\ CO_2H \end{array}$$

which is a primary alcohol as well as an α-amino acid. The *laevo*-rotatory form is obtained as one of the hydrolytic products of silk protein. The corresponding sulphur (or thiol-) compound is

Cysteine, α-amino-β-thiolpropionic acid:

$$\begin{array}{c} CH_2.SH \\ | \\ H-C-NH_2 \\ | \\ CO_2H \end{array}$$

This compound is difficult to isolate on account of the ease with which it undergoes oxidation when exposed to the atmosphere to

Cystine, di-(α-amino-β-thiopropionic acid):

$$\begin{array}{ccc} CH_2-S & -S- & CH_2 \\ | & & | \\ H-C-NH_2 & & H-C-NH_2 \\ | & & | \\ CO_2H & & CO_2H \end{array}$$

which is formed from cysteine by a reaction which may be represented:

$$HO_2C.CH(NH_2).CH_2SH + O + HSH_2C.(NH_2)HC.CO_2H$$
$$\rightarrow H_2O + HO_2C.CH(NH_2).CH_2.S.S.H_2C.(NH_2)HC.CO_2H$$

The sulphur-containing proteins and peptides have their sulphur in the cysteine unit, which is abundant in egg albumin, and in keratin, which is the insoluble protein (sclero-protein) of hair, wool and horn. Cystine is very easily isolated from clean human hair. The hair is heated on the water-bath for 24 hours with concentrated hydrochloric acid diluted with an equal volume of water. The dark-coloured solution resulting may be filtered from the small quantity of insoluble material and the filtrate *nearly* neutralised with an aqueous solution of sodium hydroxide. This still acid solution is boiled with decolorising charcoal and filtered. The excess of hydrochloric acid in the filtered solution is removed by addition of sodium acetate solution, when the following reaction takes place:

$$HCl + CH_3.CO_2Na \rightarrow CH_3.CO_2H + NaCl$$

and the cystine is precipitated, being insoluble in dilute acetic acid. The cystine which should be colourless is filtered off and washed thoroughly with water, in which it is insoluble. So obtained, cystine is a colourless crystalline substance which is *laevo*rotatory. In the dry condition or in hydrochloric acid solution it is stable; it is, however, very sensitive to alkalis and when warmed with dilute sodium hydroxide the typical reaction of sulphur in alkaline solution (sodium nitroprusside giving a violet colour) is obtained; such a solution will also give a black precipitate of lead sulphide in the presence of lead acetate solution. Having two similar asymmetric carbon atoms in its molecule, the stereochemistry of cystine is similar to that of tartaric acid (p. 393).

Phenylalanine, or β-phenyl-α-aminopropionic acid,

$$\begin{array}{c} CH_2.C_6H_5 \\ | \\ H-C-NH_2 \\ | \\ CO_2H \end{array}$$

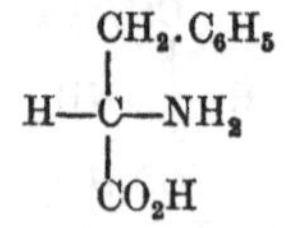

is the simplest aromatic derivative of alanine and is a hydrolytic product of casein and certain other proteins. From such sources, it is obtained in its *laevo*rotatory form, in colourless plates, m.p. 263° (decomp.). The externally compensated form has been prepared by the reduction of α-aminocinnamic acid:

$$C_6H_5.CH{=}C(NH_2).CO_2H + H_2 \rightarrow C_6H_5.CH_2.CH(NH_2).CO_2H$$

and resolved into its optically active components through its N-benzoyl derivative and subsequent hydrolysis.

Tyrosin is β-*p*-hydroxyphenyl-α-aminopropionic acid:

$$\begin{array}{ccccc} & H\ H & & & \\ & C{=}C & & & \\ & / \qquad \backslash & & & \\ HO.C & & C-CH_2 & & \\ & \backslash\backslash \qquad // & | & & \\ & C-C & | & & \\ & H\ H & | & & \\ & & H-C-NH_2 & & \\ & & | & & \\ & & CO_2H & & \end{array} \quad \text{or} \quad \begin{array}{c} CH_2.C_6H_4.OH\ (4) \\ | \\ H-C-NH_2 \\ | \\ CO_2H \end{array}$$

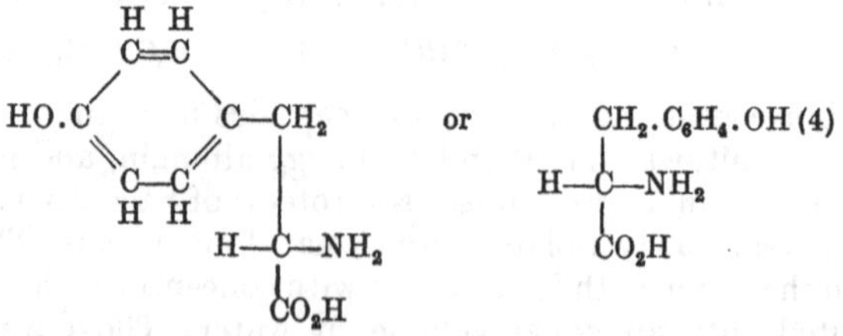

On account of its low solubility in water, tyrosine is one of the α-amino acids most easily isolated. It is obtained in the *laevo*rotatory form by the hydrolysis of many proteins, natural raw silk, casein, etc. The hydrolysis may be effected by hydrochloric or sulphuric acid. In the latter case, it is only necessary to remove the sulphuric acid by means of the requisite quantity of baryta, when the tyrosine crystallises in colourless needles on suitable evaporation of the resulting filtered solution.

The active principle of the ductless thyroid gland which can be extracted from it by using a dilute solution of barium hydroxide is known as *thyroxine*, and this can be considered to be a derivative of tyrosine or of phenylalanine. Thyroxine, as extracted, is *laevo*rotatory and the externally compensated substance has been synthesised by Harington and Barger and shown to have the following constitution:

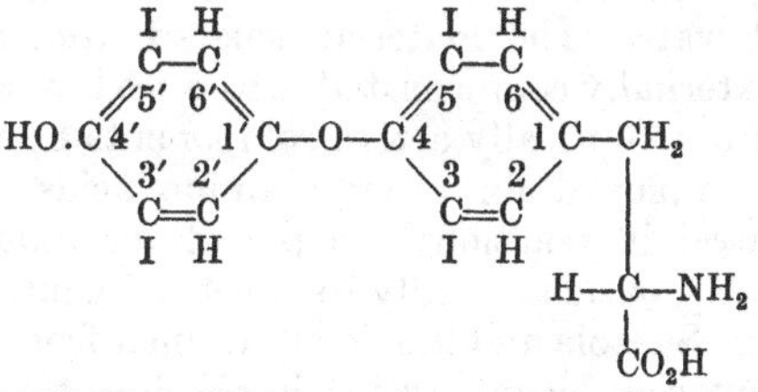

```
          I  H              I  H
          C—C               C—C
        //5' 6'\\         //5  6\\
HO.C 4'       1'C—O—C 4        1 C—CH2
        \3' 2'/           \3  2/     |
          C=C               C=C      |
          I  H              I  H     |
                                 H—C—NH2
                                     |
                                    CO2H
```

β-[3 : 5-diiodo-4-(3′ : 5′-diiodo-4′-hydroxyphenoxy)phenyl]-α-aminopropionic acid*

The synthetic material has been resolved into its optically active components and the *laevo*rotatory form shown to be three times as therapeutically active as the *dextro-* modification.†

* The iodine free substance is the *p*-hydroxyphenyl ether of tyrosine.

† Like other secretions of the endocrine organs, thyroxine plays an important part in the regulation of carbohydrate metabolism and metabolic processes generally.

Although chemically little related to thyroxine, *adrenaline* is the active principle of other ductless, the suprarenal, glands. This compound is, at the same time, a phenol, a secondary alcohol and a secondary amine (compare choline). Physiologically, it has the function of regulating the blood pressure. It is easily obtained from the suprarenal glands by extraction with a dilute solution of trichloroacetic acid in ethyl alcohol. The extract is carefully evaporated and, after addition of ammonia in slight excess, the adrenaline crystallises from the resulting solution. When pure, adrenaline is a colourless crystalline substance, m.p. 216°, readily soluble in water. As a moderately strong base (the secondary amine group is in the side chain), it forms stable salts with acids. As a phenol, it is soluble in aqueous solutions of sodium and potassium hydroxides, but it is insoluble in alkaline carbonate solutions and in an aqueous solution of ammonia.

Adrenaline is a derivative of catechol (*o*-dihydroxybenzene) and has the constitution:

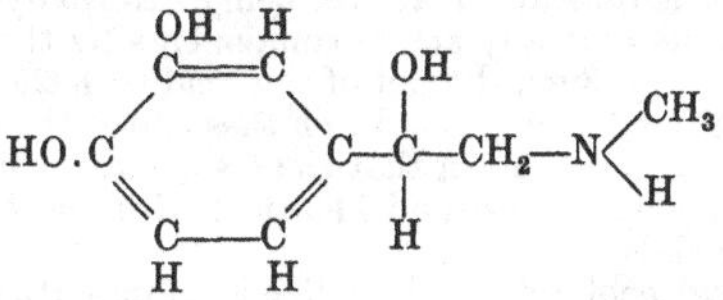

```
          OH  H
          C===C       OH
         /     \      |            CH3
HO.C            C—C—CH2—N<
         \\    //     |            H
          C———C       H
          H   H
```

4-(methylaminohydroxyethyl)-1 : 2-dihydroxybenzene

It has been synthesised by heating chloroacetyl chloride and catechol in the presence

Tryptophan, biologically, is a very important α-amino acid which was isolated by Hopkins and Cole by digestion of casein with the enzyme, *trypsin*. It has been shown to be β-indole-α-aminopropionic acid, and its constitution is represented:

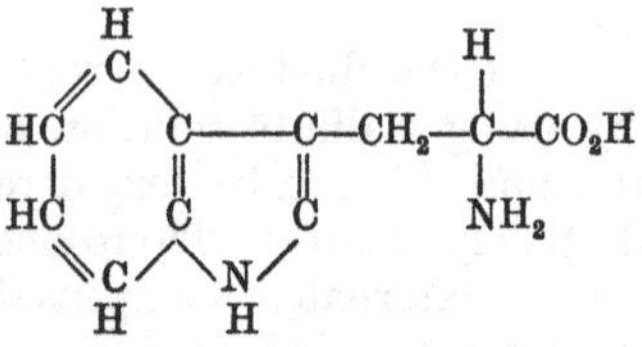

```
         H
         C                              H
      //   \                            |
    HC       C ——— C—CH₂—C—CO₂H
    |        ||    ||                  |
    HC       C     C                  NH₂
      \\   /   \  /
         C       N
         H       H
```

Tryptophan crystallises in colourless plates and is somewhat sparingly soluble in cold water. The material isolated from casein is *laevo*-rotatory. The externally compensated substance has been synthesised and resolved into its optically active components by methods analogous to those employed for other α-amino acids. Tryptophan is practically neutral in reaction; the pyrrole or indole part of the nucleus is so feebly basic as hardly to affect the neutral properties of an α-amino acid. Skatole and indole are formed from tryptophan by putrefactive bacterial decomposition in the intestine; the chemical changes which take place may be represented diagrammatically:

of phosphorus oxychloride, condensing the resulting *p*-chloroacetylcatechol with monomethylamine and reducing the keto-amine to the secondary alcohol using aluminium amalgam:

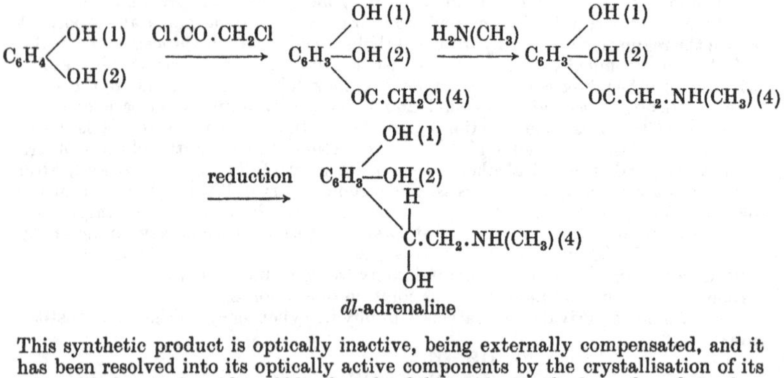

$$C_6H_4\begin{cases}OH\ (1)\\ OH\ (2)\end{cases} \xrightarrow{Cl.CO.CH_2Cl} C_6H_3\begin{cases}OH\ (1)\\ OH\ (2)\\ OC.CH_2Cl\ (4)\end{cases} \xrightarrow{H_2N(CH_3)} C_6H_3\begin{cases}OH\ (1)\\ OH\ (2)\\ OC.CH_2.NH(CH_3)\ (4)\end{cases}$$

$$\xrightarrow{\text{reduction}} C_6H_3\begin{cases}OH\ (1)\\ OH\ (2)\\ \overset{H}{\underset{OH}{C}}.CH_2.NH(CH_3)\ (4)\end{cases}$$

dl-adrenaline

This synthetic product is optically inactive, being externally compensated, and it has been resolved into its optically active components by the crystallisation of its salts with *d*-tartaric acid (p. 398), the salt of this acid with the *laevo*- base being less soluble than that with the *dextro*- base. From these salts, the optically active bases were obtained and it has been shown that the *laevo*- base is fifteen times as active therapeutically as the *dextro*- base and identical with the *l*-adrenaline extracted from the suprarenal glands.

In the form of its hydrochloride, *l*-adrenaline is an important drug. It is of great value in arresting local haemorrhage and as a prophylactic and therapeutic agent in the treatment of asthma.

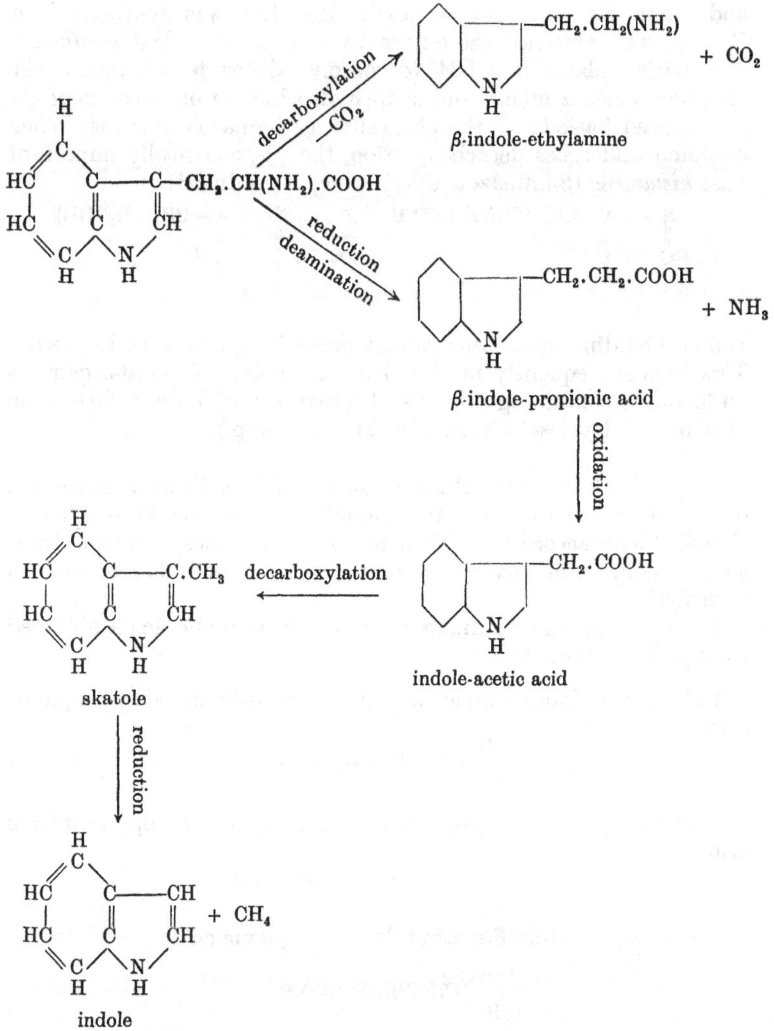

$C_8H_6N.CH_2.CH(NH_2).COOH$ (tryptophan, drawn as a structural formula)

→ decarboxylation ($-CO_2$) → $C_8H_6N—CH_2.CH_2(NH_2)$ + CO_2

β-indole-ethylamine

→ reduction, deamination → $C_8H_6N—CH_2.CH_2.COOH$ + NH_3

β-indole-propionic acid

↓ oxidation

$C_8H_6N—CH_2.COOH$

indole-acetic acid

← decarboxylation

$C_8H_6N—C.CH_3$

skatole

↓ reduction

C_8H_7N + CH_4

indole

Histidine is usually obtained by the hydrolysis of blood proteins. It is β-iminazole-alanine, and on account of the *tautomerism* of the glyoxaline nucleus (p. 307) its constitution may be written:

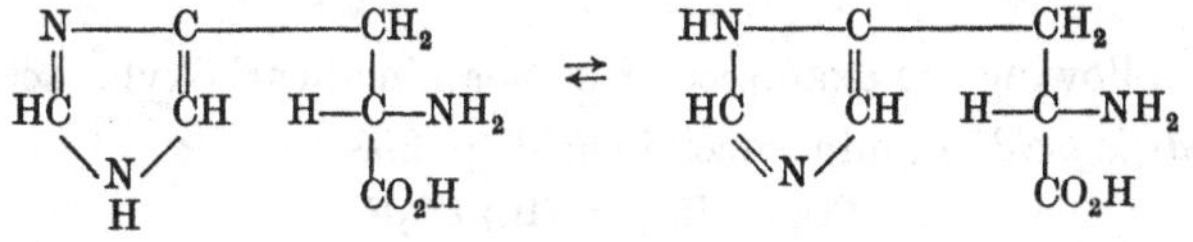

$$N{=}CH{-}NH{-}CH{=}C{-}CH_2.CH(NH_2).CO_2H \rightleftarrows HN{-}CH{=}N{-}CH{=}C{-}CH_2.CH(NH_2).CO_2H$$

and either formula may be used. Histidine was synthesised by Pyman, who obtained the externally compensated or *dl*-compound in colourless plates, m.p. 284° (decomp.). Although an α-amino acid, histidine forms a mono- and a di- hydrochloride on account of the pronounced basicity of the glyoxaline or iminazole nucleus. When histidine undergoes decarboxylation, the physiologically important base *histamine* (β-iminazole-ethylamine) is produced:

```
N——C—CH2.CH(NH2).CO2H              N——C—CH2.CH2(NH2)
‖     ‖                    ——→     ‖     ‖
HC    CH                   -CO2    HC    CH
  \  /                               \  /
   N                                  N
   H                                  H
```

Unlike histidine, histamine cannot occur in optically active forms. This base is frequently produced by the action of micro-organisms on histidine-containing proteins. Its hydrochloride crystallises from alcohol in colourless prisms, m.p. 240° (decomp.).

Only a few of the typical α-amino-monocarboxylic acids have been discussed. Some forty to fifty animal proteins have been differentiated, chiefly according to their physical properties, and from these some twenty amino acids have been isolated by suitable methods of hydrolysis.

Examples of other α-monoamino-monocarboxylic acids obtained from proteins are:

Valine, α-amino*iso*valeric acid or α-amino-ββ-dimethylpropionic acid:

$$(CH_3)_2{>}CH.CH(NH_2).CO_2H$$

Leucine, α-amino*iso*caproic acid or α-amino-β-*iso*propylpropionic acid:

$$(CH_3)_2{>}CH.CH_2.CH(NH_2).CO_2H$$

Iso*leucine*, α-amino-β-methyl-β-ethylpropionic acid:

$$\begin{matrix}CH_3\\C_2H_5\end{matrix}{>}CH.CH(NH_2).CO_2H$$

Leucine and *iso*leucine are isomeric and the latter differs from all the other α-amino acids mentioned above in having two asymmetric carbon atoms in the molecule (compare cystine, which is a di-(α-amino acid)).

The following are examples of monoaminodicarboxylic acids:

Aspartic acid, α-aminosuccinic acid (p. 358):

$$CO_2H.CH_2.CH(NH_2).CO_2H$$

The *laevo-* variety of its acid amide, *asparagine*:

$$CONH_2.CH_2.CH(NH_2).CO_2H$$

is present in beet-root, the shoots of peas, beans and vetches and in asparagus.

Glutamic acid, α-aminoglutaric acid (p. 358):

$$CO_2H.CH_2.CH_2.CH(NH_2).CO_2H$$

Aspartic acid and glutamic acid can be designated as alanine derivatives, the former being α-amino-β-carboxypropionic acid or β-carboxyalanine, and the latter α-amino-β-carboxymethylpropionic acid or β-carboxymethylalanine.

Examples of diaminomonocarboxylic acids are:

Ornithine, α : δ-diamino-*n*-valeric acid:

$$CH_2(NH_2).CH_2.CH_2.CH(NH_2).CO_2H$$

which as a derivative of alanine may be designated β-aminoethylalanine. This amino acid plays an important part in the synthesis of urea in the animal body.

Lysine, α : ε-diamino-*n*-caproic acid or γ-amino-*n*-propylalanine:

$$CH_2(NH_2).CH_2.CH_2.CH_2.CH(NH_2).CO_2H$$

The first attempts to synthesise proteins or protein-like substances by Emil Fischer were the production of *polypeptides* by condensing together different amino acid units. The simplest of these is the *dipeptide*, *glycylglycine*, which is produced by gently warming 2 : 5-diketopiperazine with hydrochloric acid:

```
      CO—CH2
     /      \
  HN          NH  +  H2O  →  H2N.CH2.CO.NH.CH2.CO2H
     \      /
      CH2—CO
```

The simplest *tripeptide*, *diglycylglycine*, was prepared by Fischer using the following reactions:

$$ClCH_2.COCl + H_2N.CH_2.CO_2H \rightarrow ClCH_2.CO.NH.CH_2.CO_2H + HCl$$

chloroacetyl chloride → N-chloroacetylglycine

$$NH_3 + ClCH_2.CO.NH.CH_2.CO_2H \rightarrow NH_2.CH_2.CO.NH.CH_2.CO_2H + HCl$$

glycylglycine

$$ClCH_2.COCl + H_2N.CH_2.CO.NH.CH_2.CO_2H \rightarrow ClCH_2.CO.NH.CH_2.CO.NH.CH_2.CO_2H + HCl$$

N-chloroacetylglycylglycine

$$NH_3 + ClCH_2.CO.NH.CH_2.CO.NH.CH_2.CO_2H \rightarrow NH_2.CH_2.CO.NH.CH_2.CO.NH.CH_2.CO_2H + HCl$$

diglycylglycine

Such reactions can be extended considerably and, by using different amino-acid units, mixed polypeptides may be obtained. Another method for the production of polypeptides is by the interaction of the acid chloride of an amino acid with either that or another amino acid. For example, the dipeptide, *alanylleucine*, was produced by the interaction of alanyl chloride and leucine, thus:

$$CH_3.CH(NH_2).C\begin{matrix}{\nearrow O}\\{\searrow Cl}\end{matrix} + H_2N-\overset{\displaystyle CH_2.CH(CH_3)_2}{\underset{\displaystyle CO_2H}{C}}-H$$

$$\rightarrow \underset{\text{alanylleucine}}{CH_3.CH(NH_2).CO.HN}-\overset{\displaystyle CH_2.CH(CH_3)_2}{\underset{\displaystyle CO_2H}{C}}-H \quad + HCl$$

This reaction is also capable of being extended, using acid chlorides of other amino acids to react with the dipeptide producing a tripeptide, and so on.*

One complex polypeptide which has been synthesised is an octadecapeptide consisting of three leucine and fifteen glycine units, having a molecular weight of 1212. In building up these polypeptides optically active amino acids have been employed. Although the synthesis of a protein identical with a naturally occurring protein is far from being realised, some polypeptides which have been isolated by the partial hydrolysis of proteins have been synthesised by the methods outlined. These synthetic polypeptides, moreover, give the same colour and biuret (p. 478) reactions as do the proteins themselves.

The simplest aromatic aminocarboxylic acids having the amino and the carboxyl groups directly attached to the benzene nucleus are

* Whether the amino-acid units exist in the protein molecule as they do in the polypeptides, or whether they are present in some similar way as in the ketopiperazines, is still not determined. From glycine and alanine, for example, has been synthesised 2 : *5-diketo-3-methylpiperazine*:

$$\begin{matrix} & CH_2-CO.\boxed{OH} & \\ H.N\boxed{H} & & \boxed{H}N.H \\ \boxed{HO}\,OC-CH(CH_3) & & \end{matrix} \rightarrow \begin{matrix} & CH_2-CO & \\ & 6 \quad 5 & \\ HN\,1 & & 4\,NH \\ & 2 \quad 3 & \\ & CO-CH(CH_3) & \end{matrix} + 2H_2O$$

This diketopiperazine has been found among the hydrolysis products of natural raw silk; but this is not necessarily proof that it exists as a unit in that protein.

```
        NH2                    NH2                    NH2
        |                      |                      |
        C                      C                      C
      //  \                  //  \                  //  \
    HC     C.COOH          HC     CH              HC     CH
    |      ||              |      ||              |      ||
    HC     CH              HC     C.COOH          HC     CH
      \\  /                  \\  /                  \\  /
        C                      C                      C
        H                      H                      |
                                                      CO2H
```

o-aminobenzoic acid or *anthranilic acid* m.p. 145°

m-*aminobenzoic acid* m.p. 173°

p-*aminobenzoic acid* m.p. 186°

Of these, the most important is the *ortho-* compound, which is generally described under the name of *anthranilic acid,*

$$C_6H_4\begin{cases} NH_2\,(1) \\ CO_2H\,(2) \end{cases}$$

and was first obtained by fusing indigo with alkali (Fritsche, 1841). When anthranilic acid is heated above its melting point it is decomposed into aniline and carbon dixoide:

$$C_6H_4\begin{cases} NH_2\,(1) \\ CO_2H\,(2) \end{cases} \rightarrow C_6H_5NH_2 + CO_2$$

It is the reduction product of *o*-nitrobenzoic acid:

$$C_6H_4\begin{cases} NO_2\,(1) \\ CO_2H\,(2) \end{cases} + 3H_2 \rightarrow C_6H_4\begin{cases} NH_2\,(1) \\ CO_2H\,(2) \end{cases} + 2H_2O$$

o-Nitrobenzoic acid is the oxidation product of *o*-nitrotoluene, which along with *p*-nitrotoluene is produced as the first nitration product of toluene:

$$C_6H_5.CH_3 \nearrow C_6H_4\begin{cases} NO_2\,(2) \\ CH_3\,(1) \end{cases} \xrightarrow[KMnO_4]{oxidation} C_6H_4\begin{cases} NO_2\,(2) \\ CO_2H\,(1) \end{cases}$$

$$C_6H_5.CH_3 \searrow C_6H_4\begin{cases} NO_2\,(4) \\ CH_3\,(1) \end{cases} \xrightarrow[KMnO_4]{oxidation} C_6H_4\begin{cases} NO_2\,(4) \\ CO_2H\,(1) \end{cases}$$

p-*Aminobenzoic acid* is the reduction product of *p*-nitrobenzoic acid. The most convenient way of obtaining m-*aminobenzoic acid* is by the reduction of *m*-nitrobenzoic acid produced by the nitration of benzoic acid:

$$C_6H_5.COOH \rightarrow C_6H_4\begin{cases} CO_2H\,(1) \\ NO_2\,(3) \end{cases} \rightarrow C_6H_4\begin{cases} CO_2H\,(1) \\ NH_2\,(3) \end{cases}$$

These three acids, like the typical α-amino acids, form salts with bases and acids but not with acetic acid. They yield esters (replacement of the hydrogen of the carboxyl group by alkyl and aryl radicles) and N-acyl derivatives (replacement of a hydrogen atom of the —NH_2 group by acyl groups such as acetyl, benzoyl, etc.). The N-acetyl derivatives of these acids can also be obtained by the oxidation of the acetyl derivatives of the corresponding *o*-, *m*- and *p*- toluidines.

Anthranilic acid is manufactured in considerable quantities as an intermediate in the commercial production of synthetic *indigo*, which has now almost completely replaced the natural product (see below).

The necessary material for the production of anthranilic acid is *phthalic anhydride*, the anhydride of the dibasic acid, *o*-dicarboxybenzene or *phthalic acid*. When phthalic acid is heated to about 220° it loses water and is converted into the anhydride:

```
      H                              H
      C                              C
    //  \                          //  \
  HC     C—CO.OH                 HC     C—CO
  |      ||             →        |      ||   >O  +  H2O
  HC     C—CO.OH                 HC     C—CO
    \\  /                          \\  /
      C                              C
      H                              H

  phthalic acid               phthalic anhydride
```

Phthalic anhydride is obtained as the distillate when phthalic acid is distilled. It crystallises in long colourless needles, m.p. 128°, b.p. 284°. The commercial product is of a high degree of purity. Phthalic anhydride is not only essential for the production of indigo but also for the production of a large class of dye-stuffs known as the phthaleins, of which phenolphthalein may be regarded as the parent substance. If the synthetic indigo is to compete successfully with the natural product, it is essential that the phthalic anhydride must be produced from cheap and easily accessible material. This is naphthalene, of which the market price is approximately 2½*d*. per lb.

Naphthalene on oxidation yields phthalic anhydride:

```
     H     H                          H
     C     C                          C
   //  \ /   \\                     //  \
 HC     C     CH                  HC     C—CO
 |      ||    |     + 9O   →      |      ||   >O  +  2CO2  +  2H2O
 HC     C     CH                  HC     C—CO
   \\  / \\  //                     \\  /
     C     C                          C
     H     H                          H
```

and this oxidation may be carried out in various ways, of which the more important are those used on the industrial scale. The older oxidation process employs sulphuric acid containing an excess of sulphuric anhydride (oleum) as the oxidising agent at a moderately high temperature using mercuric sulphate as a catalyst. The reaction may be expressed:

$$C_{10}H_8 + 9H_2SO_4 \rightarrow C_6H_4\left\langle\begin{matrix}CO\\ \\CO\end{matrix}\right\rangle O + 9SO_2 + 2CO_2 + 11H_2O$$

The naphthalene goes into solution, probably as a mixture of the sulphonic acids, at the beginning, and when the reaction is over the phthalic anhydride can be separated by filtration after addition of water. The anhydride has to be purified by distillation.

The better and modern process consists in the oxidation of naphthalene in the dry state in the presence of a suitable inorganic catalyst; vanadium pentoxide is stated to be the best for the purpose. Essentially, the process consists in passing naphthalene vapour and air over the heated catalyst and the collection of the phthalic anhydride in suitable receivers.

The phthalic anhydride is next converted into *phthalimide* by passing ammonia into the molten material:

$$C_6H_4\left\langle\begin{matrix}CO\\ \\CO\end{matrix}\right\rangle O + H_2NH \rightarrow \underset{\text{phthalimide}}{C_6H_4\left\langle\begin{matrix}CO\\ \\CO\end{matrix}\right\rangle NH} + H_2O$$

This reaction takes place rapidly and quantitatively. Phthalimide is a colourless crystalline substance, m.p. 238°. Like other compounds whose molecules contain more than one ring, it can be regarded constitutionally as derived from the fusion of rings having the grouping, $=C=C=$, common to both; in this case, phthalimide can be considered to be derived from a benzene and a partially oxidised pyrrolidine ring (or *maleimide*, m.p. 93°), and the following formulae indicate the relationships involved:

$$\begin{matrix} & H & \\ & C & \\ HC & & CH \\ HC & & CH \\ & C & \\ & H & \end{matrix}$$

$$\underset{\text{pyrrole}}{\begin{matrix}HC{=}CH\\ \| \quad\quad \| \\ HC{=}CH\end{matrix}\Big\rangle NH} \qquad \underset{\text{reduced pyrrole} = \textit{pyrrolidine}}{\begin{matrix}H_2C{-}CH_2\\ | \quad\quad | \\ H_2C{-}CH_2\end{matrix}\Big\rangle NH} \qquad \underset{\alpha\text{-}\textit{ketopyrrolidine} \text{ or } \alpha\text{-}\textit{pyrrolidone} \text{ or } n\text{-butyrolactam}}{\begin{matrix}H_2C{-}CO\\ | \quad\quad | \\ H_2C{-}CH_2\end{matrix}\Big\rangle NH} \qquad \underset{\alpha\alpha'\text{-}\textit{diketopyrrolidine} \text{ or } \textit{succinimide} \text{ (p. 293)}}{\begin{matrix}H_2C{-}CO\\ | \quad\quad | \\ H_2C{-}CO\end{matrix}\Big\rangle NH}$$

$$\underset{\text{maleimide (p. 382)}}{\begin{matrix}HC{-}CO\\ \| \quad\quad | \\ HC{-}CO\end{matrix}\Big\rangle NH}$$

The hydrogen attached to the nitrogen atom in phthalimide has acidic properties and can be replaced, for example, by potassium. *Potassium phthalimide,*

$$C_6H_4\begin{matrix} CO \\ \diagup\diagdown \\ \diagdown\diagup \\ CO \end{matrix}NK$$

is formed directly by the action of alcoholic potash; it is an important synthetic reagent.

Phthalimide is converted into anthranilic acid by warming it to about 80° for a short time with a mixture of aqueous solutions of sodium hydroxide and sodium hypochlorite. This reaction is merely a slight modification of Hofmann's reaction (p. 213) for converting an amide into an amine, having one carbon atom less in the molecule:

$$C_6H_4\begin{matrix} CO \\ \diagup\diagdown \\ \diagdown\diagup \\ CO \end{matrix}NH + NaOCl + 3NaOH \rightarrow C_6H_4\begin{matrix} NH_2 \\ CO_2Na \end{matrix} + NaCl + Na_2CO_3 + H_2O$$

When the reaction is over, the solution is neutralised with hydrochloric or sulphuric acid and the anthranilic acid finally precipitated with acetic acid. The market price of anthranilic produced by the stages indicated above, starting with naphthalene, is about 3*s*. 7*d*. per lb.; this is a typical example of the working costs of an adequately controlled chemical production of an important product.

The chief use of anthranilic acid is for the synthetic production of *indigo.* Indigo was the name given to the (Indian) coloured material derived from a variety of plants of the genus *Indigofera* growing more particularly in various parts of India, China and Central America, and from woad which grows in many parts of Europe. In the leaves of these plants, the material which is finally converted into the colouring matter is not present as such but as a *glucoside* (p. 445) known as *indican.* This glucoside, a colourless substance crystallising from water in prisms having the composition $C_{14}H_{17}O_6N.3H_2O$, may be readily obtained by extracting the leaves of the plant with acetone.

Indican can be hydrolysed by heating with dilute hydrochloric acid or, more easily, in the presence of water by the specific *enzyme* (*indemulsin*)* present in the plant and then yields the sugar *glucose* (p. 422) and *indoxyl* (p. 296):

$$C_{14}H_{17}O_6N + H_2O \rightarrow \underset{\text{glucose}}{C_6H_{12}O_6} + \underset{\text{indoxyl}}{C_8H_7ON}$$

* An enzyme may be defined as any catalytic substance having a specific action and produced by living cells. Unlike ordinary chemical catalysts, enzymes are used up or destroyed during the reaction which they are bringing about, and consequently for the particular reaction concerned to be continuous the enzyme has to be reproduced as a product of the living cell present at the time. If the living cell is not present and is not growing, the reaction will cease as soon as the original quantity of enzyme is consumed.

The indoxyl when liberated rapidly undergoes oxidation in the air and especially in the presence of the oxidising enzyme (oxidase) present to indigo:

$$2C_8H_7ON + O_2 \rightarrow C_{16}H_{10}O_2N_2 + 2H_2O$$

In actual practice, the colouring matter is obtained by cutting down the plants and placing them in wooden tanks. The plants, weighted down with stones and bamboos, are covered with water, when a vigorous fermentation, lasting for some fifteen hours, sets in by which the glucoside is hydrolysed to glucose and indoxyl, both of which remain in solution. The solution is drawn off and mechanically agitated in the presence of air, whereby the indoxyl is oxidised and the colouring matter is precipitated in glistening blue particles which are almost insoluble in water. Destructive fermentation is prevented by heating the suspension, the mixture of dye-stuffs filtered off, pressed into blocks which are broken up and dried. This is the material indigo, which is far from pure, and contains besides the blue colouring matter other colouring materials known as indirubin, indigo-yellow, indigo-brown, etc., as well as foreign material arising either as by-products from the fermentation or mechanically introduced. To avoid confusion with this impure material, the actual blue colouring matter or pure indigo is often referred to as *indigotin*, and it is this material which is now produced synthetically and which has for all practical purposes completely replaced the natural and impure product.

Pliny (79–23 B.C.) described the sublimation of what must have been impure indigo whereby the purple vapour condenses in blue crystals (indigotin). This is probably the best way of isolating indigotin from the crude natural indigo. Indigotin is practically insoluble in water and many organic solvents. It crystallises from hot acetic acid, phthalic anhydride, nitrobenzene, aniline and phenol. The crystals obtained by sublimation have a brilliant metallic lustre. Indigotin dissolves in cold concentrated sulphuric acid and is precipitated unchanged from this solution by the addition of water. When it is heated with concentrated sulphuric acid various sulphonic acids of indigotin are formed. The most important property of indigotin is that on which its use as a dye-stuff depends. It is reduced by alkaline reducing agents to *indigo-white* (p. 338), which is soluble in water to a yellow solution. This solution in contact with air undergoes oxidation, regenerating indigotin. Indigotin or pure indigo belongs to the class of *vat dyes* which are insoluble in water and yield soluble *leuco*-compounds (for example, indigo-white) when treated with reducing agents such as sodium hydrosulphite ($Na_2S_2O_4$) in alkaline solution. These vat dyes are used for the dyeing of textiles (unaffected by alkaline solutions), which are immersed in the reduced solution in the vat. The excess of solution is then removed from the fabric, which on exposure to the air retains the dye-stuff regenerated by oxidation impregnated in the fibre. The ancient Indians and Egyptians probably dyed their textiles by immersing them in the fermenting solution and then allowing the soluble indoxyl to oxidise on the material exposed to the air.

The successful commercial production of pure indigo has depended on the elucidation of the constitution of indigotin, the synthesis of the substance and the discovery of methods based on its synthesis for its production by reactions which can be successfully controlled and carried out economically on the manufacturing scale. The methods of large-scale production of a fundamental material like indigotin are continually being modified as a result of detailed chemical and economic investigation.*

The modern reactions for the production of synthetic indigotin are based on the classical work of A. von Baeyer during the years 1865 to 1880, and it was not until some ten years later that the large scale production of synthetic indigotin was made possible after further detailed and expensive investigations.

Before Baeyer began his investigations many facts concerning indigotin were known. The pure substance had been separated from the crude natural product by sublimation and its composition determined by analysis and vapour density determinations as $C_{16}H_{10}O_2N_2$. It was also known that aniline resulted from the distillation of indigotin with sodium hydroxide and, further, that on boiling indigotin and sodium hydroxide in aqueous solution together with manganese dioxide, anthranilic acid was obtained, the latter having been shown to be converted into aniline and carbon dioxide when heated (Fritsche, 1840–41). These facts indicated that the following skeletal arrangement of atoms must be present in the indigotin molecule:

```
        H
        C
      //  \
   HC       C—N=
   |        ||
   HC       C—C≡
      \\  /
        C
        H
```

* The successful adaptation of laboratory processes to large-scale production of chemical compounds frequently involves many more investigations than those which have led to their original discovery. These investigations necessarily take a different form from the latter and deal, for example, with the material for the construction of the 'plant', the design of the plant, the control of the heat exchanges during the reactions, the avoidance or the utilisation of by-products combined with economic considerations such as cost of materials, fuel or heating and labour cost, the initial cost of the plant and its depreciation, etc. On all of these the final cost of production and the market value of the material manufactured are based, which may have also to bear the expense of the original scientific investigation.

About the same time, Laurent and Erdmann obtained a reddish brown crystalline substance also as an oxidation product of indigotin. To this substance they gave the name *isatin* and they showed its composition to be $C_8H_5O_2N$. In 1865, Baeyer submitted isatin to reduction with sodium amalgam and obtained a compound, crystallising in yellow needles, to which the name *dioxindole* was given and which has the composition $C_8H_7O_2N$. He also showed that this compound on further reduction is converted into *oxindole*, which crystallises in colourless needles and which has the composition C_8H_7ON. Finally, Baeyer distilled oxindole with zinc dust and obtained indole, C_8H_7N (p. 296). The constitution of indole was known from Baeyer's synthesis, which consisted in heating *o*-nitrocinnamic acid (p. 242) with iron filings and potassium hydroxide:

$$C_6H_4(—CH{=}CH—CO.OH)(—NO_2) + 4H + 2KOH \rightarrow C_6H_4\langle{}^{CH}_{NH}\rangle CH + K_2CO_3 + 3H_2O$$

That there is a close relationship between indigotin and indole had been known for some time, since indigo is occasionally deposited from some specimens of urine on standing (p. 296) and particularly when indole has been injected subcutaneously. In addition, Baeyer showed that if indigotin be first reduced with tin and hydrochloric acid, and the product then distilled with zinc dust, indole is obtained. Working backwards, the constitution of the various products and their relationship to indigotin can be traced, thus:

indigotin —(oxidation)→ isatin —(reduction)→ dioxindole —(reduction)→ oxindole —(reduction)→ indole

indigotin —(reduction)→ indole

isatin: $C_6H_4\langle{}^{CO}_{NH}\rangle CO$

dioxindole: $C_6H_4\langle{}^{CH(OH)}_{NH}\rangle CO$

oxindole: $C_6H_4\langle{}^{CH_2}_{NH}\rangle CO$

indole: $C_6H_4\langle{}^{CH}_{NH}\rangle CH$

These investigations proved the presence of the indole skeleton arrangement of atoms in indigotin, and a comparison between the molecular formulae of these two substances afforded very strong evidence of the presence of two indole skeletons in a molecule of indigotin. Finally, it was suggested by Baeyer that the reactions of indigotin could only be satisfactorily explained if its constitutional formula were

```
        H                              H
        C                              C
     //   \                          /   \\
   HC       C——————CO  :  OC——————C       CH
   |        ||       |  :  |      ||       |
   HC       C        C==:==C      C        CH
     \\   /   \    /    :    \   /  \    //
        C        N      :      N       C
        H        H      :      H       H
```

When oxidation or vigorous reduction of indigotin takes place, the molecule is broken at the double linking of the two similar halves of the molecule, as indicated by the dotted line.

The starting-point in Baeyer's first synthesis, which confirmed the suggested constitutional formula, of indigotin was o-*nitrophenylacetic acid*,* which is converted into oxindole (the lactam of the γ-amino acid, o-aminophenylacetic acid)†:

```
     H                                  H                                H
     C                                  C                                C
  //   \                             //   \                           //   \
HC       C—CH2.CO2H               HC       C—CH2.CO2H              HC       C———————CH2
|        ||        reduction      |        ||              - H2O   |        ||        |
HC       C—NO2     ————————→      HC       C—NH2           ————→   HC       C         CO
  \\   /                            \\   /                           \\   /   \     /
     C                                  C                                C       N
     H                                  H                                H       H
                                 [o-aminophenylacetic acid]              oxindole
```

* This was synthesised starting from o-*nitrobenzyl chloride* by a series of reactions outlined as follows:

```
      CH2Cl (1)   KCN           CH2.CN (1)  hydrolysis          CH2.CO2H (1)
C6H4<            ————→   C6H4<             ————————————→  C6H4<
      NO2 (2)                   NO2 (2)                         NO2 (2)
                         o-nitrobenzyl cyanide             o-nitrophenylacetic acid
```

o-Nitrophenylacetic acid has m.p. 141° and, along with the *para-* compound (*p*-nitrophenylacetic acid, m.p. 152°), can be obtained by the nitration of phenylacetic acid (p. 234).

† The conversion of o-aminophenylacetic acid into its lactam (oxindole) takes place so readily that the acid itself cannot be isolated.

Dioxindole is the lactam of the corresponding hydroxy acid, o-*aminomandelic acid* (o-aminophenylhydroxyacetic acid):

```
      H          H                               H
      C         /                                C        H
   //   \      /                              //   \     /
 HC       C—C—OH                            HC       C———C
 |        ||   \             - H2O          |        ||   | \
 |        ||    CO2H        ————→           |        ||   |  OH
 HC       C—NH2                             HC       C    CO
   \\   /                                     \\   /   \  /
      C                                          C      N
      H                                          H      H
```

When oxindole is treated with nitrous acid it forms an *iso*nitroso compound, iso*nitrosooxindole,* due to the reaction of the hydrogen atoms of the $:CH_2$ group (attached to the $:CO$ group):

$$C_6H_4\langle CH_2,\ CO,\ NH \rangle + O{:}N{-}OH \rightarrow C_6H_4\langle C{=}N{-}OH,\ CO,\ NH \rangle + H_2O$$

*iso*nitrosooxindole

This *iso*nitroso compound, when reduced, is converted into the primary amine, *aminooxindole*:

$$C_6H_4\langle C{=}N{-}OH,\ CO,\ NH \rangle \xrightarrow{2H_2} C_6H_4\langle CH{-}NH_2,\ CO,\ NH \rangle + H_2O$$

On gentle oxidation, aminooxindole is converted into isatin which, in its tautomeric form, reacts with phosphorus pentachloride, yielding *chloroisatin* (isatin chloride). Finally, chloroisatin on reduction with zinc dust is converted into indigotin. These later stages of Baeyer's synthesis are summarised:

$$\text{aminooxindole} \xrightarrow{\text{oxidation}} \text{isatin (}C_6H_4\langle CO,\ CO,\ NH\rangle \rightleftarrows C_6H_4\langle CO,\ C(OH),\ N\rangle\text{)}$$

$$\xrightarrow{PCl_5} \text{chloroisatin (}C_6H_4\langle CO,\ CCl,\ N\rangle\text{)} \xrightarrow[\text{reduction}]{\text{Zn dust}} \text{indigotin (}C_6H_4\langle CO,\ NH\rangle C{=}C\langle CO,\ NH\rangle C_6H_4\text{)}$$

aminooxindole — isatin — chloroisatin — indigotin

Various other syntheses have been carried out subsequently. The synthesis of indigotin by the reduction of $o{:}o'$-dinitrodiphenyldiacetylene is strong additional proof that the two indole nuclei are united in indigotin as indicated (p. 244):

```
        C                C                       CO         OC
       /⫴               ⫴\                      /  \       /  \
$C_6H_4$    C—C    $C_6H_4$   reduction   $C_6H_4$    C═C    $C_6H_4$
       \                /     ———————→        \  /       \  /
        $NO_2$    $O_2N$                           NH         HN
```

Indoxyl is also a reduction product of indigotin into which it is easily converted by oxidation (p. 333). It has been synthesised and shown to be isomeric with oxindole. Indoxyl is 3-hydroxyindole; it forms yellow crystals, m.p. 85°. It is a substance exhibiting tautomerism and reacts according to either of the formula indicated:

```
      H                       H                         H
      C                       C                         C
    //  \                   //  \                     //  \
  HC     C———$CH_2$           HC     C———C(OH)          HC     C———CO
  |      ‖     |            |      ‖     ‖     ⇄     |      ‖     |
  HC     C     CO           HC     C     CH           HC     C     $CH_2$
    \\  / \   /               \\  / \   /               \\  / \   /
      C     N                   C     N                   C     N
      H     H                   H     H                   H     H
   oxindole                                            indoxyl
```

The production of indigotin oxidation of indoxyl from the hydrolysis of the glucoside, indican, is explained:

```
      H                          H
      C                          C
    //  \                      //  \
  HC     C———C(OH)           HC     C———CO       oxidation
2 |      ‖     ‖      ⇄ 2    |      ‖     |      ————————→
  HC     C     CH            HC     C     $CH_2$       $+O_2$
    \\  / \   /                \\  / \   /
      C     N                    C     N
      H     H                    H     H

          H                                   H
          C                                   C
        //  \                               /  \\
      HC     C———CO     OC———C     CH
      |      ‖     |        |     ‖      |      + $2H_2O$
      HC     C     C════C     C      CH
        \\  / \   /        \   / \  //
          C     N            N     C
          H     H            H     H
```

and the conversion of urinary indican into indigotin has been described (p. 296).

When indigotin is reduced with alkaline-reducing agents it is converted into the *leuco*-compound, *indigo-white* or *di-indoxyl* (p. 333):

```
       CO         OC                        C(OH)    (HO)C
      /  \       /  \                      //  \      /  \\
$C_6H_4$    C═══C    $C_6H_4$ + $H_2$ →  $C_6H_4$     C————C     $C_6H_4$
      \  /       \  /                      \  /        \  /
       N          N                         N            N
       H          H                         H            H
```

If sodium hydrosulphite and sodium hydroxide are used, the indigo-white is present as the disodium derivative:

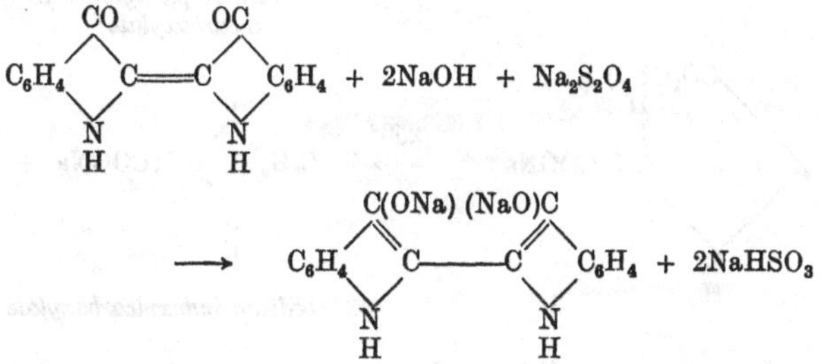

$$C_6H_4\left\langle\begin{matrix}CO\\NH\end{matrix}\right\rangle C{=}C\left\langle\begin{matrix}OC\\NH\end{matrix}\right\rangle C_6H_4 + 2NaOH + Na_2S_2O_4$$

$$\longrightarrow C_6H_4\left\langle\begin{matrix}C(ONa)\\NH\end{matrix}\right\rangle C{-}C\left\langle\begin{matrix}(NaO)C\\NH\end{matrix}\right\rangle C_6H_4 + 2NaHSO_3$$

This disodium salt is present in the solution in the vat-dyeing of textiles with indigo. When the fabric impregnated with the vat solution is exposed to air, oxidation takes place:

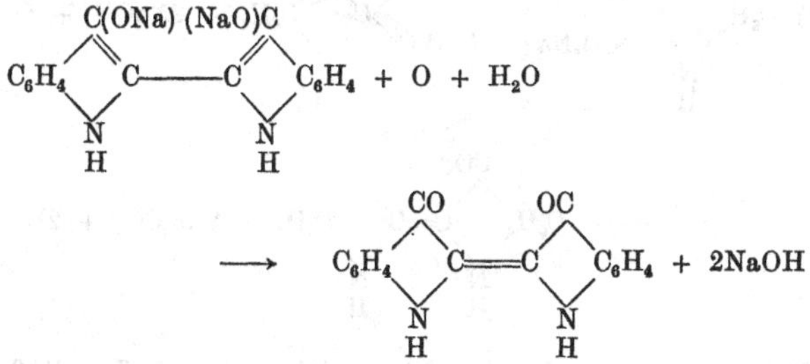

$$C_6H_4\left\langle\begin{matrix}C(ONa)\\NH\end{matrix}\right\rangle C{-}C\left\langle\begin{matrix}(NaO)C\\NH\end{matrix}\right\rangle C_6H_4 + O + H_2O$$

$$\longrightarrow C_6H_4\left\langle\begin{matrix}CO\\NH\end{matrix}\right\rangle C{=}C\left\langle\begin{matrix}OC\\NH\end{matrix}\right\rangle C_6H_4 + 2NaOH$$

leaving the indigo impregnated in the fibres.

The use of anthranilic acid as the starting material for the large-scale production of indigotin is outlined in the following reactions:

(i) $$C_6H_4\left\langle\begin{matrix}CO_2Na\\NH_2\end{matrix}\right. + HO.CH_2.SO_3Na \rightarrow C_6H_4\left\langle\begin{matrix}CO_2Na\\NH.CH_2.SO_3Na\end{matrix}\right. + H_2O$$

formaldehyde sodium bisulphite — *sodium methylanthranilate-ω-sulphonate**

(ii) $$C_6H_4\left\langle\begin{matrix}CO_2Na\\NH.CH_2.SO_3Na\end{matrix}\right. + NaCN \rightarrow C_6H_4\left\langle\begin{matrix}CO_2Na\\NH.CH_2.CN\end{matrix}\right. + Na_2SO_3$$

sodium ω-cyanomethylanthranilate

* Compounds in which a hydrogen atom on the last carbon atom of a side chain containing at least two carbon atoms is substituted are described as ω-compounds. Another example is ω-chloroacetophenone, $C_6H_5.CO.CH_2Cl$.

$$\text{(iii)}\ C_6H_4\begin{cases}CO_2Na\\ NH.CH_2.CN\end{cases} + H_2O + NaOH \rightarrow C_6H_4\begin{cases}CO_2Na\\ NH.CH_2.COONa\end{cases} + NH_3$$

*sodium N-phenylglycine-o-carboxylate**

$$\text{(iv)}\ C_6H_4\begin{cases}CO(ONa)\ \ H\\ \quad\qquad\qquad | \\ \qquad\qquad CH.COONa\\ N\\ H\end{cases} \xrightarrow{250^\circ} C_6H_4\begin{cases}CO\\ \qquad CH.COONa\\ N\\ H\end{cases} + NaOH$$

sodium indoxylcarboxylate

This fused product is dissolved in water and air passed into the resulting alkaline solution, oxidation takes place and the indigotin is precipitated:

$$\text{(v)}\ C_6H_4\begin{cases}CO\\ NH\end{cases}C\begin{cases}H\\ CO_2Na\end{cases} + \begin{cases}H\\ NaO_2C\end{cases}C\begin{cases}OC\\ NH\end{cases}C_6H_4 + 2NaOH + O_2$$

$$\longrightarrow C_6H_4\begin{cases}CO\\ NH\end{cases}C{=}C\begin{cases}OC\\ NH\end{cases}C_6H_4 + 2Na_2CO_3 + 2H_2O$$

The indigotin only requires washing with water to free it from the water-soluble sodium carbonate. It is usually made up for use in the form of a paste containing a stated amount of pure indigotin. The market price of pure indigotin is about 3*s*. 2*d*. per lb.†

* This compound in an older process was produced by the condensation of mono-chloroacetic acid and anthranilic acid in the form of their sodium salts:

$$C_6H_4\begin{cases}CO_2H\\ NH_2\end{cases} + ClCH_2.CO_2H \rightarrow C_6H_4\begin{cases}CO_2H\\ NH.CH_2.COOH\end{cases} + HCl$$

† In this method for the technical production of indigotin (pure indigo) it will be realised that naphthalene costing 2½*d*. per lb. is converted into anthranilic acid costing 3*s*. 7*d*. per lb. and this latter is converted into pure indigo costing 3*s*. 2*d*. per lb.

CHAPTER XII

STEREOISOMERISM AND OPTICAL ACTIVITY

AMONG the properties of certain compounds already described it has been pointed out that they exhibit the phenomenon of optical activity, i.e. of rotating the plane of plane-polarised light either to the right—when the compound is described as *dextro*rotatory—or to the left, the compound then being *laevo*rotatory.

After Etienne Louis Malus in 1808 had described the phenomenon of the polarisation of light, Arago (1811) noticed that certain quartz crystals are optically active and 'deflected' the plane of polarised light. This property of optical activity was also found to be exhibited by certain crystalline inorganic salts, e.g. sodium bromate, sodium chlorate, sodium periodate and the double ammonium lithium sulphate. A little later, between 1813 and 1815, J. J. Biot, to whom in his early days Louis Pasteur (p. 400) owed so much, showed that the extent of the rotation of the plane of polarised light depended on the thickness of the layer of the substance, on the temperature and on the wave-length of the light employed. Biot noticed that of plates cut in a particular way from different crystals of quartz some rotated the plane of polarisation to the right and others to the left, in each case according to the same laws. This distinguished investigator also made the discovery that certain naturally occurring organic substances, e.g. sugar, camphor, oil of turpentine, etc., exhibit optical activity either in the liquid or dissolved condition; and he pointed out the theoretical importance of these observations. Whereas in the previously recognised phenomena of the rotation of the plane of polarisation by crystals the optical activity is conditioned by the crystalline form, disappearing as soon as that form is destroyed, in the case of these organic substances this rotatory power must be inherent in the molecules of the substances themselves; for being exhibited by liquids and by substances in solution it must be dependent upon the structure which still remains, i.e. the molecule itself.

The first outstanding consequences of this pioneering work of Biot were Pasteur's investigations (1844) of the tartaric acids (p. 400), and while not in historical sequence it is useful to consider simpler compounds which were, however, not investigated in detail until later.

Scheele (1780) had isolated a lactic acid produced in the 'souring' of milk, and a lactic acid which had the same chemical reactions was isolated by Engelhardt from muscle extract in 1847; on account of its origin this latter was given the name *sarcolactic* acid. At the same time it was shown that while Scheele's (fermentation) lactic acid and sarcolactic acid were chemically identical they differed in their effect on plane polarised light, the former being optically inactive and the latter optically active and *dextro*rotatory. Starting with acetaldehyde, Wislicenus (1863) synthesised a lactic acid chemically identical with fermentation lactic acid and like it optically inactive and differing from sarcolactic acid in being optically inactive. Since these acids all had the same chemical reactions, it had to be concluded that they contained the ethylidene ($CH_3.CH:$) grouping which is also present in acetaldehyde. Finally, it was shown by Wislicenus (1873) that synthetic lactic acid (identical with fermentation lactic acid) and sarcolactic acid are both α-hydroxypropionic acids and have the constitutional formula I

$$\begin{array}{ccc} CH_3 & \qquad & CH_2OH \\ | & & | \\ C\langle^{H}_{OH} & & CH_2 \\ | & & | \\ CO_2H & & CO_2H \\ \text{I} & & \text{II} \end{array}$$

and that they have chemical properties different from hydracrylic acid (β-hydroxypropionic acid), having the constitutional formula II.

For hydroxy acids having the molecular formula $C_3H_6O_3$ only the above two constitutional formulae are possible. On account of its chemical properties, formula II had to be assigned to hydracrylic acid and formula I to the optically inactive as well as to the optically active lactic acid. Since a constitutional formula must be in keeping with the physical and chemical properties of a compound, the constitutional formula I does not account for the different behaviour towards polarised light of the optically inactive and optically active lactic acids, although it is in keeping with their chemical properties. In other words, constitutional formulae written in a plane do not explain the difference in behaviour towards plane polarised light of compounds which have identical chemical properties and therefore the same structure, and a complete understanding of the properties of compounds can only be obtained by recognising the existence of their molecules in three-dimensional space. In 1873, Wislicenus stated: 'The facts force us to explain the difference between isomeric molecules of the same structure by a different arrangement of atoms in space.' For the type of isomerism which depends on the varying space distribution of the atoms in the molecule, Wislicenus used the term 'geometrical isomerism', a name changed in 1888 by Victor Meyer to the now accepted 'stereoisomerism'.

The theory of the asymmetric carbon atom for explaining the difference in behaviour towards plane polarised light of structurally identical isomeric organic substances was announced almost simultaneously by Jacobus Henricus van 't Hoff (September, 1874) and Joseph Achille Le Bel (November, 1874). This theory is of fundamental importance in chemistry and marks the beginning of a period of rapid advancement of the science. While, for all practical purposes, the ultimate result of the expression of the theories of van 't Hoff and Le Bel is the same, it should be emphasised that Le Bel's views were based on the results of the investigations by Pasteur of the tartaric acids (p. 400), while those of van 't Hoff were based on Kekulé's theory of the quadrivalence of the carbon atom extending this theory to the spatial distribution of the four valencies of the atom.*

The most symmetrical static arrangement of the atoms in the molecule of methane is that in which the carbon atom is situated at the centre of a regular tetrahedron, the four valencies being directed to the corners of the tetrahedron at which are situated the hydrogen atoms.† This arrangement is represented in Fig. III. Such an arrangement of the atoms in space is highly symmetrical,‡ which becomes less symmetrical as the hydrogen atoms are successively replaced by other equivalent dissimilar atoms or groups

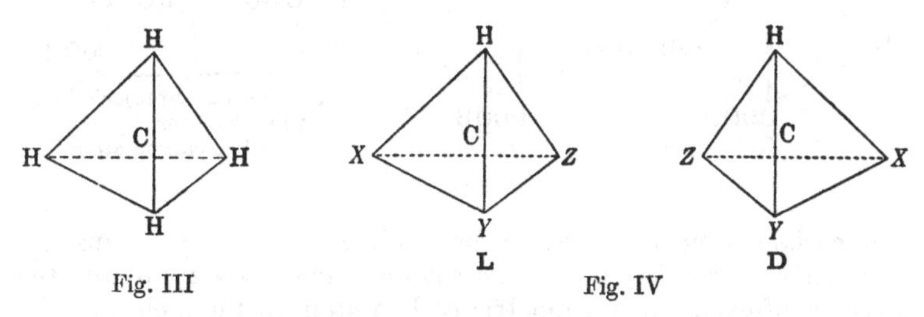

Fig. III

Fig. IV

until in a compound represented by the formula C(HXYZ), where the carbon atom is attached to four different atoms or groups, the arrangement is unsymmetrical or asymmetrical; i.e. it has no elements of symmetry (Fig. IV).

* *Dix années dans l'histoire d'une théorie*, by J. H. van 't Hoff (dedicated to Monsieur J. A. Le Bel), Rotterdam, 1887; this is the second edition of *La chimie dans l'espace*, by van 't Hoff, Rotterdam, 1875. *The arrangement of Atoms in space*, by van't Hoff, trans. by Eiloart, London, 1898. *Researches on Molecular Asymmetry*, by Louis Pasteur; Alembic Club Reprints, No. 14.

† Physical investigations have provided definite evidence that this regular tetrahedral arrangement of four similar atoms or groups round a carbon atom actually exists. This has been proved, for example, in the case of the interesting crystalline compound pentaerythritol, $C(CH_2OH)_4$.

‡ The regular tetrahedron (hemihedral form of the regular octahedron) has six diagonal planes of symmetry, four axes of threefold symmetry and three axes of twofold symmetry.

There are two, and only two, different ways of arranging tetrahedrally four dissimilar groups round a central quadrivalent atom (such as carbon) illustrated by L and D. These two arrangements correspond with an object and its mirror image or with a right and a left hand. If the arrangement represented by D were hollow and one tried to fix it over L, it would be found that the corners of D would not come into the same position as the corresponding corners of L. These mirror image arrangements are conveniently described as *non-superposable*. On the other hand, tetrahedral arrangements representing compounds CH_3X, CH_2XY or CH_2X_2 have superposable mirror images.

Such a compound as lactic acid is of the type $C(HXYZ)$, and there are two ways in which the four different groups (H—, CH_3—, HO—, —CO_2H) can be placed tetrahedrally round the central carbon atom, both arrangements having no element of symmetry and the one being the mirror image of the other. These two arrangements may be represented:

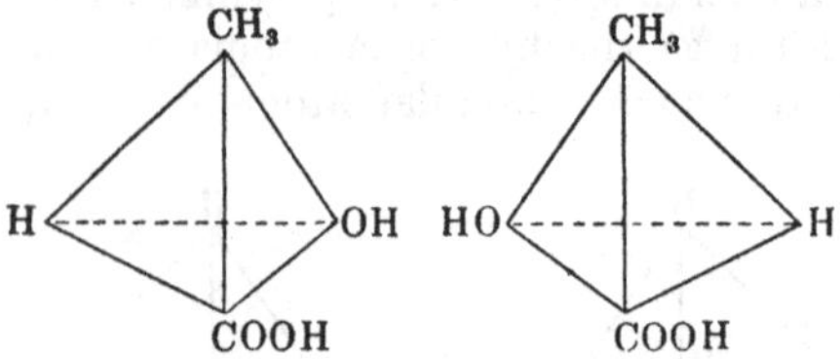

CH$_3$ / H, OH / COOH — **L**

CH$_3$ / HO, H / COOH — **D**

CH_3	CH_3
H—C—OH	HO—C—H
COOH	COOH

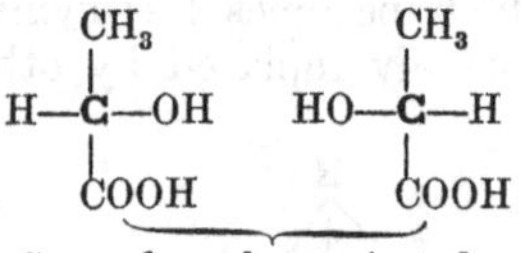

Space formulae projected on to a plane through the central carbon atom

A carbon atom attached to four different atoms or groups has frequently been referred to as an asymmetric carbon atom, and the presence of even one asymmetric carbon atom in the molecule of a compound renders the molecule asymmetric, having like the arrangement of the four different atoms or groups in lactic acid a non-superposable mirror image. This is a common feature of all naturally occurring optically active organic compounds, and the consequent asymmetry of such molecules is correlated with their optical activity. As examples of other simple naturally occurring organic compounds the following may be mentioned:

sec.-Butylcarbinol (2-methyl-*n*-butyl alcohol) occurring in fusel oil is *laevo*rotatory; alanine (α-aminopropionic acid) obtained by hydrolysis of silk fibroin is *dextro*rotatory (p. 320); malic acid (monohydroxysuccinic acid) obtained from unripe apples and especially from mountain-ash berries is *laevo*rotatory; β-hydroxybutyric acid, a product of abnormal metabolism in diabetics, is *laevo*rotatory (p. 367). The plane projection formulae of these compounds are

$$\begin{array}{c} CH_3 \\ | \\ H—C^*—C_2H_5 \\ | \\ CH_2OH \end{array} \qquad \begin{array}{c} CH_3 \\ | \\ H—C^*—NH_2 \\ | \\ COOH \end{array}$$

sec.-butylcarbinol alanine

$$\begin{array}{c} CH_2—COOH \\ | \\ H—C^*—OH \\ | \\ COOH \end{array} \qquad \begin{array}{c} CH_3 \\ | \\ H—C^*—OH \\ | \\ CH_2.COOH \end{array}$$

malic acid β-hydroxybutyric acid

In all these cases the carbon atom marked with an asterisk is attached to four different atoms or groups, the molecule in each case is therefore asymmetric. As indicated above, these naturally occurring compounds are optically active; when the compounds are synthetically prepared in the laboratory, they are optically inactive, an equimolecular mixture of the two optically active forms (*dextro-* and *laevo-*) being obtained.

Dextro-lactic acid (*d*-lactic acid) is given one of the configurations (*L* or *D*) above, and corresponding to *d*-lactic acid there should be capable of existence an *l*-lactic acid having the other configuration. A mixture of equal quantities of *l-* and *d-* lactic acids having equal and opposite specific rotatory powers† would obviously be optically

† Each optically active substance has its *specific rotatory power*, the value of which is a criterion of purity of the substance. It is known by the symbol $[\alpha]_\lambda^t$, where t represents the temperature of measurement and λ the wave-length of the light used.

The specific rotatory power of a substance is the rotatory power or angle of rotation produced by 1 gram of the optically active substance in a volume of 1 ml. when viewed through a layer of 1 dm.

For a homogeneous optically active liquid, if α denotes the angle of rotation of the plane polarised light of wave-length λ at a temperature of $t°$ C. by a layer of liquid, l dm. thick, the density of the liquid being d, the specific rotatory power is given by the expression:

$$[\alpha]_\lambda^t = \frac{\alpha}{ld}.$$

In the case of an optically active substance in solution in an optically inactive solvent, the specific rotatory power is given by the expression:

$$[\alpha]_\lambda^t = \frac{\alpha v}{lp},$$

where p represents the number of grams of the substance contained in v ml. of the solution, α and l having the same significance as before. The *molecular rotatory power*, often used for comparison of different optically active substances, is given by the relation:

$$[M]_\lambda^t = \frac{[\alpha]_\lambda^t \times M}{100},$$

where M represents the molecular weight of the substance.

From this it will be seen that the molecular rotatory power of a substance may be defined as the rotation produced by a layer of 1 mm. containing the molecular weight in grams of the optically active substance in 1 ml. This formulation of the specific rotatory power is based on the fundamental work of Biot referred to above.

inactive; this is the nature of ordinary fermentation and of synthetic lactic acid, which is therefore described as being *externally compensated* and written as *dl*-lactic acid. Similarly, the four other compounds specifically mentioned above when synthesised in the laboratory are externally compensated and described as *dl-sec.*-butylcarbinol, *dl*-alanine, *dl*-malic acid and *dl*-β-hydroxybutyric acid respectively.

The *l*- and *d*- forms of an optically active substance may be described as *enantiomers*. The enantiomeric forms are identical in all chemical properties and in all physical properties, e.g. solubility, melting point, boiling point, refractive power, etc., with the exception of the direction of rotation of the plane of polarised light; the two enantiomeric forms have, of course, equal specific rotatory powers of opposite sign.

In view of the identity of properties of the *l*- and *d*- forms of lactic acid, it is impossible to obtain them by separation of the externally compensated or *dl*- acid. If, for example, the *l*- and *d*- acids can be converted into salts, the molecules of which have configurations in space which are no longer mirror images of each other, these pairs of salts will not have identical physical properties and will differ in solubility and will be capable of being separated from each other by fractional crystallisation. Consider the case of the externally compensated lactic acid, two molecules of which can be conveniently represented as *l*A*d*A, the *l*A molecule having, of course, a configuration which is the mirror image of that of *d*A. If these two molecules be converted into salts with an optically active base (*l*B), then the two salts formed can be represented as *l*A*l*B and *d*A*l*B; such salts are usually described as *diastereoisomerides* or diastereoisomers. Such diastereoisomeric substances have configurations which are not mirror images of each other.* They will have different solubilities, and the mixture can be separated by fractional crystallisation into the two separate individuals and obtained pure. The fractional crystallisation may be prolonged if the diastereoisomeric salts differ only slightly in solubility. By liberating the acid from each of them, the pure *l*- acid and the pure *d*- acid can be obtained, and theoretically an externally compensated acid should yield half its weight of the *l*- acid and half its weight of the *d*- acid. This separation or, as it is more usually described, resolution of the externally compensated acid into its optically active components may be diagrammatically represented thus:

*l*A.*d*A externally compensated acid
+2*l*B optically active base

↙ ↘

*l*A*l*B *d*A*l*B diastereoisomeric salts
↓ ↓
*l*A *d*A optically active acids

* The mirror image of the *l*A*l*B salt would be *d*A*d*B, and the mirror image of the *d*A*l*B would be *l*A*d*B.

This same method can obviously be used for the resolution of an externally compensated base, using an optically active acid for producing the diastereoisomeric salts, and for other types of compounds; for example, an externally compensated alcohol may be converted into a mixture of diastereoisomeric esters by means of an optically active acid. This method of resolving externally compensated compounds is the most useful of the three methods originally described by Pasteur (p. 407).

Externally compensated (fermentation) lactic acid was resolved by Purdie and Walker (1892) using as the optically active base the *laevo*rotatory alkaloid strychnine. Later, Irvine (1906) showed that the *laevo*rotatory alkaloid morphine can be more conveniently employed. In this case morphine *l*-lactate (*l*A*l*B) is less soluble than the diastereoisomeric salt morphine *d*-lactate (*d*A*l*B), and from these two salts *l*-lactic acid and *d*-lactic acid, the latter identical in all respects with sarcolactic acid, were obtained in a state of purity.

The simplest organic compound which has been resolved into its optically active components is chloroiodomethane sulphonic acid, $C(H)(Cl)(I)(SO_3H)$. This compound was investigated by Pope and Read (1914) and by Read and McMath (1932); its molecule is asymmetric through the carbon atom being united to four different atoms or groups none of which contains a carbon atom. The optically active enantiomers are surprisingly stable.

The principle of the tetrahedral configuration of four different atoms or groups attached to a central atom giving rise to optical activity applies not only when the central atom is carbon, but also to other elements. Quaternary ammonium salts of the type $[Nabcd]X$, where a, b, c and d represent different monovalent hydrocarbon radicals and X a negative radical, usually chlorine, bromine or iodine, have been shown to be externally compensated and have been resolved into their optically active components. In such compounds the groups a, b, c and d in the ammonium radical are arranged tetrahedrally round the nitrogen atom and the ammonium radical is consequently asymmetric; a compound of this type is externally compensated and capable of being resolved into optically active components whatever be the nature of the negative radical in the ammonium salt. The resolution of an externally compensated ammonium halide of this type may be effected by means of the silver salt of a suitable optically active acid (represented as Ag-*d*A). The scheme of the resolution effected in a suitable medium may be represented:

$$2[Nabcd]X + 2Ag\text{-}dA \rightarrow 2AgX + l[Nabcd]dA + d[Nabcd]dA.$$

After removal of the insoluble silver halide, the two diastereoisomeric salts are separated by fractional crystallisation. From these, the optically active quaternary ammonium iodides may be obtained by treatment with potassium iodide if the iodides are less soluble than the potassium salt of the optically active acid. The first externally compensated ammonium compound of this type to be resolved into its optically active components was *dl*-methylallylphenylbenzylammonium iodide,

$$[N(CH_3)(C_3H_5)(C_6H_5)(CH_2.C_6H_5)]I$$ (Pope and Peachey, 1899).

Pope and Peachey (1900) and Smiles (1900) showed that sulphur attached to four dissimilar atoms or groups furnished externally compensated compounds, which they resolved into their optically active components. In the same year, Pope and Peachey extended this work to the quadrivalent tin compounds and Pope and Neville (1902) to the quadrivalent selenium compounds. In all these cases the tetrahedral configuration of the four dissimilar groups attached to the nitrogen, sulphur, selenium and tin atoms confers asymmetry on the respective molecular structures, making it possible for the existence of such compounds in optically active isomerides.

For many years the occurrence of optically active isomerides among carbon compounds was referred to the presence in the molecule of at least one asymmetric carbon atom, i.e. a carbon atom having four dissimilar atoms or groups attached to it. While such a tetrahedral arrangement of dissimilar atoms or groups round a carbon atom is asymmetric, the presence of such a carbon atom in a molecule renders the whole structure of the molecule asymmetric; and, as long as the occurrence of optically active isomerides was referred to the presence in the molecule of at least one asymmetric carbon atom, progress in stereochemistry was considerably restricted. The presence in the molecule of a carbon atom attached to four dissimilar groups is only one of the conditions of asymmetry and the occurrence of optically active isomers. The optical activity of all compounds exhibiting this phenomenon is better correlated with the *enantiomorphism* (i.e. having a mirror image which is non-superposable) of the molecular structure of those compounds.

In 1875, van 't Hoff showed that according to the principle of the tetrahedral configuration of groups attached to a carbon atom compound having the general formula:

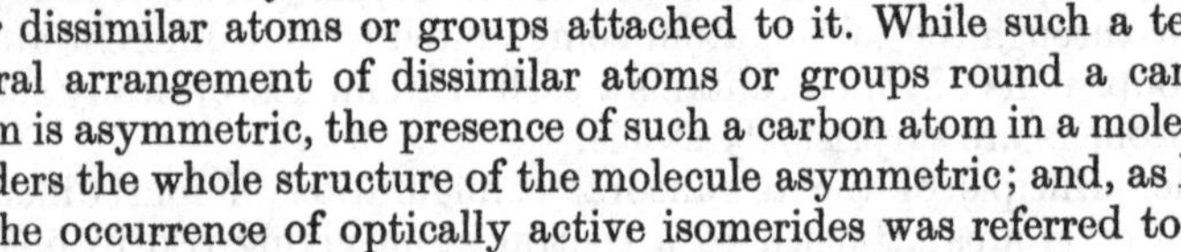

$$\begin{matrix} R' \\ R'' \end{matrix}\!\!>C\!=\!C\!=\!C\!<\!\!\begin{matrix} R''' \\ R^{\mathrm{iv}} \end{matrix} \quad \text{or}$$

R' R''' R'' R^iv

Type I

[in which the monovalent groups R' and R'' are in the same plane as each other (say, the plane of the paper) and while again the groups

R''' and R^{iv} are in the same plane as each other, one (R''') is above the plane and the other (R^{iv}) is below the plane of the paper] has a non-superposable mirror image, i.e. the molecule as a whole has an enantiomorphous configuration and therefore the compound should be capable of exhibiting optical activity either in the liquid condition or in solution, although the compound does not contain an asymmetric carbon atom. The same principle applies to the following types:

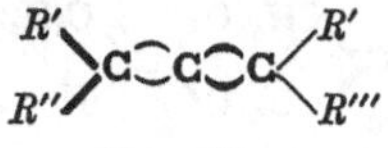

$$\begin{matrix} R' \\ R'' \end{matrix}\!\!> C{=}C{=}C <\!\!\begin{matrix} R' \\ R''' \end{matrix} \quad \text{and} \quad \begin{matrix} R' \\ R'' \end{matrix}\!\!> C{=}C{=}C <\!\!\begin{matrix} R' \\ R'' \end{matrix}$$

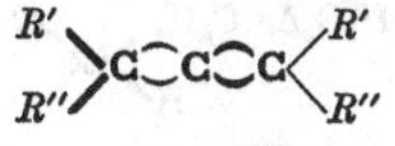

Type II Type III

Such compounds belong to the allene ($H_2C{=}C{=}CH_2$) type. Suitable derivatives of allene in the hands of many workers have proved extremely difficult to investigate from a stereochemical point of view, and therefore the result of the work of Maitland and Mills (1935) is all the more noteworthy. Maitland and Mills have been successful in obtaining a compound of type III, viz. dl-*diphenyl-di-α-naphthyl-allene,*

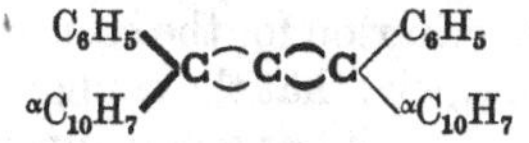

$$\begin{matrix} C_6H_5 \\ \alpha C_{10}H_7 \end{matrix}\!\!> C{=}C{=}C <\!\!\begin{matrix} C_6H_5 \\ \alpha C_{10}H_7 \end{matrix}$$

or in fuller detail

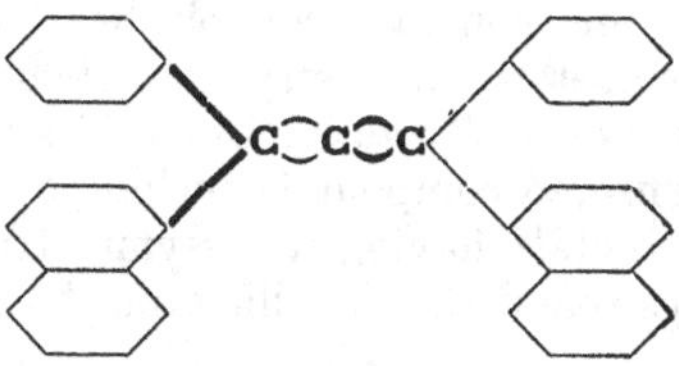

in optically active forms having equal and opposite rotatory powers: $[\alpha]_{5461} = \pm 437{\cdot}5°$.

Other compounds in which one or both of the double bonds are replaced by planar rings and which from a stereochemical standpoint are analogous to the allene type have also been investigated and have been obtained in optically active forms. Of such compounds the first investigated was *dl*-methyl*cyclo*hexylidene-4-acetic acid,

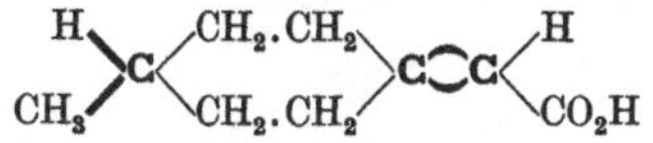

$$\begin{matrix} H \\ CH_3 \end{matrix}\!\!> C <\!\!\begin{matrix} CH_2.CH_2 \\ CH_2.CH_2 \end{matrix}\!\!> C{=}C <\!\!\begin{matrix} H \\ CO_2H \end{matrix}$$

Type II

which was synthesised and resolved into its optically active components by Perkin and Pope (1909). The *cyclo*hexyl ring in Perkin and Pope's compound can be considered equivalent to one of the double bonds in type II, and such a configuration possesses neither planes, axes nor a centre of symmetry and is completely asymmetric and the configuration has a non-superposable mirror image. In a similar manner, the following compounds

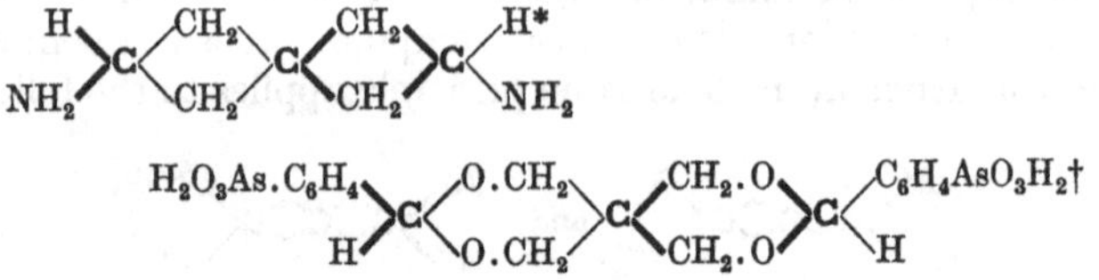

$$\begin{matrix} H \\ NH_2 \end{matrix} \rangle C \langle \begin{matrix} CH_2 \\ CH_2 \end{matrix} \rangle C \langle \begin{matrix} CH_2 \\ CH_2 \end{matrix} \rangle C \langle \begin{matrix} H^* \\ NH_2 \end{matrix}$$

$$\begin{matrix} H_2O_3As.C_6H_4 \\ H \end{matrix} \rangle C \langle \begin{matrix} O.CH_2 \\ O.CH_2 \end{matrix} \rangle C \langle \begin{matrix} CH_2.O \\ CH_2.O \end{matrix} \rangle C \langle \begin{matrix} C_6H_4AsO_3H_2\dagger \\ H \end{matrix}$$

Type III

can be regarded as belonging to type III, and these are examples of compounds which have also been synthesised and resolved into their optically active components. Unlike types I and II, the configuration represented by type III is not completely asymmetric; it possesses neither planes nor a centre of symmetry, but it possesses an axis of symmetry. Like types I and II, however, it has a non-superposable mirror image and is described as dissymmetric. This emphasises the essential criterion for the occurrence of optically active isomerides of a compound, viz. that the configuration of the molecular structure should be such that it has a non-superposable mirror image. The recognition of this general principle has led to most of the important advances in stereochemistry since 1909 and, in particular, to the resolution into optically active isomerides of numerous types of externally compensated compounds, including co-ordination compounds of various metals having a dissymmetric arrangement of atoms and/or groups round the metallic atom.‡

* *dl*-1 : 1′-diamino*spirocyclo*heptane (Pope and Janson, 1932).

† dl-*spirobis*-3 : 5-dioxan-4 : 4′-di(phenyl-*p*-arsonic acid) (Gibson and Levin, 1933) (p. 278).

‡ Externally compensated compounds having an asymmetric or dissymmetric arrangement of groups round the following elements have been resolved into their optically active components: silicon, phosphorus, arsenic, boron, beryllium, copper, zinc, cobalt, iron, aluminium, chromium, ruthenium, rhodium, iridium, platinum.

CHAPTER XIII

DIHYDRIC ALCOHOLS OR GLYCOLS AND THEIR DERIVATIVES

Glycol, Ethylene Glycol, 1:2-Dihydroxyethane

The physical properties and the preparation of *glycol* by the action of potassium permanganate in aqueous solution in the presence of sodium carbonate on ethylene have already been described (p. 36) and the reaction taking place may be summarised:

$$\begin{matrix} CH_2 \\ \| \\ CH_2 \end{matrix} + O + H_2O \rightarrow \begin{matrix} CH_2OH \\ | \\ CH_2OH \\ \text{glycol} \end{matrix}$$

Starting again from ethylene, glycol may also be synthesised by reactions which may be summarised as follows:

$$\text{(i)} \quad \begin{matrix} CH_2 \\ \| \\ CH_2 \end{matrix} + Br_2 \rightarrow \begin{matrix} CH_2Br \\ | \\ CH_2Br \\ \text{ethylene bromide} \end{matrix}$$

$$\text{(ii)} \quad \begin{matrix} CH_2Br \\ | \\ CH_2Br \end{matrix} + K_2CO_3 + H_2O \xrightarrow[\text{in presence of water}]{\text{heated for 8 hours}} \begin{matrix} CH_2OH \\ | \\ CH_2OH \end{matrix} + 2KBr + CO_2$$

$$\text{or (ii)} \quad \begin{matrix} CH_2Br \\ | \\ CH_2Br \end{matrix} + 2AgOH \rightarrow \begin{matrix} CH_2OH \\ | \\ CH_2OH \end{matrix} + 2AgBr$$

The synthesis of glycol from ethylene by the two methods is proof of its constitutional formula, which represents the compound as a di-primary alcohol; it is the simplest dihydric alcohol. Its constitutional formula indicates that it is derived from ethane by the substitution of two hydroxyl groups for two hydrogen atoms, one attached to each of the two carbon atoms. This is indicated by the systematic name 1 : 2-dihydroxyethane. The properties of glycol may readily be deduced from the properties of a simple primary alcohol, since it contains two primary alcohol groups in its molecule.

Glycol is a colourless somewhat viscous liquid having m.p. $-11{\cdot}5°$ and b.p. $197{\cdot}4°$. It has a sweet taste and is miscible with water and with alcohol in all proportions. It absorbs water from the air, being hygroscopic, and it is only sparingly soluble in ether. Like all alcohols it forms esters by interaction with acids; ethylene bromide is the normal ester of glycol and hydrobromic acid, and glycol chlorhydrin (p. 37), $\begin{array}{l}CH_2Cl\\|\\CH_2OH\end{array}$, is a half ester of glycol and hydrochloric acid.

Glycol yields a monoacetate, *glycol monoacetate*, $\begin{array}{l}CH_2O.CO.CH_3\\|\\CH_2OH\end{array}$, b.p. 182°, which is soluble in water and alcohol. *Glycol diacetate*, $\begin{array}{l}CH_2O.CO.CH_3\\|\\CH_2O.CO.CH_3\end{array}$, a colourless liquid, b.p. 186–187°, can be prepared by the action of acetic anhydride on glycol or, more conveniently, by boiling ethylene bromide, acetic acid and finely powdered potassium acetate together. It is slightly soluble in water and readily soluble in alcohol and ether.

The normal acetic esters of compounds containing 'alcoholic' hydroxyl groups are frequently used for determining the number of the latter in the parent compound. Having prepared the normal acetic ester by the action of an excess of acetic anhydride on the parent compound, this is purified, analysed and its molecular weight determined by a convenient method. By boiling a known weight of the ester with a known excess of standard alkali (usually potassium hydroxide in alcoholic solution) until all is hydrolysed, the amount of excess alkali can be determined by titration and from this the amount used for the hydrolysis of the known weight of compound calculated. From this the amount of alkali required for the hydrolysis of the molecular weight in grams of the ester is deduced. In the case of potassium hydroxide, this will be found to be $(n \times 56)$ grams of potassium hydroxide, n being the number of acetyl groups in the molecule of the ester and 56 the molecular weight of potassium hydroxide. Since each acetyl group in the ester corresponds to an 'alcoholic' group in the original compound, the number of hydroxyl groups in a molecule of the latter is determined:

$$—(OH)_n \rightarrow —(O.CO.CH_3)_n \equiv n\text{KOH}$$

In the case of glycol diacetate and glycol $n=2$, but the number of hydroxyl groups in glycol could have been determined as described above without any previous knowledge of the constitution of glycol. This supplements the proof of the constitution of glycol, since for a compound containing two carbon atoms and two hydroxyl groups to be stable one hydroxyl group must be attached to each carbon atom. In the case of more complex 'polyhydric' alcohols and their derivatives, the determination of the number of hydroxyl groups in the molecule by the above method is important evidence regarding the constitution of the compounds concerned.

The properties of the primary alcohol group indicate that the constitutions of the oxidation products of glycol must be as indicated below, the names of the various compounds being also given:

$$\underset{\text{glycol}}{\begin{matrix}CH_2OH\\|\\CH_2OH\end{matrix}} \rightarrow \underset{\substack{\text{glycollic}\\\text{aldehyde}}}{\begin{matrix}CH_2OH\\|\\CHO\end{matrix}} \rightarrow \underset{\substack{\text{glycollic}\\\text{acid}}}{\begin{matrix}CH_2OH\\|\\COOH\end{matrix}} \rightarrow \underset{\substack{\text{glyoxylic}\\\text{acid}}}{\begin{matrix}CHO\\|\\COOH\end{matrix}} \rightarrow \underset{\substack{\text{oxalic}\\\text{acid}}}{\begin{matrix}COOH\\|\\COOH\end{matrix}}$$

$$\underset{\substack{\text{glycollic}\\\text{aldehyde}}}{\begin{matrix}CH_2OH\\|\\CHO\end{matrix}} \rightarrow \underset{\text{glyoxal}}{\begin{matrix}CHO\\|\\CHO\end{matrix}} \rightarrow \underset{\substack{\text{glyoxylic}\\\text{acid}}}{\begin{matrix}CHO\\|\\COOH\end{matrix}}$$

Glycollic aldehyde, $\begin{matrix}CH_2OH\\|\\CHO\end{matrix}$, has the molecular formula $C_2H_4O_2$ and is isomeric with acetic acid. It can be obtained by oxidising glycol with hydrogen peroxide and by the interaction of monobromoacetaldehyde with baryta:

$$2\begin{matrix}CH_2Br\\|\\CHO\end{matrix} + Ba(OH)_2 \rightarrow 2\begin{matrix}CH_2OH\\|\\CHO\end{matrix} + BaBr_2$$

It has simultaneously the properties of an aldehyde and of a primary alcohol and is described as an aldehyde-alcohol. This compound has some theoretical importance, being actually the simplest 'sugar', the sugars being part of the large class of organic compounds known as carbohydrates which have the general formula $C_n(H_2O)_m$, where n and m may be equal to or different from each other. One class of the sugars and particularly part of that class known as *monosaccharides* are, like glycollic aldehyde, aldehyde-alcohols and are known as *aldoses*. Glycollic aldehyde being an aldose containing two carbon atoms in the molecule is described as an *aldodiose*. It crystallises in thin colourless crystals, m.p. 95–97°. It possesses a sweet taste, is readily soluble in water and is a strong reducing agent, giving all the typical reactions of an aliphatic aldehyde.

Glycollic acid or monohydroxyacetic acid, $\begin{matrix}CH_2OH\\|\\COOH\end{matrix}$, m.p. 80°, is most conveniently prepared by heating monochloroacetic acid (ester-acid) with water under pressure:

$$\begin{matrix}CH_2Cl\\|\\COOH\end{matrix} + HOH \rightarrow HCl + \begin{matrix}CH_2OH\\|\\COOH\end{matrix}$$

It is also obtained by the action of nitrous acid on glycine or aminoacetic acid, thus:

$$\begin{array}{l} CH_2.NH_2 \\ | \\ COOH \end{array} + ONOH \rightarrow \begin{array}{l} CH_2OH \\ | \\ COOH \end{array} + N_2 + H_2O$$

Glycollic acid is simultaneously a primary alcohol and a monobasic acid and has the properties of both these classes of substances. Like all *α-hydroxy acids* (i.e. those having an 'alcoholic' hydroxyl group on the carbon atom adjacent to the carboxyl group) when heated under reduced pressure above its melting point it yields an inner-ester by loss of two molecules of water between the alcohol and carboxyl groups of different molecules, thus:

$$\begin{array}{l} H_2C.OH \\ | \\ OC.OH \end{array} + \begin{array}{l} HO.CO \\ | \\ HO.CH_2 \end{array} \xrightarrow{-2H_2O} \begin{array}{l} H_2C—O—CO \\ | \qquad\quad | \\ OC—O—CH_2 \end{array}$$

glycollide

Glycollide has m.p. 86° and readily reverts to glycollic acid on boiling with water.

Glyoxal or diformyl, $\begin{array}{l} CHO \\ | \\ CHO \end{array}$, is a dialdehyde and therefore, as might be expected, highly reactive. It may be obtained by the oxidation of glycol with nitric acid, and when isolated is obtained in a polymerised form $[(CHO)_2]_n$ known as *polyglyoxal.* From this, the monomolecular form, having m.p. 15° and b.p. 51°, can be obtained by heating with phosphorus pentoxide. It exhibits a strong tendency to yield polymerisation products and is conveniently isolated in the form of the more stable sodium bisulphite compound:

$$\begin{array}{l} H \\ |\diagup OH \\ C—SO_3Na \\ | \\ C—SO_3Na \\ |\diagdown OH \\ H \end{array}$$

It reacts with ammonia, giving *glyoxaline* (p. 327), and with hydroxylamine to give *glyoxime*:

$$\begin{array}{l} H.C:NOH \\ \quad | \\ H.C:NOH \end{array}$$

which crystallises from water in colourless plates, m.p. 178°. A homologue of glyoxal, viz. *succinaldehyde*, $CHO.CH_2.CH_2.CHO$, has been used in the synthesis of heterocyclic compounds, of which the simplest are

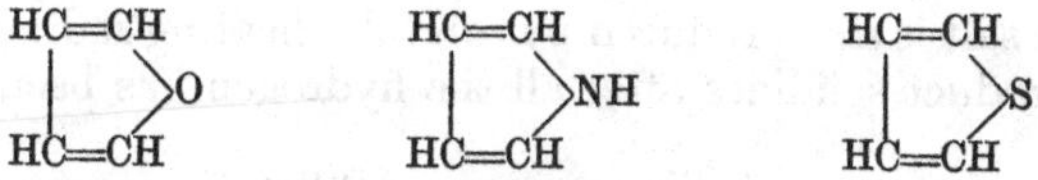

$$\begin{array}{ccc} \text{HC=CH} & \text{HC=CH} & \text{HC=CH} \\ \text{O (ring)} & \text{NH (ring)} & \text{S (ring)} \\ \text{HC=CH} & \text{HC=CH} & \text{HC=CH} \\ \text{furan} & \text{pyrrole} & \text{thiophen} \end{array}$$

Glyoxylic acid (glyoxalic acid), $\begin{array}{l}CHO \\ | \\ COOH\end{array}$, is an aldehyde-acid (compare formic acid) which when isolated has the composition $CHO.COOH+H_2O$, and it is suggested that its constitution may be represented:

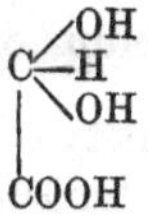

$$\begin{array}{l} \diagup OH \\ C\!-\!H \\ |\;\diagdown OH \\ | \\ COOH \end{array}$$

(compare chloral hydrate) and it is significant that the salts (but not the esters) appear to be derived from this form. It is conveniently obtained by heating dichloro- or dibromo-acetic acid with water under pressure:

$$\begin{array}{l}CHCl_2 \\ | \\ COOH\end{array} + H_2O \rightarrow \begin{array}{l}CHO \\ | \\ COOH\end{array} + 2HCl$$

It is usually a thick liquid when isolated and this crystallises in colourless prisms on standing over concentrated sulphuric acid. It yields typical aldehydic derivatives, and on boiling with alkalies it yields salts of glycollic and oxalic acids; this reaction is similar to that which formaldehyde and benzaldehyde undergo under similar conditions:

$$2\begin{array}{l}CHO \\ | \\ COOH\end{array} + H_2O \rightarrow \begin{array}{l}CH_2OH \\ | \\ COOH\end{array} + \begin{array}{l}COOH \\ | \\ COOH\end{array}$$

Glyoxylic acid is used for determining the presence of tryptophan (p. 324) both when free and combined in proteins.

Oxalic acid, $\begin{array}{l}COOH \\ | \\ COOH\end{array}$, is the simplest dibasic acid. Its sodium salt can be synthesised by the direct combination of sodium with carbon dioxide under pressure at 360°:

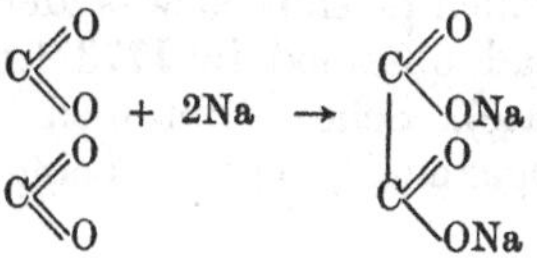

$$\begin{array}{l} C\begin{array}{l}\diagup O \\ \diagdown O\end{array} \\ C\begin{array}{l}\diagup O \\ \diagdown O\end{array} \end{array} + 2Na \rightarrow \begin{array}{l} C\begin{array}{l}\diagup O \\ \diagdown ONa\end{array} \\ | \\ C\begin{array}{l}\diagup O \\ \diagdown ONa\end{array} \end{array}$$

The sodium salt is also produced by carefully heating sodium formate until the product solidifies after all the hydrogen has been evolved:

$$\begin{matrix} H.COONa \\ H.COONa \end{matrix} \rightarrow H_2 + \begin{matrix} COONa \\ | \\ COONa \end{matrix}$$

If care is taken to avoid overheating, the hydrogen obtained from this reaction is pure and the sodium oxalate only needs recrystallisation from water. To obtain oxalic acid from the sodium salt, the latter is dissolved in water and calcium chloride added until no further precipitate is formed. The calcium oxalate, $\begin{matrix} COO \\ | \\ COO \end{matrix}\!\!>\!Ca$, is filtered off, treated with the necessary quantity of sulphuric acid and the filtrate evaporated to crystallisation after separating the sparingly soluble calcium sulphate:

$$\begin{matrix} COO \\ | \\ COO \end{matrix}\!\!>\!Ca + H_2SO_4 \rightarrow CaSO_4 + \begin{matrix} COOH \\ | \\ COOH \end{matrix}$$

Cyanogen, $\begin{matrix} CN \\ | \\ CN \end{matrix}$, is the 'nitrile' of oxalic acid, and when an aqueous solution of cyanogen is allowed to stand for some time it contains, among other products, ammonium oxalate:

$$\begin{matrix} CN \\ | \\ CN \end{matrix} + 4H_2O \rightarrow \begin{matrix} COONH_4 \\ | \\ COONH_4 \end{matrix}$$

Oxalic acid is the final oxidation product of many complex organic compounds. The technical method of preparation is by heating sawdust with potassium hydroxide in iron pans at about 200°. The potassium oxalate is extracted with water and then converted into calcium oxalate. This product is then treated with sulphuric acid and the aqueous solution of oxalic acid separated from the sparingly soluble calcium salt and treated as described above. Sucrose or cane sugar undergoes vigorous oxidation when gently warmed with fairly concentrated nitric acid. After the vigorous oxidation has subsided, the solution is neutralised and the calcium oxalate obtained and decomposed as described above. The destructive oxidation of sucrose by concentrated nitric acid is of some historical interest. In 1776 Scheele obtained, by the action of nitric acid on the sugar, a characteristic acid which he showed was identical with the impure acid which Savary had obtained in 1773 from 'wood-sorrel salt' (potassium acid oxalate). Scheele obtained the acid in a state of purity and called it 'sugar acid', the name being subsequently changed to oxalic acid.

From water, oxalic acid crystallises in monoclinic prisms, having the composition $H_2C_2O_4 . 2H_2O$. The crystals have m.p. 101° and lose water above this temperature; the anhydrous acid has m.p. 190° at which temperature decomposition sets in, the products of decomposition being as indicated:

$$(a)\quad \begin{array}{l}COOH\\ |\\ COOH\end{array} \rightarrow CO_2 + \begin{array}{l}H\\ |\\ COOH\end{array}$$

$$(b)\quad \begin{array}{l}COOH\\ |\\ COOH\end{array} \rightarrow CO + CO_2 + H_2O$$

The preparation of formic acid from oxalic acid is most conveniently carried out by heating the latter with glycerol (p. 385). When oxalic acid or its salts are warmed with concentrated sulphuric acid, carbon monoxide and carbon dioxide are evolved as indicated in equation (*b*), the sulphuric acid acting as a dehydrating agent.

Oxalic acid in the presence of sufficient dilute sulphuric acid is quantitatively oxidised at about 60° by an aqueous solution of potassium permanganate according to the equation:

$$5H_2C_2O_4 + 2KMnO_4 + 3H_2SO_4 \rightarrow 10CO_2 + 8H_2O + K_2SO_4 + 2MnSO_4.$$

This reaction finds frequent application in volumetric analysis.

The more important salts of oxalic acid are: the normal sodium salt which is anhydrous and somewhat sparingly soluble in water; the normal calcium salt, also anhydrous, which is insoluble in water and in acetic acid and frequently used for distinguishing oxalic acid from other organic acids; the acid potassium salt has the formula $\begin{array}{l}COOH\\ |\\ COOK\end{array} + 2H_2O$; the so-called potassium quadrioxalate has the composition $H_2C_2O_4 . KHC_2O_4 + 2H_2O$, is highly crystalline and is frequently used as a standard substance in volumetric analysis.

The more important esters of oxalic acid are: *dimethyl oxalate*, m.p. 51°, b.p. 162°, and *diethyl oxalate*, b.p. 186°. Both of these are colourless substances which on treatment with ammonia are converted into *oxamide*, $\begin{array}{l}CONH_2\\ |\\ CONH_2\end{array}$, which is insoluble in water. *Oxamic acid* has the constitution $\begin{array}{l}CONH_2\\ |\\ COOH\end{array}$, and is prepared by heating ammonium oxalate.

Oxalic acid is the lowest member of a homologous series of dibasic carboxylic acids having the general formula $C_nH_{2n}(COOH)_2$. Some of the acids in this series are:

$$\begin{array}{ccccc} COOH & COOH & COOH & COOH & COOH \\ | & | & | & | & | \\ CH_2 & CH_2 & CH(CH_3) & CH_2 & CH_2 \\ | & | & | & | & | \\ COOH & CH_2 & COOH & CH_2 & CH_2 \\ & | & & | & | \\ & COOH & & CH_2 & CH_2 \\ & & & | & | \\ & & & COOH & CH_2 \\ & & & & | \\ & & & & COOH \end{array}$$

malonic acid; succinic acid; methylmalonic acid; glutaric acid m.p. 97°; adipic acid m.p. 148°

Malonic acid,* $\begin{array}{c} COOH \\ | \\ CH_2 \\ | \\ COOH \end{array}$, is a direct oxidation product of the glycol, $CH_2OH.CH_2.CH_2OH$, 1:3-dihydroxypropane or trimethylene glycol, a homologue of ethylene glycol. It was first obtained by Dessaignes in 1858 by oxidising malic acid (p. 380) obtained from unripe apples; hence the name of both these acids, from *malum*, apple. It was synthesised by Kolbe and Hugo Müller in 1864 by a method which is the basis of the method of preparation and which consists in carrying out the following reactions in aqueous solution at the water-bath temperature:

$$2\begin{array}{c} Cl \\ | \\ CH_2 \\ | \\ COOH \end{array} + Na_2CO_3 \rightarrow 2\begin{array}{c} Cl \\ | \\ CH_2 \\ | \\ COONa \end{array} + CO_2 + H_2O$$

monochloroacetic acid

$$\begin{array}{c} Cl \\ | \\ CH_2 \\ | \\ COONa \end{array} + KCN \rightarrow \begin{array}{c} C{\equiv}N \\ | \\ CH_2 \\ | \\ COONa \end{array} + KCl$$

sodium cyanoacetate

$$\begin{array}{c} C{\equiv}N \\ | \\ CH_2 \\ | \\ COONa \end{array} + \begin{array}{c} HOH \\ \\ HOH \end{array} + HCl \rightarrow \begin{array}{c} COOH \\ | \\ CH_2 \\ | \\ COOH \end{array} + NH_4Cl$$

malonic acid

Malonic acid crystallises in colourless plates which are readily soluble in water and alcohol. When it is heated it begins to decompose at its melting point (132°), the following change taking place quantitatively:

* Malonic acid may be systematically described as methanedicarboxylic acid.

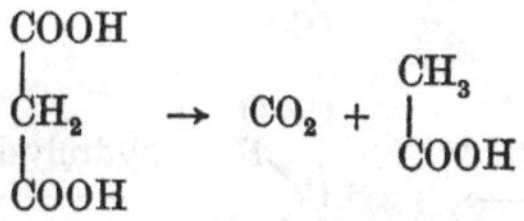

$$\begin{array}{l} COOH \\ | \\ CH_2 \\ | \\ COOH \end{array} \rightarrow CO_2 + \begin{array}{l} CH_3 \\ | \\ COOH \end{array}$$

acetic acid

This decomposition is typical of that undergone by all aliphatic acids which have two carboxyl groups attached to the same carbon atom; such dibasic acids are derivatives of malonic acid.

The most important derivative of malonic acid is its diethyl ester, which is generally obtained without actually isolating the acid. The solution of the sodium salt is evaporated, the residue mixed with absolute ethyl alcohol and the mixture treated with hydrogen chloride, one of the general methods of ester preparation. *Ethyl malonate*, $CH_2(COOC_2H_5)_2$, is a colourless liquid, b.p. 198°.

When ethyl malonate is warmed with phosphorus pentoxide it undergoes decomposition, which can be represented:

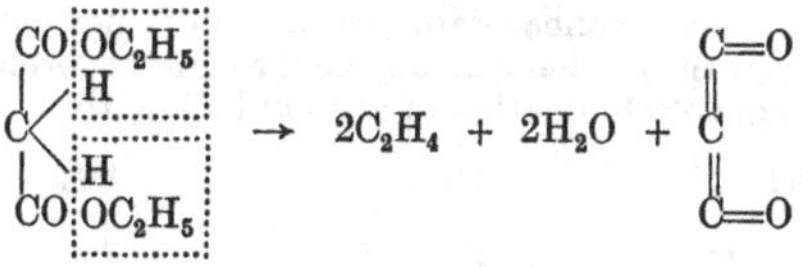

$$\begin{array}{l} CO\vdots OC_2H_5 \\ |\ \ \diagup H \\ C \\ |\ \ \diagdown H \\ CO\vdots OC_2H_5 \end{array} \rightarrow 2C_2H_4 + 2H_2O + \begin{array}{c} C{=}O \\ \| \\ C \\ \| \\ C{=}O \end{array}$$

carbon suboxide

Malonic acid also yields the interesting *carbon suboxide* (b.p. 7°) under similar conditions.

Ethyl malonate, like all substances possessing the grouping $—CO—CH_2—CO—$, behaves as a weak acid and reacts with sodium ethylate, yielding a sodium derivative. This is probably derived from the enolic form of the ester which has not been isolated (compare ethyl acetoacetate, p. 369) thus:

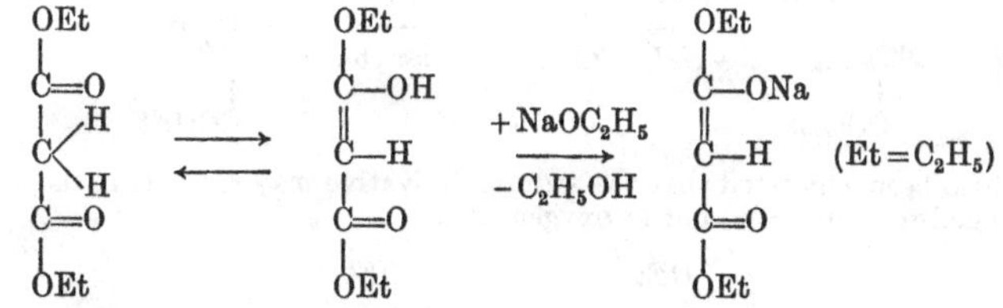

$$\begin{array}{l} OEt \\ | \\ C{=}O \\ |\ \ \diagup H \\ C \\ |\ \ \diagdown H \\ C{=}O \\ | \\ OEt \end{array} \rightleftharpoons \begin{array}{l} OEt \\ | \\ C—OH \\ \| \\ C—H \\ | \\ C{=}O \\ | \\ OEt \end{array} \xrightarrow[-C_2H_5OH]{+NaOC_2H_5} \begin{array}{l} OEt \\ | \\ C—ONa \\ \| \\ C—H \\ | \\ C{=}O \\ | \\ OEt \end{array} \quad (Et = C_2H_5)$$

ethyl malonate — enolic form — sodium ethyl malonate

This sodium derivative, not generally isolated from the alcoholic solution, reacts rapidly with alkyl and acyl halides and compounds of similar types to give monosubstituted malonic esters, thus (using methyl iodide):

```
OEt                        OEt                          
|                          |                            COOH
C—ONa                      C=O                          |  H
‖          +CH3I†          |  H       hydrolysis        C<
C—H        ——————→         C<         ——————————→       |  CH3
|          -NaI            |  CH3                       COOH
C=O                        C=O                     methylmalonic
|                          |                           acid*
OEt                        OEt
                      methylmalonic
                        ethyl ester
```

A compound of the type of the ethyl ester of methylmalonic acid will, again, react with sodium ethylate and this again will react rather more slowly with alkyl halides, etc., thus (using ethyl iodide):

* Methylmalonic acid has m.p. 130°. It is a next higher homologue to malonic acid and isomeric with succinic acid (p. 358). It is sometimes described as *iso*succinic acid. It is readily distinguished from succinic acid by its behaviour on being heated above its melting point, when it yields propionic acid and carbon dioxide.

† The mechanism of the conversion of the sodium derivatives into the ethyl ester of alkyl malonic acids and of analogous reactions in the case of ethyl aceto-acetate (p. 375) and ethyl cyanoacetate (see below) is not indicated here. Some authorities suggest that the mechanism may be through the formation of an addition compound and subsequent elimination of sodium iodide, thus

```
OEt                  OEt                   OEt
|                    |  ONa                |
C—ONa                C<                    C=O
‖         +RI        |  I       -NaI       |
|         ———→       |  H       ———→       |  H
C—H                  C<                    C<
|                    |  R                  |  R
COOEt                COOEt                 COOEt
```

Such a suggested mechanism implies the unusual addition of the alkyl iodide at the double bond. Other authorities have suggested that the reaction may proceed thus

```
OEt                  OEt                              OEt
|                    |                                |
C—ONa     +RI        C—OR      intramolec-            C=O
‖         ———→       ‖         ——————————→            |  H
C—H       -NaI       C—H       ular change            C<
|                    |                                |  R
COOEt                COOEt                            COOEt
```

It has also been suggested that the sodium derivative may react as if the sodium were linked to carbon and not to oxygen, thus

```
OEt                     OEt
|                       |
C=O        +RI          C=O
|  H       ———→         |  H
C<         -NaI         C<
|  Na                   |  R
COOEt                   COOEt
```

Such a constitution of the sodium derivative is, however, unlikely.

The present author suggests that the formation of the final product can be readily

$$EtO{-}C(=O){-}CH(CH_3){-}C(=O){-}OEt \rightleftarrows EtO{-}C(OH){=}C(CH_3){-}C(=O){-}OEt \xrightarrow[-C_2H_5OH]{+NaOC_2H_5} EtO{-}C(ONa){=}C(CH_3){-}C(=O){-}OEt \xrightarrow[-NaI]{+C_2H_5I} EtO{-}C(=O){-}C(CH_3)(C_2H_5){-}C(=O){-}OEt$$

methylethylmalonic ethyl ester

$$\xrightarrow{\text{hydrolysis}} COOH{-}C(CH_3)(C_2H_5){-}COOH$$

methylethylmalonic acid

As ethyl malonate will only form a monosodium derivative, the preparation of a disubstituted derivative of a malonic ester or acid must be carried out in two stages as indicated.

Starting with ethyl malonate it is thus possible to prepare mono- and di- substituted malonic esters and the corresponding acids. Moreover, since the latter, like malonic acid, each have two carboxyl groups attached to the same carbon atom, they will undergo a change similar to that of malonic acid on being heated above their melting points, thus:

$$COOH{-}CH(CH_3){-}COOH \xrightarrow{-CO_2} CH_3{-}CH_2{-}COOH\ ;\qquad COOH{-}C(C_2H_5)(CH_3){-}COOH \xrightarrow{-CO_2} CH_3{-}CH_2{-}CH(CH_3){-}COOH$$

methylmalonic acid; propionic acid; methylethyl-malonic acid; methylethylacetic acid or α-methyl-*n*-butyric acid

understood if its production be regarded as due to the collision of the reacting molecules and would represent the reaction in some such way as

$$EtO{-}C(ONa){=}CH{-}COOEt \longleftrightarrow I{-}R \longrightarrow EtO{-}C(=O){-}CH(R){-}COOEt + NaI$$

This suggestion also affords an explanation of the fact that the formation of the dialkyl derivative proceeds more slowly than the formation of the monoalkyl compound. In the former case effective collision of the reacting molecules for production of the final compound would be to some extent prevented by the steric hindrance afforded by the alkyl group already attached to the carbon atom.

The above reactions indicate the general method of preparing homologues of malonic acid and homologues of acetic acid.*

The next homologues to glycollic acid or hydroxyacetic acid are the two isomeric monohydroxypropionic acids which differ in the position of the alcoholic hydroxyl group, thus:

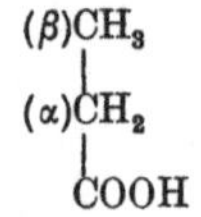

$$\begin{array}{c} (\beta)CH_3 \\ | \\ (\alpha)CH_2 \\ | \\ COOH \end{array}$$

propionic acid

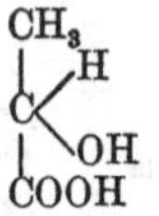

$$\begin{array}{l} CH_3 \\ |\quad\;\; H \\ C\langle \\ |\quad\;\; OH \\ COOH \end{array}$$

α-hydroxypropionic acid or *lactic acid*

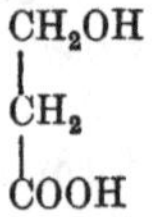

$$\begin{array}{c} CH_2OH \\ | \\ CH_2 \\ | \\ COOH \end{array}$$

β-hydroxypropionic acid or *hydracrylic acid*

Lactic acid may be considered to be an oxidation product of propylene glycol (homologue of ethylene glycol) obtained from propylene

* Although it does not contain the —CO—CH_2—CO— grouping, ethyl cyanoacetate undergoes similar reactions, the —CN grouping being like the —CO group sufficiently electro-negative in character to render the compound weakly acidic. The reactions in this case may be briefly outlined:

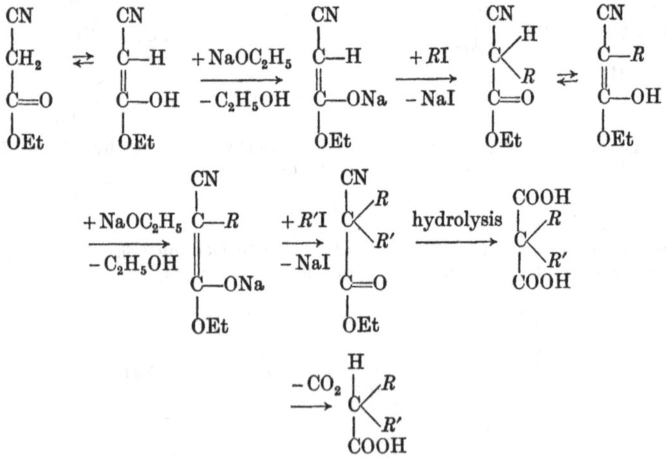

$$\begin{array}{c} CN \\ | \\ CH_2 \\ | \\ C{=}O \\ | \\ OEt \end{array} \rightleftarrows \begin{array}{c} CN \\ | \\ C{-}H \\ \| \\ C{-}OH \\ | \\ OEt \end{array} \xrightarrow[-C_2H_5OH]{+NaOC_2H_5} \begin{array}{c} CN \\ | \\ C{-}H \\ \| \\ C{-}ONa \\ | \\ OEt \end{array} \xrightarrow[-NaI]{+RI} \begin{array}{c} CN \\ |\;\; H \\ C\langle \\ |\;\; R \\ C{=}O \\ | \\ OEt \end{array} \rightleftarrows \begin{array}{c} CN \\ | \\ C{-}R \\ \| \\ C{-}OH \\ | \\ OEt \end{array}$$

$$\xrightarrow[-C_2H_5OH]{+NaOC_2H_5} \begin{array}{c} CN \\ | \\ C{-}R \\ \| \\ C{-}ONa \\ | \\ OEt \end{array} \xrightarrow[-NaI]{+R'I} \begin{array}{c} CN \\ |\;\; R \\ C\langle \\ |\;\; R' \\ C{=}O \\ | \\ OEt \end{array} \xrightarrow{\text{hydrolysis}} \begin{array}{c} COOH \\ |\;\; R \\ C\langle \\ |\;\; R' \\ COOH \end{array}$$

$$\xrightarrow{-CO_2} \begin{array}{c} H \\ |\;\; R \\ C\langle \\ |\;\; R' \\ COOH \end{array}$$

homologue of acetic acid

Ethyl cyanoacetate may thus be, and frequently is, employed in such synthetic work instead of ethyl malonate.

in ways similar to those in which ethylene glycol is obtained from ethylene, thus:

$$CH_3{-}CH{=}CH_2 \xrightarrow[\text{potassium permanganate}]{\text{oxidation with alkaline}} \underset{\text{propylene glycol}}{CH_3{-}CH(OH){-}CH_2OH} \xrightarrow{\text{oxidation}} \underset{\text{lactic acid}}{CH_3{-}CH(OH){-}COOH}$$

$$CH_3{-}CH{=}CH_2 \xrightarrow{+Br_2} \underset{\text{propylene bromide}}{CH_3{-}CHBr{-}CH_2Br} \xrightarrow{\text{hydrolysis}} \text{propylene glycol}$$

Hydracrylic acid may be similarly considered as an oxidation product of the isomeric glycol derived from propane, viz. 1 : 3-dihydroxypropane, thus:

$$\underset{\text{1 : 3-dihydroxypropane}}{CH_2OH{-}CH_2{-}CH_2OH} \xrightarrow{\text{oxidation}} \underset{\text{hydracrylic acid}}{CH_2OH{-}CH_2{-}COOH}$$

The complete direct oxidation product of 1 : 3-dihydroxypropane would be malonic acid.

Lactic acid has been obtained by the following methods:

(i) By the hydrolysis of α-chloro- and α-bromo- propionic acids (produced by the direct halogenation in sunlight of propionic acid):

$$CH_3{-}CH_2{-}COOH + Cl_2(Br_2) \xrightarrow[-HBr]{-HCl} CH_3{-}CH(Cl(Br)){-}COOH \xrightarrow{\text{hydrolysis}} CH_3{-}CH(OH){-}COOH$$

(ii) By the action of nitrous acid on alanine (p. 320):

$$CH_3{-}CH(NH_2){-}COOH + ONOH \rightarrow CH_3{-}CH(OH){-}COOH + N_2 + H_2O$$

(iii) By the hydrolysis of the cyanohydrin of acetaldehyde (lactonitrile) produced by the direct addition of hydrogen cyanide to acetaldehyde:

$$CH_3.CHO + HCN \longrightarrow CH_3.CH(OH).C{\equiv}N$$

acetaldehyde cyanohydrin

$$CH_3.CH(OH).C{\equiv}N + 2H_2O \xrightarrow[\text{of mineral acids}]{\text{boiling in presence}} CH_3.CH(OH).COOH + NH_3$$

This last method of obtaining lactic acid is a direct verification of the constitution of the substance. The acid prepared by the above methods is, like the acid generally obtained in the 'souring' of milk, optically inactive, being externally compensated (p. 346). It should be designated *dl*-lactic acid. On the other hand, the lactic acid obtained from expressed muscle, and which has long been known as *sarcolactic acid*, is *dextro*rotatory and is designated *d*-lactic acid.

Lactic acid is prepared on the technical scale by fermentation of sugar solutions. Its presence in sour milk is due to the decomposition under the agency of enzymes (p. 332) of a sugar, lactose (p. 452), which is isomeric with cane sugar or sucrose (p. 453) and which has the molecular formula $C_{12}H_{22}O_{11}$. The lactose is first hydrolysed to the simpler sugars glucose (p. 422) and galactose (p. 433):

$$\underset{\text{lactose}}{C_{12}H_{22}O_{11}} + H_2O \rightarrow \underset{\text{glucose}}{C_6H_{12}O_6} + \underset{\text{galactose}}{C_6H_{12}O_6}$$

and then these two sugars are converted into lactic acid by a reaction which in its simplest form may be represented:

$$2C_6H_{12}O_6 \rightarrow \underset{\text{lactic acid}}{4C_3H_6O_3}$$

This latter reaction, however, does not take into account the formation of any intermediate or by-products. The conversion of the sugars into lactic acid is induced by enzymes produced by the lactic acid bacillus, *B. acidi lactici*, which is present in decaying cheese, and the reactions proceed at temperatures not far removed from 37°. The 'fermentation' stops if free acid is allowed to accumulate, and to prevent this it is convenient to have an excess of zinc carbonate present so that the acid is neutralised as soon as it is formed, producing the sparingly soluble (in neutral solutions) zinc lactate. The solution remains practically neutral and the reactions proceed until the whole of the sugar disappears. The zinc lactate, $(CH_3.CHOH.CO_2)_2Zn + 3H_2O$, is separated in the usual way, and from it the free acid can be obtained by careful treatment with the calculated quantity of mineral acid.

Fermentation or *dl*-lactic acid, a colourless oil, is readily soluble in water, alcohol and ether, and its separation from an aqueous solution by means of ether requires many extractions. It forms a normal sodium salt, $CH_3.\overset{*}{C}HOH.COONa$, and this when heated with metallic sodium yields the disodium derivative, $CH_3.CH(ONa).COONa$, corresponding to an alcoholate; the calcium salt, $(CH_3.CHOH.CO_2)_2Ca + 5H_2O$, finds extensive use as a therapeutic agent.

When *dl*-lactic acid is treated with hydrobromic acid, *dl*-α-bromopropionic acid is produced by esterification:

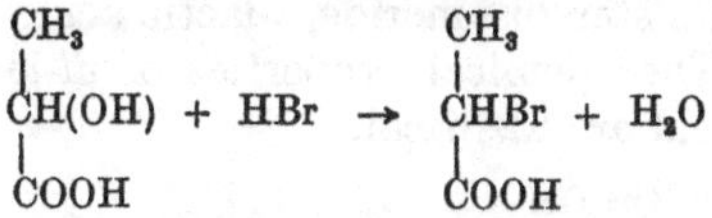

$$CH_3.CH(OH).COOH + HBr \rightarrow CH_3.CHBr.COOH + H_2O$$

Being an α-hydroxy acid, lactic acid behaves similarly to glycollic acid on being heated (distilled at ordinary pressure) and yields *lactide*, its inner-ester:

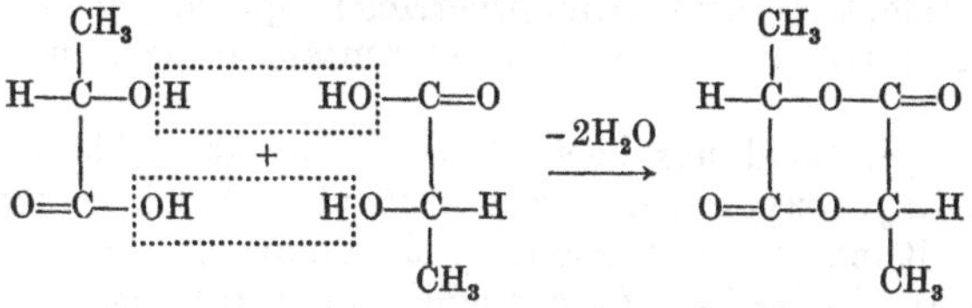

$$\begin{matrix} CH_3 \\ | \\ H{-}C{-}O\,\vdots\, H \qquad HO\,\vdots{-}C{=}O \\ | \qquad + \qquad | \\ O{=}C{-}\vdots\,OH \qquad H\,\vdots\,O{-}C{-}H \\ | \\ CH_3 \end{matrix} \xrightarrow{-2H_2O} \begin{matrix} CH_3 \\ | \\ H{-}C{-}O{-}C{=}O \\ | \qquad | \\ O{=}C{-}O{-}C{-}H \\ | \\ CH_3 \end{matrix}$$

lactide, m.p. 125°

The behaviour of lactic acid when heated with sulphuric acid may be used for its identification. With dilute sulphuric acid, it is decomposed into acetaldehyde and formic acid; with concentrated sulphuric acid the formic acid is further decomposed, carbon monoxide being evolved:

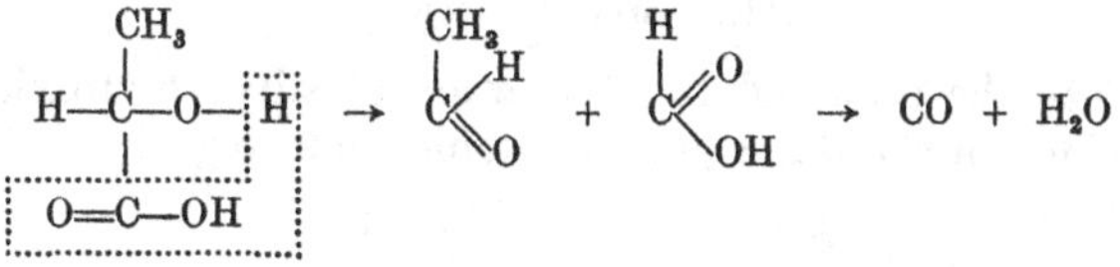

$$\begin{matrix} CH_3 \\ | \\ H{-}C{-}O{-}\vdots H \\ | \\ O{=}C{-}OH \end{matrix} \rightarrow CH_3.CHO + H.COOH \rightarrow CO + H_2O$$

On oxidation with potassium permanganate, lactic acid (a secondary alcohol-acid) is converted into the simplest ketone-acid, *pyruvic acid** (acetylformic acid):

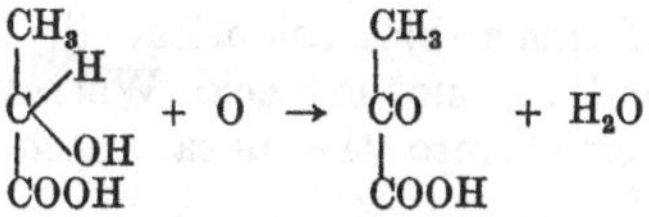

$$CH_3.CH(OH).COOH + O \rightarrow CH_3.CO.COOH + H_2O$$

pyruvic acid

* Pyruvic acid is also known as *pyroracemic acid*, on account of the fact that it may be prepared by distilling tartaric acid and racemic acid with potassium hydrogen sulphate. Pyruvic acid (m.p. 9°, b.p. 165°) may be synthesised starting with acetyl chloride as follows:

$$CH_3.CO.Cl \xrightarrow{KCN} CH_3.CO.CN \xrightarrow[+2H_2O]{\text{hydrolysis}} CH_3.CO.COOH$$

Pyruvic acid yields derivatives characteristic of a ketone and also of a typical acid.

Pyruvic acid and lactic acids may be intermediate products in the fermentation of glucose into ethyl alcohol and carbon dioxide (p. 468).

The stereoisomerism of lactic acid and the resolution of externally compensated or *dl*-lactic acid into its optically active components are described on p. 346. *d*-Lactic acid obtained in the resolution of *dl*-lactic acid is identical in all respects with sarcolactic acid. It differs only from its stereoisomeride, *l*-lactic acid, in the sign of its rotatory power. The chemical properties of *dl*-lactic acid, *d*-lactic acid and *l*-lactic acid are identical.

Hydracrylic acid, $\begin{matrix} CH_2OH \\ | \\ CH_2 \\ | \\ COOH \end{matrix}$, β-hydroxypropionic acid, differs from its isomeride, lactic acid, in not occurring in optically active forms; its molecule is symmetrical, it does not contain an asymmetric carbon atom.

Hydracrylic acid has been shown to be an oxidation product of 1 : 3-dihydroxypropane or trimethyleneglycol. Starting with ethylene, it may be synthesised by the following series of reactions, in which the hydrolysis of a cyanide to an acid is again utilised:

$$\begin{matrix} CH_2 \\ \| \\ CH_2 \end{matrix} \xrightarrow{HOCl} \underset{\text{glycol chlorhydrin or ethylene chlorhydrin}}{\begin{matrix} CH_2OH \\ | \\ CH_2 \\ | \\ Cl \end{matrix}} \xrightarrow{KCN} \underset{\text{hydroxyethyl cyanide}}{\begin{matrix} CH_2OH \\ | \\ CH_2 \\ | \\ CN \end{matrix}} \xrightarrow{\text{hydrolysis}} \underset{\text{hydracrylic acid}}{\begin{matrix} CH_2OH \\ | \\ CH_2 \\ | \\ COOH \end{matrix}}$$

It may also be obtained by the action of silver hydroxide (moist silver oxide) on the β-halogenopropionic acids, e.g.:

$$\begin{matrix} CH_2Cl \\ | \\ CH_2 \\ | \\ COOH \end{matrix} + AgOH \rightarrow \begin{matrix} CH_2OH \\ | \\ CH_2 \\ | \\ COOH \end{matrix} + AgCl$$

Hydracrylic acid forms a thick colourless oil. It is simultaneously a primary alcohol and a monobasic acid. When it is heated, it loses water and is converted into the unsaturated acid, *acrylic acid* (p. 227):

$$\begin{matrix} (\beta)\,CH_2OH \\ | \\ (\alpha)\,CH_2 \\ | \\ COOH \end{matrix} \xrightarrow{-H_2O} \begin{matrix} CH_2 \\ \| \\ CH \\ | \\ COOH \end{matrix}$$

This behaviour on heating is characteristic of β-hydroxycarboxylic acids. Under similar conditions, the isomeric α-hydroxypropionic acid, lactic acid, is converted into the 'inner-ester' or lactide; consequently, the anhydrides of both these acids are unknown. Acrylic acid is the lowest member of the series of unsaturated acids having the general formula, $C_nH_{2n-1}.COOH$, of which oleic acid is a member.

A homologue of lactic acid and of hydracrylic acid is *β-hydroxy-*n-*butyric acid*:

$$\begin{array}{l} CH_3 \\ | \\ C\!\diagup^{H}_{\diagdown OH} \\ | \\ CH_2 \\ | \\ COOH \end{array}$$

the molecule of which, like that of lactic acid, is asymmetrical—the β-carbon atom being united to four different atoms or groups. The compound is therefore capable of existing in optically active forms —*dextro-* and *laevo-* rotatory—and in the optically inactive (*dl-*) or externally compensated form. The externally compensated acid has been synthesised and resolved into the *dextro-* and *laevo-* rotatory isomerides by the fractional crystallisation of the quinine salts. As a β-hydroxy acid, it behaves like hydracrylic acid on being heated, undergoing loss of water as indicated: the product of the dehydration being the unsaturated β-methylacrylic acid, which, as an ethylene derivative of the type $\begin{array}{c} CHX \\ \| \\ CHY \end{array}$, is capable of existing in stereoisomeric (*cis-* and *trans-*) forms, thus:

$$\begin{array}{ccc} \begin{array}{c} CH_3 \\ | \\ \boxed{HO}\!-\!C\!-\!H \\ | \\ \boxed{H}\!-\!C\!-\!H \\ | \\ COOH \end{array} & \xrightarrow{-H_2O} & \begin{array}{c} CH_3\!-\!C\!-\!H \\ \| \\ HOOC\!-\!C\!-\!H \\ \textit{cis-}\text{crotonic acid} \end{array} \\ \begin{array}{c} CH_3 \\ | \\ \boxed{HO}\!-\!C\!-\!H \\ | \\ \boxed{H}\!-\!C\!-\!COOH \\ | \\ H \end{array} & \xrightarrow{-H_2O} & \begin{array}{c} CH_3\!-\!C\!-\!H \\ \| \\ H\!-\!C\!-\!COOH \\ \textit{trans-}\text{crotonic acid} \end{array} \end{array} \Bigg\} \begin{array}{c} \beta\text{-methylacrylic acids} \\ \text{(p. 228)} \end{array}$$

l-β-Hydroxybutyric acid frequently occurs along with acetoacetic acid (*v.* below) and acetone in the urine of patients suffering from diabetes mellitus.

The direct oxidation product of the externally compensated or optically active β-hydroxybutyric acids will be a homologue of pyruvic acid:

$$\begin{array}{c} CH_3 \\ | \\ C(H)(OH) \\ | \\ CH_2 \\ | \\ COOH \end{array} \xrightarrow{\text{oxidation}} \begin{array}{c} CH_3 \\ | \\ CO \\ | \\ CH_2 \\ | \\ COOH \end{array}$$

acetoacetic acid

This may be described as *β-keto*-n-*butyric acid* or *acetylacetic acid*: it is more usually termed *acetoacetic acid*.

Free acetoacetic acid or its sodium salt, which may be obtained from the ethyl ester, ethyl acetoacetate (p. 369), by hydrolysis with cold dilute sodium hydroxide, is not very important chemically. From a biochemical standpoint, it is of great importance. It occurs in 'diabetic' urine, together with acetone and, frequently, *l*-β-hydroxybutyric acid; the acetone is formed from acetoacetic acid or its sodium salt by decomposition:

$$CH_3.CO.CH_2.COOH \rightarrow CH_3.CO.CH_3 + CO_2$$

$$CH_3.CO.CH_2.COONa + H_2O \rightarrow CH_3.CO.CH_3 + NaHCO_3$$

l-β-Hydroxybutyric acid, acetoacetic acid and acetone are usually described collectively as the 'acetone bodies', and their production in the animal organism is explained in the following way.

During digestion, fats are hydrolysed in the intestine to glycerol and fatty acids and the latter are normally oxidised to carbon dioxide and water. This oxidation takes place in stages which may be represented thus:

$$\underset{\text{stearic acid}}{C_{15}H_{31}.CH_2.CH_2.COOH} + O_2 \rightarrow \underset{\beta\text{-ketostearic acid}}{C_{15}H_{31}.CO.CH_2\ COOH} + H_2O$$

$$C_{15}H_{31}.CO.CH_2.COOH + 2O_2 \rightarrow \underset{\text{palmitic acid}}{C_{15}H_{31}.COOH} + 2CO_2 + H_2O$$

$$\underset{\text{palmitic acid}}{C_{13}H_{27}.CH_2.CH_2.COOH} + O_2 \rightarrow \underset{\beta\text{-ketopalmitic acid}}{C_{13}H_{27}.CO.CH_2.COOH} + H_2O$$

$$C_{13}H_{27}.CO.CH_2.COOH + 2O_2 \rightarrow C_{13}H_{27}.COOH + 2CO_2 + H_2O$$

n-myristic acid
m.p. 53·8°, b.p. 220·5°

and so on to the final stages:

(*a*) $\underset{n\text{-butyric acid}}{CH_3.CH_2.CH_2.COOH} + O_2 \rightarrow CH_3.CO.CH_2.COOH + H_2O$

β-ketobutyric acid,
i.e. acetoacetic acid

and (*b*) $CH_3.CO.CH_2.COOH + 4O_2 \rightarrow 4CO_2 + 3H_2O$

In the case of the diabetic, the oxidation stops at reaching stage (*a*) and the acetoacetic acid may be excreted or partially reduced to *l*-β-hydroxybutyric acid or partially decomposed to acetone:

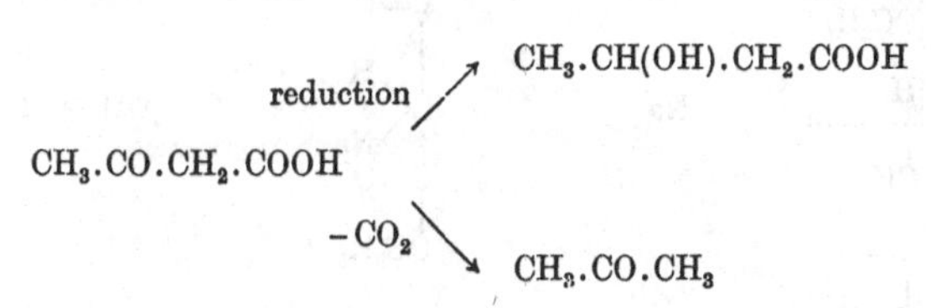

Being closely related to each other constitutionally, acetone and acetoacetic acid have many reactions in common. They both yield iodoform under appropriate and similar conditions, and dilute solutions of each when treated with a freshly prepared dilute aqueous solution of sodium nitroprusside and then made alkaline, by addition of a little sodium hydroxide, yield a deep red colour. Unlike acetone, however, a dilute solution of acetoacetic acid when treated with a dilute solution of ferric chloride yields a characteristic claret-red colour. This reaction will detect 1 part of acetoacetic acid in 100,000 parts of the solution. A dilute solution of acetoacetic acid (or of its sodium salt) may be obtained by mixing 13 grams of the ethyl ester (*v.* below) with 5 litres 0·02*N* sodium hydroxide and allowing the mixture to stand for 48 hours at the ordinary temperature; such a solution will be found convenient for investigating the reactions of acetoacetic acid and comparing them with those which take place with 'diabetic' urine under similar conditions.

From a chemical point of view, the ethyl ester, *ethyl acetoacetate*, is the most important derivative of acetoacetic acid. It was discovered by Geuther (1863), and is prepared by the action of finely divided sodium on ethyl acetate containing a little ethyl alcohol. Hydrogen is evolved, the sodium dissolves and a pasty colourless precipitate—a mixture of sodium ethylate and the sodium derivative of ethyl acetoacetate—separates. When the mixture is acidified with acetic acid, an oil separates which after isolation is distilled under reduced pressure. The product, ethyl acetoacetate, has b.p. 181°. It is a colourless liquid and has a small solubility in water; it is readily soluble in the usual organic solvents.

If the constitutional formula of acetoacetic acid is, as indicated above, $CH_3.CO.CH_2.COOH$, the constitutional formula of ethyl acetoacetate would be $CH_3.CO.CH_2.COOC_2H_5$; since the compound contains the grouping —CO—CH_2—CO—, the formation of a sodium derivative is readily understood (compare p. 359). The production of ethyl acetoacetate as described above might be explained:

$$\begin{array}{c} CH_3 \\ | \\ C{=}O \\ | \\ \boxed{\begin{array}{l} OC_2H_5 \\ H \end{array}} \\ | \\ CH_2 \\ | \\ C{=}O \\ | \\ OC_2H_5 \end{array} + Na \rightarrow \begin{array}{c} CH_3 \\ | \\ C{=}O \\ | \\ C\begin{array}{l} \diagup H \\ \diagdown Na \end{array} \\ | \\ C{=}O \\ | \\ OC_2H_5 \end{array} + C_2H_5OH + H$$

$$\begin{array}{c} CH_3 \\ | \\ C{=}O \\ | \\ C\begin{array}{l} \diagup H \\ \diagdown Na \end{array} \\ | \\ COOC_2H_5 \end{array} + CH_3.COOH \rightarrow \begin{array}{c} CH_3 \\ | \\ C{=}O \\ | \\ CH_2 \\ | \\ COOC_2H_5 \end{array} + CH_3.COONa$$

ethyl acetoacetate
keto-formula

This explanation does not take into account the necessity for the presence of ethyl alcohol before the reaction proceeds. This difficulty was surmounted by explaining the course of the reactions, thus:

$$CH_3{-}C\begin{array}{l} \diagup\!\!\diagup O \\ \diagdown OC_2H_5 \end{array} + NaOC_2H_5 \rightarrow CH_3{-}\overset{\displaystyle ONa}{\overset{\diagup}{C}}{-}OC_2H_5 \; \begin{array}{l} \\ \diagdown OC_2H_5 \end{array}$$

intermediate compound

$$CH_3{-}\overset{\displaystyle ONa}{\overset{\diagup}{C}}\boxed{\begin{array}{l} {-}OC_2H_5 \quad + \quad H{-} \\ \diagdown OC_2H_5 \qquad\quad H \diagup \end{array}}\overset{\displaystyle H}{\overset{\diagdown}{C}}.C\begin{array}{l} \diagup\!\!\diagup O \\ \diagdown OC_2H_5 \end{array}$$

$$\rightarrow CH_3{-}C(ONa){=}CH.COOC_2H_5 + 2C_2H_5OH$$

$$\begin{array}{c} CH_3 \\ | \\ C{-}ONa \\ \| \\ CH \\ | \\ COOC_2H_5 \end{array} + CH_3.COOH \rightarrow \begin{array}{c} CH_3 \\ | \\ C{-}OH \\ \| \\ CH \\ | \\ COOC_2H_5 \end{array} + CH_3.COONa$$

ethyl acetoacetate
enol-formula

These two modes of explanation of the formation of ethyl acetoacetate give rise to two different formulae of the substance, known as the keto- and enol- formula, respectively (from ketone and unsaturated alcohol).

Actually, it has long been known that ethyl acetoacetate reacts as though it possessed both constitutions. It reacts as a ketone with hydroxylamine and with phenylhydrazine.* It reacts as an alcohol in yielding an acetyl derivative, in giving a violet colour with ferric chloride (compare acetoacetic acid) and by decolorising a solution of bromine, the last reaction indicating an unsaturated compound necessitated by the enolic formula. For some time it was disputed as to what was the actual constitution of ethyl acetoacetate, but the two views of the constitution were reconciled by assuming that the compound could react either as a ketone or an unsaturated alcohol according to the conditions, and that ethyl acetoacetate constituted a typical example of tautomerism. Later, it was shown that although pure ethyl acetoacetate as ordinarily prepared behaves as a homo geneous substance with a definite boiling point, it is actually a mixture of the two forms in equilibrium and that ordinary ethyl acetoacetate should be represented thus:

* The oxime readily loses ethyl alcohol and is transformed into a *hetero*cyclic compound, 3-*methyl isoxazolone* (m.p. 169–170°), thus:

```
                 HO                                  O
                   \                               /  1 \
C2H5O.OC            N                          OC 5      2 N
       |            ||    →   C2H5OH   +         |          ||
       |            ||                           | 4      3 ||
    H2C ———————— C.CH3                        H2C ———————— C.CH3
```

The phenylhydrazone, similarly, readily loses ethyl alcohol and is converted into the *hetero*cyclic compound, 1-*phenyl*-3-*methyl*-5-*pyrazolone*, m.p. 127° (the first pyrazole derivative to be described):

```
                                                    C6H5
                                                     |
              H.N.C6H5                               N
                   \                               /  1 \
C2H5O.OC            N                          OC 5      2 N
       |            ||    →   C2H5OH   +         |          ||
       |            ||                           | 4      3 ||
    H2C ———————— C.CH3                        H2C ———————— C.CH3
```

This pyrazolone on being heated with methyl iodide and methyl alcohol under pressure is converted into the hydriodide of *antipyrine*, from which the base itself is obtained by treatment with alkali. Antipyrine, m.p. 113°, is a well-known febrifuge. Its constitution is:

```
      C6H5
       |
       N
     /  1 \
  OC 5     2 N.CH3        i.e. 1-phenyl-2 : 3-dimethyl-5-pyrazolone
   |         |
   | 4     3 |
  HC ======= C.CH3
```

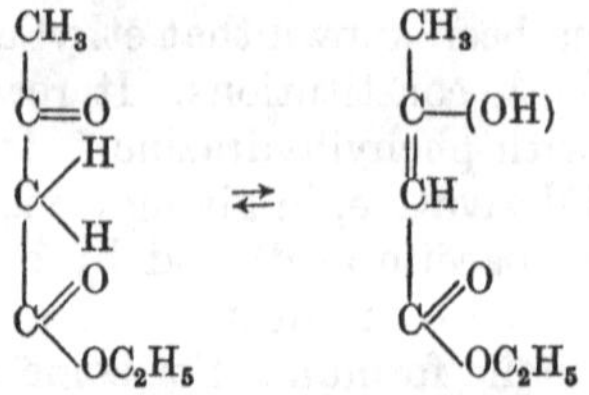

$$
\begin{array}{ccc}
CH_3 & & CH_3 \\
| & & | \\
C{=}O & & C{-}(OH) \\
| & & \| \\
C{<}^{H}_{H} & \rightleftarrows & CH \\
| & & | \\
C{\overset{O}{\ll}}_{OC_2H_5} & & C{\overset{O}{\ll}}_{OC_2H_5}
\end{array}
$$

By cooling a solution of pure ethyl acetoacetate in an indifferent solvent, such as ether or ligroin, to −78°, Knorr (1911) obtained colourless crystals of the ester which, if kept at or below that temperature, could be separated and preserved without change. These crystals do not give a violet coloration with ferric chloride or combine with bromine and are the crystals of the pure keto (or, as it is sometimes described, the 'β-form') of ethyl acetoacetate,

$$CH_3.CO.CH_2.COOC_2H_5$$

The mother liquor from which the crystals are separated continues to show unsaturated properties and to give a violet colour with ferric chloride at temperatures below −78°. By treating the sodium derivative of ethyl acetoacetate with hydrogen chloride, Knorr obtained the enolic form. He separated it in a pure state by distillation in high vacuum in quartz (not glass) vessels. The enol- ('α-') form has a higher refractive index than the keto- ('β-') form, but it has not been obtained in the crystalline condition. The enol- or 'α-' form should theoretically exist in *cis-* and *trans-* stereoisomeric forms:

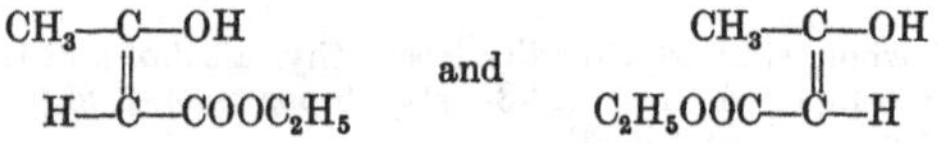

$$
\begin{array}{c}
CH_3{-}C{-}OH \\
\| \\
H{-}C{-}COOC_2H_5
\end{array}
\quad \text{and} \quad
\begin{array}{c}
CH_3{-}C{-}OH \\
\| \\
C_2H_5OOC{-}C{-}H
\end{array}
$$

and, provided it is kept below −78° and uncontaminated, it remains as the enol- or 'α-' form and reacts accordingly.

As the temperature of the pure keto- or 'β-' form is allowed to rise from −78° to the ordinary temperature, it soon yields a violet colour with ferric chloride, exhibits the properties of an unsaturated compound and slowly acquires the properties of pure ethyl acetoacetate as ordinarily prepared. Similarly, the enol- or 'α-' form when allowed to reach the ordinary temperature assumes the physical and chemical properties of ordinary pure ethyl acetoacetate; this must, therefore, be an equilibrium mixture of the two forms which are interconvertible, thus:

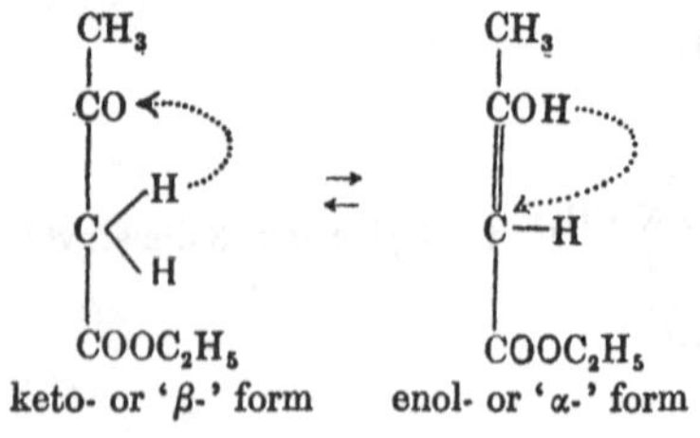

$$
\begin{array}{ccc}
CH_3 & & CH_3 \\
| & & | \\
CO & & COH \\
| & & \| \\
C{<}^{H}_{H} & \rightleftarrows & C{-}H \\
| & & | \\
COOC_2H_5 & & COOC_2H_5
\end{array}
$$

keto- or 'β-' form enol- or 'α-' form

by the wandering of a hydrogen atom as indicated. To describe this type of equilibrium between isomeric substances which are interconvertible, Lowry (1899) introduced the convenient term *dynamic isomerism*. A substance exhibiting tautomerism is capable of reacting in two ways according to conditions of reaction, whereas dynamic isomerism implies the possibility that the two tautomeric forms of the substance are actually capable of being isolated.

By a careful examination of the physical properties of pure ethyl acetoacetate as ordinarily prepared, it has been shown that the substance is an equilibrium mixture containing about 93 per cent. of the keto- or 'β-' form and about 7 per cent. of the enol- or 'α-' form.

The nature of ordinary ethyl acetoacetate can readily be demonstrated. A saturated aqueous solution of the ester is prepared and a small quantity of this, diluted considerably with water, is placed in a suitable large glass vessel.* On addition of a small quantity of a freshly prepared aqueous solution of ferric chloride, a violet colour is developed owing to the presence of the enolic form. On treatment with an aqueous solution of bromine, this violet colour is discharged, but develops again on standing. On addition of more bromine water, the violet colour is again discharged and again reappears after some little time and can be discharged by addition of more aqueous solution of bromine. The phenomena can be repeated until finally the violet colour does not reappear. Bearing in mind that the enolic form is an unsaturated compound and will combine directly with bromine, these changes are explained as indicated:

$$CH_3{-}CO{-}CH_2{-}COOC_2H_5 \rightleftarrows CH_3{-}C(OH){=}CH{-}COOC_2H_5 \xrightarrow{+Br_2} CH_3{-}C(OH)(Br){-}CH(Br){-}COOC_2H_5 \xrightarrow{-HBr}$$

colour with $FeCl_3$ (enol form)

93 % (keto form) + 7 % (enol form) = ordinary ethyl acetoacetate

$$CH_3{-}CO{-}CH(Br){-}COOC_2H_5 \rightleftarrows CH_3{-}C(OH){=}C(Br){-}COOC_2H_5 \xrightarrow{+Br_2} CH_3{-}C(OH)(Br){-}CBr_2{-}COOC_2H_5 \xrightarrow{-HBr} CH_3{-}CO{-}CBr_2{-}COOC_2H_5$$

$CH_3COCH(Br)COOC_2H_5$: no colour with $FeCl_3$; $CH_3C(OH){=}C(Br)COOC_2H_5$: colour with $FeCl_3$; $CH_3COCBr_2COOC_2H_5$: no colour with $FeCl_3$, ethyl $\alpha\alpha$-dibromoacetoacetate

* The use of a dilute aqueous solution makes it unnecessary to use large quantities of the saturated aqueous solution of bromine: otherwise the bulk of the solution becomes inconveniently large.

It will be noticed that, under the conditions of the reaction, the addition of bromine only affects the enolic form; as soon as this form is removed as indicated more ketonic form is changed into the enolic form until, finally, the whole is converted into *ethyl αα-dibromoaceto-acetate* or *ethyl αα-dibromo-β-keto*-n-*butyrate*, which cannot change into an enolic form and which, therefore, does not give a violet colour with ferric chloride. The solution becomes opalescent owing to the low solubility of the liquid dibromo- ester.

Ethyl acetoacetate forms other metallic derivatives: e.g. the copper derivative is obtained by shaking an alcoholic solution of the ester with a solution of cupric acetate. If the constitution of the sodium derivative is as indicated above, to the copper derivative might be assigned the constitution:

```
CH3               CH3
 |                 |
 C—O—Cu—O—C   *
 ||                ||
 C—H               C—H
 |                 |
COOC2H5          COOC2H5
```

* It has been shown, particularly by the stereochemical investigations of W. H. Mills and others, that the constitution of the copper derivative should be represented thus:

```
      OC2H5              CH3
       |                  |
       C══O       O──C
      /     ↘   /     \\
   HC         Cu         CH
     \\      /  ↖       /
       C──O       O══C
       |                  |
      CH3               OC2H5
```

in which by the 'co-ordination' of the oxygen atom of the >C=O of the carbethoxy group to the copper atom a 'chelate' structure is built up. This conception of the structure of this typical metallic derivative of ethyl acetoacetate is not only in keeping with the results of stereochemical investigations but is also in keeping with the electronic theory of valency. The constitution of the copper derivative of ethyl acetoacetate also corresponds with that of the inorganic compound, cupric diammino chloride, which is represented thus:

```
   Cl      Cl
     \    /
      Cu
    ↗    ↖
 H3N      NH3
```

This compound is a non-electrolyte in which two ammonia molecules are co-ordinated to the copper atom. The "chelate" structure of ethyl acetoacetate and of its sodium derivative are similarly represented:

```
      OC2H5                  OC2H5
       |                      |
       C══O                   C══O
      /     ↘                /     ↘
   HC         H           HC         Na
     \\      /              \\      /
       C──O                   C──O
       |                      |
      CH3                    CH3
```

When ethyl acetoacetate is heated with an alcoholic solution of sodium hydroxide or a concentrated aqueous solution of sodium hydroxide, it yields acetic acid (as sodium acetate):

$$\begin{array}{c}CH_3\\|\\CO\\ \cdots|\cdots\\CH_2\\|\\COOC_2H_5\end{array} \rightleftarrows \begin{array}{c}CH_3\\|\\C{-}OH\\ \cdots\|\cdots\\CH\\|\\COOC_2H_5\end{array} + \begin{array}{c}HONa\\ \\HONa\end{array} \rightarrow 2\begin{array}{c}CH_3\\|\\COONa\end{array} + C_2H_5OH$$

When it is heated with a dilute aqueous solution of sodium hydroxide, ethyl acetoacetate yields acetone:

$$\begin{array}{c}CH_3\\|\\CO\\|\\CH_2\\ \cdots|\cdots\\COOC_2H_5\end{array} \rightleftarrows \begin{array}{c}CH_3\\|\\C{-}OH\\\|\\CH\\ \cdots|\cdots\\COOC_2H_5\end{array} + H_2O + NaOH \rightarrow \begin{array}{c}CH_3\\|\\CO\\|\\CH_3\end{array} + C_2H_5OH + NaHCO_3$$

These two modes of decomposition of ethyl acetoacetate are known as 'acid hydrolysis' and 'ketonic hydrolysis' respectively.

While the sodium derivative of ethyl acetoacetate must be derived from the enolic form, the ketonic form contains, like ethyl malonate, the grouping —CO—CH_2—CO—, and, like ethyl malonate (and ethyl cyanoacetate), ethyl acetoacetate is an important synthetic reagent for the preparation of substituted acetone and substituted acetic acid derivatives. In the simplest and typical cases, to the ethyl acetoacetate mixed with an excess of ethyl alcohol is added an equivalent amount of sodium dissolved in ethyl alcohol; an alkyl halide (RX) or other suitable compound is then added, the reaction being completed by heating. The reactions taking place may be represented:

$$\begin{array}{c}CH_3\\|\\C{=}O\\|\\C\langle^{H}_{H}\\|\\COOC_2H_5\end{array} \rightleftarrows \begin{array}{c}CH_3\\|\\C{-}OH\\\|\\C{-}H\\|\\COOC_2H_5\end{array} \xrightarrow[-C_2H_5OH]{+NaOC_2H_5} \begin{array}{c}CH_3\\|\\C{-}ONa\\\|\\C{-}H\\|\\COOC_2H_5\end{array} \xrightarrow[-NaX]{+RX}$$

$$\begin{array}{c}CH_3\\|\\C{=}O\\|\\C\langle^{H}_{R}\\|\\COOC_2H_5\end{array} \rightleftarrows \begin{array}{c}CH_3\\|\\C{-}OH\\\|\\C{-}R\\|\\COOC_2H_5\end{array}$$

The product is a monosubstituted derivative of ethyl acetoacetate which is isolated after acidification with acetic acid. By ketonic hydrolysis a monosubstituted derivative of acetone may be obtained:

$$CH_3-CO-CH(R)-COOC_2H_5 \rightleftarrows CH_3-C(OH)=C(R)-COOC_2H_5 + H_2O \rightarrow CH_3-C(=O)-CH_2R + CO_2 + C_2H_5OH$$

By acid hydrolysis a monosubstituted derivative of acetic acid is produced:

$$CH_3-C(=O)-CH(R)-COOC_2H_5 \rightleftarrows CH_3-C(OH)=C(R)-COOC_2H_5 + H_2O \rightarrow CH_3-COOH + CH_2R-COOH + C_2H_5OH$$

To obtain disubstituted acetones and disubstituted acetic acids, the monoalkyl substituted ethyl acetoacetate is treated in a similar manner with alcohol, sodium and a suitable halide ($R'X$) as indicated·

$$CH_3-C(=O)-CH(R)-COOC_2H_5 \rightleftarrows CH_3-C(OH)=C(R)-COOC_2H_5 \xrightarrow[-C_2H_5OH]{+NaOC_2H_5} CH_3-C(ONa)=C(R)-COOC_2H_5 \xrightarrow[-NaX]{+R'X} CH_3-C(=O)-C(R)(R')-COOC_2H_5$$

The product, a disubstituted derivative of ethyl acetoacetate, on ketonic hydrolysis yields a disubstituted acetone:

$$CH_3-CO-C(R)(R')-COOC_2H_5 + H_2O \rightarrow CH_3-CO-CH(R)(R') + CO_2 + C_2H_5OH$$

and, by acid hydrolysis, a disubstituted acetic acid:

$$\begin{array}{l} CH_3 \\ | \\ CO \\ \cdots|\cdots\cdots \\ | \\ C\begin{smallmatrix} R \\ R' \end{smallmatrix} \\ | \\ COOC_2H_5 \end{array} + 2H_2O \rightarrow \begin{array}{l} CH_3 \\ | \\ COOH \end{array} + \begin{array}{l} C\begin{smallmatrix} H \\ R \\ R' \end{smallmatrix} \\ | \\ COOH \end{array} + C_2H_5OH$$

Higher acids of the oxalic acid series may be conveniently synthesised, starting from ethyl acetoacetate; for example, the synthesis of succinic acid is outlined:

$$\begin{array}{l} CH_3 \\ | \\ C{=}O \\ | \\ C\begin{smallmatrix} H \\ H \end{smallmatrix} \\ | \\ COOC_2H_5 \end{array} \rightleftarrows \begin{array}{l} CH_3 \\ | \\ C{-}OH \\ \| \\ CH \\ | \\ COOC_2H_5 \end{array} \xrightarrow[-C_2H_5OH]{+NaOC_2H_5} \begin{array}{l} CH_3 \\ | \\ C{-}ONa \\ \| \\ CH \\ | \\ COOC_2H_5 \end{array} \xrightarrow[-NaCl]{\substack{+ClCH_2.COOC_2H_5 \\ \text{ethyl monochloroacetate}}}$$

$$\begin{array}{l} CH_3 \\ | \\ C{=}O \\ | \\ C\begin{smallmatrix} H \\ CH_2.COOC_2H_5 \end{smallmatrix} \\ | \\ COOC_2H_5 \end{array} \rightleftarrows \begin{array}{l} CH_3 \\ | \\ C{-}OH \\ \| \\ C.CH_2.COOC_2H_5 \\ | \\ COOC_2H_5 \end{array} \xrightarrow[+2H_2O]{\text{acid hydrolysis}}$$

$$\begin{array}{l} CH_3 \\ | \\ COOH \end{array} + \begin{array}{l} COOH \\ | \\ CH_2 \\ | \\ CH_2 \\ | \\ COOH \end{array} + C_2H_5OH$$

succinic acid

By using ethyl β-chloropropionate, $Cl.CH_2.CH_2.COOC_2H_5$, in a similar series of reactions, the next homologue to succinic acid, viz. glutaric acid, $HOOC.CH_2.CH_2.CH_2.COOH$, is obtained (compare the analogous reactions of ethyl malonate and of ethyl cyanoacetate).

The monosubstituted derivatives of ethyl acetoacetate have properties analogous to those of ethyl acetoacetate itself, while the disubstituted derivatives react as typical ketones.

Succinic acid. The product obtained by distilling amber (Latin: *succinum*) has been known since the middle of the sixteenth century. It was identified by Lémery in 1675 as an acid and since that time has been known as succinic acid.

This homologue of oxalic acid can be considered to be a direct oxidation product of the glycol, 1 : 4-dihydroxybutane:

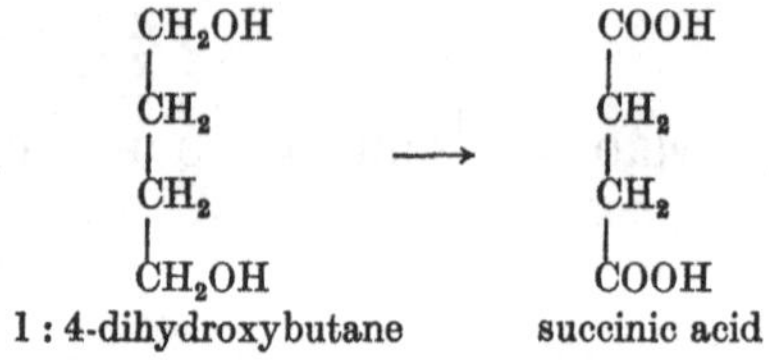

$$\begin{array}{c}CH_2OH\\|\\CH_2\\|\\CH_2\\|\\CH_2OH\end{array}\longrightarrow\begin{array}{c}COOH\\|\\CH_2\\|\\CH_2\\|\\COOH\end{array}$$

1 : 4-dihydroxybutane   succinic acid

It has been synthesised by the following series of reactions, which affords proof of the constitution assigned to the compound:

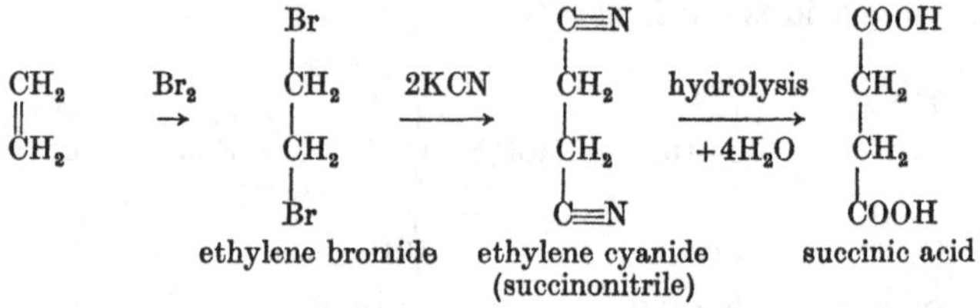

$$\begin{array}{c}CH_2\\\|\\CH_2\end{array}\xrightarrow{Br_2}\begin{array}{c}Br\\|\\CH_2\\|\\CH_2\\|\\Br\end{array}\xrightarrow{2KCN}\begin{array}{c}C\equiv N\\|\\CH_2\\|\\CH_2\\|\\C\equiv N\end{array}\xrightarrow[+4H_2O]{\text{hydrolysis}}\begin{array}{c}COOH\\|\\CH_2\\|\\CH_2\\|\\COOH\end{array}$$

ethylene bromide   ethylene cyanide (succinonitrile)   succinic acid

The synthesis of succinic acid starting from ethyl acetoacetate has been outlined. Starting from ethyl malonate (p. 359), the acid may be synthesised by a series of reactions, briefly indicated:

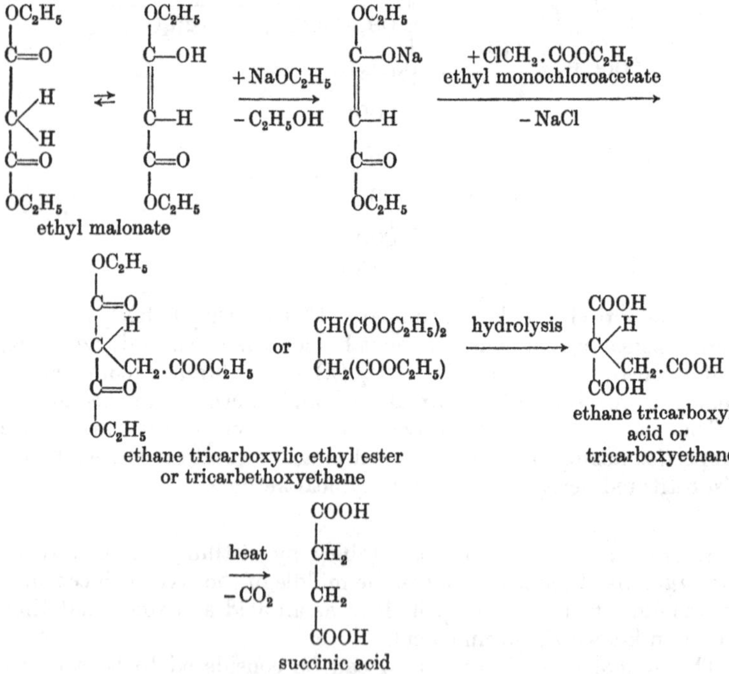

$$\begin{array}{l}OC_2H_5\\|\\C{=}O\\|\\C{<}^{H}_{H}\\|\\C{=}O\\|\\OC_2H_5\end{array}\rightleftarrows\begin{array}{l}OC_2H_5\\|\\C{-}OH\\\|\\C{-}H\\|\\C{=}O\\|\\OC_2H_5\end{array}\xrightarrow[-C_2H_5OH]{+NaOC_2H_5}\begin{array}{l}OC_2H_5\\|\\C{-}ONa\\\|\\C{-}H\\|\\C{=}O\\|\\OC_2H_5\end{array}\xrightarrow[-NaCl]{+ClCH_2.COOC_2H_5 \text{ ethyl monochloroacetate}}$$

ethyl malonate

$$\begin{array}{l}OC_2H_5\\|\\C{=}O\\|\\C{<}^{H}_{CH_2.COOC_2H_5}\\|\\C{=}O\\|\\OC_2H_5\end{array}\text{ or }\begin{array}{l}CH(COOC_2H_5)_2\\|\\CH_2(COOC_2H_5)\end{array}\xrightarrow{\text{hydrolysis}}\begin{array}{l}COOH\\|\\C{<}^{H}_{CH_2.COOH}\\|\\COOH\end{array}$$

ethane tricarboxylic ethyl ester or tricarbethoxyethane   ethane tricarboxylic acid or tricarboxyethane

$$\xrightarrow[-CO_2]{\text{heat}}\begin{array}{c}COOH\\|\\CH_2\\|\\CH_2\\|\\COOH\end{array}$$

succinic acid

Succinic acid crystallises from hot water in anhydrous colourless prisms; it is a convenient standard acid in volumetric analysis. It has m.p. 185° and, at its boiling point (235°), it loses water and is converted into *succinic anhydride*:*

$$\begin{matrix} H_2C-C(=O)-O-H \\ | \\ H_2C-C(=O)-O-H \end{matrix} \xrightarrow{-H_2O} \begin{matrix} H_2C-C{=}O \\ | \quad\quad \rangle O \\ H_2C-C{=}O \end{matrix}$$

succinic anhydride
m.p. 119·6°, b.p. 261°

Succinic acid yields well-defined salts and the ammonium salt on being heated loses water and ammonia and is converted into *succinimide*, succinamide being an intermediate product, thus:

$$\begin{matrix} H_2C-C(=O)-ONH_4 \\ | \\ H_2C-C(=O)-ONH_4 \end{matrix} \xrightarrow{-2H_2O} \begin{matrix} H_2C-C(=O)-NH_2 \\ | \\ H_2C-C(=O)-NH_2 \end{matrix} \xrightarrow{-NH_3} \begin{matrix} H_2C-C{=}O \\ | \quad\quad \rangle NH \\ H_2C-C{=}O \end{matrix}$$

succinamide   succinimide

Succinimide is a colourless crystalline compound, m.p. 126°, b.p. 288°.† Like other compounds containing the —CO—NH—CO—

* *Furan* (p. 355), b.p. 31·5° at 756 mm., may be considered to be a direct reduction product of succinic anhydride, thus:

$$\begin{matrix} H_2C-C{=}O \\ | \quad\quad \rangle O \\ H_2C-C{=}O \end{matrix} \text{ or } \begin{matrix} HC{=}C(OH) \\ | \quad\quad \rangle O \\ HC{=}C(OH) \end{matrix} \xrightarrow[+2H_2-2H_2O]{\text{reduction}} \begin{matrix} HC{=}CH \\ | \quad\quad \rangle O \\ HC{=}CH \end{matrix} \text{ or } \begin{matrix} O \\ HC \quad\quad CH \\ \| \quad\quad\quad \| \\ HC-CH \end{matrix}$$

alternative formula of succinic anhydride   furan

Furan is a typical *hetero*cyclic compound, i.e. a compound having a ring structure and at least one of the atoms forming the ring being dissimilar from the others.

† Like succinic anhydride, succinimide may be considered the 'parent' substance of certain *hetero*cyclic compounds; in this case one of the 'ring' atoms is nitrogen. *Pyrrole* (p. 293) is obtained by reduction (distillation with zinc dust) and *pyrrolidone* (m.p. 24·6°, b.p. 245°) and *pyrrolidine* (p. 293) are also reduction products of succinimide, thus:

$$\begin{matrix} CH_2-CO \\ | \quad\quad \rangle NH \\ CH_2-CO \end{matrix} \text{ or } \begin{matrix} HC{=}C(OH) \\ | \quad\quad \rangle NH \\ HC{=}C(OH) \end{matrix} \xrightarrow[+2H_2-2H_2O]{\text{distillation with zinc dust}} \begin{matrix} HC{=}CH \\ | \quad\quad \rangle NH \\ HC{=}CH \end{matrix}$$

succinimide   alternative formula   pyrrole

alternative formula $\downarrow +2H_2 \; -H_2O$ pyrrole $\downarrow +2H_2$

$$\begin{matrix} HC{=}C(OH) \\ | \quad\quad \rangle NH \\ H_2C-CH_2 \end{matrix} \text{ or } \begin{matrix} H_2C-CO \\ | \quad\quad \rangle NH \\ H_2C-CH_2 \end{matrix} \xrightarrow{+2H_2-H_2O} \begin{matrix} H_2C-CH_2 \\ | \quad\quad \rangle NH \\ H_2C-CH_2 \end{matrix}$$

pyrrolidone   pyrrolidine

grouping, succinimide has acidic properties, in that it yields metallic derivatives by the replacement of the hydrogen atom of the $>$NH group by such metals as sodium, potassium, silver, etc.

Succinic acid undergoes direct halogenation with chlorine and bromine, yielding monohalogenated and dihalogenated succinic acids:

$$\begin{array}{l} COOH \\ | \\ CH_2 \\ | \\ CH_2 \\ | \\ COOH \end{array} \xrightarrow{+X_2-HX} \begin{array}{l} COOH \\ | \\ CHX \\ | \\ CH_2 \\ | \\ COOH \end{array} \xrightarrow{+X_2-HX} \begin{array}{l} COOH \\ | \\ CHX \\ | \\ CHX \\ | \\ COOH \end{array} \quad (X=Cl,\ Br)$$

In general, the properties of these halogenated succinic acids are similar to those of succinic acid. The stereoisomerism of the monohalogenated succinic acids is similar to that of lactic acid (p. 345), since they are compounds of the type *Cabcd*; and the stereoisomerism of the dihalogenated acids is similar to that of the tartaric acids (p. 393), they being compounds of the type *Cabc.Cabc*. dl-*Monochlorosuccinic acid* has m.p. 152° and its *anhydride* has m.p. 41°. The corresponding bromo-compounds have m.p. 160° and 31° respectively.

When *dl*-monochlorosuccinic acid or the corresponding bromo-compound is treated with moist silver oxide or with dilute aqueous sodium hydroxide, it is converted into dl-*monohydroxysuccinic acid* or dl-*malic acid*:

$$\begin{array}{l} COOH \\ | \\ CHX \\ | \\ CH_2 \\ | \\ COOH \end{array} + AgOH \rightarrow \begin{array}{l} COOH \\ |\ \diagup H \\ C \\ |\ \diagdown OH \\ CH_2 \\ | \\ COOH \end{array} + AgBr$$

dl-*Malic acid* forms deliquescent colourless crystals, m.p. 130°. It is identified by its highly crystalline monoammonium salt, $C_4H_5O_5(NH_4).H_2O$. The projection formula of malic acid is

$$\begin{array}{c} OH \\ | \\ H—C—COOH \\ | \\ H—C—COOH \\ | \\ H \end{array}$$

and its stereoisomerism is similar to that of lactic acid. *dl*-Malic acid is conveniently resolved into its optically active components by the fractional crystallisation of its salts with the alkaloid cinchonine (*l*-base), the *lAl*B salt being less soluble than its diastereoisomeride (*dAl*B) in aqueous solution.

l-*Malic acid* occurs in unripe apples (the acid is known in German as *Äpfelsäure*), in gooseberries (Scheele, 1785) and in mountain-ash berries.

dl-malic acid is obtained by the partial reduction of *dl*-tartaric acid with hydriodic acid; *d*-tartaric and *l*-tartaric acids are converted into *d*- and *l*- malic acids respectively under the same treatment:

$$\begin{array}{l} COOH \\ | \\ C\langle^{H}_{OH} \\ | \\ C\langle^{H}_{OH} \\ | \\ COOH \end{array} \quad + \quad \begin{array}{c} HI \\ \\ HI \end{array} \quad \rightarrow \quad \begin{array}{l} COOH \\ | \\ C\langle^{H}_{OH} \\ | \\ C\langle^{H}_{H} \\ | \\ COOH \end{array} \quad + H_2O + I_2$$

When it is heated to 140–150°, malic acid loses water and is converted into a dibasic unsaturated acid, $C_2H_2(COOH)_2$, known as *fumaric acid*. When it is heated rapidly to 180°, it is converted not only into fumaric acid but also, by further loss of water, into a bstance, *maleic anhydride*, $C_2H_2\langle\begin{array}{l} C{=}O \\ \quad\; \rangle O \\ C{=}O \end{array}$. On prolonged boiling in dilute sodium hydroxide solution, malic acid is converted into (the sodium salt of) fumaric acid. Further, by digesting maleic anhydride with water, *maleic acid* isomeric with fumaric acid is produced. Again, when fumaric acid is heated above 200°, it is converted into maleic anhydride and maleic acid is changed directly into fumaric acid by heating in a sealed tube at 200°. These changes may be outlined:

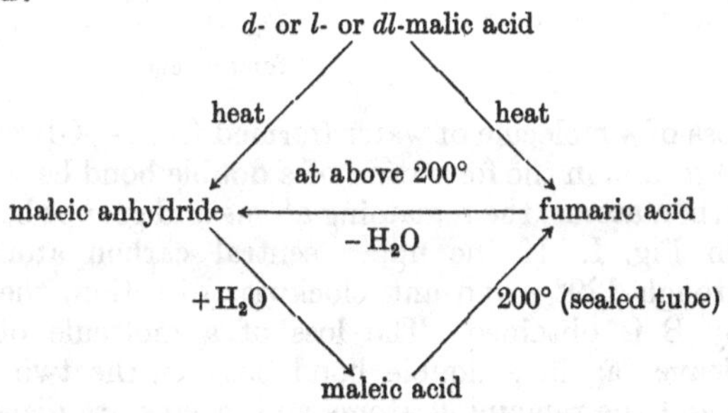

Fumaric acid is very sparingly soluble in cold water and crystallises from hot water in small colourless prisms; whereas maleic acid crystallises in large colourless prisms and has a high solubility in cold water. These two isomeric acids are therefore easily distinguished and, as indicated above, they can each be converted into the other. Fumaric acid combines directly with hydrogen chloride, yielding *dl*-monochlorosuccinic acid, $HOOC.CHCl.CH_2.COOH$ (p. 380).

The loss of a molecule of water from a molecule of malic acid is usefully considered stereochemically. If the constitution be represented by the conventional spring models, the atoms or groups attached to the central carbon atoms, being at the angles of a regular tetrahedron, are arranged as in A (the dotted lines representing valency directions away from the observer).

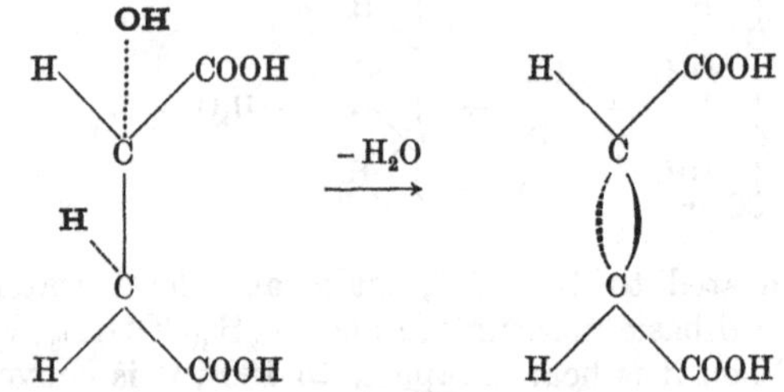

OH
H COOH
C
H
C
H COOH
A

$-H_2O$ →

H COOH
C
C
H COOH
Fig. I
maleic acid

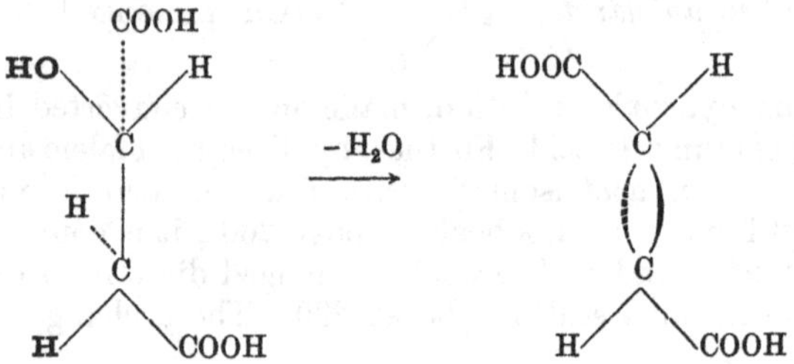

COOH
HO H
C
H
C
H COOH
B

$-H_2O$ →

HOOC H
C
C
H COOH
Fig. II
fumaric acid

In this case, loss of a molecule of water (formed from —OH and —H in heavy type) results in the formation of a double bond between the two central carbon atoms, the remaining atoms and groups becoming disposed as in Fig. I. If the upper central carbon atom in A be rotated through 120° in an anti-clockwise direction, the model represented by B is obtained. The loss of a molecule of water as indicated leaves again a double bond between the two central carbon atoms and the remaining atoms and groups are disposed as in Fig. II. Thus, on the assumption of the possibility of free rotation about the single bond joining two carbon atoms, it is possible to derive from malic acid the constitutional formulae of two isomeric unsaturated dicarboxylic acids, which are ethylene derivatives of the type $\begin{matrix}CHX\\ \Vert\\ CHX\end{matrix}$ or $\begin{matrix}Cab\\ \Vert\\ Cab\end{matrix}$ (p. 41). That acid which yields an anhydride (maleic anhydride) the more easily is assumed to be the one having the two carboxyl groups on the same side of the double bond and this constitution is assigned to maleic acid. The other constitutional formula is assigned to fumaric acid.

Maleic anhydride, whose constitutional formula is represented:

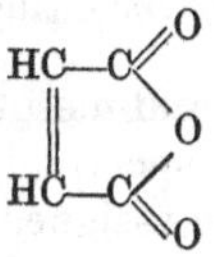

HC—C(=O)
‖   >O
HC—C(=O)

has m.p. 53° and is formed from both acids as indicated. It is also produced when these acids are submitted to the action of dehydrating agents, e.g. acetyl chloride, phosphorus pentoxide, etc.

Fumaric and maleic acids are typical unsaturated dibasic acids. They are further distinguished by their behaviour on oxidation with potassium permanganate in alkaline solution. The reaction taking place is similar to the formation of glycol from ethylene; fumaric acid is converted into externally compensated tartaric acid (racemic acid) and maleic acid is converted into internally compensated tartaric acid or mesotartaric acid (p. 396). These changes may be conveniently represented diagrammatically:

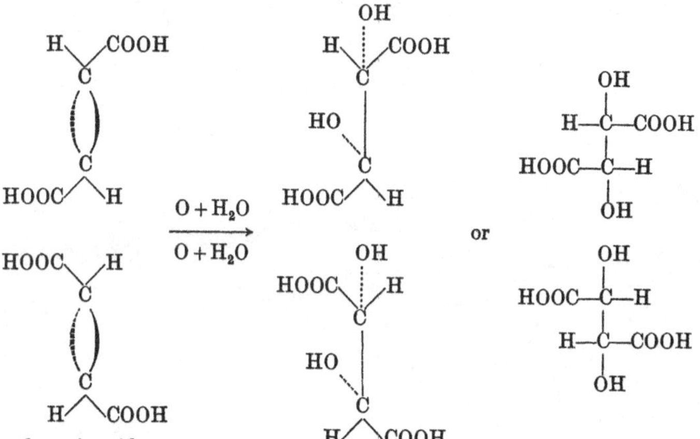

H, COOH / C / C / HOOC, H — fumaric acid

HOOC, H / C / C / H, COOH

$\xrightarrow[O+H_2O]{O+H_2O}$

(OH) H, COOH / C / HO / C / HOOC, H

(OH) HOOC, H / C / HO / C / H, COOH

or

OH
H—C—COOH
HOOC—C—H
OH

OH
HOOC—C—H
H—C—COOH
OH

(projection formulae)

d- and *l*- tartaric acids = racemic acid, i.e. externally compensated or *dl*- tartaric acid

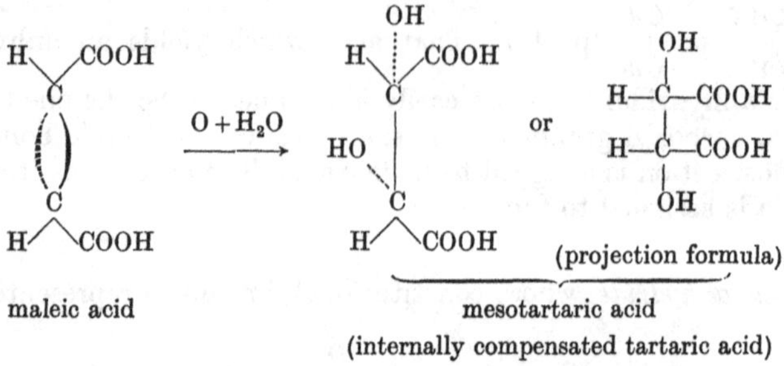

$$\text{maleic acid} \xrightarrow{O + H_2O} \text{mesotartaric acid}$$

H—C—COOH (OH above)
H—C—COOH (OH below)

(projection formula)

maleic acid

mesotartaric acid
(internally compensated tartaric acid)

The production of racemic and mesotartaric acids by the oxidation of fumaric and maleic acids respectively is important additional confirmation of the constitutions assigned to the latter two substances.

CHAPTER XIV

GLYCEROL, ERYTHRITOL AND THE TARTARIC ACIDS

GLYCEROL, 1:2:3-TRIHYDROXYPROPANE

By replacing one hydrogen atom attached to each of the two carbon atoms in the molecule of ethane by hydroxyl, the constitutional formula of glycol or dihydroxyethane, the simplest aliphatic dihydric alcohol, is deduced. Similarly, by replacing one hydrogen atom attached to each of the three carbon atoms in the molecule of propane by hydroxyl, the constitutional formula of glycerol or trihydroxy-propane, the simplest trihydric alcohol, is derived, thus:

$$\begin{array}{l} CH_3 \\ | \\ CH_3 \end{array} \qquad \begin{array}{l} CH_2OH \\ | \\ CH_2OH \end{array}; \qquad \begin{array}{l} CH_3 \\ | \\ CH_2 \\ | \\ CH_3 \end{array} \qquad \begin{array}{l} CH_2OH \\ | \\ CHOH \\ | \\ CH_2OH \end{array}$$

glycol — glycerol or trihydroxypropane

Glycerol has already been referred to in some detail as a product of the hydrolysis of fats, the chief constituents of which are fatty acid esters of glycerol (p. 220). Glycerol is obtained technically by carrying out the hydrolysis of fats by superheated steam. From the resulting mixture, the insoluble fatty acids separate on cooling and the glycerol is obtained from the aqueous solution by distillation under reduced pressure:

$$\begin{array}{l} CH_2O.OCR \\ | \\ CHO.OCR \\ | \\ CH_2O.OCR \end{array} + 3H_2O \rightarrow 3RCO.OH + \begin{array}{l} CH_2OH \\ | \\ CHOH \\ | \\ CH_2OH \end{array}$$

The name of the great French chemist, Chevreul, is indissolubly connected with the chemistry of glycerol. It was he who introduced (1813) the name and recognised the chief constituents of oils and fats as esters of glycerol. The synthesis of glycerol by the following series of reactions affords proof of its constitution (Friedel and Silva, 1872):

$$\underset{\text{acetic acid}}{CH_3.COOH} \rightarrow \underset{\text{calcium acetate}}{(CH_3.CO.O)_2Ca} \xrightarrow{\text{heat}} \underset{\text{acetone}}{CH_3.CO.CH_3} \xrightarrow{\text{reduction}} \underset{i\text{-propyl alcohol}}{CH_3.CH(OH).CH_3} \xrightarrow{\text{conc. } H_2SO_4} \underset{\text{propylene}}{CH_3.CH{=}CH_2}$$

$$\xrightarrow{+Cl_2} \underset{\text{propylene chloride, 1:2-dichloropropane}}{CH_3.CHCl.CH_2Cl} \xrightarrow[-HCl]{+Cl_2} \underset{\text{1:2:3-trichloropropane}}{CH_2Cl.CHCl.CH_2Cl}$$

$$CH_2Cl.CHCl.CH_2Cl \xrightarrow[-3AgCl]{+3AgOH} CH_2OH.CHOH.CH_2OH$$

$$CH_2Cl.CHCl.CH_2Cl \xrightarrow[-3AgCl]{+3CH_3.COOAg} \underset{\text{glyceryl triacetate (triacetin)}}{CH_2O.OC.CH_3.CHO.OC.CH_3.CH_2O.OC.CH_3} \xrightarrow{\text{hydrolysis}} CH_2OH.CHOH.CH_2OH$$

Glycerol is a viscous colourless liquid possessing a sweet taste. It has m.p. 17°, but remains supercooled much below this temperature. At its boiling point under ordinary pressure (290°) it undergoes slight decomposition, but it can readily be distilled under reduced pressure. It is highly hygroscopic and is miscible with water and with alcohol in all proportions; it is insoluble in ether.

When it is heated with such substances as concentrated sulphuric acid, potassium hydrogen sulphate, phosphorus pentoxide and anhydrous magnesium sulphate, glycerol is converted into acrolein or acrylic aldehyde (p. 226) by loss of two molecules of water:

$$CH_2OH.CH(OH).CH(OH)H \rightarrow CH_2{=}CH.CHO + 2H_2O$$

The production of the unsaturated aldehyde having a characteristic odour is a convenient method of identifying both combined and free glycerol. Glycerol dissolves alkalis and many metallic oxides, forming compounds analogous to the alcoholates.

The most important derivatives of glycerol are its numerous esters. Those with fatty acids, such as butyric, palmitic and stearic acids, and with the unsaturated oleic acid are the chief constituents of oils and fats. If, in these esters, the three acyl groups are different from each other, i.e. the esters are of the type:

$$\begin{array}{l} CH_2OR \\ | \\ CHOR' \\ | \\ CH_2OR'' \end{array}$$

they may exhibit optical activity, the individual ester being a compound of the general type *Cabcd*. Lecithin (p. 260) is a compound of this type and is optically active:

$$\begin{array}{l} CH_2O.CO.C_{17}H_{35} \\ | \\ CHO.CO.C_{17}H_{33} \\ | \qquad\quad \diagup OH \\ CH_2O.PO \\ \qquad\quad \diagdown \\ \qquad\qquad O.CH_2.CH_2.N\begin{matrix} \diagup CH_3 \\ -CH_3 \\ \diagdown CH_3 \end{matrix}\Bigg] OH \end{array}$$

Optically active esters of glycerol may also be of types A and B, whereas isomeric esters of types C and D cannot exhibit optical activity:

A	B	C	D
CH_2OR	CH_2OR	CH_2OH	CH_2OR
$CHOH$	$CHOR$	$CHOR$	$CHOH$
CH_2OH	CH_2OH	CH_2OH	CH_2OR

Glycerol esters may thus be of five types and, in the case of the esters with hydrochloric acid, all five compounds are known. These are:

CH_2Cl	CH_2OH	CH_2Cl	CH_2Cl	CH_2Cl
$CHOH$	$CHCl$	$CHOH$	$CHCl$	$CHCl$
CH_2OH	CH_2OH	CH_2Cl	CH_2OH	CH_2Cl
α-chlorhydrin b.p. 139° at 18 mm.	β-chlorhydrin b.p. 146° at 18 mm.	αα′-dichlorhydrin b.p. 174°	αβ-dichlorhydrin b.p. 183°	glyceryl trichloride, 1 : 2 : 3-trichloropropane b.p. 158°

Much of the glycerol produced technically is converted into the ester with nitric acid, *glyceryl trinitrate* or, as it is frequently wrongly described, *nitroglycerine*. For the preparation elaborate precautions are necessary on account of the highly explosive nature of the compound; the method consists in the very slow addition of pure glycerol to a well-cooled mixture of concentrated nitric and sulphuric acids, allowing the mixture to stand, separating the ester which forms the upper layer and washing it thoroughly with water and drying it with a neutral drying agent. The reaction may be expressed:

$$\begin{array}{l} CH_2OH \\ | \\ CHOH \\ | \\ CH_2OH \end{array} + \begin{array}{l} HONO_2 \\ \\ HONO_2 \\ \\ HONO_2 \end{array} \rightarrow \begin{array}{l} CH_2O.NO_2 \\ | \\ CHO.NO_2 \\ | \\ CH_2O.NO_2 \end{array} + 3H_2O$$

glyceryl trinitrate

Glyceryl trinitrate is a colourless oil which explodes with great violence when heated rapidly or on percussion. If the carbon and hydrogen in glyceryl trinitrate are converted into carbon dioxide and water respectively and the nitrogen is liberated as such, it will be noticed that the compound already contains more than sufficient oxygen for the purpose of oxidising the carbon and hydrogen. The fact that the compound on hydrolysis yields glycerol and a nitrate proves that the compound is a nitric acid ester and not a nitro-compound. Glyceryl trinitrate absorbed in kieselguhr forms a plastic material, which is the explosive dynamite. Glyceryl trinitrate is also an essential constituent of smokeless powder.

Glyceryl monophosphate,

$$\begin{array}{l} CH_2OH \\ | \\ CHOH \\ | \\ CH_2O.PO(OH)_2 \end{array}$$

prepared by warming glycerol with orthophosphoric acid at a temperature not exceeding 105°, is used in medicine, particularly in the form of the so-called *glycerophosphates*. Glyceryl monophosphate reacts as a dibasic acid and theoretically is capable of existing in optically active forms. The so-called glycerophosphates are the salts of glyceryl monophosphate, e.g. calcium glycerophosphate,

$$CH_2OH.CHOH.CH_2O.POO_2Ca.$$

Magnesium, iron (ferric), potassium, sodium, quinine and strychnine salts are among those which have been used medicinally.

By heating glycerol with anhydrous oxalic acid, formic acid and allyl alcohol are obtained according to the conditions employed. At lower temperatures the chief product is formic acid, while allyl alcohol is chiefly formed at higher temperatures. The products are obtained by the formation and decomposition of glycerol monooxalate, glycerol dioxalate and glycerol monoformate, and the reactions taking place may be outlined thus:

$$\begin{array}{ccccccc} & & & & & & CH_2OH \\ & & & & & & | \\ & & & & & & CHOH \\ & & & & & & | \\ & & & & & & CH_2OH \\ & & & & & & \uparrow -CO \\ CH_2OH & + \begin{array}{c} COOH \\ | \\ COOH \end{array} & & CH_2O.OC.COOH & & & CH_2O.OC.H \\ | & & & | & \xrightarrow{-CO_2} & & | \\ CHOH & \xrightarrow[-H_2O]{} & & CHOH & & & CHOH \\ | & & & | & & & | \\ CH_2OH & & & CH_2OH & & & CH_2OH \\ & & & \text{glycerol monoxalate} & & & \text{glycerol monoformate} \end{array}$$

$$\text{glycerol monoxalate} \xrightarrow{-H_2O} \text{glycerol dioxalate}; \qquad \text{glycerol monoformate} \xrightarrow{\begin{array}{c} COOH \\ | \\ COOH \end{array}} \text{glycerol monoxalate}$$

$$\begin{array}{ccccccc} & CH_2 & & CH_2O\,.\,OC & & CH_2O.OC.COOH & \\ & \| & & | \quad\;\; | & & | & \\ 2CO_2 + & CH & \longleftarrow & CH.O.OC & \xleftarrow{-H_2O} & CHOH & + \; H.COOH \\ & | & & | & & | & \text{formic acid} \\ & CH_2OH & & CH_2OH & & CH_2OH & \\ & \text{allyl alcohol} & & \text{glycerol dioxalate} & & \text{glycerol monoxalate} & \end{array}$$

When glycerol is heated with hydriodic acid—the first product is *i*-propyl iodide (compare p. 392) and the final product is allyl iodide —the reactions taking place may be outlined:

$$\begin{array}{lcl} \begin{array}{l} \quad\quad\quad\quad\; HI \\ CH_2\,OH \quad\; HI \\ | \\ CH\;\;OH \;+\; HI \\ | \\ CH_2\,OH \quad\; HI \\ \quad\quad\quad\quad\; HI \end{array} & \longrightarrow & \begin{array}{l} CH_3 \\ |\;\;\diagup H \\ C \\ |\;\;\diagdown I \\ CH_3 \end{array} \;+\; 2I_2 \;+\; 3H_2O \end{array}$$

$$\begin{array}{ccccc} CH_2OH & & CH_2I & & CH_2 \\ | & & | & & \| \\ CHOH \;+\; 3HI & \longrightarrow & CHI & \xrightarrow{-I_2} & CH \\ | & & | & & | \\ CH_2OH & & CH_2I & & CH_2I \\ & & \text{glyceryl iodide} & & \text{allyl iodide} \\ & & \text{unstable} & & \text{b.p. } 101^\circ \end{array}$$

As the molecule of glycerol contains two primary and one secondary alcohol groups, the constitutional formulae of the direct oxidation products of glycerol which are theoretically capable of existing can be readily deduced, thus:

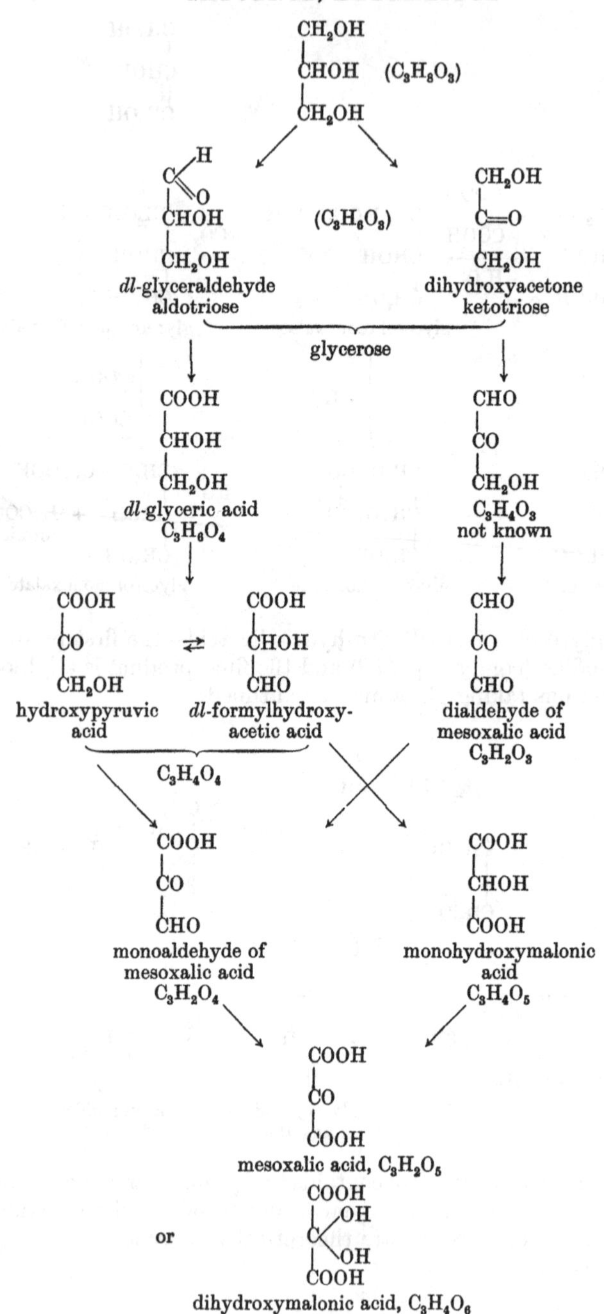

CH_2OH
$CHOH$
CH_2OH
($C_3H_8O_3$)
H
C
O
CHOH
CH_2OH
($C_3H_6O_3$)
CH_2OH
C=O
CH_2OH
dl-glyceraldehyde
aldotriose
dihydroxyacetone
ketotriose
glycerose
COOH
CHOH
CH_2OH
dl-glyceric acid
$C_3H_6O_4$
CHO
CO
CH_2OH
$C_3H_4O_3$
not known
COOH
CO
CH_2OH
hydroxypyruvic acid
COOH
CHOH
CHO
dl-formylhydroxy-acetic acid
$C_3H_4O_4$
CHO
CO
CHO
dialdehyde of mesoxalic acid
$C_3H_2O_3$
COOH
CO
CHO
monoaldehyde of mesoxalic acid
$C_3H_2O_4$
COOH
CHOH
COOH
monohydroxymalonic acid
$C_3H_4O_5$
COOH
CO
COOH
mesoxalic acid, $C_3H_2O_5$
or
COOH
C
OH
OH
COOH
dihydroxymalonic acid, $C_3H_4O_6$

When glycerol is subjected to partial oxidation with dilute nitric acid or bromine water it is converted into *glycerose*, which is chiefly a mixture of the two isomeric substances, *dl*-glyceraldehyde and dihydroxyacetone, which are also isomeric with lactic acid. dl-*Glyceraldehyde* crystallises in colourless needles, m.p. 138° (bimolecular form); it is highly soluble in water and yields derivatives through its aldehyde and alcohol groups. It is a strong reducing agent and its chief interest lies in the fact that it belongs to the same group of compounds as glycolaldehyde or glycollic aldehyde (p. 353), i.e. it is an aldehyde-alcohol and is a sugar, an aldose, and, having three carbon atoms in the molecule, is an aldotriose. *Dihydroxyacetone*, m.p. 75°, has a sweet taste and is highly soluble in water. It is a ketone-alcohol and forms derivatives through both these groupings. Dihydroxyacetone is the simplest type of another class of sugar, viz. the *ketone-alcohols* or *ketoses*, and having three carbon atoms in the molecule is described as a ketotriose. Glyceraldehyde and dihydroxyacetone are simple monosaccharides belonging to the general class of carbohydrates, having the general formula $C_m(H_2O)_n$; in this particular case, $m=n=3$.

When glycerose is allowed to stand in dilute aqueous alkaline solution, it undergoes an interesting condensation to a six carbon sugar, $C_6H_{12}O_6$, known as α-acrose or *dl*-fructose, from which both the well-known glucose (aldohexose) and fructose (ketohexose) have been obtained (p. 444); this condensation may be represented:

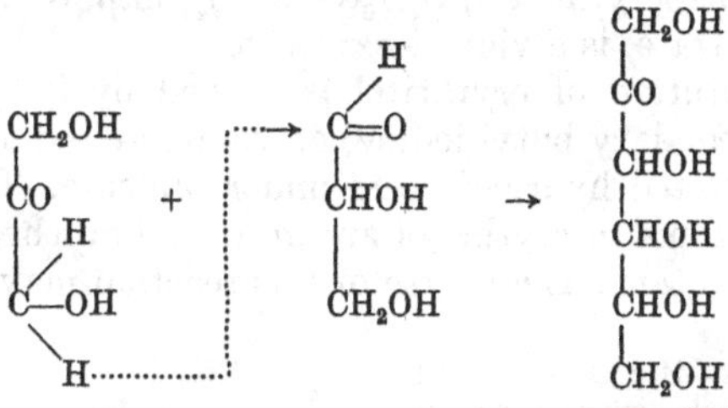

```
                                          CH2OH
                         H                  |
                        /                   CO
CH2OH            ....→ C=O                  |
  |              :     |                   CHOH
  CO       +     :    CHOH        →         |
  |  H           :     |                   CHOH
  | /            :    CH2OH                 |
  C—OH           :                         CHOH
   \             :                          |
    H............:                         CH2OH
```

α-acrose, $C_6H_{12}O_6$,
dl-fructose

Tetrahydric Alcohols—The Erythritols

The compound derived from *n*-butane by replacing one hydrogen atom attached to each of the four carbon atoms by a hydroxyl group has the constitutional formula indicated:

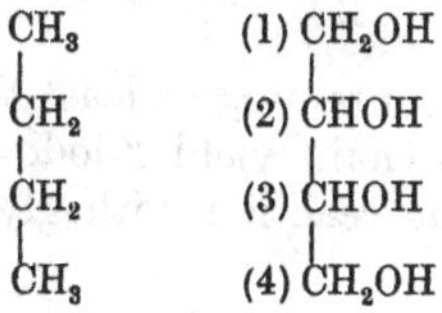

```
CH3        (1) CH2OH
 |              |
CH2        (2) CHOH
 |              |
CH2        (3) CHOH
 |              |
CH3        (4) CH2OH
```

It is thus a tetrahydric alcohol having two primary alcohol and two secondary alcohol groups. This is the formula of the compound *erythritol*, whose molecule contains two asymmetric carbon atoms (numbered 2 and 3).

On account of the symmetry of the molecule, the following stereoisomerides of erythritol are possible: d-*erythritol* (*dextro*rotatory), l-*erythritol* (*laevo*rotatory), externally compensated or dl-*erythritol* (optically inactive, but capable of being resolved into *d*- and *l*-erythritols) and internally compensated or meso*erythritol* (optically inactive and incapable of being resolved into optically active isomerides). All compounds of the general type $CXYZ.CXYZ$ (where X, Y and Z represent different monovalent atoms or groups), of which the disubstituted succinic acids and, in particular, the dihydroxy succinic acids or tartaric acids are important examples, exhibit the same type of stereoisomerism, which will be discussed in detail in the case of the tartaric acids.

Optically inactive, externally compensated or dl-*erythritol* occurs free in the alga *Protococcus vulgaris*. It is a colourless crystalline substance, m.p. 126°; it possesses a sweet taste and is readily soluble in water, sparingly soluble in ethyl alcohol and insoluble in ether. It yields esters, for example *tetraacetyl erythritol*,

$$CH_2O.OC.CH_3.(CHO.OC.CH_3)_2.CH_2O.OC.CH_3$$

m.p. 85° (by the action of acetic anhydride), and the *tetranitrate* (misnamed nitroerythritol), $C_4H_6(O.NO_2)_4$, m.p. 61°. The latter, like glyceryl trinitrate, is a violent explosive.

The constitution of erythritol is proved by its yielding 2-iodobutane (*n*-secondary butyl iodide) on complete reduction by heating with concentrated hydriodic acid under pressure. This proves that the carbon atoms in erythritol are in an unbranched chain as they are in 2-iodobutane. The course of this reaction may be indicated:

$$\begin{array}{lll} CH_2\,OH & H\,I & \\ | & HI & \\ CH\,OH & H\,I & \\ | & HI & \\ | & + & \\ CH\,OH & H\,I & \\ | & & \\ CH_2\,OH & H\,I & \\ & HI & \end{array} \rightarrow \begin{array}{l} CH_3 \\ | \\ CH_2 \\ | \\ | \\ CHI \\ | \\ CH_3 \end{array} + 3I_2 + 4H_2O$$

All polyhydric alcohols containing at least three carbon atoms in an unbranched or straight chain yield 2-iodo- straight chain paraffin derivatives (compare the reaction with glycerol, p. 389), and this reaction is used for demonstrating the configuration of the carbon atoms both in the polyhydric alcohols themselves and in the sugars which, as will be pointed out, are converted into them by gentle reduction with sodium and alcohol. Since erythritol yields a tetra-acetyl derivative, it must contain four hydroxyl groups, each of which in such a stable compound must be attached to a different carbon atom. These facts, together with the constitutional formulae of the direct oxidation products of erythritol, indicate that its constitutional formula must be that indicated above.

The first oxidation product of erythritol is called erythrose, which is a mixture of the two compounds:

	I		II
(1)	CHO	(1)	CH_2OH
(2)	CHOH	(2)	CO
(3)	CHOH	(3)	CHOH
(4)	CH_2OH	(4)	CH_2OH

Compound I is an aldotetrose, which having two asymmetric carbon atoms (numbered 2 and 3) is capable of existing in $2^2=4$ optical isomers, two being *dextro*rotatory and two being *laevo*rotatory. Compound II is a ketotetrose, which has only one asymmetric carbon atom (numbered 3) and like lactic acid is capable of existing in a *dextro*rotatory and a corresponding *laevo*rotatory form. These two compounds are isomeric, each having the molecular formula, $C_4H_8O_4$, and are 'four carbon sugars'.

The Tartaric Acids

Among the direct oxidation products of erythritol is a dihydroxy-dibasic acid having the constitutional formula indicated:

CH_2OH		(1)	COOH
CHOH	→	(2)	CHOH
CHOH		(3)	CHOH
CH_2OH		(4)	COOH

Such a dibasic acid has the same number of isomerides as erythritol, from which it differs in having carboxyl groups in place of the primary alcohol groups. The carbon atoms numbered 2 and 3 each have four different groups attached, as indicated:

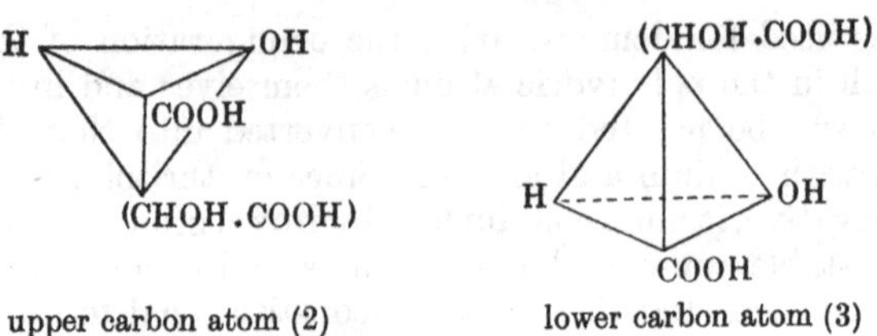

upper carbon atom (2) lower carbon atom (3)

The atoms and groups attached to carbon atom 2 can be arranged round the carbon atom in only two ways, I and II, thus:

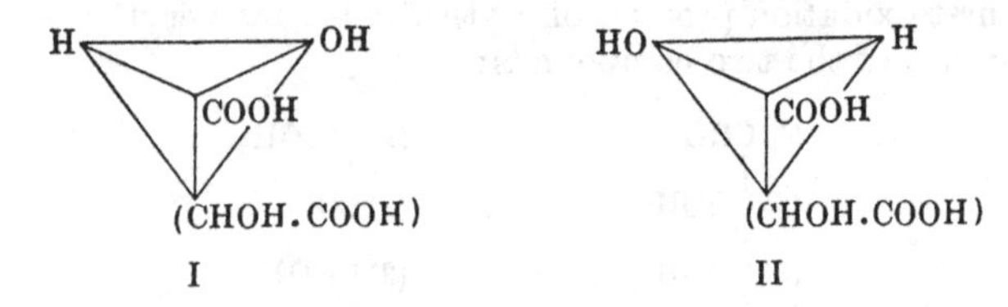

These two possible arrangements are mirror images of each other. Similarly there are two possible arrangements of the four different atoms and groups attached to carbon atom 3, as indicated in III and IV:

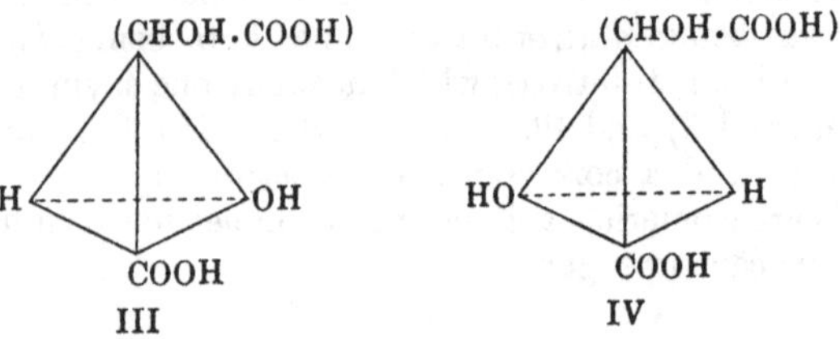

which arrangements of atoms and groups are again mirror images of each other. The whole constitution of the compound can be built up (*a*) by combining I and III, (*b*) by combining II and IV, (*c*) by combining I and IV and (*d*) by combining II and III, thus:

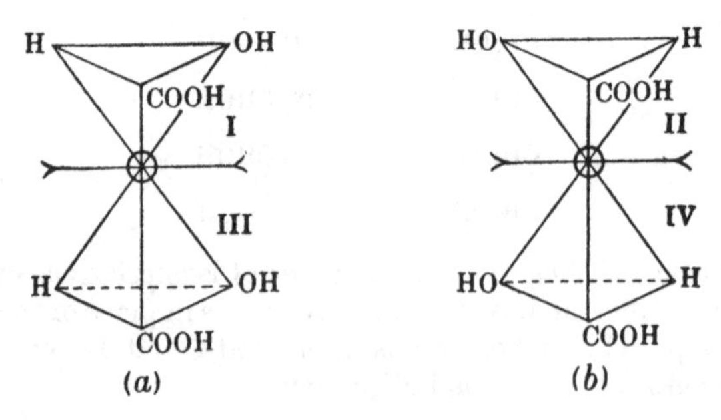

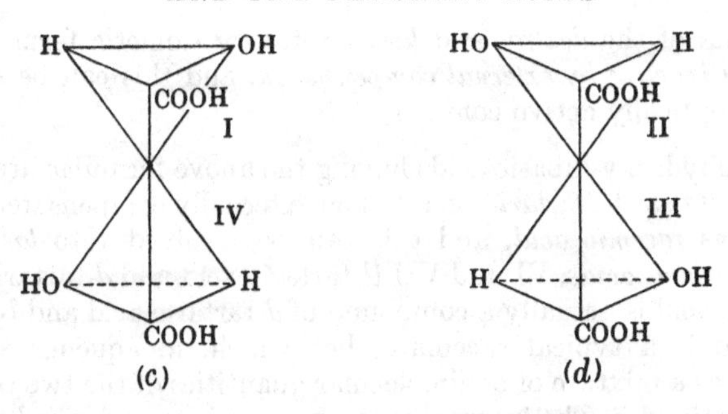

It will be noticed that the upper and lower halves of (*a*) and (*b*) are the mirror images of each other, i.e. each of these has a plane of symmetry passing through the point of joining of the two tetrahedra (⊛). Actually (*a*) is identical with (*b*), so that both these models represent one compound which is *internally compensated* and which can be represented concisely by V:

```
                   COOH                    COOH
   COOH             |                       |                    COOH
    |          H----|----OH        HO-------|-------H             |
H---C---OH          |                       |                HO---C---H
    |               |          or           |                     |
H---C---OH          |                       |                HO---C---H
    |          H----|----OH        HO-------|-------H             |
   COOH             |                       |                    COOH
                   COOH                    COOH
                                 V
```

A compound represented by the constitutional formulae (*a*) or (*b*) or V cannot be obtained in optically active forms because it has a plane of symmetry; it may be described as being *optically inactive by internal compensation*.

On the other hand, both the models (*c*) and (*d*) are devoid of symmetry, the model (*c*) is the mirror image of (*d*) and these can be represented concisely by the projection formulae, VI and VII:

```
                   COOH                    COOH
   COOH             |                       |                    COOH
    |          H----|----OH        HO-------|-------H             |
H---C---OH          |                       |                HO---C---H
    |               |                       |                     |
HO--C---H           |                       |                 H---C---OH
    |          HO---|----H         H--------|-------OH            |
   COOH             |                       |                    COOH
                   COOH                    COOH
              VI                                       VII
```

Models (*c*) and (*d*) and formulae VI and VII represent the constitution of optically active (*laevo-* and *dextro-*) isomeric forms of the compound having equal and opposite rotatory powers. A mixture of equal quantities of the *dextro-* and *laevo-* rotatory isomeric forms will be *optically inactive by external compensation*, and this can be resolved into its optically active components.

The dihydroxy-dibasic acids having the above formulae are known collectively as the *tartaric acids**. The externally compensated acid is known as *racemic acid*, and this can be resolved into *laevo-* and *dextro- tartaric acids*, VI and VII (*l*-tartaric acid and *d*-tartaric acid). Racemic acid is actually a compound of *d*-tartaric acid and *l*-tartaric acid and is a typical racemate, but which, in aqueous solution, behaves as a mixture of equimolecular quantities of the two optically active antipodes. The internally compensated isomeride is known as *mesotartaric acid*, V, and this, for the reason stated, cannot be resolved into optically active isomerides or antipodes.

The stereoisomerism of erythritol is similar to that of the tartaric acids; there is an externally compensated form yielding, on resolution, *d-* and *l-* erythritols, and also an internally compensated isomeride which cannot be resolved. All compounds of the general formula, $\begin{array}{l}CXYZ\\ |\\ CXYZ\end{array}$ (where X, Y and Z represent different monovalent atoms or groups), have stereoisomerides analogous in constitution to those of the tartaric acids. On the other hand, if one of the groups attached to the carbon atoms is different from all of the other five groups, as in a compound of the type $\begin{array}{l}CXYZ\\ |\\ CXYW\end{array}$, the compound is capable of yielding $2^2=4$ optically active isomerides, two *dextro*rotatory isomerides and two corresponding *laevo*rotatory isomerides, for each *dextro*rotatory isomeride there being a *laevo*rotatory isomeride having an equal and opposite optical rotation. This is so, for example, in the case of the aldotetrose, $CHO.CHOH.CHOH.CH_2OH$, the configuration of the four isomeric forms of which may be represented by the projection formulae, VIII and IX:

VIII		IX	
CHO	CHO	CHO	CHO
H—C—OH	HO—C—H	HO—C—H	H—C—OH
H—C—OH	HO—C—H	H—C—OH	HO—C—H
CH_2OH	CH_2OH	CH_2OH	CH_2OH
d	*l*	*d*	*l*

* The isomerism of diallyl tetrabromide, 1:2:5:6-tetrabromohexane (p. 54) would be explained by this compound occurring in externally compensated (theoretically resolvable into optically active forms) and internally compensated isomerides.

From the above, it will be realised that when either of the optically active sugars represented by formula VIII is oxidised to the corresponding dibasic acid, the latter will be optically inactive, internally compensated, mesotartaric acid; if no racemisation takes place in the process, the oxidation of the optically active sugars represented by formula IX will result in the production of the optically active tartaric acids.

The sparingly soluble acid potassium salt of *d*-tartaric acid, COOH.CHOH.CHOH.COOK, occurs in the form of crystalline crusts, *argol*, deposited by grape juice during fermentation. From this, *d*-tartaric acid is extracted (Scheele, 1769) and racemic acid is obtained as a by-product in the extraction. Mesotartaric acid is obtained from *d*-tartaric acid by heating it at 165° for some 48 hours (Pasteur).

Kekulé, Perkin (Sir W. H.) and Duppa (1861) synthesised racemic and mesotartaric acids starting from succinic acid obtained from amber. Since succinic acid can be obtained starting from ethylene which can be synthesised from carbon and hydrogen, the synthesis of racemic and mesotartaric acids may be regarded as complete:

```
                                              COOH              COOH
                                               |                 |
                                            H—C—Br   AgOH     H—C—OH
                                               |     ——→         |
                                            H—C—Br            H—C—OH
                                               |                 |
                                              COOH              COOH
                                         mesodibromo-       mesotartaric
                                         succinic acid          acid
                                        ↗
  COOH                     COOH        / bromination
   |                        |         /
H—C—H   bromination     H—C—Br
   |    ——————→             |
H—C—H                   H—C—H
   |                        |
  COOH                     COOH        \
                                        ↘     COOH              COOH
                                               |                 |
                                            H—C—Br   AgOH     H—C—OH
                                               |     ——→         |
                                           Br—C—H             HO—C—H
                                               |                 |
                                              COOH              COOH
                                    externally compensated     racemic
                                    dibromosuccinic acid        acid
```

The conversion of fumaric acid and maleic acid into racemic and mesotartaric acids respectively by oxidation has already been described (p. 383). Racemic and mesotartaric acids may also be synthesised starting from glyoxal, converting this into the dicyanohydrin and hydrolysing the latter compound:

$$\begin{array}{c}\text{H}\\ \diagup\\ \text{C=O}\\ |\\ \text{C=O}\\ \diagdown\\ \text{H}\end{array}
\begin{array}{l}\nearrow^{\text{HCN}}\\ \\ \\ \searrow_{\text{HCN}}\end{array}
\begin{array}{c}
\begin{array}{c}\text{CN}\\ |\\ \text{H—C—OH}\\ |\\ \text{H—C—OH}\\ |\\ \text{CN}\end{array} \xrightarrow{\text{hydrolysis}} \begin{array}{c}\text{COOH}\\ |\\ \text{H—C—OH}\\ |\\ \text{H—C—OH}\\ |\\ \text{COOH}\end{array}\\[2ex]
\begin{array}{c}\text{CN}\\ |\\ \text{H—C—OH}\\ |\\ \text{HO—C—H}\\ |\\ \text{CN}\end{array} \xrightarrow{\text{hydrolysis}} \begin{array}{c}\text{COOH}\\ |\\ \text{H—C—OH}\\ |\\ \text{HO—C—H}\\ |\\ \text{COOH}\end{array}
\end{array}$$

Racemic acid (referred to by Pasteur as 'paratartaric acid') crystallises in large colourless rhombic prisms which have the composition $C_4H_6O_6.H_2O$ and which effloresce in dry air. When anhydrous it has m.p. 206°, and profound decomposition sets in. Its solubility in water is less than that of *d*-tartaric acid.

Racemic acid forms typical salts and esters such as the dimethyl and diethyl esters ($CO_2R.CHOH.CHOH.CO_2R$), and these esters again form diacetyl and dibenzoyl derivatives, which are prepared from them by the action of acetyl and benzoyl chlorides; the diacetyl derivative of dimethyl racemate, $CO_2Me.CHOAc.CHOAc.CO_2Me$, has m.p. 86° and the production of such a compound is proof of the presence of two carboxyl groups and two secondary alcohol (: CHOH) groups in the molecule of racemic acid.

Racemic acid and its derivatives are, of course, optically inactive; being externally compensated tartaric acid, racemic acid is resolvable into *d*- and *l*- tartaric acids (*v.* below).

Mesotartaric acid is more soluble in water than racemic acid; it forms derivatives, salts and esters, similar to those of racemic acid, but the corresponding derivatives of the two acids have different physical properties, such as solubility, melting points, etc. Mesotartaric acid cannot be resolved into optically active components, since its molecule possesses a plane of symmetry and may be described as internally compensated.

d-*Tartaric acid* is extracted commercially from argol. The crude acid potassium salt may be converted into the sparingly soluble neutral calcium *d*-tartrate and the free acid liberated from this by treatment with the calculated quantity of sulphuric acid. After separating the calcium sulphate, the aqueous solution of *d*-tartaric acid is carefully evaporated. During this evaporation some racemisation takes place, resulting in the production of a certain amount of racemic acid. *d*-Tartaric acid is the ordinary commercial tartaric acid; it has m.p. 167–170°.

Its acid potassium salt, $HO_2C.CHOH.CHOH.CO_2K$, is sparingly soluble in water and constitutes a distinguishing test for the potassium ion. The normal potassium salt, $KO_2C.CHOH.CHOH.CO_2K + \frac{1}{2}H_2O$, is readily soluble in water.

Being a dibasic acid, *d*-tartaric acid forms a large variety of simple and mixed salts, many of which have been obtained in massive crystals. *Rochelle salt*, discovered in 1672 by Pierre Seignette, a pharmacist of La Rochelle, is the mixed sodium potassium *d*-tartrate having the composition $NaO_2C.CHOH.CHOH.CO_2K + 4H_2O$. It is a constituent of Fehling's solution used for the detection of reducing sugars. Unlike most salts of *d*-tartaric acid, its massive crystals show little tendency to effloresce under ordinary atmospheric conditions. Potassium antimonyl *d*-tartrate, usually known as *tartar emetic*, has the composition $KO_2C.CHOH.CHOH.CO_2(SbO) + \frac{1}{2}H_2O$; it is prepared by boiling acid potassium tartrate with antimony oxide and water and allowing the hot filtered solution to crystallise. It crystallises in massive colourless crystals which rapidly effloresce under normal atmospheric conditions. It finds important applications in medicine and, particularly, in the treatment of bilharziasis by intravenous injection of its aqueous solution. The mixed sodium ammonium *d*-tartrate, $NaO_2C.CHOH.CHOH.CO_2NH_4 + 4H_2O$, is of special theoretical interest and will be referred to below. The numerous salts, esters and other derivatives of *d*-tartaric acid have been the subject of many chemical and physico-chemical investigations.

The crystals of *d*-tartaric acid are themselves optically active (and, curiously enough, *laevo*rotatory), while its aqueous solutions are *dextro*rotatory. The specific rotation, $[\alpha]_D$, at 20° of a 20 per cent. aqueous solution, is $+11{\cdot}98°$. The optical activity of aqueous solutions of *d*-tartaric acid was first described by Biot in 1815.

When *d*-tartaric acid is added under proper conditions to a mixture of concentrated nitric and sulphuric acids, *dinitrotartaric acid*, $HO_2C.CH(ONO_2).CH(ONO_2).CO_2H$, is produced. This compound, being actually an ester of nitric acid, is better described as *d*-tartaric acid dinitrate. This compound can be separated from the acid mixture, and when treated with water at a low temperature it decomposes producing *dioxytartaric acid*, $HO_2C.C(OH)_2.C(OH)_2.CO_2H$, which can be conveniently separated in the form of its sparingly soluble disodium salt. This sodium salt is used in the preparation of certain dye-stuffs. Dioxytartaric acid is a direct oxidation product of *d*-tartaric acid and its stereoisomerides. Its production may be regarded as a proof of the existence of two secondary alcohol groups (: CHOH) in the molecule of these compounds, thus:

$$\begin{array}{c} CO_2H \\ | \\ CHOH \\ | \\ CHOH \\ | \\ CO_2H \end{array} \rightarrow \begin{array}{c} CO_2H \\ | \\ C(OH)_2 \\ | \\ C(OH)_2 \\ | \\ CO_2H \end{array} \text{ or } \begin{array}{c} CO_2H \\ | \\ CO \\ | \\ CO \\ | \\ CO_2H \end{array} + 2H_2O$$

(compare chloral and chloral hydrate).

Pasteur's classical work on the tartaric acids, which may be considered as the basis of stereochemistry, started with the crystallographic investigation of the salts of *d*-tartaric acid. As part of his training in crystallography, Pasteur repeated the work on the crystal forms of tartaric (*d*-tartaric acid) and paratartaric (racemic acid) acids and their salts which had been published by de la Provostaye in 1841, and he found that the latter had failed to note the occurrence of 'hemihedral' faces on all the crystals of the tartrates examined.*

* Hemihedrism of crystals is intimately associated with enantiomorphism, and the following elementary treatment of this subject is largely based on the teaching and writings of the author's friend, the late Dr T. V. Barker, and acknowledgment is made to Messrs Macmillan for permission to reproduce the various figures taken from Barker's contribution to vol. II of Roscoe and Schorlemmer's *Treatise on Chemistry*, 1913.

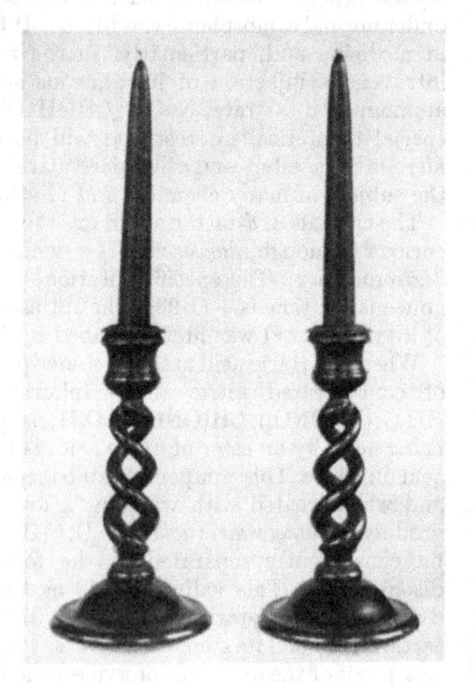

Enantiomorphism, i.e. the existence of bodies which can exist in two non-superposable forms which are mirror images of each other, is not uncommon among natural objects. The commonest example is that of the human hands; the mirror image of a right hand appears to be a left hand, and while the right and left hands are similar, they are easily distinguished from each other; they are non-superposable and enantiomorphous. The common hop plant climbs in a left-handed spiral; the mirror image of such a plant would be one which appears to climb in a right-handed spiral which would be non-superposable on its object. A left-handed and right-handed spiral are enantiomorphous arrangements. Similarly, the arrangement of the multiple fruits in the pineapple is that of a right-handed spiral which also would have a non-superposable mirror image. The antique pair

Further, Mitscherlich in 1844 had examined the crystals of sodium ammonium *d*-tartrate and of sodium ammonium racemate and had declared that these two substances 'have the same chemical composition, the same crystalline form with the same angles, the same

of carved oak candlesticks, of which a photograph is shown, exhibit enantiomorphism, the individual candlesticks being mirror images of each other and neither can be superposed on the other.

The geometrical study of crystals has shown that certain classes (eleven out of the thirty-two crystal classes known) can exist in enantiomorphous forms, and, further, that those crystals which exhibit enantiomorphism only possess ordinary axes of symmetry, and that to exhibit enantiomorphism, they must be devoid of the following three elements of symmetry: plane, centre and alternating axis of symmetry. The principle of enantiomorphism may be illustrated in the case of crystals of magnesium sulphate, $MgSO_4 . 7H_2O$, which belong to the orthorhombic system (having three planes of symmetry and three twofold axes of symmetry of unequal length at right angles to each other). The simplest form of crystal of magnesium sulphate is illustrated in Fig. I.

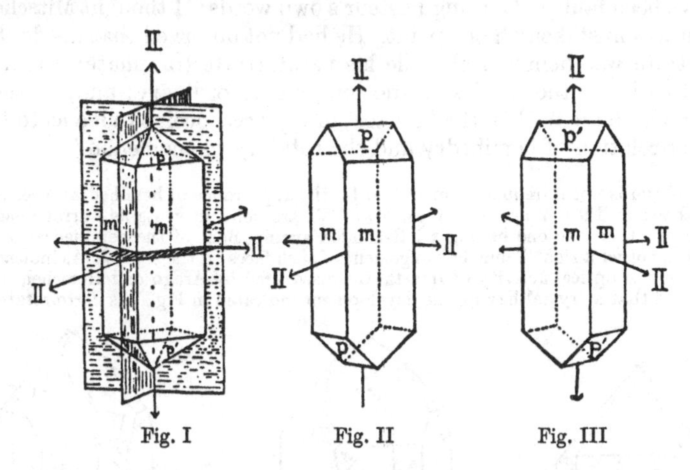

Fig. I Fig. II Fig. III

The development of the crystals may be of a less symmetrical character than that indicated by Fig. I, and two kinds of these less-developed crystals are possible and known (Figs. II and III). In Fig. II, only half of the eight terminal faces of Fig. I are developed, the remaining half of the eight terminal faces being developed in Fig. III. This is what is known as a hemihedral development of the terminal faces, and as a consequence the three planes of symmetry in Fig. I are absent in Figs. II and III, both of which, however, still possess the three twofold axes of symmetry of Fig. I. Figs. II and III are enantiomorphous, the one being the non-superposable mirror image of the other. Crystals of magnesium sulphate of types II and III are optically active (one being *dextro-* and the other *laevo-* rotatory), but the aqueous solution of each of these is optically inactive.

The earliest known example of enantiomorphism amongst crystals was that of quartz (SiO_2), which furnishes an interesting example of hemihedrism. In their simplest form, quartz crystals have the habit indicated in Fig. IV, but such crystals, in spite of their appearance, have no centre or planes of symmetry (as is shown by etched figures and other crystallographic properties). Such hemihedral and enantiomorphous crystals of quartz are well known. In the case of quartz, the enantiospecific weight (density), the same double refraction and consequently the same inclination in their optical axes. When dissolved in water, their refraction is the same. But the dissolved tartrate deviates the plane of polarisation, while the paratartrate (racemate) is indifferent. ... Yet, here the nature and number of the atoms, their arrangement and distances, are the same in the two substances compared.' In view of the optical rotatory power of the tartrates previously discovered by Biot, Pasteur was led to believe that there might be a relationship between the hemihedrism of the tartrates and their property of rotating the plane of polarised light. On the other hand, Mitscherlich had shown that in crystal form and all other respects the tartrates and racemates were completely identical except in the matter of their behaviour towards polarised light. Since substances having properties not completely identical must be dissimilar, the only alternative left to Pasteur was to conclude that Mitscherlich's observations must have been faulty. Quoting Pasteur's own words: 'I thought Mitscherlich was mistaken on one point. He had not observed that his double tartrate was hemihedral while his paratartrate (racemate) was not. If this is so, the results are no longer extraordinary; and, further, I should have in this, the best test of my preconceived idea as to the inter-relation of hemihedry and the rotatory phenomenon.'

morphism is not infrequently exhibited by the appearance of hemihedral faces (x), as shown in Figs. V and VI. Figs. V and VI are non-superposable mirror images of each other, the one having a left-handed arrangement of the hemihedral faces and the other a right-handed arrangement of such faces on the crystal. As indicated above, the optical activity of quartz was discovered by Arago and Herschel, who showed that a crystal having the development indicated in Fig. V is *dextro*rotatory

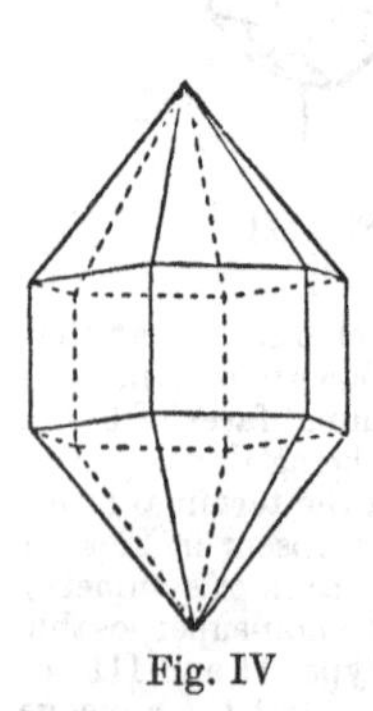

Fig. IV

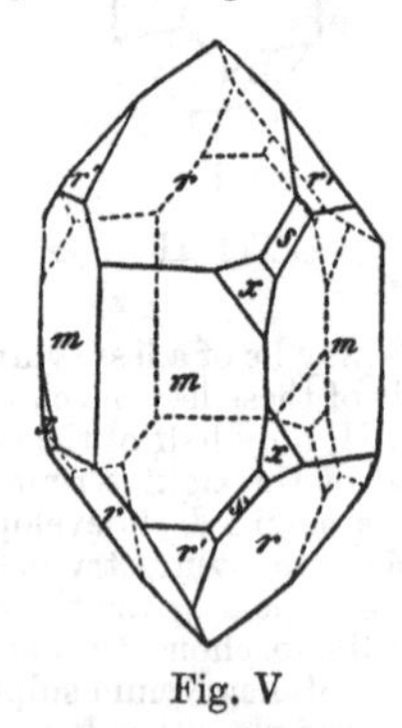

Fig. V

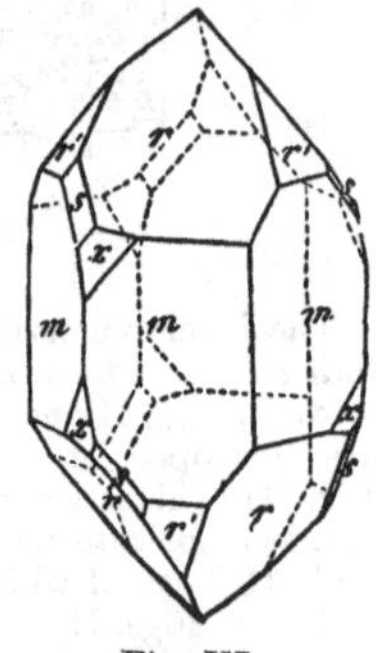

Fig. VI

while that having the development indicated by Fig. VI is *laevo*rotatory. Even if the quartz crystal possesses no hemihedral faces, it is always optically active; but whether it is *dextro-* or *laevo-* rotatory has to be determined experimentally. Like magnesium sulphate, quartz is only optically active in the crystalline condition. Thus, the association of enantiomorphism of crystals with their optical activity was fully recognised before Pasteur commenced his investigations of the tartaric acids and their salts.

'I hastened therefore to reinvestigate the crystalline form of Mitscherlich's two salts. I found, as a matter of fact, that the tartrate was hemihedral, like all the other tartrates which I had previously studied, but, strange to say, the paratartrate (racemate) was hemihedral also. Only, the hemihedral faces which in the tartrate were all turned the same way, were in the paratartrate (racemate) inclined sometimes to the right and sometimes to the left. In spite of the unexpected character of this result, I continued to follow up my idea. I carefully separated the crystals which were hemihedral to the right from those hemihedral to the left and examined their solutions separately in the polarising apparatus. I then saw with no less surprise than pleasure that the crystals hemihedral to the right deviated the plane of polarisation to the right, and that those hemihedral to the left deviated it to the left; and when I took an equal weight of each of the two kinds of crystals, the mixed solution was indifferent towards the light in consequence of the neutralisation of the two equal and opposite deviations.'

Pasteur then summarises this method of resolution of racemic acid into its optically active components: 'Thus, I start with paratartaric (racemic) acid; I obtain in the usual way the double paratartrate of soda and ammonia; and the solution of this deposits, after some days, crystals all possessing exactly the same angles and the same aspect. To such a degree is this the case that Mitscherlich, the celebrated crystallographer, in spite of the most minute and severe study possible, was not able to recognise the smallest difference. And yet the molecular arrangement in one set is entirely different from that in the other. The rotatory power proves this, as does also the mode of symmetry of the crystals. The two kinds of crystals are isomorphous and isomorphous with the corresponding tartrate. But the isomorphism presents itself with a hitherto unobserved peculiarity; it is the isomorphism of an asymmetric crystal with its mirror image. This comparison expresses the fact exactly; indeed, if, in a crystal of each kind, I imagine the hemihedral faces prolonged till they meet, I obtain two symmetrical tetrahedra, inverse, and which cannot be superposed, in spite of the perfect identity of all their respective parts. From this I was justified in concluding that by crystallisation of the double paratartrate of soda and ammonia (sodium ammonium racemate) I had separated two symmetrically isomorphous atomic groups, which are intimately united in paratartaric (racemic) acid. Nothing is easier than to show that these two species of crystals represent two distinct salts from which two different acids can be extracted.'

The two kinds of crystals which Pasteur obtained by allowing sodium ammonium racemate to crystallise are shown in Figs. I and II, which illustrate the enantiomorphous nature of the crystals and the positions of their hemihedral faces (*p*). Further, Figs. III and IV indicate the enantiomorphous tetrahedra obtained by carrying out Pasteur's suggestion of extending the hemihedral faces in space.

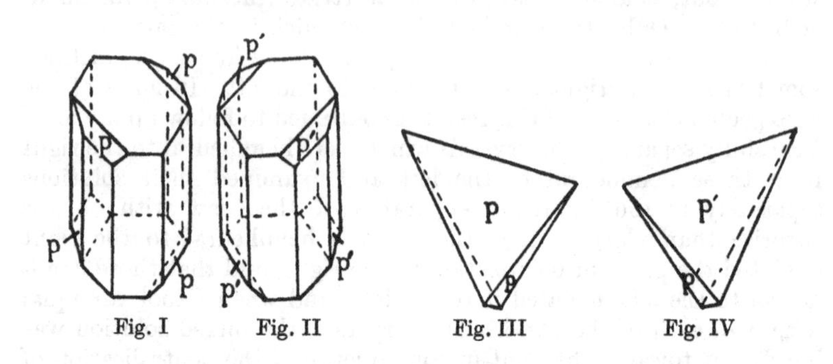

Fig. I Fig. II Fig. III Fig. IV

Pasteur's announcement of his resolution of racemic acid was received not without scepticism. Again, in Pasteur's own words: 'The announcement of the above facts naturally placed me in communication with Biot, who was not without some doubts regarding their accuracy. Being charged with giving an account of them to the Academy, he made me come to him and repeat before his eyes the decisive experiment. He handed over to me some paratartaric (racemic) acid which he had himself previously studied with particular care and which he had found to be perfectly indifferent to polarised light. I prepared the double salt in his presence with soda and ammonia, which he had likewise desired to provide. The liquid was set aside for slow evaporation in one of his rooms. When it had furnished about 30 to 40 grams of crystals, he asked me to call at the *Collège de France* in order to collect them and isolate before him, by recognition of their crystallographic character, the right and left crystals, requesting me to state once more whether I really affirmed that the crystals which I should place at his right hand would deviate to the right, and the others to the left. This done, he told me that he would undertake the rest. He prepared the solutions with carefully measured quantities, and when ready to examine them in the polarising apparatus, he once more invited me to come into his room. He first placed in the apparatus the more interesting solution, that which ought to deviate to the left. Without even making a measurement, he saw by the appearance of the tints of the two images, ordinary and extraordinary, in the analyser, that there was a strong deviation to the left. Then very visibly affected, the illustrious old man took me by the arm and said: "My dear child, I have loved science so much throughout my life that this touches my very heart" (Mon cher enfant, j'ai tant aimé les sciences dans ma vie que cela me fait battre le cœur).'*

In one respect, the success of Pasteur's resolution of racemic acid via the double sodium ammonium salt was due to a happy chance, for other investigators failed to obtain Pasteur's result. Pasteur happened to allow his solutions to crystallise at temperatures below 27°, and failures to repeat his results were largely due to the solutions being allowed to crystallise at temperatures higher than 27°. Investigations of the transition points have shown that below 27° the *d*- and *l*- sodium ammonium tartrates are stable and at 27° an equimolecular mixture of these salts is transformed into sodium ammonium racemate which, again, at 35° undergoes decomposition into the two salts sodium racemate and ammonium racemate. These changes in aqueous solution may be represented:

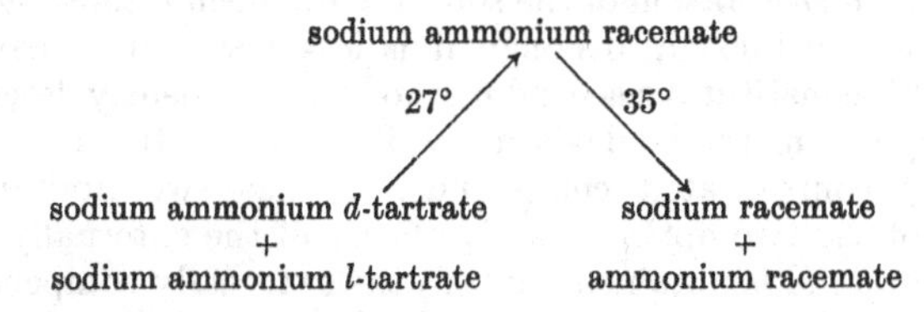

From the neutral sodium ammonium *d*- and *l*- tartrates, the pure optically active acids are obtained through the sparingly soluble (in neutral solution) calcium salts by decomposing the latter with the calculated quantity of sulphuric acid as indicated:

$$\underset{NaNH_4C_4H_4O_6.4H_2O}{\text{Na.NH}_4\ d\text{- or } l\text{-tartrate}} \xrightarrow{CaCl_2} \underset{CaC_4H_4O_6.H_2O}{\text{Ca } d\text{- or } l\text{-tartrate}} \xrightarrow{H_2SO_4} \underset{C_4H_6O_6}{d\text{- or } l\text{-tartaric acid}}$$

This is Pasteur's first method for the resolution of externally compensated compounds which, although of great importance in itself, is generally of limited application. Apart from its being a method of effecting such a resolution, the theoretical consequences of this work were far-reaching. The relationship between enantiomorphism of crystal structure and optical activity was emphasised in a striking manner. Further, since the optical activity of the tartaric acids is manifest not only in their crystals but also when the crystalline form is destroyed (i.e. when the crystals are dissolved in water), Pasteur concluded that the enantiomorphous arrangement of the atoms constituting the individual molecules must persist both in the crystal and when the form of the latter no longer exists. It was on these classical investigations of Pasteur that Le Bel formulated his

* From *Researches on Molecular Asymmetry*, by Louis Pasteur (1860), Alembic Club Reprints, No. 14.

theory of the asymmetric molecule in 1874, which together with the theory of van 't Hoff, announced almost at the same time, became the basis of stereochemistry.

Following his investigations on the resolution of racemic acid through the crystallisation of the sodium ammonium salt, Pasteur showed that the optically active tartaric acids can be differentiated by the action upon them of moulds and yeasts and, particularly, *Penicillium glaucum*. When *Penicillium* is allowed to grow in an aqueous solution of ammonium racemate, the solution soon becomes *laevo*rotatory, and this *laevo*rotation gradually increases to a maximum at which point there is no ammonium *d*-tartrate present in the solution, all having been destroyed. If the action of the organism be stopped at this stage (by boiling the solution), pure *l*-tartaric acid can be isolated. If, however, the action of the organism be allowed to continue the *laevo*rotation of the solution diminishes, showing that *l*-tartrate is also attacked, although it is less susceptible than the *d*-tartrate. This constitutes a second method—conveniently described as the biological method—devised by Pasteur for the resolution of externally compensated compounds. It has two limitations: (i) only one of the two optically active forms of the externally compensated compound is obtained, and (ii) the externally compensated compound must be non-poisonous to the 'resolving' organism and must not inhibit its growth. While this method of resolving externally compensated compounds is rarely employed, it was the beginning of the work which made Pasteur the greatest genius in medicine of his time. It is remarkable that Pasteur's investigations in crystallography and the relationships of optically active isomerides led to his energy being devoted to the elucidation of those infections of plants and animals with which his name is universally associated and which has made him one of the greatest benefactors of humanity.

The third method introduced by Pasteur for the resolution of externally compensated acids has been described in detail in the case of the resolution of *dl*-lactic acid. In the case of racemic acid, the resolving agent is the alkaloid, *cinchonine*, which is a *dextro*rotatory diacid base having the molecular formula $C_{19}H_{22}ON_2$. When racemic acid and an equimolecular quantity of cinchonine are dissolved in sufficient hot water, the solution on cooling deposits the less soluble cinchonine *l*-tartrate, the mother liquor containing the whole of the cinchonine *d*-tartrate and some of the less soluble diastereoisomeride. After one crystallisation from water the cinchonine *l*-tartrate is pure, and from this pure *l*-tartaric acid is obtained by decomposing the cinchonine salt. By a careful fractionation of the mixture of salts left in the mother liquor, pure cinchonine *d*-tartrate can be obtained and the *d*-tartaric acid separated from this in the usual manner. Briefly, the resolution of racemic acid may be outlined:

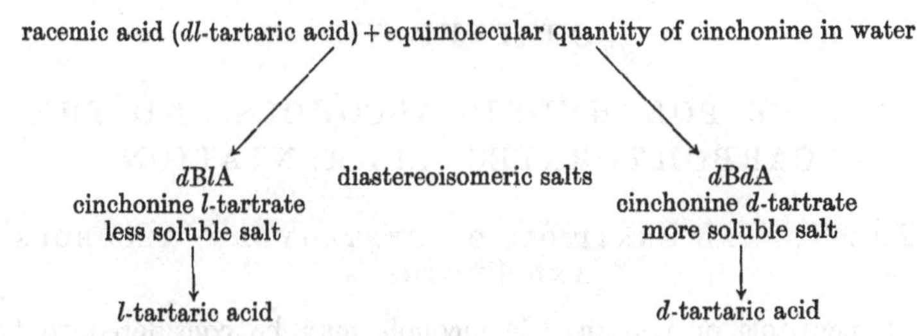

This method is the basis of most methods now used for the resolution of externally compensated compounds. An externally compensated base is resolved by the fractional crystallisation of its diastereoisomeric salts with an optically active acid; an externally compensated alcohol may be resolved by converting it into the diastereoisomeric esters using an optically active acid. In principle these methods are the same, and various devices have been introduced from time to time for the more effective and easy separation of the diastereoisomeric compounds.

l-*Tartaric acid* is identical in chemical properties with its stereoisomeride, *d*-tartaric acid. Its hemihedral crystals belong to the monosymmetric system and are enantiomorphous with those of *d*-tartaric acid. Its rotatory power is equal and opposite in sign to that of *d*-tartaric acid, and in all other respects the physical properties of the two substances are identical. The derivatives of *l*-tartaric acid which are stereoisomerides of analogous derivatives of *d*-tartaric acid are prepared in a similar manner, have identical chemical properties and the only difference in their physical properties is in regard to the sign of their rotatory powers. The diastereoisomerides of these two acids have identical chemical properties, but differ in all physical properties, including rotatory power, crystalline form and solubility.

CHAPTER XV

HIGHER POLYHYDRIC ALCOHOLS AND THE CARBOHYDRATES; FERMENTATION

THE NORMAL PENTITOLS OR PENTAHYDRIC ALCOHOLS AND PENTOSES

THE pentitols or pentahydric alcohols may be considered to be oxidation products of and derived from *n*-pentane by substitution of a hydroxyl group for one hydrogen atom attached to each of the five carbon atoms:

$$\begin{array}{lcl} CH_3 & \quad & CH_2OH \\ | & & | \\ CH_2 & & (2)\ CHOH \\ | & & | \\ CH_2 & & (3)\ CHOH \\ | & & | \\ CH_2 & & (4)\ CHOH \\ | & & | \\ CH_3 & & CH_2OH \end{array}$$

On account of the possibility of there being a plane of symmetry through carbon atom 3, there are capable of existing two optically active isomers (one *dextro-* and the corresponding *laevo-* rotatory form) and two optically inactive and internally compensated isomerides. These isomeric pentitols are generally obtained by reduction of the aldopentoses (5-carbon-aldehyde sugars), which are discussed below. An inactive pentitol, *adonitol*, having the constitution

$$\begin{array}{ccccccc} & & H & & H & & H & \\ & & | & & | & & | & \\ HO.CH_2 & — & C & — & C & — & C & — CH_2.OH \\ & & | & & | & & | & \\ & & OH & & OH & & OH & \end{array}$$

occurs naturally in *Adonis vernalis*. It is obtained by the reduction of *l*-ribose (p. 410).

The most important derivatives of the pentitols are compounds having the general formula, $C_5H_{10}O_5$, and are their simplest oxidation products. These compounds are sugars (pentoses) belonging to the two isomeric classes:

$$\begin{array}{lcl} (1)\ CHO & \quad & (1)\ CH_2OH \\ | & & | \\ (2)\ CHOH & & (2)\ CO \\ | & & | \\ (3)\ CHOH & & (3)\ CHOH \\ | & & | \\ (4)\ CHOH & & (4)\ CHOH \\ | & & | \\ (5)\ CH_2OH & & (5)\ CH_2OH \\ \text{aldopentoses} & & \text{ketopentoses} \end{array}$$

These pentoses, on reduction, are converted into the pentitols, and this is the usual way of preparing the latter compounds.

All sugars which react as though they contain either of the terminal groupings, O:CH—CH.OH— (aldoses) or CH_2OH—CO— (ketoses), behave similarly with phenylhydrazine, yielding *osazones*. In all cases, only the terminal groupings mentioned are concerned in osazone formation. The reaction is carried out by mixing the sugar in aqueous solution with rather more than three molecular proportions of freshly purified phenylhydrazine (or phenylhydrazine hydrochloride and an equivalent quantity of sodium acetate) and an excess of acetic acid and heating the mixture in a boiling water-bath until the osazone crystallises from the hot solution or on cooling. The reactions taking place are:

in the case of the aldoses,

$$\text{(i)}\quad \begin{array}{l} \text{H} \\ \diagup \\ \text{C=O} \\ | \\ \text{CHOH} \\ | \end{array} + H_2N.N\begin{array}{l}\diagup H \\ \diagdown C_6H_5\end{array} \rightarrow \begin{array}{l} \text{H} \\ \diagup \\ \text{C=N.N}\begin{array}{l}\diagup H \\ \diagdown C_6H_5\end{array} \\ | \\ \text{CHOH} \\ | \end{array} + H_2O$$

phenylhydrazone, capable of existing in two stereoisomeric forms of which only one is usually isolated

$$\text{(ii)}\quad \begin{array}{l} \text{H} \\ \diagup \\ \text{C=N.N}\begin{array}{l}\diagup H \\ \diagdown C_6H_5\end{array} \\ | \\ \text{CHOH} \\ | \end{array} + H_2N.N\begin{array}{l}\diagup H \\ \diagdown C_6H_5\end{array} \overset{*}{\rightarrow} \begin{array}{l} \text{H} \\ \diagup \\ \text{C=N.N}\begin{array}{l}\diagup H \\ \diagdown C_6H_5\end{array} \\ | \\ \text{C=O} \\ | \end{array} + NH_3 + C_6H_5NH_2$$

(intermediate product)

$$\text{(iii)}\quad \begin{array}{l} \text{H} \\ \diagup \\ \text{C=N.N}\begin{array}{l}\diagup H \\ \diagdown C_6H_5\end{array} \\ | \\ \text{C=O} \\ | \end{array} + H_2N.N\begin{array}{l}\diagup H \\ \diagdown C_6H_5\end{array} \rightarrow \begin{array}{l} \text{H} \\ \diagup \\ \text{C=N.N}\begin{array}{l}\diagup H \\ \diagdown C_6H_5\end{array} \\ | \\ \text{C=N.N}\begin{array}{l}\diagup H \\ \diagdown C_6H_5\end{array} \\ | \end{array} + H_2O$$

osazone, also capable of existing in stereoisomeric forms of which only one is usually isolated

in the case of the ketoses,

$$\text{(i)}\quad \begin{array}{l} CH_2OH \\ | \\ \text{C=O} \\ | \end{array} + H_2N.N\begin{array}{l}\diagup H \\ \diagdown C_6H_5\end{array} \rightarrow \begin{array}{l} CH_2OH \\ | \\ \text{C=N.N}\begin{array}{l}\diagup H \\ \diagdown C_6H_5\end{array} \\ | \end{array} + H_2O$$

phenylhydrazone

* In reaction (ii), the phenylhydrazine acts as an oxidising agent, being reduced to ammonia and aniline, the neighbouring primary and secondary alcohol groups being oxidised at the same time to aldehyde (O : CH—) and ketone (: CO) groups respectively. The reduction of phenylhydrazine may be represented:

$$H_2N.NHC_6H_5 + H_2 \rightarrow NH_3 + C_6H_5NH_2$$

$$\text{(ii)}\quad \begin{array}{c} \text{H} \\ | \\ \text{H—C—OH} \\ | \\ \text{C=N.N}\!\begin{array}{l}\text{H}\\ \text{C}_6\text{H}_5\end{array} \\ | \end{array} + \text{H}_2\text{N.N}\!\begin{array}{l}\text{H}\\ \text{C}_6\text{H}_5\end{array} \xrightarrow{*} \begin{array}{c} \text{H} \\ | \\ \text{C=O} \\ | \\ \text{C=N.N}\!\begin{array}{l}\text{H}\\ \text{C}_6\text{H}_5\end{array} \\ | \end{array} + \text{NH}_3 + \text{C}_6\text{H}_5\text{NH}_2$$

$$\text{(iii)}\quad \begin{array}{c} \text{H} \\ | \\ \text{C=O} \\ | \\ \text{C=N.N}\!\begin{array}{l}\text{H}\\ \text{C}_6\text{H}_5\end{array} \\ | \end{array} + \text{H}_2\text{N.N}\!\begin{array}{l}\text{H}\\ \text{C}_6\text{H}_5\end{array} \rightarrow \begin{array}{c} \text{H} \\ | \\ \text{C=N.N}\!\begin{array}{l}\text{H}\\ \text{C}_6\text{H}_5\end{array} \\ | \\ \text{C=N.N}\!\begin{array}{l}\text{H}\\ \text{C}_6\text{H}_5\end{array} \\ | \\ \textit{osazone} \end{array} + \text{H}_2\text{O}$$

Derivatives of phenylhydrazine may be, and frequently are, used to yield corresponding osazones, provided that the hydrazine contains the $\begin{array}{l}\text{H}\\ \text{H}\end{array}\!\!>\!\text{N—}$ grouping. The osazones are yellow compounds and are used for identifying the sugars and establishing their stereochemical relationships.

The aldopentoses contain three asymmetric carbon atoms (numbered 2, 3 and 4) and there are consequently $2^3=8$ optically active aldopentoses occurring in four pairs of *dextro-* and *laevo-* rotatory isomerides, there being for each *dextro*rotatory aldopentose a *laevo*-rotatory aldopentose having an equal and opposite rotatory power, the constitution of the one being the mirror image of the other. Adopting the simple projection formulae, as used in the case of the tartaric acids, the constitution of these compounds may be written:

$$\begin{array}{c} \text{CHO} \\ | \\ \text{H—C—OH} \\ | \\ \text{H—C—OH} \\ | \\ \text{H—C—OH} \\ | \\ \text{CH}_2\text{OH} \\ \text{I} \end{array} \qquad \begin{array}{c} \text{CHO} \\ | \\ \text{HO—C—H} \\ | \\ \text{H—C—OH} \\ | \\ \text{H—C—OH} \\ | \\ \text{CH}_2\text{OH} \\ \text{II} \end{array} \qquad \begin{array}{c} \text{CHO} \\ | \\ \text{H—C—OH} \\ | \\ \text{HO—C—H} \\ | \\ \text{H—C—OH} \\ | \\ \text{CH}_2\text{OH} \\ \text{III} \end{array} \qquad \begin{array}{c} \text{CHO} \\ | \\ \text{HO—C—H} \\ | \\ \text{HO—C—H} \\ | \\ \text{H—C—OH} \\ | \\ \text{CH}_2\text{OH} \\ \text{IV} \end{array}$$

The ketopentoses have not been so extensively studied as the isomeric aldopentoses and are not discussed in the present work. The general formula given above, however, shows that carbon atoms 3 and 4 are asymmetric and therefore $2^2=4$ optically active isomeric ketopentoses are theoretically capable of existing in two pairs of *dextro-* and *laevo-* rotatory compounds.

The constitution of the four either *dextro-* or *laevo-* rotatory aldopentoses, *arabinose*, *ribose*, *xylose* and *lyxose*, may be deduced from the following considerations:

* See footnote on p. 409.

(i) Arabinose and ribose yield identical osazones and therefore the configuration of carbon atoms 3 and 4 must be identical in each. Therefore arabinose and ribose must be represented by either formulae I and II or III and IV.

(ii) On oxidation, arabinose yields an optically active dibasic acid and ribose and xylose yield optically inactive dibasic acids. Compounds II and IV obviously yield optically active dibasic acids and compounds I and III yield optically inactive dibasic acids. Therefore, arabinose is either II or IV, ribose and xylose must be I and III and the remaining one, lyxose, must be either IV or II.

(iii) When hydrocyanic acid is added to an aldopentose and the resulting cyanide is hydrolysed, two stereoisomeric tetrahydroxy-*n*-valeric acids are produced by the resulting introduction of another asymmetric carbon atom, thus:

CHO–CHOH–CHOH–CHOH–CH_2OH —HCN→ CN–(H–C–OH)–CHOH–CHOH–CHOH–CH_2OH —hydrolysis→ COOH–(H–C–OH)–CHOH–CHOH–CHOH–CH_2OH —oxidation→ COOH–(H–C–OH)–CHOH–CHOH–CHOH–COOH

CHO–CHOH–CHOH–CHOH–CH_2OH —HCN→ CN–(HO–C–H)–CHOH–CHOH–CHOH–CH_2OH —hydrolysis→ COOH–(HO–C–H)–CHOH–CHOH–CHOH–CH_2OH —oxidation→ COOH–(HO–C–H)–CHOH–CHOH–CHOH–COOH

intermediate cyanide; stereoisomeric tetrahydroxy-*n*-valeric acids; stereoisomeric dibasic acids

On direct oxidation, these two stereoisomeric tetrahydroxy-*n*-valeric acids yield two stereoisomeric dibasic acids having the configurations indicated. When arabinose is submitted to this treatment, it yields a mixture of two acids both of which are optically active; on the other hand, of the two acids given by lyxose one is optically active and the other optically inactive. This proves that lyxose must have constitution IV. This is shown:

```
      CHO                   COOH                       COOH
      |                     |                          |
  HO—C—H                H—C—OH                     HO—C—H
      |                     |                          |
  HO—C—H       →       HO—C—H          and         HO—C—H
      |                     |                          |
   H—C—OH              HO—C—H                      HO—C—H
      |                     |                          |
      CH2OH              H—C—OH                      H—C—OH
   IV                       |                          |
                            COOH                       COOH
                    optically inactive         optically active
                    \_______________________________________/
                                  dibasic acids
```

Therefore, arabinose must have constitution II; ribose must have constitution I and xylose must have constitution III. These are detailed:

```
     CHO             CHO              CHO              CHO
     |               |                |                |
  H—C—OH         HO—C—H           H—C—OH          HO—C—H
     |               |                |                |
  H—C—OH          H—C—OH          HO—C—H          HO—C—H
     |               |                |                |
  H—C—OH          H—C—OH           H—C—OH           H—C—OH
     |               |                |                |
     CH2OH           CH2OH            CH2OH            CH2OH
  ribose          arabinose         xylose           lyxose
```

These configurations have been assigned to the *d*- forms of the above sugars and the *l*- forms must have configurations which are mirror images of these. It may be emphasised here that the signs *d*- and *l*- in the sugar group do not necessarily imply a *dextro*- and *laevo*-rotatory power respectively; these letters, in the sugar group, imply stereochemical relationships. Those sugars having the configuration of the hydrogen atom and the hydroxyl group attached to the carbon atom adjacent to the primary alcohol group (—CH_2OH) disposed as indicated above belong, by convention, to the *d*- series.

The above aldopentoses yield respectively the pentahydric alcohols indicated:

```
     d-ribose          d-arabinose       d-xylose           d-lyxose
        ↓                  ↓                 ↓                  ↓
    (1) CH2OH              CH2OH             CH2OH              CH2OH
        |                  |                 |                  |
(2)  H—C—OH           HO—C—H             H—C—OH           HO—C—H
        |                  |                 |                  |
(3)  H—C—OH            H—C—OH           HO—C—H            HO—C—H
        |                  |                 |                  |
(4)  H—C—OH            H—C—OH            H—C—OH            H—C—OH
        |                  |                 |                  |
    (5) CH2OH              CH2OH             CH2OH              CH2OH
     adonitol,          d-arabitol         xylitol,           d-arabitol
 optically inactive—                  optically inactive—
     internally                           internally
    compensated                          compensated
```

As shown, the reduction products of *d*-arabinose and *d*-lyxose are identical.

The aldopentoses are readily distinguished from the more important hexoses by the fact that, unlike the latter, they do not undergo fermentation by yeasts. The aldopentoses when heated with dilute hydrochloric acid are converted into *furfural* (a colourless oil, b.p. 162°, having properties analogous to those of the typically aromatic aldehyde, benzaldehyde) by elimination of water, thus:

```
       H         H
       |         |
  HO — C ——————— C — OH                HC——CH
       |         | / H                 ‖    ‖
  H  — C         C — CHO      →        HC   C.CHO   + 3H₂O
      / \       /                        \ /
     H   OH   OH                          O
                                       furfural*
```

Furfural yields a characteristic derivative with phloroglucinol, a litharge-coloured powder having the composition $C_{11}H_{12}O_6$, which is used in the quantitative estimation of aldopentoses. The qualitative test for aldopentoses consists in heating them with dilute hydrochloric acid containing phloroglucinol, when a cherry-red coloration is produced.

dl-Arabinose occurs in the urine in the somewhat rare pathological condition known as *pentosuria*.

The aldopentoses have been described above as possessing 'open'-aldehyde formulae. Actually such an open formula is more adequately represented in some such way as

```
              H  OH
              | /
              C
        H   / 5 \ H
        |  /          O
        C 4          C⟨
        |             1 \ H      (p. 421)
       HO   H     HO
             \   /
              C ——— C
              | 3   | 2
              OH    H
```

It has been shown by Haworth and co-workers that while the aldopentoses react as aldehydes as indicated, their more stable forms

* Compare furan (p. 379).

possess a closed ring constitution and are given 'pyranose'* formulae which, in the case of arabinose, is

```
        H                                    H
        |                                    |
        C5 ----------- O                     C5 ----------- O
   H   /|               \    H          H   /|               \    OH
   |  / H                \   |          |  / H                \   |
   C4                     C1     and    C4                     C1
   |  \ H           OH   /   |          |  \ H           OH   /   |
  OH   \|           |   /    OH        OH   \|           |   /    H
        C3 ---------C2                       C3 ---------C2
        |           |                        |           |
        OH          H                        OH          H
```

The significance of this constitution will be made more clear in the case of glucose (p. 426).

The *n*-Hexitols (Normal Hexahydric Alcohols) and the Hexoses

The normal hexahydric alcohols are oxidation products of *n*-hexane, thus:

```
CH3             (1) CH2OH
 |                   |
CH2             (2) CHOH
 |                   |
CH2             (3) CHOH
 |              .........|.........
CH2             (4) CHOH
 |                   |
CH2             (5) CHOH
 |                   |
CH3             (6) CH2OH
```

n-hexane, C_6H_{14} *n*-hexahydric alcohols, $C_6H_{14}O_6$

and they can be considered to be derived from *n*-hexane by the replacement of six hydrogen atoms, one attached to each carbon atom, by hydroxyl groups.

The above general formula of the hexahydric alcohols contains four asymmetric carbon atoms (2, 3, 4 and 5). The molecule may have a plane of symmetry as indicated, and according to the theory of van 't Hoff there must be capable of existence ten compounds having such a general formula. Of these, eight are optically active, occurring in four pairs of *dextro-* and *laevo-* rotatory isomerides, and

* I.e. a sugar related to *pyran*, a *hetero*cyclic compound having the constitution:

```
      HC ------ O
     //          \
   HC             CH
     \           //
      CH2 ------ CH
```

two are optically inactive by internal compensation. These ten hexahydric alcohols are known and all yield 2-iodohexane on reduction with hydriodic acid (p. 392). These alcohols may be obtained by reduction of the hexoses (*v.* below) in aqueous solution with sodium amalgam:

$$\underset{\text{hexose}}{C_6H_{12}O_6} + H_2 \rightarrow \underset{\text{hexahydric alcohol}}{C_6H_{14}O_6}$$

and the constitution of the hexahydric alcohol can be deduced from that of the aldohexose from which, under suitable conditions, it is the only product produced by this reaction. These alcohols are colourless crystalline substances; they possess a sweet taste and are readily soluble in water.

On account of their relationship with the four more important hexoses, three hexahydric alcohols, viz. *mannitol*, *sorbitol* and *dulcitol*, have special interest.

d-*Mannitol* (colourless needles, m.p. 166°) occurs in manna, the evaporated sap of the manna tree, from which it may be obtained by extraction with and crystallisation from alcohol. It is the direct reduction production of the aldohexose *d*(+)-mannose*, and it is produced along with *d*-sorbitol by the reduction of the ketohexose, *d*(−)-fructose.

d-*Sorbitol* is present in mountain-ash berries. It crystallises in colourless needles containing one molecule of water of crystallisation and the anhydrous compound has m.p. 110°. It is obtained by the reduction of *d*(+)-glucose in aqueous solution with sodium amalgam.

Dulcitol, one of the two hexahydric alcohols which are inactive by internal compensation, is obtained by the reduction of *d*(+)-galactose. It crystallises in monoclinic prisms, m.p. 188·5°.

The relationships between these three hexahydric alcohols and the above four hexoses are summarised:

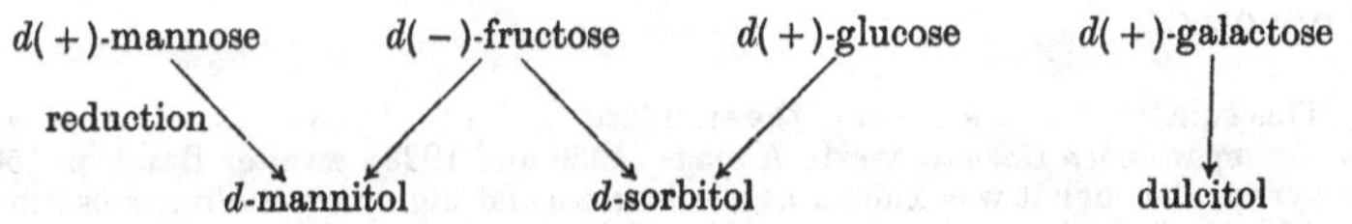

As far as carbon atoms 2, 3, 4, 5 and 6 are concerned, the configuration of the hexahydric alcohols is the same as that of the aldohexoses, from which they are obtained by reduction with sodium amalgam, in which process the only change is the conversion of the terminal aldehyde group into a primary alcohol group.

The hexaacetyl derivatives are easily prepared by treatment of

* The significance of this terminology will be made clear in what follows; "*d*(+)-mannose" indicates the aldohexose mannose belonging by configuration to the "*d*" series (p. 412) and having a *dextro*- or + rotatory power.

the hexahydric alcohol with an excess of acetic anhydride in the presence of a little zinc chloride. These hexaacetyl derivatives are colourless crystalline compounds and serve for the identification of the particular alcohol. *Hexaacetyl*-d-*mannitol* has m.p. 121°; *hexa-acetyl*-d-*sorbitol* has m.p. 99°;* *hexaacetyldulcitol* has m.p. 171°. These esters of acetic acid have the general formula $C_6H_8(O.OC.CH_3)_6$.

The Aldohexoses and Ketohexoses or Monosaccharides

The sugars having the general formula $C_6H_{12}O_6$ constitute a group of compounds containing the well-known glucose (formerly known as dextrose) and fructose (formerly known as laevulose).† These compounds are of two types:

aldohexoses	ketohexoses
(1) CHO	(1) CH_2OH
(2) CHOH	(2) CO
(3) CHOH	(3) CHOH
(4) CHOH	(4) CHOH
(5) CHOH	(5) CHOH
(6) CH_2OH	(6) CH_2OH

They may be regarded as the simplest direct oxidation products of the hexahydric alcohols and they are described generally as monosaccharides, being hydrolysis products of the more complicated carbohydrates, disaccharides, trisaccharides and polysaccharides. The hydrolysis of di-, tri- and poly-saccharides may be briefly represented:

* This substance is described in the standard book of reference [*Beilsteins Handbuch der organischen Chemie*, vierte Auflage (1920 and 1929), zweiter Band, p. 150] as a syrup, although it was known as a colourless and highly crystalline substance, m.p. 99°, in 1912. It has been prepared and further described in the literature by later investigators. A specimen prepared by the present author by acting upon $d(+)$-sorbitol with an excess of acetic anhydride in the presence of a little zinc chloride was obtained as a colourless crystalline substance, m.p. 99°, having $[\alpha]^{20°}_{5461} = +3{\cdot}35°$ in 0·5 per cent. solution in ethyl alcohol. The substance was analysed and gave the following result: 3·795 mg. gave 6·950 mg. CO_2 and 2·060 mg. H_2O; C=49·95, H=6·03 per cent.; calculated for $C_{18}H_{26}O_{12}$, C=49·8, H=6·0 per cent. (compare pp. 7, 512).

† The old names of dextrose and laevulose were assigned on account of the common forms of these sugars being *dextro*- and *laevo*- rotatory respectively. These names have become unsuitable and confusing chiefly on account of the isolation of a *laevo*rotatory form of glucose and a *dextro*rotatory form of fructose.

$$C_{12}H_{22}O_{11} + H_2O \rightarrow C_6H_{12}O_6 + C_6H_{12}O_6$$
disaccharides

$$C_{18}H_{32}O_{16} + 2H_2O \rightarrow C_6H_{12}O_6 + C_6H_{12}O_6 + C_6H_{12}O_6$$
trisaccharides

$$(C_6H_{10}O_5)_n + nH_2O \rightarrow nC_6H_{12}O_6$$
polysaccharides

Aldohexoses

The simplest formula by which these compounds can be represented is

$$CHO.CHOH.CHOH.CHOH.CHOH.CH_2OH$$

and the substantial accuracy of this is demonstrated by the following considerations:

(i) Being aldehydes, they, like all aldehydes, yield by direct oxidation acids containing the same number of carbon atoms as the original compound. All aldohexoses on direct oxidation are converted into monobasic and dibasic acids having the general formulae $HOOC.(CHOH)_4.CH_2OH$ and $HOOC.(CHOH)_4.COOH$ respectively.

(ii) They each contain five hydroxyl groups in the molecule, yielding in each case a pentaacetyl derivative.

(iii) The presence of the unbranched chain of carbon atoms in the molecule is proved by each aldohexose yielding by reduction a hexahydric alcohol; this hexahydric alcohol on further reduction with hydriodic acid is converted into 2-iodohexane (p. 392)

$$CH_3.CH_2.CH_2.CH_2.CHI.CH_3$$

(iv) They are reducing agents and each yields an osazone, the mechanism of osazone formation being that indicated in the case of the aldopentoses (p. 409), and the general formula of the phenyl osazones is

$$\begin{array}{l} H \\ | \\ C{=}N.N\langle^{H}_{C_6H_5} \\ | \\ C{=}N.N\langle^{H}_{C_6H_5} \\ | \\ (CHOH)_3 \\ | \\ CH_2OH \end{array}$$

(v) The general formula shows the presence in the molecule of four asymmetric carbon atoms, and since there is no symmetry in the molecule there should be capable of existence $2^4 = 16$ optically active isomerides, occurring in eight pairs of *dextro-* and *laevo-*rotatory isomerides, for each *dextro*rotatory aldohexose there should be an aldohexose having an equal and opposite rotatory power; all these sixteen optically active isomeric aldohexoses have been isolated.

The constitution of the sixteen aldohexoses is deduced from that of the eight aldopentoses. An aldohexose may be considered to be derived from an aldopentose by the insertion of another :CHOH group between the terminal aldehyde group and the adjacent :CHOH group in the aldopentose. Since the H— and HO— of the inserted :CHOH group can occupy two positions (either right or left of the carbon chain in the straight projection formula), two aldohexoses can thus be derived from each aldopentose.*

From arabinose, two aldohexoses, *glucose* and *mannose*, are derived:

```
                                                              H
                                                              |      H
                                                              C=N.N<
     CHO†            CHO                                      |      C6H5
     |               |                                        |      H
 H—C—OH          HO—C—H                                       C=N.N<
     |               |                                        |      C6H5
HO—C—H          HO—C—H      both of which yield the       HO—C—H
     |               |       same phenylosazone               |
 H—C—OH           H—C—OH     having the constitution:      H—C—OH
     |               |                                        |
 H—C—OH           H—C—OH                                   H—C—OH
     |               |                                        |
    CH2OH           CH2OH                                    CH2OH
     I               II
```

* The conversion of an aldose containing n carbon atoms into one containing $n+1$ carbon atoms is carried out by the series of reactions (Kiliani) which is represented diagrammatically:

$OHC.(CHOH)_{n-2}.CH_2OH$ or $OHC.CHOH.CHOH.(CHOH)_{n-4}.CH_2OH$

$\xrightarrow{HCN}$ $CN.CHOH.CHOH.CHOH.(CHOH)_{n-4}.CH_2OH$

$\xrightarrow{\text{hydrolysis}}$ $HOOC.CHOH.CHOH.CHOH.(CHOH)_{n-4}.CH_2OH$

$\xrightarrow{-H_2O}$ $OC.CHOH.CHOH.CH.(CHOH)_{n-4}.CH_2OH$ (lactone) [OC and CH joined through —O—]

$\xrightarrow[\text{sodium amalgam}]{\text{reduction}}$ $OHC.CHOH.CHOH.CHOH.(CHOH)_{n-4}.CH_2OH$

or $OHC.(CHOH)_{n-1}.CH_2OH$

† For convenient reference the carbon atoms in the aldohexoses are always designated thus: $OHC.CHOH.CHOH.CHOH.CHOH.CH_2OH$ (carbons numbered 1, 2, 3, 4, 5, 6).

From ribose, the two derived aldohexoses, *allose* and *altrose*, are:

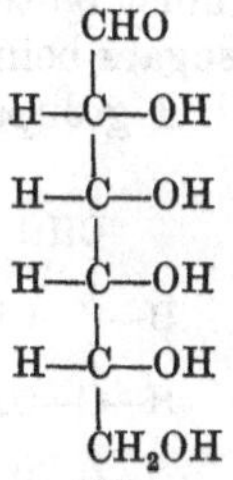

$$\begin{array}{ccc}
\mathrm{CHO} & \mathrm{CHO} & \\
\mathrm{H-C-OH} & \mathrm{HO-C-H} & \\
\mathrm{H-C-OH} & \mathrm{H-C-OH} & \\
\mathrm{H-C-OH} & \mathrm{H-C-OH} & \\
\mathrm{H-C-OH} & \mathrm{H-C-OH} & \\
\mathrm{CH_2OH} & \mathrm{CH_2OH} & \\
\mathrm{III} & \mathrm{IV} &
\end{array}$$

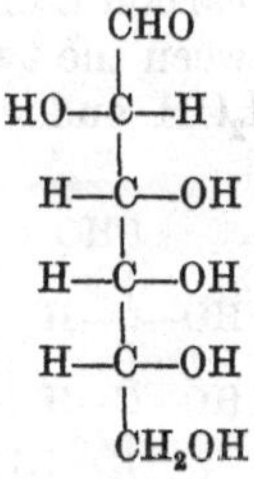

the phenylosazone from both having the constitution:

$$\begin{array}{c}
\mathrm{H-C{=}N.N{<}^{H}_{C_6H_5}} \\
\mathrm{C{=}N.N{<}^{H}_{C_6H_5}} \\
\mathrm{H-C-OH} \\
\mathrm{H-C-OH} \\
\mathrm{H-C-OH} \\
\mathrm{CH_2OH}
\end{array}$$

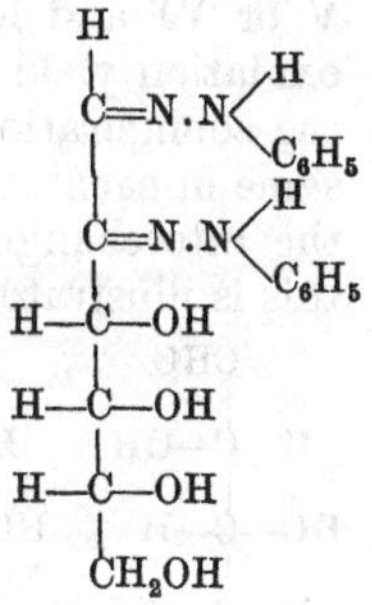

From xylose are derived the two aldohexoses, *gulose* and *idose*:

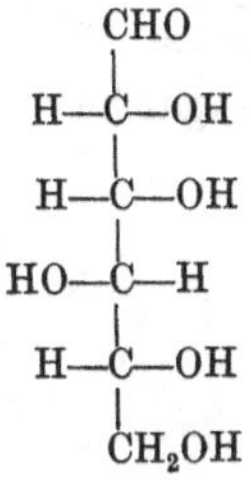

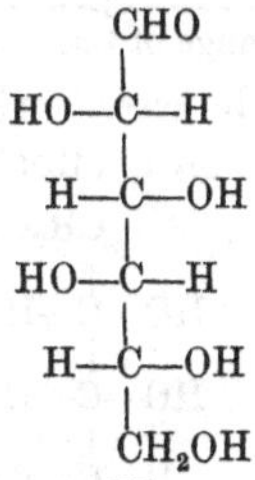

$$\begin{array}{cc}
\mathrm{CHO} & \mathrm{CHO} \\
\mathrm{H-C-OH} & \mathrm{HO-C-H} \\
\mathrm{H-C-OH} & \mathrm{H-C-OH} \\
\mathrm{HO-C-H} & \mathrm{HO-C-H} \\
\mathrm{H-C-OH} & \mathrm{H-C-OH} \\
\mathrm{CH_2OH} & \mathrm{CH_2OH} \\
\mathrm{V} & \mathrm{VI}
\end{array}$$

the phenylosazone derived from both having the constitution:

$$\begin{array}{c}
\mathrm{H-C{=}N.N{<}^{H}_{C_6H_5}} \\
\mathrm{C{=}N.N{<}^{H}_{C_6H_5}} \\
\mathrm{H-C-OH} \\
\mathrm{HO-C-H} \\
\mathrm{H-C-OH} \\
\mathrm{CH_2OH}
\end{array}$$

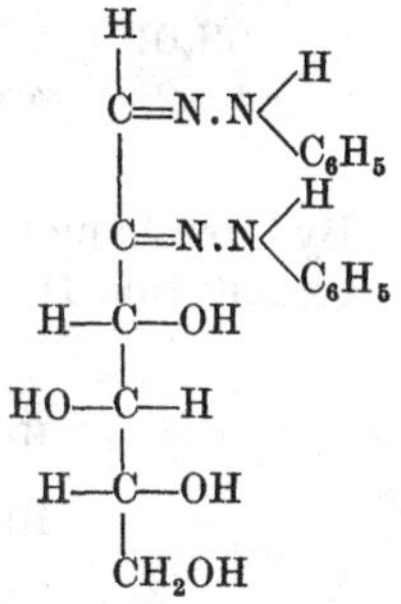

From lyxose are derived the two aldohexoses, *galactose* and *talose*:

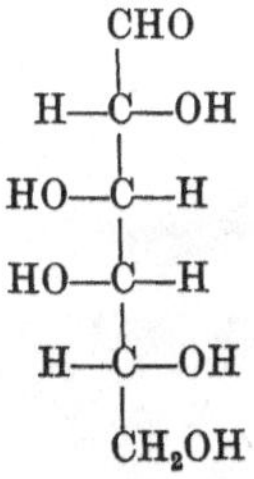

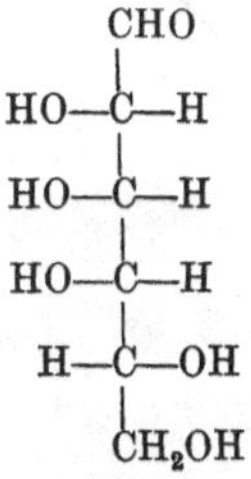

$$\begin{array}{cc}
\mathrm{CHO} & \mathrm{CHO} \\
\mathrm{H-C-OH} & \mathrm{HO-C-H} \\
\mathrm{HO-C-H} & \mathrm{HO-C-H} \\
\mathrm{HO-C-H} & \mathrm{HO-C-H} \\
\mathrm{H-C-OH} & \mathrm{H-C-OH} \\
\mathrm{CH_2OH} & \mathrm{CH_2OH} \\
\mathrm{VII} & \mathrm{VIII}
\end{array}$$

the phenylosazone derived from each of these has the constitution:

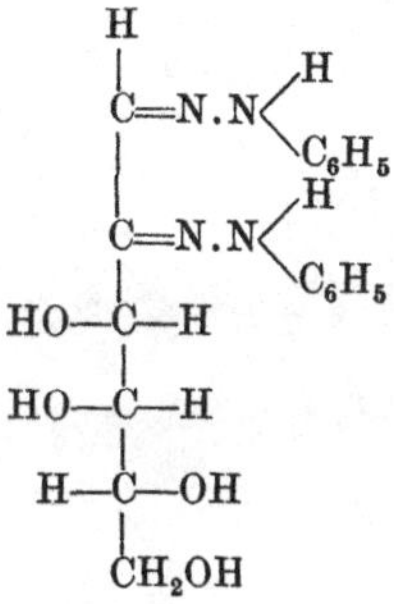

$$\begin{array}{c}
\mathrm{H-C{=}N.N{<}^{H}_{C_6H_5}} \\
\mathrm{C{=}N.N{<}^{H}_{C_6H_5}} \\
\mathrm{HO-C-H} \\
\mathrm{HO-C-H} \\
\mathrm{H-C-OH} \\
\mathrm{CH_2OH}
\end{array}$$

The other eight aldohexoses have constitutions which are represented respectively as mirror images of each of the above compounds represented by formulae I–VIII.

The constitutions of glucose, mannose, gulose and idose are determined by the following considerations:

Glucose must be either I or II, mannose either II or I, gulose either V or VI and idose either VI or V. Now, glucose and gulose on oxidation yield the same dibasic acid (*saccharic acid*) and therefore the configuration of the four asymmetric carbon atoms must be the same in each case, the only difference between the two sugars being the interchanging of the terminal —CH_2OH and —CHO groups; this is illustrated:

$$\begin{array}{rcl} & CHO & \\ & | & \\ H- & C & -OH \\ & | & \\ HO- & C & -H \\ & | & \\ H- & C & -OH \\ & | & \\ H- & C & -OH \\ & | & \\ & CH_2OH & \\ & \text{I} & \end{array} \rightarrow \begin{array}{rcl} & COOH & \\ & | & \\ H- & C & -OH \\ & | & \\ HO- & C & -H \\ & | & \\ H- & C & -OH \\ & | & \\ H- & C & -OH \\ & | & \\ & COOH & \\ & \text{saccharic acid} & \end{array} \leftarrow \begin{array}{rcl} & CH_2OH & \\ & | & \\ H- & C & -OH \\ & | & \\ HO- & C & -H \\ & | & \\ H- & C & -OH \\ & | & \\ H- & C & -OH \\ & | & \\ & CHO & \\ & & \end{array} \leftarrow \left[\begin{array}{rcl} & CHO & \\ & | & \\ HO- & C & -H \\ & | & \\ HO- & C & -H \\ & | & \\ H- & C & -OH \\ & | & \\ HO- & C & -H \\ & | & \\ & CH_2OH & \end{array}\right] \quad \begin{array}{rcl} & CHO & \\ & | & \\ H- & C & -OH \\ & | & \\ H- & C & -OH \\ & | & \\ HO- & C & -H \\ & | & \\ H- & C & -OH \\ & | & \\ & CH_2OH & \\ & \text{V} & \end{array}$$

mirror image of V

interchange of terminal groups

By interchanging the terminal groups in the case of the sugar having constitution II no change in the constitution is effected:

$$\begin{array}{rcl} & CHO & \\ & | & \\ HO- & C & -H \\ & | & \\ HO- & C & -H \\ & | & \\ H- & C & -OH \\ & | & \\ H- & C & -OH \\ & | & \\ & CH_2OH & \\ & \text{II} & \end{array} \qquad \begin{array}{rcl} & CH_2OH & \\ & | & \\ HO- & C & -H \\ & | & \\ HO- & C & -H \\ & | & \\ H- & C & -OH \\ & | & \\ H- & C & -OH \\ & | & \\ & CHO & \\ & & \end{array}$$

$$\searrow \qquad \swarrow$$

$$\begin{array}{rcl} & COOH & \\ & | & \\ HO- & C & -H \\ & | & \\ HO- & C & -H \\ & | & \\ H- & C & -OH \\ & | & \\ H- & C & -OH \\ & | & \\ & COOH & \end{array}$$

mannosaccharic acid

Consequently to glucose must be assigned constitution I and to the sugar gulose which on oxidation yields the same dibasic acid as glucose must be assigned constitution V. To mannose, in the case of which interchanging of the terminal groups causes no change in the constitution, must be assigned constitution II.

The constitutions of glucose and gulose having been determined, the constitutions of mannose and idose also follow from the fact that the latter yield osazones identical with those of glucose and gulose respectively.

The constitution of allose must be represented by III, because allose yields an optically inactive (internally compensated) dibasic acid on oxidation and similarly an optically inactive hexahydric alcohol on reduction. The constitution of altrose must be IV, and this sugar yields the same osazone as allose. Similarly, the constitution of galactose must be VII, this sugar also yielding an optically inactive (internally compensated) dibasic acid (*mucic acid*) and an optically inactive hexahydric alcohol (dulcitol) on reduction. Galactose having constitution VII, that of talose must be VIII, since these two sugars yield the same osazone.

Among the sugars the signs *d*- and *l*- do not necessarily imply the sign of the rotatory power as in general stereochemistry. These signs imply stereochemical relationships*. The constitutional formula for the common *dextro*rotatory form of glucose (*d*-glucose) is chosen arbitrarily (although it has been proved to be correct) as that indicated by I and hence that of its optical and *laevo*rotatory isomeride (*l*-glucose) is the mirror image of I. *d*- and *l*-Glucose can be considered as the 'key' substances of the aldohexoses and it is usual to regard as belonging to the *d*- series all compounds derivable from *d*-glucose by simple reactions, regardless of the sign of their rotatory power. Thus *d*-arabinose and *d*-fructose (p. 437) (obtainable from *d*-glucose) belong to the *d*- series although they are *laevo*rotatory. Any ambiguity may be avoided by using the prefix *d*- or *l*- to denote the series, followed by (+) or (−) to indicate the sign of the rotatory power; thus, ordinary glucose would be described as *d*(+)-glucose, ordinary fructose as *d*(−)-fructose and galactose as *d*(+)-galactose.

It may be emphasised that a combination of carbon atoms in an unbranched chain is not a straight line as represented (for convenience) in the above formulae for the aldohexoses and aldopentoses, etc. The tetrahedral configuration of groups attached to a carbon atom necessitates that the combination of at least five such atoms takes approximately the form of a regular pentagon, and it is much more in keeping with the actual spatial configuration to represent the 'open' constitutional formula of glucose diagrammatically thus:

* This further emphasizes that all sugars having the hydrogen atom and hydroxyl group attached to the carbon atom adjacent to the terminal primary alcohol disposed in the linear projection formula thus:

$$\begin{array}{c} | \\ H\text{—}C\text{—}OH \\ | \\ CH_2OH \end{array}$$

belong to the *d*- series (p. 412).

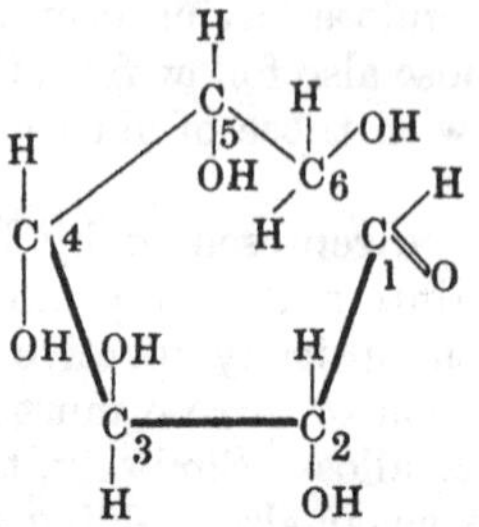

H
C 5, H, OH, OH, C 6, H, H
H, C 4, C 1, H, O
OH, OH, H
C 3 —— C 2
H, OH

compare

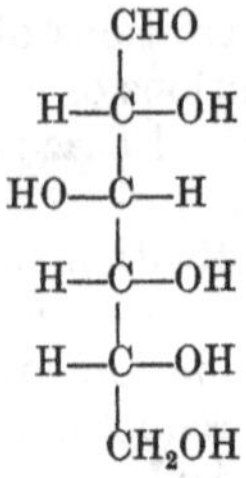

CHO
H—C—OH
HO—C—H
H—C—OH
H—C—OH
CH_2OH

in which the open ring structure is regarded as being at right angles to the plane of the paper, the thickened edges being towards the observer. The hydrogen atoms and hydroxyl groups are then above and below the plane of the partial ring; those which are on the left side of the carbon chain in the straight formula being above the plane of the ring and those on the right side of the carbon chain being below the plane of the ring. In such a representation, the carbon atom 6 of the $—CH_2OH$ group almost coincides with the carbon atom 1 of the —CHO group. It is hardly necessary to point out that the 'open' formulae of the other hexoses and of the pentoses may be usefully represented in a similar manner.

d-*Glucose*, *d*(+)-glucose, grape sugar, formerly described as dextrose, occurs together with fructose in sweet fruits and honey. It is found combined in certain natural compounds belonging to the class known as glucosides which on hydrolysis by means of mineral acids or specific enzymes (p. 446) yield glucose and at least one other substance. Glucose is a constituent of the blood, in which it occurs on the average to the extent of 0·1 per cent. in normal cases. The glucose content of the blood is markedly higher in cases of diabetes mellitus, when it also occurs in the urine.

Glucose may be prepared from sucrose or cane sugar, which on hydrolysis yields an equimolecular mixture of glucose and fructose:

$$\underset{\text{sucrose}}{C_{12}H_{22}O_{11}} + H_2O \rightarrow \underset{\text{glucose}}{C_6H_{12}O_6} + \underset{\text{fructose}}{C_6H_{12}O_6}$$

the glucose being the less soluble in alcohol containing a little water.

For the preparation, 40 grams of powdered sucrose is added to a mixture of 120 c.c. of 90 per cent. ethyl alcohol and 5 c.c. of concentrated hydrochloric acid and the mixture heated at 40–50° for 2 hours with occasional stirring and then allowed to cool. Glucose separates from the mixture, especially after introduction of a small crystal of glucose; the glucose may be recrystallised from ethyl alcohol diluted with half its volume of water.

Glucose may also be prepared from starch by hydrolysis effected by heating with very dilute sulphuric acid under pressure:

$$\underset{\text{starch}}{(C_6H_{10}O_5)_n} + nH_2O \rightarrow \underset{\text{glucose}}{nC_6H_{12}O_6}$$

The resulting solution is neutralised with barium carbonate, heating with decolorising charcoal, filtered and evaporated under reduced pressure.

Glucose is the hydrolysis product of maltose (p. 451), one molecule of the latter yielding two molecules of glucose:

$$\underset{\text{maltose}}{C_{12}H_{22}O_{11}} + H_2O \rightarrow \underset{\text{glucose}}{C_6H_{12}O_6} + \underset{\text{glucose}}{C_6H_{12}O_6}$$

From alcohol or concentrated aqueous solution at 30°, glucose crystallises in colourless needles, m.p. 146°, having the composition $C_6H_{12}O_6$; it also crystallises from cold water in colourless plates, m.p. 86°, having the composition $C_6H_{12}O_6 . H_2O$. On reduction, it is converted into sorbitol (p. 415), and on oxidation with bromine water it is converted into the monobasic *gluconic acid*; on oxidation with nitric acid it is converted into *saccharic acid*:

```
       CHO
       |
    H—C—OH
       |
   HO—C—H
       |
    H—C—OH
       |
    H—C—OH
       |
       CH2OH
  d(+)-glucose
```

↗

```
       COOH
       |
    H—C—OH
       |
   HO—C—H
       |
    H—C—OH
       |
    H—C—OH
       |
       CH2OH
 d-gluconic acid
```

⟶

```
       COOH
       |
    H—C—OH
       |
   HO—C—H
       |
    H—C—OH
       |
    H—C—OH
       |
       COOH
 d-saccharic acid
```

↘

```
       CHO
       |
    H—C—OH
       |
   HO—C—H
       |
    H—C—OH
       |
    H—C—OH
       |
       COOH
 d-glucuronic acid
```

Glucuronic acid is an oxidation product of glucose formed in the human body, especially after the administration of chloral hydrate as a soporific. Chloral hydrate is probably reduced to trichloroethyl alcohol, which then reacts with glucose (as do other alcohols, e.g. methyl alcohol, p. 426) in such a manner that the $—CH_2OH$ of the sugar molecule is left free to undergo oxidation. The compound (*urochloralic acid*) so formed may then be excreted and the glucuronic acid obtained from this by hydrolysis (compare p. 450). Each of these acids, direct oxidation products of glucose, have, like glucose, six carbon atoms in the molecule.

Glucose, like all hexoses, is a reducing agent and will reduce an ammoniacal solution of silver oxide or silver nitrate to silver, which is usually obtained in the form of a mirror on the walls of the vessel. Glucose and all the hexoses reduce alkaline solutions of cupric hydroxide to cuprous oxide, and this reaction is the basis of volumetric methods for the estimation of reducing sugars (p. 462).

As an aldehyde, glucose reacts with phenylhydrazine, yielding a phenylhydrazone which occurs in two isomeric forms (m.p. 160° and 141° respectively), both of which have rotatory powers opposite in sign to that of the sugar. With excess of phenylhydrazine it yields phenylglucosazone, occurring in two isomeric forms (m.p. 145° and 205°, respectively), and these again have rotatory powers opposite in sign to that of the sugar. The phenylosazones obtained from *d*(+)-glucose, *d*(+)-mannose and *d*(−)-fructose are identical. Through its osazone *d*(+)-glucose is converted into *d*(−)-fructose (p. 440).

Glucose, when treated with hydroxylamine, yields the corresponding oxime (*glucoseoxime*, m.p. 137°), theoretically capable of existing in two isomeric forms. This compound is of interest since, through it, glucose may be degraded to arabinose. Omitting intermediate stages, this typical conversion of an aldohexose to an aldopentose may be outlined thus:

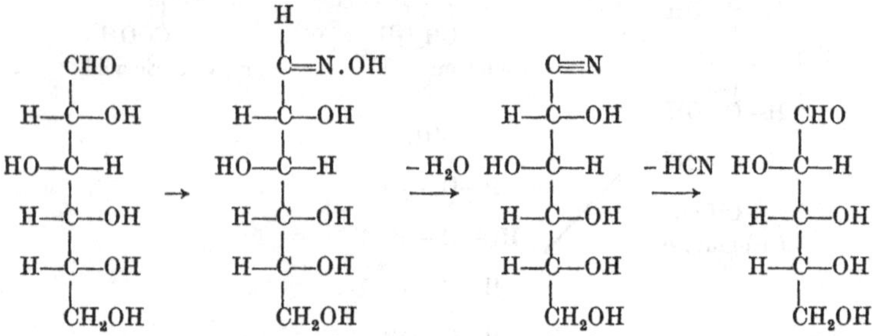

$$
\begin{array}{c}
CHO \\
H\text{—}C\text{—}OH \\
HO\text{—}C\text{—}H \\
H\text{—}C\text{—}OH \\
H\text{—}C\text{—}OH \\
CH_2OH \\
d(+)\text{-glucose} \\
\text{aldohexose}
\end{array}
\rightarrow
\begin{array}{c}
H \\
C\text{=}N.OH \\
H\text{—}C\text{—}OH \\
HO\text{—}C\text{—}H \\
H\text{—}C\text{—}OH \\
H\text{—}C\text{—}OH \\
CH_2OH \\
d\text{-glucoseoxime}
\end{array}
\xrightarrow{-H_2O}
\begin{array}{c}
C\equiv N \\
H\text{—}C\text{—}OH \\
HO\text{—}C\text{—}H \\
H\text{—}C\text{—}OH \\
H\text{—}C\text{—}OH \\
CH_2OH \\
\text{nitrile of} \\
d\text{-gluconic acid}
\end{array}
\xrightarrow{-HCN}
\begin{array}{c}
CHO \\
HO\text{—}C\text{—}H \\
H\text{—}C\text{—}OH \\
H\text{—}C\text{—}OH \\
CH_2OH \\
d(-)\text{-arabinose} \\
\text{aldopentose}
\end{array}
$$

The close relationship between glucose, mannose and (as will be further emphasised later, p. 440) fructose is shown by the interconversion of these sugars in aqueous solution containing small quantities of alkali. When an aqueous solution of glucose is treated with alkali, the rotatory power of the solution falls and, on attaining equilibrium, the solution is found to contain glucose, mannose and fructose, thus:

$$\text{glucose} \rightleftarrows \text{mannose} \rightleftarrows \text{fructose}$$

The solution also contains small amounts of other products and an equilibrium between these three sugars is set up whichever sugar is used as the starting material. The interconversion of these three sugars is effected with weak alkalies and organic bases and even with disodium hydrogen phosphate.

The general method for the conversion of an aldose into another aldose in which the configuration of the hydrogen atom and hydroxyl group attached to carbon atom 2 is reversed (compare glucose and mannose), a reversible process known as *epimerisation*, is illustrated by the conversion of glucose into mannose, thus:

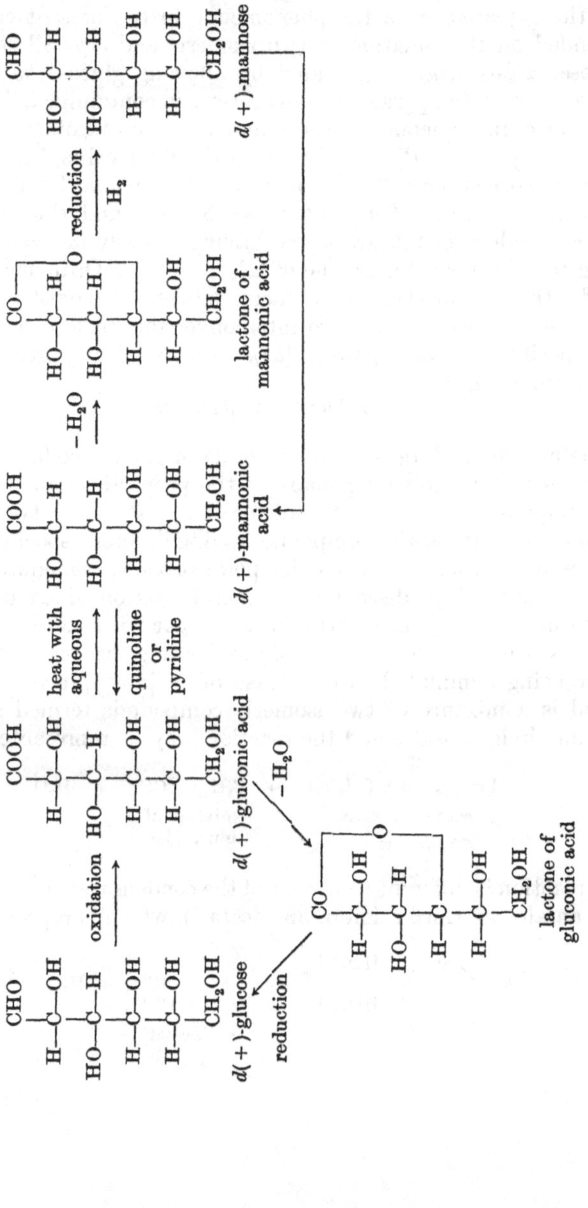

```
   CHO                       COOH                           COOH
H—C—OH                    H—C—OH     heat with          HO—C—H
HO—C—H    oxidation      HO—C—H      aqueous            HO—C—H
H—C—OH    ———————→        H—C—OH     ————————→           H—C—OH
H—C—OH                    H—C—OH     ←————————           H—C—OH
   CH2OH                     CH2OH   quinoline              CH2OH
d(+)-glucose          d(+)-gluconic acid  or          d(+)-mannonic
                                     pyridine               acid

      CO————                  −H2O                       CHO
   HO—C—H    |               ———→    reduction        HO—C—H
   HO—C—H    O                       ————————→        HO—C—H
    H—C——————                           H2             H—C—OH
    H—C—OH                                             H—C—OH
      CH2OH                                              CH2OH
  lactone of                                         d(+)-mannose
  mannonic acid

d(+)-gluconic acid —(−H2O)→ lactone of gluconic acid —(reduction)→ d(+)-glucose

      CO————
    H—C—OH   |
   HO—C—H    O
    H—C——————
    H—C—OH
      CH2OH
  lactone of
  gluconic acid

(d(+)-mannose —→ d(+)-mannonic acid)
```

If the rotatory power of a cold aqueous solution of ordinary glucose is observed as soon as possible after preparing the solution, it is found to diminish to almost half its initial value on standing, reaching a final and constant value of $[\alpha]_D = +52{\cdot}5°$. The attainment of the final constant value of the optical rotatory power may be hastened by boiling the solution for a few minutes or by the addition of a very small quantity of aqueous ammonia. This alteration of the rotatory power of a solution on standing is known as *mutarotation* (Lowry), and the explanation of the phenomenon in the present case finally depended on the isolation of two isomeric and crystalline forms of glucose. α-*d*-Glucose is prepared by allowing glucose to crystallise at the ordinary temperature from acetic acid containing a little water. This form in aqueous solution undergoes mutarotation and has initially $[\alpha]_D = +110°$, diminishing to the final value, $[\alpha]_D = +52{\cdot}5°$. When glucose is crystallised from pure acetic acid at higher temperatures another form, β-*d*-glucose, is obtained and this in aqueous solution undergoes mutarotation, having initially $[\alpha]_D = +17{\cdot}5°$ and rising to a final constant value of $[\alpha]_D = +52{\cdot}5°$. Ordinary glucose is chiefly the α- variety. These facts indicate the existence of two varieties of glucose which are interconvertible in aqueous solution, the specific rotatory power, $[\alpha]_D = +52{\cdot}5°$, being given by the equilibrium mixture:

$$\alpha\text{-glucose} \rightleftarrows \beta\text{-glucose}$$

The unbranched, open chain, formula of glucose so far used, while satisfactory for explaining many of the properties and reactions of the compound, does not account for the existence of two interconvertible varieties of the compound having different specific rotatory powers and is therefore, from this point of view, inadequate.

E. Fischer in 1893 discovered a general reaction which all reducing sugars undergo, a condensation of the sugar with methyl alcohol in the presence of hydrogen chloride by heating for some time at 65°, water being eliminated. In the case of ordinary glucose, the compound is a mixture of two isomeric compounds termed α- and β-monomethylglucosides and the reaction may be represented:

$$\underset{\text{glucose}}{C_6H_{12}O_6} + CH_3OH \rightarrow \underset{\substack{\text{mixture of}\\ \text{glucosides}}}{C_6H_{11}O_6.CH_3} + H_2O$$

This reaction is different from that of the condensation of an aldehyde with an alcohol (formation of an 'acetal'), which is represented:

$$R.C\begin{cases}H\\ O\end{cases} + \begin{matrix}HO.CH_3\\ HO.CH_3\end{matrix} \rightarrow \underset{\text{'acetal'}}{R.C\begin{cases}H\\ OCH_3\\ OCH_3\end{cases}} + H_2O$$

The two monomethylglucosides (known as 'α-' and 'β-' respectively) have been obtained in a state of purity as colourless crystalline substances; their physical constants are:

α-monomethylglucoside, m.p. 165–166°, $[\alpha]_D = +159°$

β-monomethylglucoside, m.p. 105°, $[\alpha]_D = -34°$

Unlike glucose, they are not reducing agents, they do not undergo mutarotation and do not possess aldehydic properties. To explain their formation and properties, Fischer assigned to them a ring structure and, in accordance with present views, their formation may be represented thus:

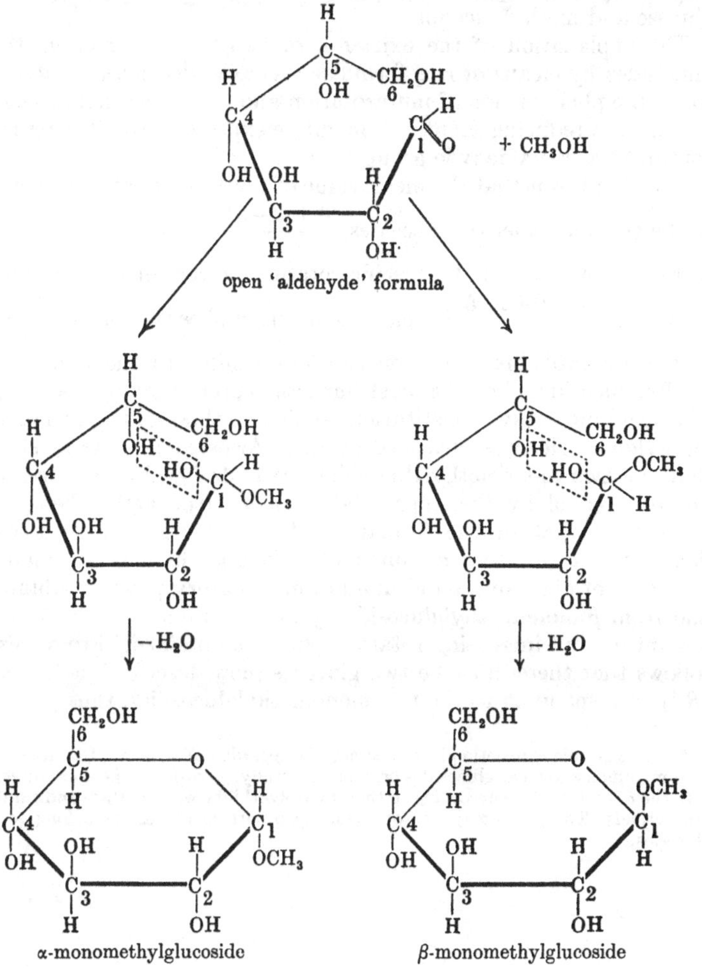

$+ CH_3OH$

open 'aldehyde' formula

$-H_2O$ $-H_2O$

α-monomethylglucoside β-monomethylglucoside

The difference between the isomeric forms of the compound is in the relative positions of the methoxy (—OCH_3) group to the plane of the ring. In the α-glucoside the methoxy group is attached to carbon atom 1 on the same side of the ring as the —OH group on the neighbouring carbon atom 2. In the β-glucoside the methoxy group is on the side of the ring opposite from that of the —OH group attached to carbon atom 2. α-Monomethylglucoside is hydrolysed by the enzyme, *maltase*, and the β- compound is hydrolysed by the enzyme, *emulsin*.* These two compounds are also hydrolysed by boiling with dilute hydrochloric acid, and whether hydrolysed in this way or by the respective enzymes the products of hydrolysis are glucose and methyl alcohol.

The explanation of the existence of two isomeric monomethylglucosides by means of ring formulae was extended later to account for (i) the phenomenon of mutarotation and (ii) glucose not acting as rapidly as a reducing agent as it might be expected to do if it actually contained a free aldehyde group.

It is significant that the mean value of the specific rotatory powers of the two monomethylglucosides, $\frac{+159-34}{2} = +62{\cdot}5°$, is very close to the mean value of the specific rotatory powers of the two forms of glucose, $\frac{+110+17{\cdot}5}{2} = +63{\cdot}7°$, and this implies that the two forms of glucose probably have constitutions similar to the two monomethylglucosides; but the most important proof that the two forms of glucose must have constitutions similar to those of the two monomethylglucosides was furnished by E. F. Armstrong. Having shown that the two monomethylglucosides are hydrolysed to glucose and methyl alcohol by the appropriate enzyme (the action being the reversal of that of the formation of the glucosides), Armstrong demonstrated that from α-monomethylglucoside by means of maltase a glucose of high initial and diminishing rotatory power is obtained and from β-monomethylglucoside by means of emulsin a glucose of low initial and increasing rotatory power is obtained. From this it follows that there must be two glucoses (now described as 'α-' and 'β-') corresponding to the two monomethylglucosides, thus:

* These simple synthetic glucosides may be regarded as typical of a large class of compounds known as glucosides or, more generally, *glycosides*. These are divisible into the α- and β- series according to their hydrolysability with maltase and emulsin respectively. The great majority of naturally occurring glucosides belong to the β- series.

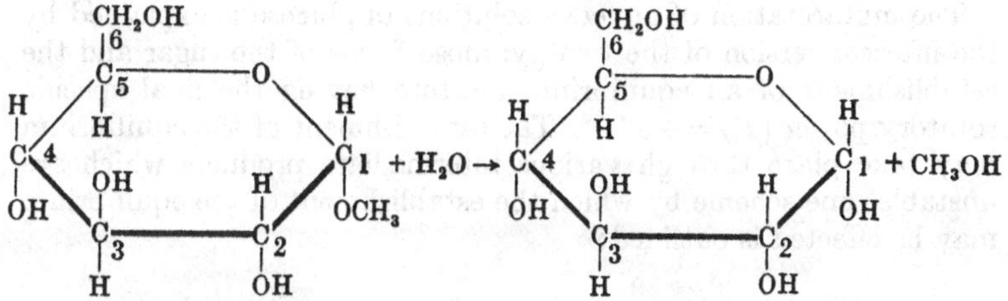

α-monomethylglucoside, $[\alpha]_D = +159°$ $+ H_2O \rightarrow$ α-glucose, $[\alpha]_D = +110°$ $+ CH_3OH$

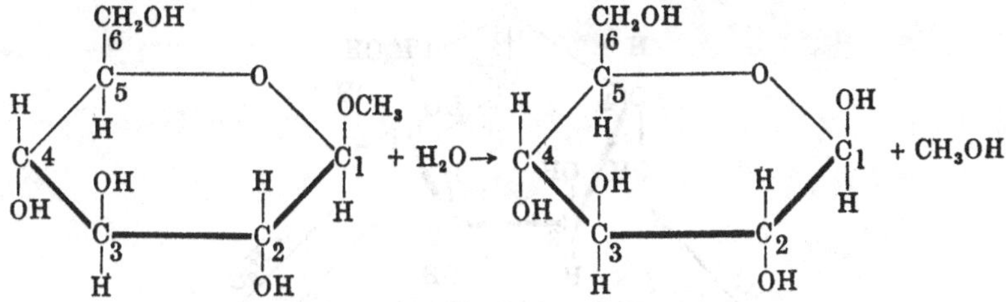

β-monomethylglucoside, $[\alpha]_D = -34°$ $+ H_2O \rightarrow$ β-glucose, $[\alpha]_D = +17{\cdot}5°$ $+ CH_3OH$

The above formulae for the α- and β- monomethylglucosides and the α- and β- forms of glucose are based on a ring structure containing five carbon atoms and an oxygen atom. The simplest compound having a constitution built up in this way is *pyran* (1 : 2-pyran), and

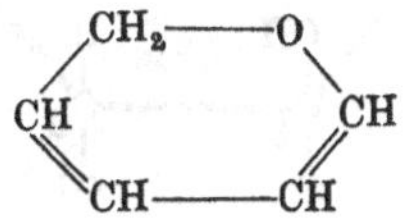

1 : 2-pyran 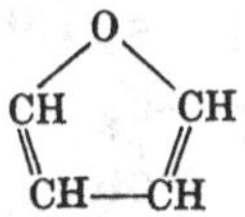 furan 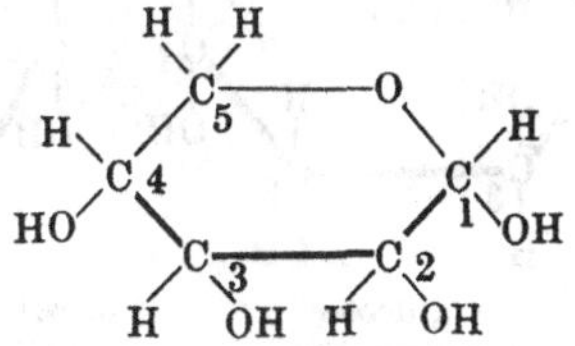 pentose

the 'pyranose' formula for the most stable forms of glucose and of other monosaccharides has been amply demonstrated by the investigations of Haworth and his collaborators.* At the same time, other ring formulae such as that based on the constitution of *furan* are possible and, in many cases, the 'furanose' forms which are less stable and more easily oxidisable than the pyranose forms are known. As indicated above reduction and hydroxylation of 1 : 2-pyran will theoretically yield a tetrahydrotetrahydroxypyran which can be regarded as the simplest normal sugar, i.e. a pentose such as arabinose, etc. (p. 413).

* For a fuller account of this work, see *The Constitution of the Sugars*, by W. N. Haworth (Arnold, 1929); *The Carbohydrates*, by E. F. and K. F. Armstrong (Longmans, Green and Co., 1934).

The mutarotation of aqueous solutions of glucose is explained by the interconversion of the two pyranose forms of the sugar and the establishment of an equilibrium mixture having the final specific rotatory power $[\alpha]_D = +52{\cdot}5°$. The establishment of the equilibrium may take place through various intermediate products which are unstable; one scheme by which the establishment of the equilibrium may be effected is outlined:

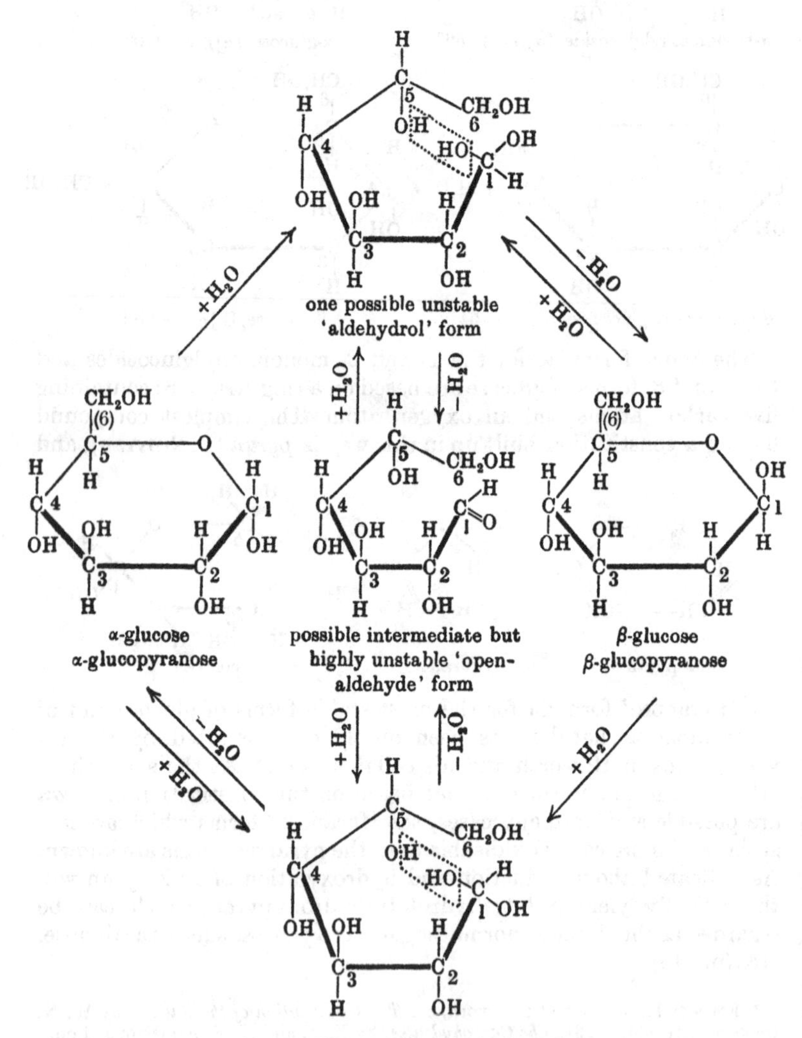

It should be noted that when ring formation from the 'aldehydrol' forms takes place the —CH_2OH group attached to carbon atom 5 has to take a position above the plane of the ring. This will be made clear if the constitution of the monosaccharides is studied using suitable 'spring' models.

While the chief constituents of an aqueous solution of glucose in equilibrium are α- and β- glucose, it is probable that the solution may also contain in small quantities other more chemically active isomeric forms of the sugar, among which may be the two isomeric furanose forms (compare α- and β- glucose) having the formulae:

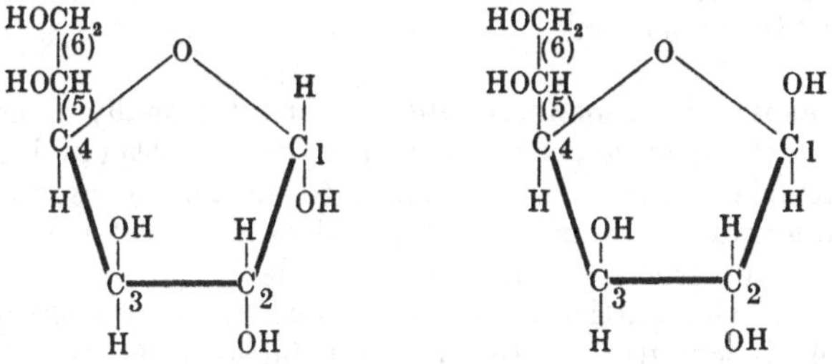

α-glucofuranose

β-glucofuranose

This mixture of the two furanose forms may be the less stable and more easily oxidisable γ-glucose which has not yet been isolated, although its two isomeric tetramethyl derivatives have been described.

The reducing power of glucose, as shown by its action on solutions of cupric salts, is almost equal to that of mannose, galactose and fructose. The reduction of glucose to sorbitol by means of sodium amalgam has been described (p. 415).

l(−)-Glucose has chemical properties identical with those of its optical isomeride, the ordinary form of the sugar. In its physical properties (and those of its derivatives), *l*-glucose only differs from *d*-glucose (and its corresponding derivatives) in the sign of the optical rotatory power. α- and β- *l*-glucoses have pyranose formulae which are mirror images of the corresponding α- and β- *d*-glucoses and have specific rotatory powers which are equal to but opposite in sign to those of the latter.

The ring formulae for the isomeric forms of glucose have the free —CH_2OH outside the ring; hence it is easy to understand the formation of *glucuronic acid* under suitable conditions. Glucuronic acid derived from β-glucopyranose would have the constitution:

```
              COOH
               |
               C————————O
        H    / |         \    OH
        |   /  H          \   |
        C                   C
        | \    OH     H    / |
       OH  \   |      |   /  H
            C—————————C
            |         |
            H         OH
```

As a result of many investigations, it has been established that all the aldohexoses exist in stable α- and β- pyranose forms and less stable α- and β- furanose forms corresponding to the isomeric forms of glucose. Two other aldohexoses, *mannose* and *galactose*, will be described in some detail.

d-*Mannose*, *d*(+)-mannose, differs constitutionally from *d*(+)-glucose in the relative positions of the hydrogen atom and hydroxyl group attached to carbon atom 2 next to the aldehyde group (in the 'open-aldehyde' formula). *d*-Mannose does not occur free in nature but is derived from certain polysaccharides known as *mannans* or *mannosans*. One particular such source of mannose is the so-called vegetable ivory: fruit of the Tagua palm growing extensively in Southern Rhodesia. Mannose may be readily obtained by hydrolysing the vegetable ivory by warming with moderately concentrated sulphuric acid, removing the acid with barium carbonate, evaporating the solution to a syrupy consistency from which the crystalline sugar is obtained by addition of acetic acid. *d*-Mannose has been obtained from tubercle bacilli.

Mannose crystallises in colourless rhombic prisms, which have a curious sweet changing to a bitter taste. This is remarkably different from the sweetness of glucose and fructose, which are so closely related to mannose.

The phenylosazone from maltose is identical with the phenylosazone obtained from glucose and from fructose. Maltose is readily identified by its sparingly soluble phenylhydrazone,

```
          H    H    OH   OH     H
          |    |    |    |     /        H
HOCH2.C————C————C————C————C=N.N<
          |    |    |    |              C6H5
          OH   OH   H    H
```

which separates very readily when phenylhydrazine is added to a solution of maltose in acetic acid.

The aqueous solution of mannose undergoes mutarotation. When the aqueous solution of *d*(+)-mannose reaches equilibrium, it has $[\alpha]_D = +14{\cdot}6°$ and mannose must exist in isomeric forms similar to those of glucose. The more stable α- and β- mannopyranoses have the following constitutions:

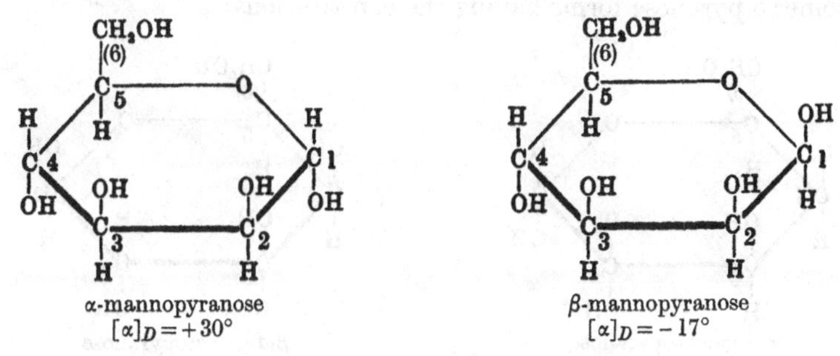

α-mannopyranose
$[\alpha]_D = +30°$

β-mannopyranose
$[\alpha]_D = -17°$

and the corresponding mannofuranoses are represented:

α-mannofuranose

β-mannofuranose

The reducing power of mannose as shown by its action on solutions of cupric salts is almost equal to that of glucose, galactose and fructose. Reduction of *d*-mannose to *d*-mannitol has already been referred to (p. 415).

d-*Galactose*, *d*(+)-galactose, occurs combined with *d*-glucose in lactose or milk sugar. The galactosides belong like the glucosides to the class of glycosides and yield galactose on hydrolysis. Galactose may be described as the sugar of the brain, where it occurs in combination in the cerebrosides which occur in nerve and brain tissue. When galactose was first isolated from this source it was described under the name cerebrose.

Galactose can be readily obtained from lactose by hydrolysis with sulphuric acid:

$$\underset{\text{lactose}}{C_{12}H_{22}O_{11}} + H_2O \rightarrow \underset{d(+)\text{-glucose}}{C_6H_{12}O_6} + \underset{d(+)\text{-galactose}}{C_6H_{12}O_6}$$

Galactose is purified by crystallisation from aqueous methyl alcohol and from acetic acid; it is less soluble than glucose.

In aqueous solution, galactose undergoes mutarotation and there is evidence that several (three or possibly four) isomeric forms are present in the solution. As in other cases, the most stable are the isomeric pyranose forms having the constitutions:

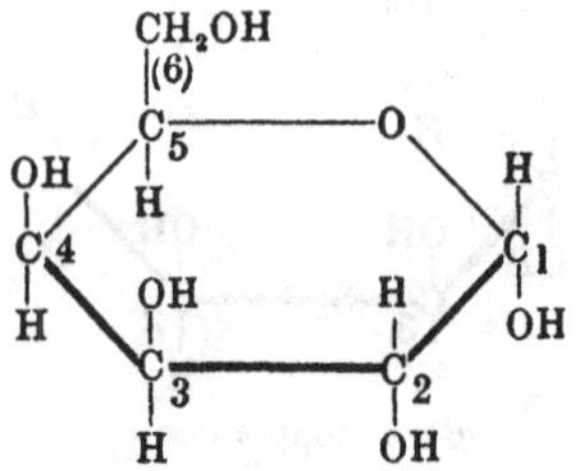

α-d-galactopyranose
$[\alpha]_D = +145°$

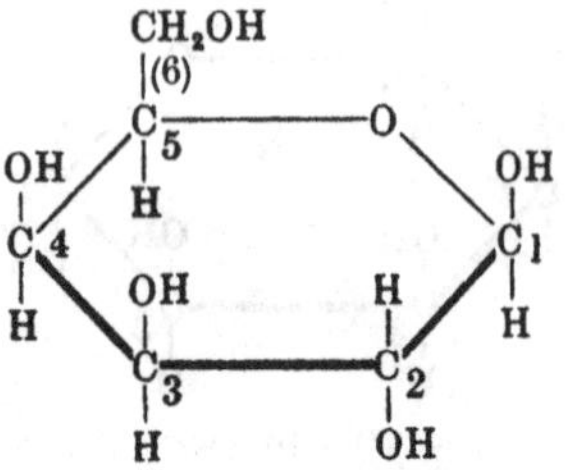

β-d-galactopyranose
$[\alpha]_D = +54°$

Both of these have been isolated and evidence for the existence in aqueous solution of at least one of the furanose forms has been obtained. The furanose forms have the constitutions indicated (compare the furanose formulae of glucose):

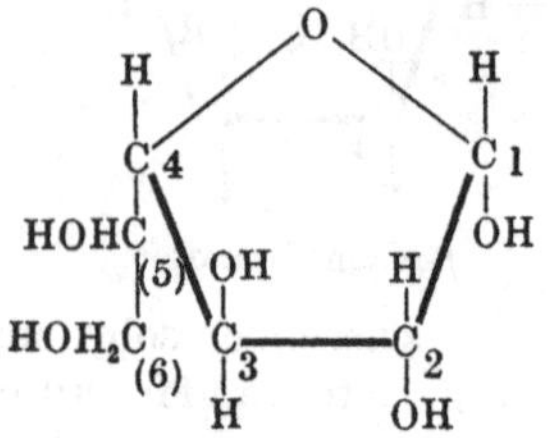

α-galactofuranose

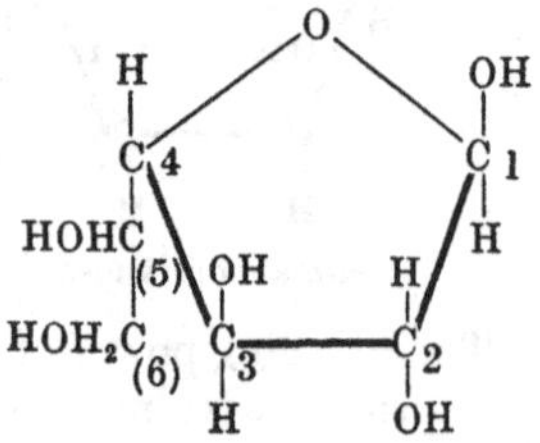

β-galactofuranose

The constitution of galactose follows from (i) its being derived from lyxose, which determines the configuration of the hydrogen atoms and hydroxyl groups attached to carbon atoms (3), (4) and (5); (ii) on reduction, it yields the inactive hexahydric alcohol, dulcitol, which determines the configuration of the hydrogen atom and hydroxyl group attached to carbon atom (2); (iii) on oxidation, it yields the inactive dibasic acid, mucic acid, which also determines the configuration of the hydrogen atom and hydroxyl group attached to carbon atom (2). Galactose yields a phenylosazone identical with that derived from talose.

The reducing power of galactose as determined by its action on solutions of cupric salts is practically the same as that of the other monosaccharides.

Ketohexoses

The ketohexoses are isomeric with the aldohexoses. Their simplest general formula:

$$\begin{array}{l} (1)\ CH_2OH \\ \quad\ \ | \\ (2)\ CO \\ \quad\ \ | \\ (3)\ CHOH \\ \quad\ \ | \\ (4)\ CHOH \\ \quad\ \ | \\ (5)\ CHOH \\ \quad\ \ | \\ (6)\ CH_2OH \end{array}$$

implies

(i) that the carbon atoms are in an unbranched chain. This is proved by each of them yielding on reduction in aqueous solution with sodium amalgam two isomeric hexahydric alcohols whose constitutions may be represented, thus:

$$\begin{array}{c} CH_2OH \\ | \\ H\text{—}C\text{—}OH \\ | \\ (CHOH)_3 \\ | \\ CH_2OH \end{array} \quad \text{and} \quad \begin{array}{c} CH_2OH \\ | \\ HO\text{—}C\text{—}H \\ | \\ (CHOH)_3 \\ | \\ CH_2OH \end{array}$$

On complete reduction with hydriodic acid these alcohols all yield 2-iodohexane and, therefore, as in the case of the aldohexoses, the carbon atoms in the ketohexoses must be in an unbranched chain;

(ii) the presence of five hydroxyl groups in the molecule, which is proved by each of them yielding pentaacyl (pentaacetyl, pentabenzoyl, etc.—esters of acetic, benzoic, etc. acids) derivatives;

(iii) that they are ketones and the carbonyl or keto group is adjacent to one of the primary alcohol groups. This is proved by each of them on oxidation yielding two acids, one containing two carbon atoms and the other four carbon atoms in their molecules. Such a scission of the molecule in direct oxidation implies the presence of the ketonic group (compare definition of a ketone as a substance which on oxidation yields acids each of which contains fewer carbon atoms in the molecule than the original substance) and the position of the ketonic group is determined by the number of carbon atoms in the molecules of the acids produced, thus:

$$\begin{array}{c} CH_2OH \\ | \\ CO \\ \cdots\cdots|\cdots\cdots \\ CHOH \\ | \\ CHOH \\ | \\ CHOH \\ | \\ CH_2OH \end{array} \quad \xrightarrow{O_2} \quad \begin{array}{c} CH_2OH \\ | \\ COOH \\ \\ COOH \\ | \\ CHOH \\ | \\ CHOH \\ | \\ CH_2OH \end{array}$$

The two acids indicated are glycollic (hydroxyacetic) acid and trihydroxybutyric acid (containing two asymmetric carbon atoms and, therefore, capable of existing in four optically active—two *dextro*rotatory and two *laevo*rotatory—forms). The nature of the acids obtained depends, of course, on the conditions of oxidation, since the immediate oxidation products indicated above may undergo further oxidation, e.g. to oxalic and tartaric acids.

(iv) That there are three asymmetric carbon atoms in the molecule of each of them, carbon atoms (3), (4) and (5). Therefore, there should be capable of existence $2^3 = 8$ optically active isomerides occurring in four *dextro*rotatory and *laevo*rotatory pairs, to each *dextro*rotatory compound there being a *laevo*rotatory isomeride having an equal and opposite rotatory power. Actually, these eight optically active ketohexoses are known, four of which are *dextro*rotatory and four *laevo*-rotatory.

The number of isomeric ketohexoses is half that of the isomeric aldohexoses. The possible configurations of the hydrogen atoms and hydroxyl groups attached to carbon atoms (3), (4) and (5) are indicated in the case of eight of the aldohexoses (p. 418), the other eight having configurations which are mirror images of these. It will be seen that there are correspondingly four different possible configurations for the ketohexoses, since the configurations of the hydrogen atoms and hydroxyl groups attached to carbon atoms (3), (4) and (5) in the case of each pair of aldohexoses (e.g. glucose and mannose) are the same. This implies that there is one ketohexose corresponding to glucose and mannose, a second one corresponding to allose and altrose, a third to gulose and idose and a fourth to galactose and talose. Or regarded in another way, the isomerism of the ketohexoses will be analogous to that of the aldopentoses, carbon atoms (2), (3) and (4) in the latter becoming carbon atoms (3), (4) and (5) in the former.

The ketohexoses yield phenylhydrazones (which may exist in stereoisomeric forms) and osazones, or, more particularly, phenylosazones (which also may exist in stereoisomeric forms) having the general formulae:

$$\begin{array}{l} CH_2OH \\ | \\ C{=}N{-}N\begin{cases} H \\ C_6H_5 \end{cases} \\ | \\ (CHOH)_3 \\ | \\ CH_2OH \end{array} \quad \text{and} \quad \begin{array}{l} \quad H \\ \; / \\ C{=}N.N\begin{cases} H \\ C_6H_5 \end{cases} \\ | \\ C{=}N.N\begin{cases} H \\ C_6H_5 \end{cases} \\ | \\ (CHOH)_3 \\ | \\ CH_2OH \end{array}$$

It is therefore obvious that the osazones of ketohexoses will be identical with the corresponding osazones of aldohexoses in which the configurations of hydrogen atoms and hydroxyl groups attached to carbon atoms (3), (4) and (5) in each class of sugar are the same. For example, fructosazone is identical with the osazone produced from glucose and from mannose. Consequently, the simplest formula of fructose must be as indicated, the constitutional formulae of glucose and mannose being given for comparison:

$$
\begin{array}{ccc}
\begin{array}{c}
CH_2OH \\
| \\
CO \\
| \\
HO{-}C{-}H \\
| \\
H{-}C{-}OH \\
| \\
H{-}C{-}OH \\
| \\
CH_2OH
\end{array}
&
\begin{array}{c}
CHO \\
| \\
H{-}C{-}OH \\
| \\
HO{-}C{-}H \\
| \\
H{-}C{-}OH \\
| \\
H{-}C{-}OH \\
| \\
CH_2OH
\end{array}
&
\begin{array}{c}
CHO \\
| \\
HO{-}C{-}H \\
| \\
HO{-}C{-}H \\
| \\
H{-}C{-}OH \\
| \\
H{-}C{-}OH \\
| \\
CH_2OH
\end{array}
\\
d(-)\text{-fructose} & d(+)\text{-glucose} & d(+)\text{-mannose}
\end{array}
$$

Ordinary *fructose* which, on account of its being *laevo*rotatory, was formerly known as *laevulose* should be systematically described as *d*(−)-fructose to indicate its close relationship with ordinary glucose. It is the most important ketohexose. It occurs along with glucose in fruit juices and honey, and, combined with glucose, it occurs in sucrose or cane sugar. Inulin, which is the reserve food material in dahlia tubers and Jerusalem artichokes (compare starch in potatoes, etc.), yields fructose alone on hydrolysis.

Fructose may be prepared using sucrose as the starting material. Sucrose (300 grams) is dissolved in water (1200 ml.) and to this solution is added concentrated hydrochloric acid (7 ml.) and the mixture, occasionally shaken, heated in boiling water for 2 hours. The solution is then diluted to 3000 ml. and calcium hydroxide (200 grams) added with thorough stirring which is continued until the calcium fructosate is precipitated, calcium glucosate remaining in solution. The calcium fructosate is filtered off, suspended in water and decomposed by stirring the suspension with the calculated quantity of oxalic acid. The solution is filtered from calcium oxalate and the filtrate evaporated under reduced pressure until the fructose crystallises. The reactions involved in the preparation are:

(i) Hydrolysis of the sucrose,

$$
\underset{\text{sucrose}}{C_{12}H_{22}O_{11}} + H_2O \rightarrow \underset{d(+)\text{-glucose}}{C_6H_{12}O_6} + \underset{d(-)\text{-fructose}}{C_6H_{12}O_6}
$$

This is frequently known as 'inversion' of the sucrose. Sucrose is *dextro*rotatory and on hydrolysis the mixture becomes *laevo*rotatory, since fructose has a *laevo*rotatory power larger than the *dextro*rotatory power of glucose; the mixture of equimolecular quantities of glucose and fructose produced by the inversion of cane sugar is frequently described as 'invert sugar'.

(ii) Neutralisation of the hydrochloric acid by calcium hydroxide and the conversion of glucose and fructose into their calcium hydroxide derivatives. The latter may be regarded as being formed thus:

$$C_6H_{11}O_5.OH + Ca(OH)_2 \rightarrow C_6H_{11}O_5.O.Ca(OH) + H_2O$$

(iii) Decomposition of the sparingly soluble calcium fructosate by the calculated quantity of oxalic acid:

$$2C_6H_{11}O_5.O.Ca(OH) + H_2C_2O_4 \rightarrow 2C_6H_{12}O_6 + CaC_2O_4$$

The decomposition of calcium fructosate may also be effected by passing carbon dioxide into its aqueous suspension, the reaction being:

$$C_6H_{11}O_5.O.Ca(OH) + CO_2 \rightarrow C_6H_{12}O_6 + CaCO_3$$

This reaction must, however, be carried out in the boiling solution to prevent the formation of the soluble calcium bicarbonate which, although unstable, might cause the resulting fructose to be contaminated with calcium carbonate.

Fructose may readily be obtained from inulin (dahlia tubers being a convenient source) by warming this with dilute sulphuric acid on a boiling water-bath for about an hour. The resulting solution is neutralised with barium carbonate and the filtrate evaporated under reduced pressure until crystallisation of the fructose commences.

The crude fructose obtained in either of the above ways may be purified by recrystallisation from pure ethyl alcohol, but it crystallises much less easily than glucose. In aqueous solution fructose exhibits mutarotation and this is accounted for in a similar way to the mutarotation of glucose. Actually, an aqueous solution of fructose contains not only the α- and β- pyranose forms (α- and β-fructopyranose) but also other isomerides. The pyranose forms can be derived from the unbranched ketone formula of fructose as shown on opposite page.

The formulae of α- and β- glucopyranose are given for comparison.

It should be noted that the ketonic carbon atom in fructose is (2), whereas the aldehydic carbon atom in glucose is (1). When an aqueous solution of fructose has reached equilibrium, it has $[\alpha]_D = -92°$. Ordinary fructose is chiefly β-fructopyranose.

The more active and oxidisable form of fructose is the so-called γ-fructose, which may have a furanose structure of which there are two possible α- and β- modifications. These may be represented:

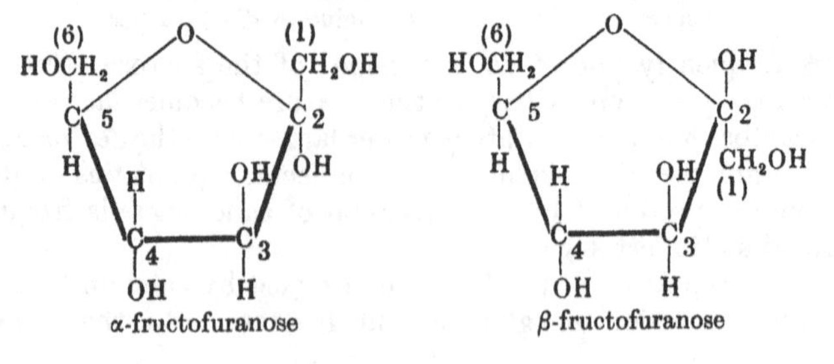

α-fructofuranose β-fructofuranose

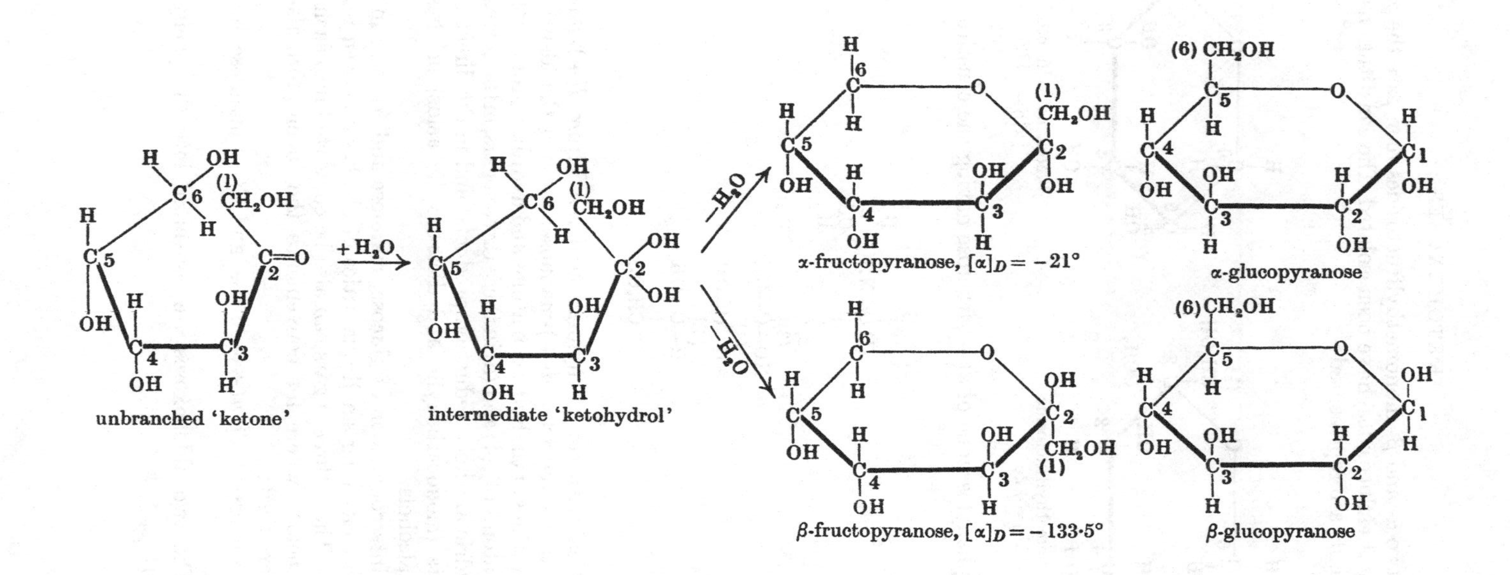
unbranched 'ketone'
$+H_2O$
intermediate 'ketohydrol'
$-H_2O$
$-H_2O$
α-fructopyranose, $[\alpha]_D = -21°$
β-fructopyranose, $[\alpha]_D = -133{\cdot}5°$
α-glucopyranose
β-glucopyranose

The two α- and β- monomethylfructosides (compare the α- and β-monomethylglucosides) have been isolated. The constitutions of these compounds are represented:

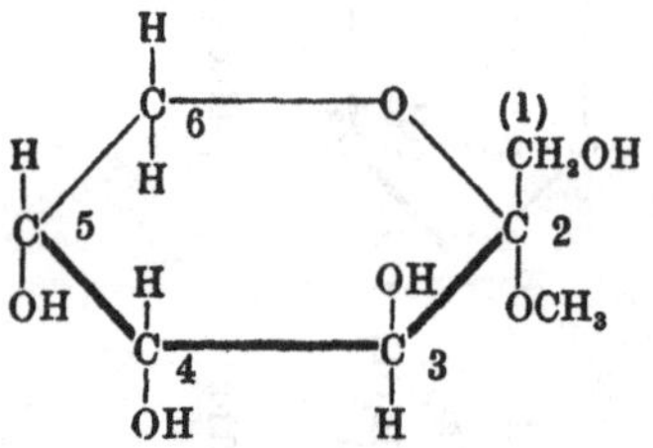

```
          H                                            H
          |                                            |
          C6 ---------O     (1)                        C6 ---------O
   H    / |            \    CH2OH               H    / |            \    OCH3
   |   /  H             \   |                   |   /  H             \   |
   C 5                   C 2                    C 5                   C 2
   |  \  H         OH   /   |                   |  \  H         OH   /   |
   OH  \ |         |   /    OCH3                OH  \ |         |   /    CH2OH
         C4 ------ C 3                                C4 ------ C 3      (1)
         |         |                                  |         |
         OH        H                                  OH        H
```

α-monomethylfructoside
$[\alpha]_D = +26°$

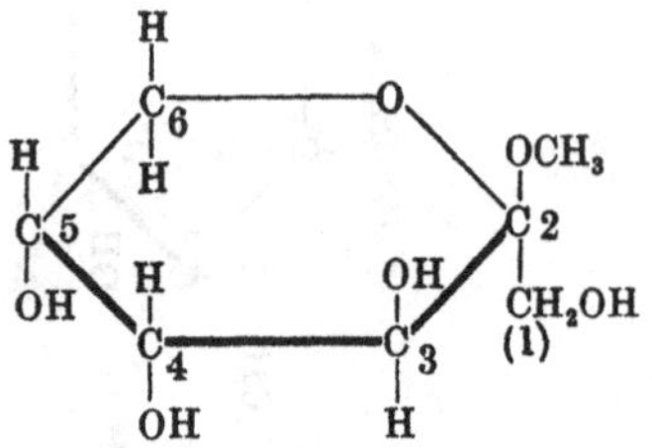

β-monomethylfructoside
$[\alpha]_D = -172°$

The phenylosazone of $d(-)$-fructose having the constitution:

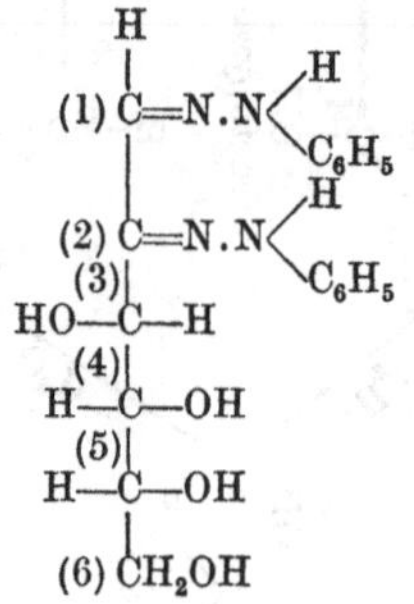

$$
\begin{array}{l}
\qquad\quad \mathrm{H} \\
\qquad\quad | \\
(1)\ \mathrm{C{=}N.N} \begin{smallmatrix} \mathrm{H} \\ \mathrm{C_6H_5} \end{smallmatrix} \\
\qquad\quad | \\
(2)\ \mathrm{C{=}N.N} \begin{smallmatrix} \mathrm{H} \\ \mathrm{C_6H_5} \end{smallmatrix} \\
\qquad (3)| \\
\mathrm{HO{-}C{-}H} \\
\qquad (4)| \\
\mathrm{H{-}C{-}OH} \\
\qquad (5)| \\
\mathrm{H{-}C{-}OH} \\
\qquad\quad | \\
(6)\ \mathrm{CH_2OH}
\end{array}
$$

is identical with the phenylosazone derived from $d(+)$-glucose and with that derived from $d(+)$-mannose, proving the identity of the configurations of the hydrogen atoms and hydroxyl groups attached to carbon atoms (3), (4) and (5) in these three sugars, in the corresponding hexahydric alcohols (reduction products) and in the direct oxidation products (monocarboxylic and dicarboxylic acids) of these hexahydric alcohols.

The interconversion of glucose, mannose and fructose in aqueous solution containing small quantities of alkalies has been referred to (p. 424). These three sugars can also be converted into each other as shown below, where for convenience the 'open' formulae of the sugars are used:

(*a*) Conversion of $d(+)$-glucose and $d(+)$-mannose into $d(-)$-fructose.

(i) The two aldohexoses are converted into the phenylosazone (p. 409):

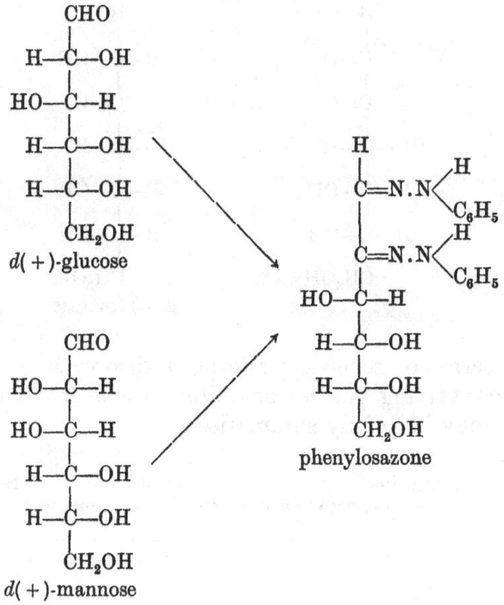

$$\begin{array}{c} CHO \\ | \\ H-C-OH \\ | \\ HO-C-H \\ | \\ H-C-OH \\ | \\ H-C-OH \\ | \\ CH_2OH \end{array}$$

d(+)-glucose

$$\begin{array}{c} CHO \\ | \\ HO-C-H \\ | \\ HO-C-H \\ | \\ H-C-OH \\ | \\ H-C-OH \\ | \\ CH_2OH \end{array}$$

d(+)-mannose

$$\longrightarrow \quad \begin{array}{c} H \\ | \\ C{=}N.N\langle^{H}_{C_6H_5} \\ | \\ C{=}N.N\langle^{H}_{C_6H_5} \\ | \\ HO-C-H \\ | \\ H-C-OH \\ | \\ H-C-OH \\ | \\ CH_2OH \end{array}$$

phenylosazone

(ii) The phenylosazone is hydrolysed by treatment with concentrated hydrochloric acid for a short time at 45° when it is converted into the *osone*,* phenylhydrazine hydrochloride being eliminated:

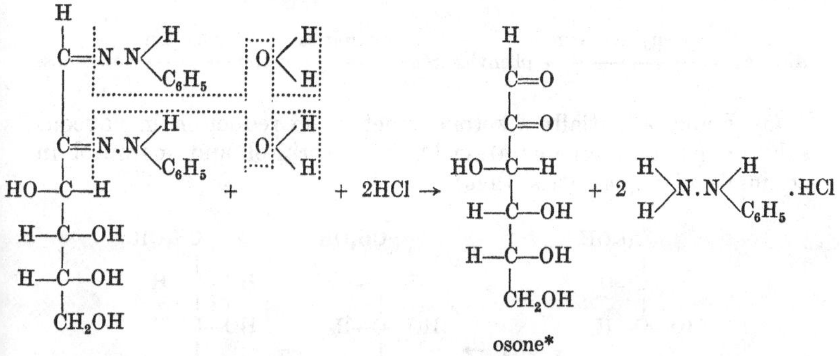

$$\begin{array}{c} H \\ | \\ C{=}N.N\langle^{H}_{C_6H_5} \\ | \\ C{=}N.N\langle^{H}_{C_6H_5} \\ | \\ HO-C-H \\ | \\ H-C-OH \\ | \\ H-C-OH \\ | \\ CH_2OH \end{array} + \begin{array}{c} O\langle^{H}_{H} \\ \\ O\langle^{H}_{H} \end{array} + 2HCl \rightarrow \begin{array}{c} H \\ | \\ C{=}O \\ | \\ C{=}O \\ | \\ HO-C-H \\ | \\ H-C-OH \\ | \\ H-C-OH \\ | \\ CH_2OH \end{array} + 2\ {}^{H}_{H}\rangle N.N\langle^{H}_{C_6H_5}.HCl$$

osone*

(iii) The osone—an oxidation product of the three sugars—on gentle reduction (zinc and acetic acid) is converted into fructose, the keto group remaining unchanged:

* This osone may be designated *glucosone*, *mannosone* or *fructosone* if it is desired to indicate from which sugar it is derived. The configuration of the osone indicated in the text is that of the osone belonging to the *d*- series and normally designated d(–)-*glucosone*. The l(+)-*glucosone* obtained from *l*(–)-glucose, *l*(–)-mannose and *l*(+)-fructose is also known.

$$\begin{array}{c} \mathrm{H} \\ | \\ \mathrm{C{=}O} \\ | \\ \mathrm{C{=}O} \\ | \\ \mathrm{HO{-}C{-}H} \\ | \\ \mathrm{H{-}C{-}OH} \\ | \\ \mathrm{H{-}C{-}OH} \\ | \\ \mathrm{CH_2OH} \end{array} + \mathrm{H_2} \rightarrow \begin{array}{c} \mathrm{H} \\ | \\ \mathrm{H{-}C{-}OH} \\ | \\ \mathrm{C{=}O} \\ | \\ \mathrm{HO{-}C{-}H} \\ | \\ \mathrm{H{-}C{-}OH} \\ | \\ \mathrm{H{-}C{-}OH} \\ | \\ \mathrm{CH_2OH} \\ d(-)\text{-fructose} \end{array}$$

The above series of reactions affords a theoretically quantitative method of converting glucose or mannose into fructose; the series of reactions may be briefly summarised thus:

$$\left.\begin{array}{c}\text{Glucose}\\ \text{and/or}\\ \text{Mannose}\end{array}\right\} \xrightarrow{\text{phenylhydrazine}} \text{phenylglucosazone} \xrightarrow{\text{hydrolysis}} \text{glucosone} \xrightarrow{\text{reduction}} \text{fructose}$$

(fructose $\xrightarrow{\text{phenylhydrazine}}$ phenylglucosazone)

Or, more generally, the method for transforming aldohexoses into ketohexoses, having the same configuration of the hydrogen atoms and hydroxyl groups attached to carbon atoms (3), (4) and (5), is according to the scheme:

$$\text{Aldohexose} \xrightarrow{\text{phenylhydrazine}} \text{phenylosazone} \xrightarrow{\text{hydrolysis}} \text{osone} \xrightarrow{\text{reduction}} \text{ketohexose}$$

(*b*) Being potentially a ketone, fructose on reduction in aqueous solution (sodium amalgam) yields both sorbitol and mannitol in equimolecular quantities, thus:

$$2\begin{array}{c} \mathrm{CH_2OH} \\ | \\ \mathrm{C{=}O} \\ | \\ \mathrm{HO{-}C{-}H} \\ | \\ \mathrm{H{-}C{-}OH} \\ | \\ \mathrm{H{-}C{-}OH} \\ | \\ \mathrm{CH_2OH} \\ d(-)\text{-fructose} \end{array} + 2\mathrm{H_2} \rightarrow \begin{array}{c} \mathrm{CH_2OH} \\ | \\ \mathrm{H{-}C{-}OH} \\ | \\ \mathrm{HO{-}C{-}H} \\ | \\ \mathrm{H{-}C{-}OH} \\ | \\ \mathrm{H{-}C{-}OH} \\ | \\ \mathrm{CH_2OH} \\ d(+)\text{-sorbitol} \end{array} + \begin{array}{c} \mathrm{CH_2OH} \\ | \\ \mathrm{HO{-}C{-}H} \\ | \\ \mathrm{HO{-}C{-}H} \\ | \\ \mathrm{H{-}C{-}OH} \\ | \\ \mathrm{H{-}C{-}OH} \\ | \\ \mathrm{CH_2OH} \\ d(+)\text{-mannitol} \end{array}$$

Since hexahydric alcohols are converted into aldohexoses by a general series of reactions which may be summarised thus:

CH_2OH		COOH		C=O (ring O)		CHO
CHOH		CHOH		CHOH		CHOH
CHOH	oxidation →	CHOH	$-H_2O$ →	CHOH	reduction →	CHOH
CHOH		CHOH		H—C (ring O)		CHOH
CHOH		CHOH		CHOH		CHOH
CH_2OH		CH_2OH		CH_2OH		CH_2OH
hexahydric alcohol		hexonic acid		lactone of hexonic acid		aldohexose

the method of conversion of fructose into equal quantities of glucose and mannose (theoretically, 50 per cent. of each calculated on the fructose being obtained) is illustrated:

CH_2OH	
C=O	
HO—C—H	
H—C—OH	reduction →
H—C—OH	
CH_2OH	
d(−)-fructose	

CH_2OH		COOH		O=C (ring O)		CHO
H—C—OH		H—C—OH		H—C—OH		H—C—OH
HO—C—H	oxida- →	HO—C—H	$-H_2O$ →	HO—C—H	re- →	HO—C—H
H—C—OH	tion	H—C—OH		H—C (ring O)	duction	H—C—OH
H—C—OH		H—C—OH		H—C—OH		H—C—OH
CH_2OH		CH_2OH		CH_2OH		CH_2OH
d(+)-sorbitol		*d*(+)-gluconic acid		gluconolactone		*d*(+)-glucose

CH_2OH		COOH		O=C (ring O)		CHO
HO—C—H		HO—C—H		HO—C—H		HO—C—H
HO—C—H	oxida- →	HO—C—H	$-H_2O$ →	HO—C—H	re- →	HO—C—H
H—C—OH	tion	H—C—OH		H—C (ring O)	duction	H—C—OH
H—C—OH		H—C—OH		H—C—OH		H—C—OH
CH_2OH		CH_2OH		CH_2OH		CH_2OH
d(+)-mannitol		*d*(+)-mannonic acid		mannonolactone		*d*(+)-mannose

These are general reactions and are applicable for the conversion of any ketohexose into equimolecular quantities (theoretically) of two aldohexoses in which the configuration of the hydrogen atoms and hydroxyl groups attached to carbon atoms (3), (4) and (5) is the same in the three sugars.

The reducing power of fructose in aqueous solution as estimated quantitatively by its action on standard solutions of cupric salts is almost the same as that of glucose, etc.

As a class, the monosaccharides have certain general reactions which are frequently used for their qualitative identification, although these reactions do not serve for the identification of the particular monosaccharide. Three of such reactions are:

(i) On warming with alkalies, aqueous solutions of the monosaccharides yield yellowish brown solutions and ultimately resinification takes place (Moore's test).

(ii) Any monosaccharide when heated in the dry state yields furfural, which may be identified by its action on paper moistened with an aqueous solution of aniline acetate held in the escaping vapours, the paper developing a red coloration.

(iii) Molisch's test for all carbohydrates (including starch): a few drops of an alcoholic solution of α-naphthol (Molisch's reagent) is added to about 1 ml. of an aqueous solution of the carbohydrate and a layer of concentrated sulphuric acid is run under it by pouring the latter down the side of the sloping test-tube. A purple ring is formed at the junction of the layers either at once or after a short time. This purple coloration is due to a hydroxymethylfurfural derivative.

Synthesis of Monosaccharides

The synthesis of glucose and of fructose has been effected starting from glycerol and from formaldehyde. As these two compounds have been systematically synthesised, the synthesis of these two monosaccharides and of the sugars which can be derived from them may be regarded as complete.

Glycerol, on careful oxidation, yields a mixture of glyceraldehyde and dihydroxyacetone (a mixture frequently described as glycerose, p. 390). This mixture when treated with alkalies condenses, yielding a ketohexose known as *α-acrose* which is largely *dl*-fructose. This condensation may be illustrated:

$$\begin{array}{ccccc}
\mathrm{HO.HCH} & & \mathrm{C{\overset{H}{\lessdot}}O} & & \mathrm{HOHC{-\!-\!-}CHOH} \\
| & & | & & |\qquad\quad | \\
\mathrm{CO} & + & \mathrm{CHOH} & \longrightarrow & \mathrm{CO}\qquad \mathrm{CHOH} \\
| & & | & & |\qquad\quad | \\
\mathrm{CH_2OH} & & \mathrm{CH_2OH} & & \mathrm{CH_2OH}\quad \mathrm{CH_2OH} \\
\text{dihydroxyacetone} & & \text{glyceraldehyde} & & \alpha\text{-acrose} \\
 & \text{glycerose} & & & dl\text{-fructose}
\end{array}$$

From the *dl*-fructose, the optically active glucoses and fructoses are obtained by reactions which have been indicated and by processes of resolution of externally compensated monobasic acids.

The synthesis starting from formaldehyde depends on the production from the latter by the action of lime water of a complex mixture known as *formose*. A stage in the condensation may be illustrated as a type of aldol condensation (p. 158), thus:

$$H_2{:}C{:}O + H.CHO + H.CHO + H.CHO + H.CHO + H.CHO$$
$$\rightarrow HOH_2C.CHOH.CHOH.CHOH.CHOH.C{\begin{matrix}H\\ \| \\ O\end{matrix}}$$

From formose, α-acrose has been isolated and the production of a ketohexose from the preliminary aldohexose may be due to the usual interconversion of monosaccharides in the presence of alkalies in aqueous solution (p. 424). The complex nature of formose may be assumed to be due to the same interconversion.

Glycosides

The glycosides constitute a large class of substances which are derivatives of the simple sugars, chiefly the monosaccharides. The term glycoside is applied to those substances which when hydrolysed by acids or enzymes yield a sugar and one or more other products which are not sugars. Many of these compounds occur in plants and correspond in structure to the simple synthetic monomethylglucosides. The simplest type of glycoside may be given the general formula:

$$CH_2OH.\underbrace{CH.CHOH.CHOH.CHOH.CH}_{O}.OR$$

where R represents the radicle split off from the sugar during hydrolysis, forming a compound having the general formula R.OH. While glycoside is the general term applied to this class of substances, the term glucoside (formerly applied to the whole class) is now restricted to those which, on hydrolysis, yield glucose. Frequently, the non-sugar substance obtained by the hydrolysis of a glycoside is referred to as the *aglycone*.

The largest class of naturally occurring glycosides are glucose derivatives and β-glucosides. The β-glucosides may be given the general formula:

```
              (6) CH2OH
                   |
                   C5 ---------- O
          H       /  |             \       OR
          |      /   H              \      |
          C4                          C1
          |  \   OH          H      /   |
          OH  \  |           |     /    H
                 C3 ---------- C2
                 |             |
                 H             OH
```

corresponding to the pyranose formula of β-glucose and to the formula of β-monomethylglucoside. Glucosides having the furanose structure are unknown in nature, but several have been prepared synthetically.

The glycosides may occur in the fruit, bark and roots of the appropriate plants and, in each case, the plant tissue also contains the hydrolysing enzyme, but in cells different from those containing the glycosides. The enzyme only gains access to the glycoside when the tissue is destroyed by maceration. A large number of hydrolysing enzymes are known, but the commonest is *emulsin*, which occurs in the bitter almond; it is the specific hydrolysing enzyme for β-glycosides, and hydrolysis by emulsin is used for assigning a particular glycoside to the β- class.

As a class, the glycosides are colourless crystalline solids and generally *laevo*rotatory. They may be obtained by extracting the plant tissue with alcohol or water, but, in the latter case, the enzyme should be first destroyed by heating to a temperature of about 65°. Apart from being hydrolysed by enzymes, glycosides are all hydrolysed by heating with dilute mineral acids, although at very different rates. Such a hydrolysis, whether by enzyme or mineral acid, may be represented in the case of a simple typical glycoside as:

$$\underbrace{CH_2OH.\overbrace{CH.(CHOH)_3.CH}^{\text{---- O ----}}.OR}_{\beta\text{-glycoside}} + H_2O \rightarrow \underbrace{CH_2OH.\overbrace{CH.(CHOH)_3.CHOH}^{\text{---- O ----}}}_{\text{monosaccharide }(\beta\text{-pyranose form})} + \underbrace{R.OH}_{\text{aglycone}}$$

A reaction of this type is reversible. In practice, enzyme hydrolysis (or enzyme action generally) goes almost to completion and the synthetic action of enzymes has been realised experimentally in only a few cases.

An enzyme (see p. 332) may be defined as a substance produced by the living cell which has, apart from the cell, a specific catalytic action. Enzymes are classified according to the type of reaction in which they play the part of catalysts and emulsin is a typical hydrolytic enzyme, hydrolysing β-glucosides; *maltase*, a hydrolytic enzyme found in barley malt, has no action on β-glycosides but hydrolyses α-glycosides. Unlike inorganic catalysts, enzymes are destroyed during the reaction in which they play the part of catalysts. Like true catalysts, enzymes have their conditions for optimum activity. Their activity increases with increasing temperature, but the temperature limits of activity are narrowly defined since the enzymes themselves are unstable and are usually destroyed at 60–65°. At 0° they are inactive and their temperature of optimum activity is about 37–45°, destruction of the enzyme beginning at about 45°.

The chief experimental difficulty in studying enzyme action is due to the fact that no enzyme has yet been obtained as a chemical individual and consequently any reaction in which an enzyme preparation is used as the catalyst is necessarily attended by side reactions. The greater number of enzymes are hydrolytic agents, but many act as catalysts in other types of reaction. All digestive processes are enzyme actions and the biochemical production from glucose of ethyl alcohol, lactic acid, etc. are also examples of enzyme action.*

Many naturally occurring glycosides (particularly glucosides) have for long been used for medicinal purposes, their therapeutic action being due to the aglycone (or aglucone) which they yield on hydrolysis. Such glycosides are arbutin and salicin.

Arbutin, $C_{12}H_{16}O_7.\frac{1}{2}H_2O$, is obtained as a colourless crystalline substance having a bitter taste from the leaves of the bearberry (*Arbutus uva ursi*); when anhydrous, it has m.p. 187°. Its methyl derivative, *methylarbutin* (m.p. 176°), always occurs with it. Arbutin is *laevo*rotatory, having $[\alpha]_D = -63{\cdot}5°$, and is a β-glucoside, being hydrolysed by emulsin and by mineral acids to glucose and hydroquinone, *p*-dihydroxybenzene or quinol:

$$C_{12}H_{16}O_7 + H_2O \rightarrow C_6H_{12}O_6 + C_6H_4(OH)_2$$

Methylarbutin is similarly, but more readily, hydrolysed to glucose and the monomethyl ether of hydroquinone. Hydroquinone is an antiseptic and may be used in the treatment of chronic cystitis; it is non-poisonous. Arbutin has been synthesised and its constitution may be represented:

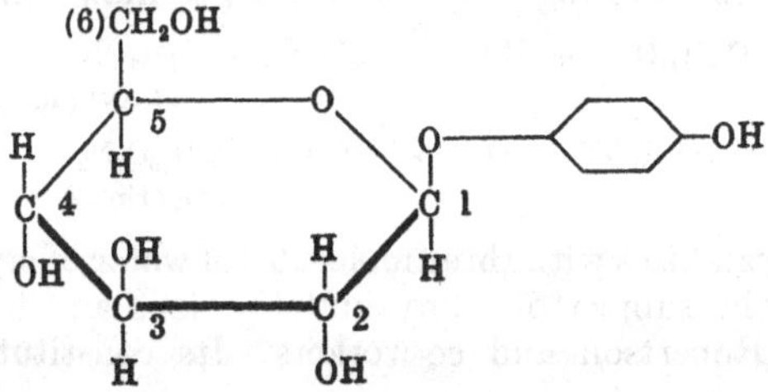

```
        (6)CH2OH
           |
           C-----------O
     H    / 5           \       O---------< C6H4 >--OH
     |   /  |            \      |
     C 4    H             \     C 1
     |  \                  \   /|
    OH   \  OH         H    \ / H
          \ |          |     /
           C-----------C
           3           2
           |           |
           H           OH
```

Salicin, $C_{13}H_{18}O_7$, is a colourless crystalline substance, m.p. 200°, obtained from *Salix fragilis*, *S. purpurea* and other species of willow. It also occurs in the flower buds of meadow sweet (*Spirea ulmaria*). It has long been used as an antipyretic and in the treatment of rheumatism.

* The following are recommended books concerning enzymes and enzyme action: *Enzymes*, by J. B. S. Haldane, M.A., F.R.S. (Longmans, Green and Co.); *The Nature of Enzyme Action*, by Sir W. M. Bayliss, D.Sc., F.R.S. (Longmans, Green and Co.); *Alcoholic Fermentation*, by Sir A. Harden, D.Sc., F.R.S. (Longmans, Green and Co.).

It is a *laevo*rotatory ($[\alpha]_D = -65°$) β-glucoside, being hydrolysed by emulsin to glucose and *o*-hydroxybenzyl alcohol, salicylic alcohol or saligenin:

$$C_{13}H_{18}O_7 + H_2O \rightarrow C_6H_{12}O_6 + C_6H_4\begin{cases}OH\\CH_2OH\end{cases}$$

The leaves and twigs of the willow contain the specific enzyme, *salicase*, which hydrolyses the salicin. The therapeutic action of salicin depends on the oxidation of the salicylic alcohol in the blood to salicylic acid. The constitution of salicin may be represented:

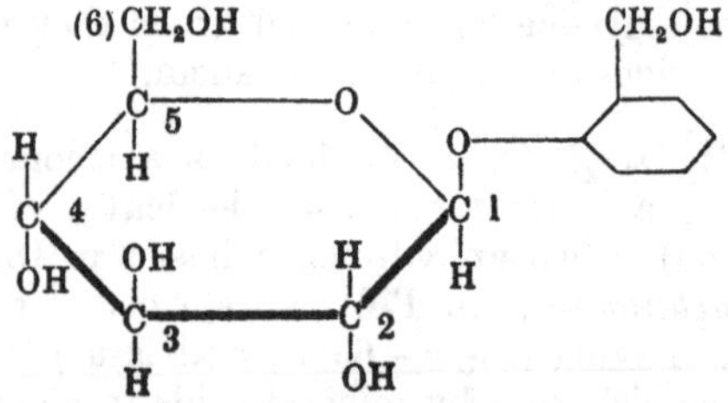

Indican, $C_{14}H_{17}O_6N$, is a β-glucoside which occurs in the leaves of various species of indigofera and in woad (*Isatis tinctoria*) and is readily extracted from the leaves by means of acetone. Indican is the source of natural indigo, which has now been displaced by the artificially produced material (indigotin, p. 334). The specific enzyme which occurs in the plant is *indemulsin*, which hydrolyses the indican more rapidly than emulsin. The indican is hydrolysed to glucose and indoxyl (p. 338) and the latter is oxidised (the oxidation being promoted by an oxidising enzyme, *oxidase*) to indigotin:

$$C_{14}H_{17}O_6N + H_2O \rightarrow C_6H_{12}O_6 + \underset{\text{indoxyl (aglucone)}}{C_8H_7ON}$$

$$2C_8H_7ON + O_2 \rightarrow 2H_2O + \underset{\text{indigotin}}{C_{16}H_{10}O_2N_2}$$

Indican crystallises with three molecules of water of crystallisation. It has been the subject of many investigations and has been synthesised by Robertson and co-workers. Its constitution may be represented:

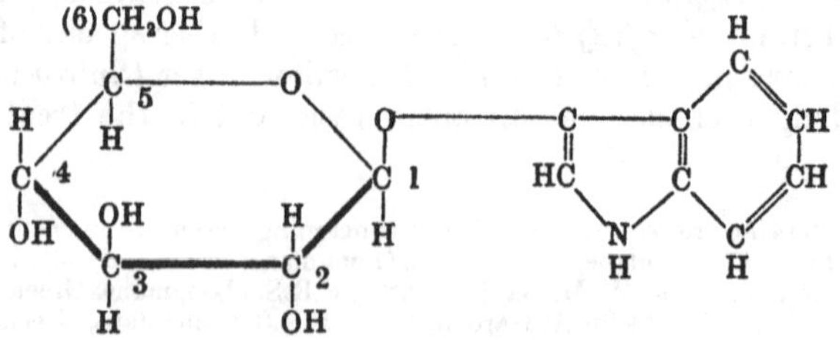

Amygdalin is a different type of glycoside. It is one of the cyanophoric glycosides which on hydrolysis yield hydrocyanic acid. It is a colourless crystalline substance having a bitter taste and soluble in water; it is found in the bitter almond and in the kernels of apricots, peaches and plums. Amygdalin is a β-glycoside, since it is hydrolysed by emulsin which also occurs in the bitter almond. It has $[\alpha]_D = -38{\cdot}3°$.

When amygdalin is hydrolysed either with emulsin or with acids, it yields two molecules of glucose, benzaldehyde and hydrocyanic acid, the general reaction being represented:

$$\underset{\text{amygdalin}}{C_{20}H_{27}O_{11}N} + 2H_2O \rightarrow C_6H_5.CHO + HCN + 2C_6H_{12}O_6$$

As a result of many investigations and chiefly those of Haworth and co-workers who have effected the synthesis of amygdalin, this glycoside has been shown to be a derivative of the somewhat uncommon disaccharide, *gentiobiose*, the benzaldehyde and hydrogen cyanide being combined in the glycoside as they are in mandelonitrile. Amygdalin may be described as mandelonitrile β-glucose-6-β-glucoside.

Amygdalin has been hydrolysed by an enzyme, *amygdalase*, which is present in yeast extract, to glucose and mandelonitrile-β-glucoside:

$$C_{20}H_{27}O_{11}N + H_2O \rightarrow C_6H_{12}O_6 + \underset{\text{mandelonitrile-}\beta\text{-glucoside}}{C_6H_5.CH(CN).O.C_6H_{11}O_5}$$

whereas emulsin from bitter almonds, and which also contains amygdalase, hydrolyses amygdalin completely as indicated above.

Amygdalin may also be described as the gentiobioside of mandelonitrile or the β-glucoside of mandelonitrile-β-glucoside. Its constitution may be represented:

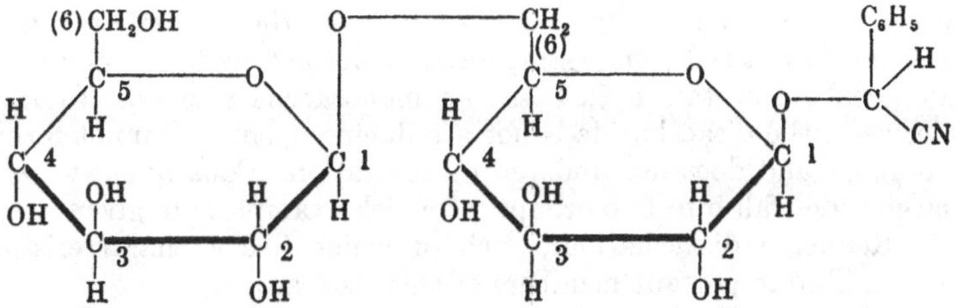

Other naturally occurring glycosides are the yellow anthoxanthin pigments and the most interesting blue, violet and red anthocyanin pigments of flowers, which have been investigated by Robinson and co-workers.

Although not strictly glycosides, the derivatives of glucuronic acid (p. 423) are closely related to the glucosides; they may be considered as oxidation products of the glucosides and are frequently described as 'paired glucuronates'. They are produced after administration of such substances as chloral, camphor, menthol, phenol, etc. After administration of chloral, the latter is reduced to trichloroethyl alcohol which is excreted as *urochloralic acid*, having the probable constitution:

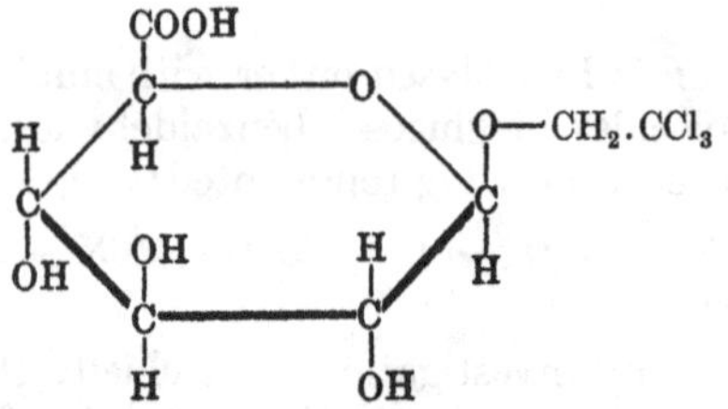

```
              COOH
               |
               C————————O
         H   / |         \   O—CH2.CCl3
         |  /  H          \  |
         C                   C
         |  \  OH       H  / |
        OH   \ |        | /  H
               C————————C
               |        |
               H        OH
```

These 'paired glucuronates' on hydrolysis yield glucuronic acid which, like glucose, is a strong reducing agent.

DISACCHARIDES

These carbohydrates, having the general formula $C_{12}H_{22}O_{11}$, are closely related to the glycosides, since on hydrolysis by means of acids or appropriate enzymes they yield two molecules of monosaccharides, thus:

$$C_{12}H_{22}O_{11} + H_2O \rightarrow C_6H_{12}O_6 + C_6H_{12}O_6$$

In this case the aglycone is one of the monosaccharides, whereas in the typical glycosides the aglycone is not a carbohydrate.

One of the constituent monosaccharides in a disaccharide is always an aldohexose and the functional aldehyde group (carbon atom, 1) is substituted as in the monomethylglucosides. The aldehyde or ketone group (carbon atom, 2) of the second constituent monosaccharide may remain functional or not. In the former case, the disaccharide is a reducing agent, forms a characteristic osazone and undergoes mutarotation just as the monosaccharides do. In the latter case, the disaccharide is not a reducing agent, it cannot form an osazone and does not undergo mutarotation. Consequently, the disaccharides fall into two groups, of which examples are given:

(i) Reducing disaccharides, each of which has a characteristic osazone. Two important members of this class are:

Maltose or malt sugar, which on hydrolysis yields two molecules of glucose, and

Lactose or milk sugar, which on hydrolysis yields a molecule of glucose and a molecule of galactose.

Gentiobiose (p. 449) which, like maltose, yields two molecules of glucose on hydrolysis is also a reducing disaccharide.

(ii) Non-reducing disaccharides which cannot yield osazones. The most important member of this class is:

Sucrose, cane sugar or beet sugar, which on hydrolysis yields a molecule of glucose and a molecule of fructose.

The isomerism among the disaccharides is extensive, as can be seen from the following considerations:

(*a*) There are several ways in which two molecules of a monosaccharide having the pyranose structure can theoretically be condensed together;

(*b*) The monosaccharide constituents may be the same or different from each other;

(*c*) One or both of the monosaccharide constituents may have the furanose structure.

Many disaccharides have been synthesised.

The disaccharides may occur naturally or they may be obtained by the hydrolysis of glycosides or of polysaccharides. All the disaccharides have eight free hydroxyl groups in the molecule and they all yield octaacyl (octaacetyl, octabenzoyl, etc., i.e. esters of acetic, benzoic, etc. acids) derivatives and octaalkyl (octamethyl, etc., i.e. ethers) derivatives; the hydrogen atoms of the hydroxyl groups in the former case are replaced by acyl ($CH_3.CO$—, etc.) groups (acetylation, etc.) and, in the latter case, by alkyl (CH_3—, etc.) groups (alkylation).

Maltose is the disaccharide obtained by the partial hydrolysis of starch; complete hydrolysis of starch yields glucose. The starch is hydrolysed to maltose and *dextrin* by the enzyme, *diastase*, present in germinated barley or malt. This hydrolysis may be expressed:

$$\underset{\text{starch}}{(C_6H_{10}O_5)_n} + H_2O \rightarrow \underset{\text{maltose}}{C_{12}H_{22}O_{11}} + \underset{\text{dextrin}}{(C_6H_{10}O_5)_{n-2}}$$

Maltose may be obtained also by the incomplete hydrolysis of starch by acids and also by the action of diastase on glycogen or liver starch.

Maltose is readily soluble in water and crystallises from it in colourless needles having the composition $C_{12}H_{22}O_{11}.H_2O$. In aqueous solution, maltose undergoes mutarotation and the specific rotatory power when equilibrium is attained is $[\alpha]_D = +136°$. This mutarotation indicates the existence of two forms of maltose (α- and β-). The octaacetyl derivatives of these two forms have been prepared and they have the following constants:

$$\alpha\text{-, m.p. } 125°, [\alpha]_D = +122{\cdot}7°,$$
$$\beta\text{-, m.p. } 159°, [\alpha]_D = +62{\cdot}6°,$$

and they both have the general composition $C_{12}H_{14}O_3(O.CO.CH_3)_8$. When carefully prepared, the phenylosazone:

$$\begin{array}{l} \quad\quad\quad\quad\; C(H)=N.N(H)(C_6H_5) \\ \quad\quad\quad\quad\; | \\ C_{10}H_{19}O_9—C=N.N(H)(C_6H_5) \end{array}$$

which crystallises out as a hot aqueous solution cools, forms characteristic needle-shaped crystals which may be differentiated microscopically from those of glucosazone and lactosazone.

Maltose reduces Fehling's solution and other solutions of cupric salts, but its reducing power is much less than that of glucose. After hydrolysis (by hot dilute hydrochloric acid) the reducing power is much greater, since two molecules of glucose are produced from one molecule of maltose:

$$\underset{\text{maltose}}{C_{12}H_{22}O_{11}} + H_2O \rightarrow \underset{\text{glucose}}{C_6H_{12}O_6} + \underset{\text{glucose}}{C_6H_{12}O_6}$$

After this hydrolysis the specific rotatory power of the solution is less than that before hydrolysis, since the specific rotatory power of glucose is less than that of maltose.

Maltose is hydrolysed by maltase and not by emulsin; maltose is therefore an α-glucoside, α-glucose being obtained from it initially on hydrolysis. The structure of maltose has been determined and α-maltose has been shown to be a combination of two α-glucose residues, thus:

```
      (6) CH2OH                                 (6) CH2OH
           |                                         |
           C-----------O                             C-----------O
   H     / | 5          \     H           H        / | 5          \     H
   |    /  H             \    |           |       /  H             \    |*
   C 4                    C 1             C 4                       C 1
   |    OH          H     |               |      OH          H      |
   OH    \          |     |               |        \         |      OH
          C---------C 2   +------O--------+         C--------C 2
          | 3       |                               | 3      |
          H         OH                              H        OH
```

and is described as α-glucose-4-α-glucoside. The reducing group in maltose is indicated (*).

Lactose is present in the milk of all mammals and has not been found in the vegetable kingdom. It occurs to the extent of 6–8 per cent. in human milk, cow's milk containing generally about half this amount. The sugar is prepared by evaporation of whey obtained in the preparation of cheese. The lactose readily crystallises, and after recrystallisation from water is obtained as a colourless microcrystalline powder.

In aqueous solution, lactose undergoes mutarotation, the specific rotatory power of the solution in equilibrium being $[\alpha]_D = +55{\cdot}3°$. This mutarotation affords proof of the existence of two forms of lactose (α- and β-) and the octaacetyl derivatives of these two forms have been isolated and have the following constants:

α-, m.p. 152°, $[\alpha]_D = +53{\cdot}6°$,
β-, m.p. 90°, $[\alpha]_D = -4{\cdot}7°$,

both having the general composition $C_{12}H_{14}O_3(O.CO.CH_3)_8$. The crystalline form of phenyllactosazone (lactose phenylosazone) is characteristic and when viewed microscopically may serve for the partial identification of the sugar. The general composition of the phenylosazone is

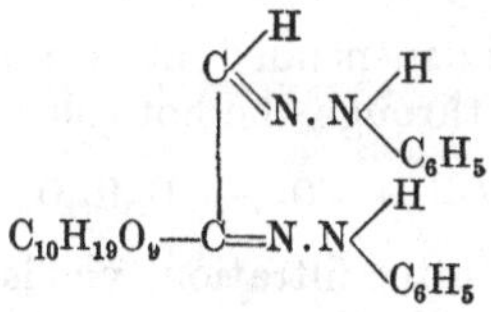

$$\begin{array}{l} \quad\quad\quad\; C\!\!\begin{array}{l}\diagup H \\ \diagdown\!\!\!= N.N\!\begin{array}{l}\diagup H\\ \diagdown C_6H_5\end{array}\end{array} \\ \quad\quad\quad\;\; | \\ C_{10}H_{19}O_9\text{—}C{=}N.N\!\begin{array}{l}\diagup H\\ \diagdown C_6H_5\end{array}\end{array}$$

Lactose, like maltose, is a reducing agent and its reducing power is greatly increased after hydrolysis, when a molecule of lactose yields a molecule of glucose and a molecule of galactose:

$$\underset{\text{lactose}}{C_{12}H_{22}O_{11}} + H_2O \rightarrow \underset{\text{glucose}}{C_6H_{12}O_6} + \underset{\text{galactose}}{C_6H_{12}O_6}$$

Lactose is hydrolysed by the enzyme emulsin and not by maltase. It is therefore a β-glycoside (β-galactoside). The constitution of β-lactose has been determined by Haworth and is represented thus:

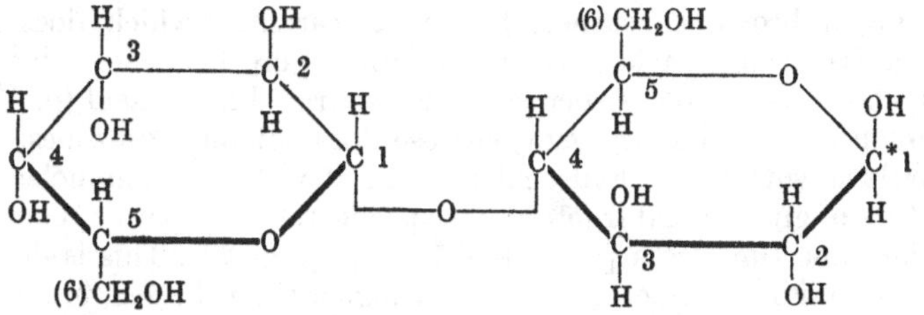

β-galactose unit β-glucose unit

where the usual constitutional formula of β-galactose (p. 434) has, for convenience, been turned through 180°. Lactose is systematically named β-glucose-4-β-galactoside. The reducing group in lactose is marked (*).

Sucrose, the typical non-reducing disaccharide, is also known as saccharose, cane sugar and beet sugar. It is widely distributed in the vegetable kingdom. Expressed juice from the sugar cane contains about 20 per cent. of sucrose, beetroot juice contains somewhat less and smaller quantities occur in other plants, such as maple and birch.

The method of extraction of the sugar consists in neutralising the plant juice with milk of lime and boiling to precipitate protein material. The solution is then treated with carbon dioxide to remove all calcium and with sulphur dioxide for decolorisation. The boiled and filtered solution is then evaporated under reduced pressure until extensive crystallisation begins. The non-crystalline material (molasses) may be made to yield more crystalline sucrose by treatment with calcium or strontium hydroxide, when the sparingly soluble so-called calcium or strontium saccharate is formed:

$$C_{12}H_{21}O_{10}.OH + Ca(OH)_2 \rightarrow C_{12}H_{21}O_{10}.O.CaOH + H_2O$$

The solid is separated, suspended in water and decomposed by passing carbon dioxide through the hot solution:

$$C_{12}H_{21}O_{10}.O.CaOH + CO_2 \rightarrow C_{12}H_{21}O_{10}.OH + CaCO_3$$

The resulting solution, after filtration, yields crystalline sucrose on evaporation under reduced pressure.

In contrast with other sugars, sucrose crystallises extremely readily. This is connected with the fact that it is not a reducing agent and does not undergo mutarotation in aqueous solution. Consequently only one form of cane sugar exists, and it only yields one highly crystalline octaacetyl derivative, $C_{12}H_{14}O_3(O.CO.CH_3)_8$, m.p. 69°, $[\alpha]_D = +59{\cdot}6°$.

Sucrose has a high solubility in water and this solution has $[\alpha]_D = +66{\cdot}5°$. At 160° it forms a glassy mass, frequently known as barley sugar, which slowly crystallises. At about 200°, it loses water, forming a brown substance, known as caramel, which does not crystallise. Oxalic acid was first obtained pure by Scheele by the destructive oxidation of sucrose by concentrated nitric acid (p. 356).

Not being a reducing agent, sucrose does not form osazones. On hydrolysis with dilute hydrochloric acid, it yields an equimolecular mixture of glucose and fructose, the specific rotatory power changing at the same time from $[\alpha]_D = +66{\cdot}5°$ to $[\alpha]_D = -20°$. This is due to fructose having a *laevo*rotatory power higher than the *dextro*rotatory power of glucose. For this reason, the hydrolysis of sucrose is often described as *inversion* and the equimolecular mixture of glucose and fructose as *invert sugar*:

$$\underset{\text{sucrose}}{C_{12}H_{22}O_{11}} + H_2O \rightarrow \underbrace{\underset{\text{glucose}}{C_6H_{12}O_6} + \underset{\text{fructose}}{C_6H_{12}O_6}}_{\text{invert sugar}}$$

This production of invert sugar from sucrose is brought about by an enzyme present in yeasts, moulds and many plants and also in bees; this enzyme is termed *invertase* or *sucrase*. Honey is a mixture of glucose and fructose. It is only after inversion that sucrose can undergo fermentation.

Since sucrose can be readily obtained chemically pure and it easily undergoes quantitative hydrolysis to glucose and fructose, it is the most convenient standardising reagent for solutions of cupric salts used in the volumetric estimation of reducing sugars (p. 465). Being a non-reducing sugar, sucrose is unaffected by alkalies and is readily differentiated from reducing sugars by what is known as Moore's test—behaviour of reducing sugars with alkali in aqueous solution. Like all carbohydrates, sucrose responds to Molisch's test.

The determination of the constitution of sucrose has proved to be one of great difficulty and has been the subject of much controversy. In spite of statements to the contrary, some of which have, unfortunately, found their way into the literature, sucrose has not yet (1936) been synthesised. All attempts to repeat the so-called syntheses have proved unsuccessful; at the best, isomeric disaccharides have only been obtained. Sucrose is both a glucoside and a fructoside in which the monosaccharide residues are united in such a way as to destroy both the potential aldehyde (carbon atom, 1, in the case of the glucose residue) and ketone (carbon atom, 2, in the case of the frucrose residue) groups to give a product which is not a reducing agent and which contains eight hydroxyl groups. The fructose obtained from sucrose immediately on hydrolysis is the labile γ-fructose having the furanose structure, i.e. fructofuranose, and this reverts after formation to fructopyranose. The chief constituents in the solution after hydrolysis and when the solution has reached equilibrium are the pyranose forms of the two monosaccharides; but all the forms of glucose and fructose which occur in their simple solutions are also present, even if some of them are only in small amounts. From its behaviour towards maltase, sucrose is regarded as an α-glucoside, and from its behaviour towards other enzymes it is, for the time being, also regarded as a β-fructoside. The formulation of sucrose by Haworth as α-glucopyranose-β-fructofuranose gives the most satisfactory formula for this sugar in the light of our present knowledge. The constitution is graphically represented:

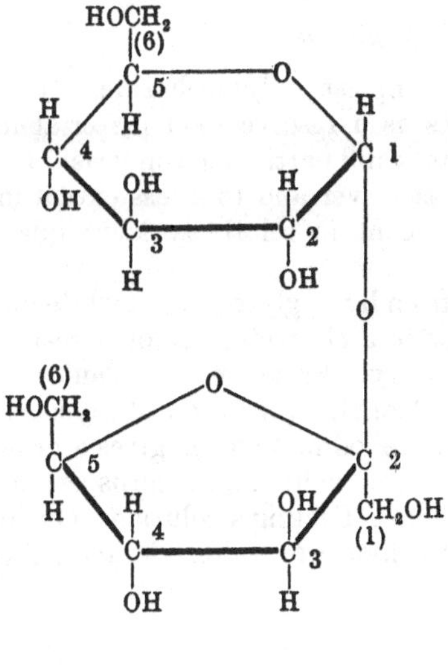

α-glucopyranose unit

β-fructofuranose unit

A number of tri- and tetra- saccharides have been described and investigated, but the more important remaining members of the carbohydrate group are the *polysaccharides*, which include *starch*, *inulin* and *cellulose*.

POLYSACCHARIDES

Previous references to polysaccharides, and particularly to starch, have been chiefly concerned with their being sources of monosaccharides, which they yield on hydrolysis. In general, the polysaccharides are non-crystalline substances of high molecular weight yielding colloidal solutions. They are thus difficult to obtain in a state of chemical purity. Recent elaborate investigations have shown that most polysaccharides consist of long regular chains of monosaccharide units connected with each other through an oxygen atom, as are the two monosaccharide units in a disaccharide.

The polysaccharides are conveniently classified according to the monosaccharides which they yield on hydrolysis. Thus, the polysaccharides having the general formula $(C_6H_{10}O_5)_n$ are divided into:

Glucosans (yielding glucose) such as *glycogen*, *starch*, *dextrin* and *cellulose*.

Fructosans (yielding fructose), the most important one being *inulin*.

Mannosans (yielding mannose) and *galactosans* (yielding galactose).

Other polysaccharides are known which consist wholly or partly of pentose units. In the former case they have the general formula $(C_5H_8O_4)_n$ and are described as *pentosans*.

Glycogen

Glycogen is probably the simplest polysaccharide. It is closely related to starch and serves as a reserve food material in animal organs; it is also found in plants and particularly in yeast. In animals, glycogen is chiefly found in the liver and to a less extent in muscle. Oysters and other molluscs contain relatively large quantities of glycogen.

When suitably extracted from liver, glycogen is a white amorphous powder soluble in cold water to a characteristic opalescent colloidal solution which is *dextro*rotatory. From such a solution it is reprecipitated by addition of alcohol. A solution of glycogen treated with a small quantity of a dilute iodine solution gives a brownish red coloration which disappears on warming and returns as the solution cools. Glycogen does not reduce Fehling's solution. On hydrolysis by boiling with dilute hydrochloric or sulphuric acid, glycogen is quantitatively converted into glucose, and this solution, after neutralisation, reduces Fehling's solution and other solutions of cupric salts. The production of glycogen from monosaccharides carried in the portal blood stream to the liver is brought about by an enzyme, *glycogenase*, and glycogen, like starch, is hydrolysed under the influence of ptyalin of the saliva to maltose, the precursor of glucose on complete hydrolysis. Glycogen of muscle is the source of *dextro*- or sarcolactic acid produced during muscle contraction.

Glycogen has been investigated by Haworth and co-workers, who conclude that the molecule of glycogen consists of from twelve to eighteen α-glucopyranose units arranged similarly to the two α-glucose units in maltose and the molecular weight of glycogen is between 2000 and 3000. The molecular structure of the glycogen molecule may be represented thus:

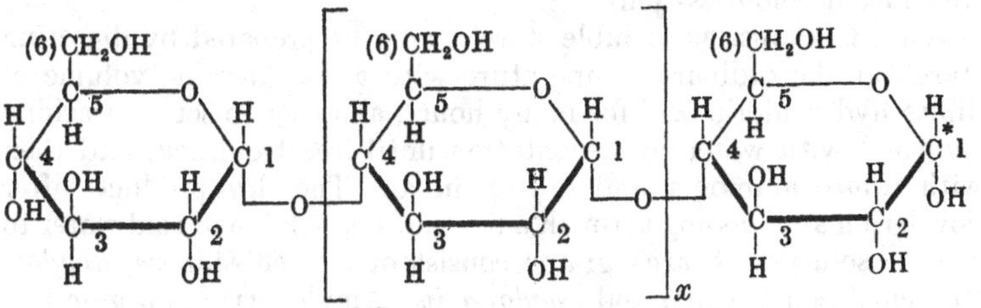

For glycogen, $x = 10$–16. For starch, x may be about 23.

Assuming the presence of eighteen α-glucose units in the molecule, the complete hydrolysis of glycogen is represented:

$$\underset{\text{glycogen}}{(C_6H_{10}O_5)_{18}} + 18H_2O \rightarrow \underset{\text{glucose}}{18C_6H_{12}O_6}$$

* Glycogen and starch not being reducing agents, the terminal grouping cannot be as indicated by the asterisk, since carbon atom 1 in the right-hand end unit is potentially aldehydic as in the aldohexoses. A non-reducing material would be represented if one α-glucopyranose unit was united to the rest of the molecule through carbon atoms 1, thus:

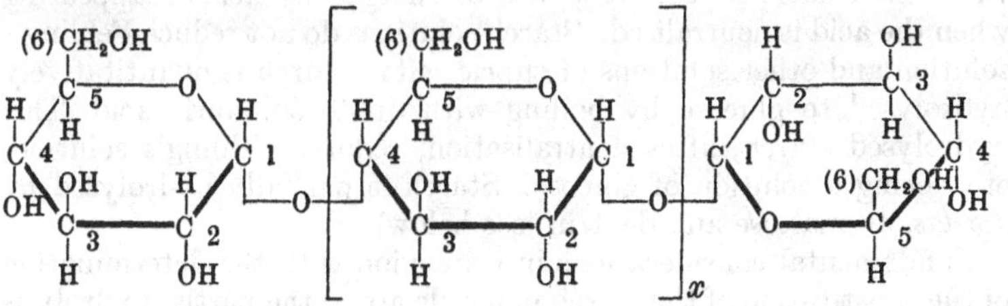

Another non-reducing material would be represented if the =CH.OH group marked with an asterisk is changed to =CO (oxidation); this would only reduce the molecular weight by two units.

Starch and dextrin

Starch is present chiefly in seeds and tubers of plants. It is also present in the green leaf and in the fruits. Dried seeds may contain as much as 70 per cent. by weight of starch and potatoes as much as 30 per cent. Microscopically, starch grains possess an organised structure by which the source of the starch may be determined.

Starch grains form a white powder, insoluble in cold water. If boiled with water, the grains swell and burst, yielding an opalescent colloidal solution frequently described as 'starch paste'. A convenient way of preparing a solution of starch is to macerate the starch with a small quantity of cold water and pour this suspension into a large volume of almost boiling water. According to the starch content, the 'paste' may, at the ordinary temperature, be a limpid solution or a translucent colourless jelly.

What is known as 'soluble starch' may be prepared by digesting starch at the ordinary temperature with a considerable volume of dilute hydrochloric acid for many hours, allowing to settle, washing the solid with water by decantation until free from acid and then with dilute ammonia and drying in air. The dry product, after powdering and passing through a fine sieve, is soluble in cold water to a clear solution. Starch grains consist of two substances, *amylose* (the chief constituent) and *amylopectin*. Amylopectin is a gum-like substance which swells up without dissolving in water and the gelatinisation of starch solutions is probably due to this constituent. Constitutionally, there appears to be no difference between the two constituents except that amylopectin appears to have a higher molecular weight than amylose.

Starch paste and solutions of soluble starch are *dextro*rotatory and give the following reactions. The starch is precipitated by the addition of alcohol. With small quantities of dilute iodine solution a blue colour develops; this colour disappears in hot solutions and reappears when the solutions cool; it is also discharged by acids reappearing when the acid is neutralised. Starch solutions do not reduce Fehling's solution and other solutions of cupric salts. Starch is quantitatively hydrolysed to glucose by boiling with dilute sulphuric acid. This hydrolysed starch, after neutralisation, reduces Fehling's solution, etc., being a solution of glucose. Starch is partially hydrolysed by *diastase* to maltose and dextrin (see below).

Fundamental considerations in connexion with the determination of the constitution of the starch molecule are (i) the partial hydrolysis (degradation) of starch to maltose and dextrin, represented in outline, thus:

$$\underset{\text{starch}}{(C_6H_{10}O_5)_n} + H_2O \rightarrow \underset{\text{maltose}}{C_{12}H_{22}O_{11}} + \underset{\text{dextrin}}{(C_6H_{10}O_5)_{n-2}}$$

(ii) the quantitative hydrolysis of starch to glucose by boiling with dilute sulphuric acid, represented in outline, thus:

$$\underset{}{(C_6H_{10}O_5)_n} + nH_2O \rightarrow \underset{\text{glucose}}{nC_6H_{12}O_6}$$

and (iii) the close relationship structurally between maltose, glycogen, dextrin and starch. Haworth and co-workers as a result of long investigations, particularly, as in the case of glycogen, of the methyl derivatives (methyl ethers), have suggested that the starch molecule consists of a regular chain of α-glucopyranose units united as in maltose and glycogen, there being about 25 such units corresponding to a molecular weight for starch of about 5000. It has still to be determined whether the small amounts of combined phosphorus and silica present in all varieties of starch do or do not have some significant value as a factor in the chemical constitution of the polysaccharide. It may be pointed out that the very small amounts of phosphorus and silicon present do not appear to have any stoichiometric relationships with the chemical molecule as a whole. If starch be represented by the provisional general formula $(C_6H_{10}O_5)_{25}$ its degradation (hydrolysis) to maltose and dextrin and to glucose may be represented, respectively, thus:

$$\underset{\text{starch}}{(C_6H_{10}O_5)_{25}} + H_2O \rightarrow \underset{\text{maltose}}{C_{12}H_{22}O_{11}} + \underset{\text{dextrin}}{(C_6H_{10}O_5)_{23}}$$

$$(C_6H_{10}O_5)_{25} + 25H_2O \rightarrow \underset{\text{glucose}}{25C_6H_{12}O_6}$$

When prepared from starch by the action of diastase, *dextrin* is a colourless powder, soluble in water, the solution being 'gummy' and *dextro*rotatory. Unlike starch, it has slight reducing properties and reacts with alkalies similarly to reducing sugars. It differs also from starch in giving a reddish brown coloration (not very different from that of iodine solutions) with dilute iodine solution. Like starch, dextrin is readily hydrolysed to glucose by boiling with dilute sulphuric acid. The molecule of dextrin must contain two α-glucopyranose units less than the molecule of starch, but the structural formula must also account for the small but definite reducing properties of the substance.

Cellulose

Cellulose is the chief constituent of the cell walls of plants in which other substances (lignin, etc.) are encrusted. The fibres of the cotton plant, hemp and flax are particularly rich in cellulose, from which 'pure' cellulose is conveniently prepared. Good quality filter paper is a fairly 'pure' form of cellulose.

Cellulose is insoluble in water and most solvents. It is decomposed, i.e. hydrolysed, by heating with water under pressure. Cellulose is generally stable towards alkalies. With strong solutions of alkali hydroxides, the cellulose fibre undergoes a curious thickening and gelatinisation. Treatment of cotton whilst under tension with strong alkali renders the surface translucent and this is the basis of 'mercerisation' (Mercer, 1844). Cellulose when treated with alkali hydroxides combines with carbon disulphide and forms a mixture of 'cellulose xanthates', compounds of the general formula *R*O.CS.SNa; these compounds when treated with water swell up, yielding colloidal solutions which decompose in air precipitating cellulose in a viscous form. This is the basis of the 'viscose' artificial silk process.

Cellulose becomes gelatinous and ultimately dissolves in aqueous solutions of zinc chloride. It dissolves in an ammoniacal solution of cupric hydroxide (Schweitzer's reagent), and this solution when acidified with dilute sulphuric acid reprecipitates the cellulose, again in a viscous form. This is the basis of another process for the production of artificial silk.

The esters of cellulose are of industrial importance. The esters of nitric acid (the so-called 'nitrocelluloses', compare 'nitroglycerine', p. 387) are the chief constituents of collodion, celluloid (tri- and tetranitrates) and guncotton (hexanitrate) and are prepared by treating cellulose with nitric and sulphuric acid mixtures under various conditions. Cellulose nitrates can be 'denitrated' with suitable reagents, e.g. sodium hydrosulphide, and are converted into a viscous form of cellulose, which can be expressed into threads. This is the basis of de Chardonnet's artificial silk process (1885). The cellulose esters of acetic acid are prepared by the action on cellulose of acetic acid or acetic anhydride in the presence of sulphuric acid; they constitute another variety of artificial silk ('rayon').

Cellulose is quantitatively hydrolysed to glucose by heating with dilute sulphuric acid. The essential difference between starch and cellulose is that the former is built of α-glucopyranose units and the latter of β-glucopyranose units; the molecule of cellobiose, the disaccharide corresponding to cellulose just as maltose does to starch, consists of two such β-glucopyranose units. Haworth and co-workers have shown that there are probably 200 β-glucopyranose units in the cellulose molecule and that these units are united as indicated, the constitutional formula of β-cellobiose being given for comparison:

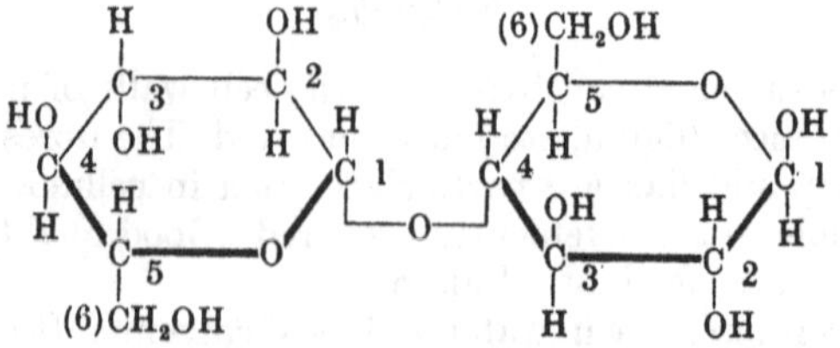

β-cellobiose

cellulose

This means that the molecular weight of cellulose is of the order of 32,000. On the assumption of 200 β-glucopyranose units in the molecule, the complete hydrolysis of cellulose may be represented:

$$\underset{\text{cellulose}}{(C_6H_{10}O_5)_{200}} + 200H_2O \rightarrow \underset{\text{glucose}}{200C_6H_{12}O_6}$$

Inulin

Inulin, the polysaccharide occurring in dahlia tubers and in the Jerusalem artichoke, is a convenient source of *d*(−)-fructose (p. 437). It is a colourless powder which swells up and dissolves in hot water, yielding a clear solution which is *laevo*rotatory.

Like starch it is precipitated from its aqueous solution by alcohol. Its aqueous solution yields a brown coloration with a dilute aqueous solution of iodine. All preparations have a slight but definite reducing action towards Fehling's solution, etc. Inulin is very easily and quantitatively hydrolysed to fructose by boiling with dilute mineral acids.

The molecule of inulin appears to consist of some thirty fructofuranose units, the probable structure being conveniently represented, thus:

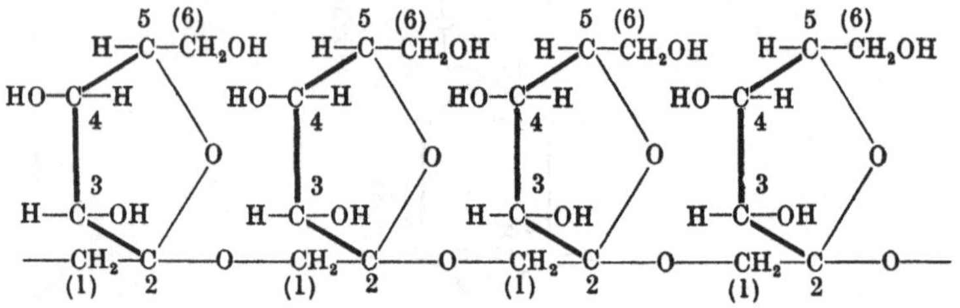

```
     5 (6)             5 (6)             5 (6)             5 (6)
   H—C—CH2OH         H—C—CH2OH         H—C—CH2OH         H—C—CH2OH
HO—C—H    \      HO—C—H    \      HO—C—H    \      HO—C—H    \
   |4      O        |4      O        |4      O        |4      O
   |3     /         |3     /         |3     /         |3     /
 H—C—OH  /        H—C—OH  /        H—C—OH  /        H—C—OH  /
——CH2—C——O————CH2—C——O————CH2—C——O————CH2—C——O——
  (1)  2           (1)  2           (1)  2           (1)  2
```

The molecular weight of inulin is of the order of 5000. On the basis of the above figures, the hydrolysis of inulin to fructose may be briefly represented, thus:

$$\underset{\text{inulin}}{(C_6H_{10}O_5)_{30}} + 30H_2O \rightarrow \underset{d(-)\text{-fructose}}{30C_6H_{12}O_6}$$

Estimation of sugars

The estimation of sugars, and particularly glucose, is important from industrial and medical points of view. If the nature of the sugar and its specific rotatory power under standard conditions are known and the sugar is in colourless aqueous solution, the simplest method of estimation is by determining the rotatory power of the solution by means of the polarimeter and calculating the concentration, c (in grams per 100 ml. of the solution) from the relation:

$$c = \frac{\alpha . 100}{[\alpha] . l},$$

where α represents the observed rotatory power of the solution, $[\alpha]$ the known specific rotatory power and l the length in decimetres of the polarimeter tube, i.e. length of layer of the solution (p. 345). The rotatory power is determined using the same wave-length and at the same temperature as the already known specific rotatory power of the sugar.

In actual practice, it frequently happens that aqueous solutions of sugars may not be colourless and, in these cases, the sugar is estimated by reactions which depend on its being a reducing agent. These methods can be employed for estimating reducing sugars directly; non-reducing sugars and polysaccharides are estimated by these methods after hydrolysis and neutralisation of the resultant solution. Lactose and maltose (reducing disaccharides) may either be estimated directly or after hydrolysis, the latter method giving the more accurate result. Pentoses may also be estimated directly, or, if other carbohydrates be present, they can be estimated by converting them into furfural, which is then combined with phloroglucinol and the resulting compound (p. 413) weighed directly.

The chemical reaction or reactions concerned depend on the reduction of alkaline solutions of cupric salts to cuprous compounds by the sugar, which is thereby oxidised at the same time. The oxidation of reducing sugars by alkaline solutions of cupric salts was first investigated by Hermann von Fehling (1812–1885), who showed that under conditions devised by him (referred to later as *standard* conditions) the following relationships hold in the case of glucose:

$$C_6H_{12}O_6 \equiv 5CuSO_4.5H_2O \equiv 5CuO \equiv 5Cu''$$

Since the nature of the oxidation products of glucose, etc. in the conditions studied by Fehling are not accurately known, it is impossible to state the reactions between reducing sugars and alkaline solutions of cupric salts in the form of equations. The method of estimation is therefore an arbitrary one and is only accurate when the *standard* conditions are maintained throughout the estimation. The standard conditions are (i) the concentration of the sugar, reckoned as glucose, must be between 0·5 and 1·0 per cent., (ii) the solution of the sugar must be added to the gently boiling solution of the cupric salt and further sugar solution added only after the sugar previously added has all been oxidised. The end of the reaction is determined by the disappearance of the blue colour of the original solution of the cupric salt.

Fehling's reagent (1849) consists of two solutions *A* and *B*:

A containing 69·28 grams of $CuSO_4.5H_2O$ dissolved in water and made up to 1 litre.

B containing 346 grams of sodium potassium tartrate (Rochelle salt, $NaKC_4H_4O_6.4H_2O$) and 130 grams of sodium hydroxide dissolved in water and made up to 1 litre.

For the estimation, exactly equal volumes of *A* and *B* are mixed together and 10 ml. of the reagent are equivalent to 0·05 gram of glucose. If, after trial, less than 5 ml. of the solution of sugar of unknown strength are required for 10 ml. of Fehling's solution, the sugar solution must be diluted accordingly.

Using Fehling's solution, cuprous oxide is precipitated, and at the end-point the supernatant solution is colourless. The precipitation of the cuprous oxide tends to make the end-point obscure and various devices have been suggested for overcoming this difficulty. Using Fehling's unmodified method, it is convenient to use starch-potassium iodide as indicator towards the end of the reaction for determining the presence of unchanged cupric salt remaining in the solution.

Another method, devised by Pavy, depends on diluting Fehling's reagent to a known amount with a concentrated aqueous solution of ammonia. In this case, the cuprous oxide remains in solution and the solution becomes colourless; but this modification of Fehling's method has the disadvantage that ammonia is lost during the determination and the end-point is rarely as clear as may be desired.

Other modifications of Fehling's method are due to Gerrard and to Bertrand. Gerrard made use of potassium cyanide for preventing the precipitation of the cuprous oxide. Bertrand has accurately applied Mohr's method, which consists in filtering off the precipitated cuprous oxide and estimating the latter by dissolving it in a solution of ferric sulphate made acid with sulphuric acid and titrating the ferrous sulphate produced by the reaction:

$$Cu_2O + Fe_2(SO_4)_3 + H_2SO_4 = 2CuSO_4 + 2FeSO_4 + H_2O$$

by means of a standard aqueous solution of potassium permanganate.

The most convenient method for the estimation of glucose, etc., and which is employed very largely in clinical medical practice, is that devised by Benedict. This depends on the precipitation of colourless cuprous thiocyanate and the production of a colourless or (in some cases) a pale greenish coloured solution above the precipitate. Benedict's solution contains 18 grams of $CuSO_4.5H_2O$, 200 grams of $Na_2CO_3.10H_2O$, 200 grams of sodium citrate and 125 grams of potassium thiocyanate, together with 0·25 gram of potassium ferrocyanide made up with water to 1 litre; it is important that the reagents should be as pure as possible. In the estimation, 25 ml. of the solution, to which some 5–10 grams of anhydrous sodium carbonate and a little porous pot, to promote steady ebullition, are added, is kept gently boiling in a suitable wide test-tube and the sugar solution (diluted if necessary so as to have a concentration of 0·5 to 1·0 per cent.) added gradually from a burette until the solution is decolorised. If the Benedict's solution contains the above quantity of copper sulphate, it has been shown that 25 ml. of the solution are equivalent to 0·05 gram of glucose, but it is always necessary to standardise the solution against a solution of glucose of known concentration (see below).

The reducing power of the monosaccharides are not quite the same; for example, 10 ml. of Fehling's reagent are equivalent to 0·05 gram of glucose, 0·0511 gram of galactose, 0·0514 gram of fractose and to 0·0431 gram of mannose. In actual practice, these four monosaccharides are generally regarded as having the same reducing power as glucose.

Owing to the difficulty in obtaining it pure, glucose is not convenient for standardising alkaline solutions of cupric salts. Pure sucrose is readily available and is always used for this purpose. A weighed quantity (say, 1·75 grams) of pure sucrose is dissolved in about 40 ml. of water and to this is added 25 ml. of 0·5*N* HCl. The mixture is heated on a boiling water-bath for 15 minutes, when hydrolysis is complete, and then neutralised by means of the requisite quantity of solid sodium carbonate. The cooled mixture is transferred quantitatively to a standard flask and made up to 250 ml. with water. According to the equation:

$$C_{12}H_{22}O_{11} + H_2O = C_6H_{12}O_6 + C_6H_{12}O_6$$

the solution contains 1·842 grams of reducing sugar (equal quantities of glucose and fructose) considered as glucose and the solution is assumed to contain 0·7368 gram of glucose per 100 ml. Such a solution is used for the standardisation of the solutions of the cupric salts in terms of the amount of glucose in grams equivalent to or oxidised by a stated volume of the solution.

Fermentation of Hexoses

The production of ethyl alcohol and carbon dioxide from glucose by the action of enzymes produced by yeast (saccharomyces, of which there are many varieties) has been referred to under the term 'fermentation'. Only four hexoses, viz. *d*(+)-glucose, *d*(+)-mannose, *d*(+)-galactose and *d*(−)-fructose, which are all naturally occurring substances, undergo the change mentioned. Galactose is, however, fermented by only some of the yeasts, all of which ferment the other three sugars. The mechanism of the alcoholic fermentation of these three sugars must be identical, and it will be realised that the configuration of hydrogen atoms and hydroxyl groups attached to carbon atoms (3), (4), (5) and (6) is the same in these three sugars. The reaction taking place has previously (p. 96) been expressed simply as:

$$C_6H_{12}O_6 \rightarrow 2C_2H_5OH + 2CO_2$$

The process of alcoholic fermentation has been and is being studied by many investigators. The work of Pasteur on fermentation began in 1857 and may be said to be the foundation of our knowledge of the subject. In 1860 Pasteur concluded that 'alcoholic fermentation is an act correlated with the life and organisation of the yeast cells, not with the death or putrefaction of the cells, any more than it is a phenomenon of contact, in which case the transformation of sugar would be accomplished in the presence of the ferment without yielding up to it or taking from it anything.' In these words he rejected some of the views of the process held up to that time.

The next great advance was made by E. Buchner in 1897, who succeeded in extracting from yeast a liquid which in the complete absence of the yeast cells was capable of effecting the transformation of sugar into ethyl alcohol and carbon dioxide; and Pasteur was only in error in that he believed the presence of the actual living cells to be necessary for fermentation to take place. The yeast juice prepared according to Buchner's method or as modified by later workers is a thick opalescent brownish liquid which is faintly acid and almost optically inactive. It contains generally rather more than 1 per cent. of nitrogen and leaves an ash amounting to rather less than 2 per cent. It is important to realise that the juice always contains small amounts of mineral phosphates or pyrophosphates. This extract is capable of bringing about the fermentation of the sugar slowly for a time, after which it loses its fermenting power; at this stage the fermenting substance in the yeast extract has disappeared. To the fermenting substance in the extract, Buchner gave the name *zymase* which, although not isolated, was recognised as belonging to the class of organic catalysts or enzymes. It is specifically classified as a *sucroclastic enzyme*. If a mixture of alcohol and ether be added to the yeast extract the precipitate produced, even after separation and drying, is capable of inducing fermentation.

Much later, Harden and Young (1905) showed that the zymase is composed of two parts. By filtering the yeast juice through a special gelatin filter, they separated it into two parts—the residue remaining in the filtering apparatus and the filtrate; they further showed that neither the residue nor the filtrate is capable of bringing about the fermentation. When the residue and the filtrate are mixed, the resulting solution brings about the fermentation in the normal manner. Obviously, Buchner's zymase contains at least two components, and the zymase system may be briefly described thus:

Zymase (often described as the *fermenting complex*)

residue in filter (often described as *apozymase*)	filtrate (which has been termed *cozymase*)

Actually, in the fermenting complex there are several enzymes which play a part in the fermenting process.

The fermentation induced by fresh yeast juice is slow, and Harden and Young showed that on addition of soluble phosphate (Na_2HPO_4) the rate of fermentation rapidly increases to as much as twenty times the original rate and then falls again to something like, but generally higher than, the original. This indicates that phosphates play an essential part in the fermentation process, and so-called *hexose phosphates*, which are esters, have been isolated from the fermenting solutions. Harden further showed that carbon dioxide and ethyl alcohol are produced as the hexose phosphates are formed by such reactions as (*a*) and (*b*):

$$(a)\quad 3C_6H_{11}O_5.OH + 2HO.PO(ONa)_2 \rightarrow \underset{\text{hexose monophosphate}}{2C_6H_{11}O_5.O.PO(ONa)_2} + 2H_2O + 2CO_2 + 2C_2H_5OH$$

$$(b)\quad 2C_6H_{10}O_4(OH)_2 + 2HO.PO(ONa)_2 \rightarrow \underset{\text{hexose diphosphate}}{C_6H_{10}O_4\begin{cases}O.PO(ONa)_2\\O.PO(ONa)_2\end{cases}} + 2H_2O + 2CO_2 + 2C_2H_5OH$$

As the fermentation slackens, the concentration of free phosphate in the solution increases, pointing to hydrolysis of the hexose phosphates by reactions (*c*) and (*d*). The hydrolysis of the hexose diphosphate may, of course, proceed in two stages:

$$(c)\quad C_6H_{11}O_5.O.PO(ONa)_2 + H_2O \rightleftarrows \underset{\text{hexose}}{C_6H_{11}O_5.OH} + Na_2HPO_4$$

$$(d)\quad C_6H_{10}O_4\begin{cases}O.PO(ONa)_2\\O.PO(ONa)_2\end{cases} + 2H_2O \rightleftarrows \underset{\text{hexose}}{C_6H_{10}O_4(OH)_2} + 2Na_2HPO_4$$

The formation and hydrolysis of hexose phosphates are catalysed by an enzyme, *hexosephosphatase*, present in the fermenting complex. The above reactions imply that fermentation depends on the supply of phosphates and that synthesis and hydrolysis of the hexose phosphates proceed *pari passu* during the fermentation process. Probably, the hexose regenerated by hydrolysis is in a more active

(? furanose) form than the normal hexose. The presence in small amounts of inorganic phosphates in yeast juice explains the slow fermentation produced by the latter in the absence of added phosphates. Owing to the interconversion of glucose, mannose and fructose in solution in the presence of phosphates (p. 425), the same hexose phosphates are probably produced whichever of these sugars is initially present.

This outline of the conditions governing the production of carbon dioxide and ethyl alcohol from the specific hexoses does not explain by what stages the final products are reached. At the time of Pasteur's investigations it was known that besides carbon dioxide and ethyl alcohol small quantities of other compounds are formed during alcoholic fermentation; the compounds then recognised are glycerol, succinic acid and the higher alcohols present in fusel oil (p. 112). Later acetaldehyde was also detected, and Neuberg (1918) showed that the amount of acetaldehyde (and of glycerol) was increased considerably by the addition of sodium sulphite when once fermentation has become well established (*v.* below). Succinic acid and fusel oil do not arise from the hexose (*v.* below).

Suggestions concerning the stages of breakdown of the hexose during fermentation must take into account the various products which have been isolated either in the normal process or by carrying out the process in the presence of reagents which are known to 'fix' the possible intermediate substances. In addition to glycerol and acetaldehyde, it has been demonstrated that pyruvic acid (p. 365) and the monomethyl derivative of glyoxal (p. 354) or methylglyoxal are formed as intermediate products, and it is significant that pyruvic acid, methylglyoxal and glycerol all have three carbon atoms in the molecule. The series of reactions suggested by Neuberg is:

(i) $$C_6H_{12}O_6 - 2H_2O \rightarrow \underbrace{\underset{\text{keto}}{2CH_3.CO.C\langle^{H}_{O}} \rightleftarrows \underset{\text{enol}}{2CH_2:C(OH).C\langle^{H}_{O}}}_{\text{methylglyoxal}}$$

(ii) $$\left.\begin{array}{l} CH_2:C(OH).C\langle^{H}_{O} + H_2O \\ CH_2:C(OH).C\langle^{H}_{O} + \begin{array}{c}H\ \ H\\ \diagdown\diagup \\ O\end{array} \end{array}\right\} \rightarrow \begin{array}{l} \underset{\text{glycerol}}{CH_2OH.CHOH.CH_2OH} \\ \quad + \\ CH_2:C(OH).COOH \rightarrow \underset{\text{pyruvic acid}}{CH_3.CO.COOH} \end{array}$$

(iii) $$CH_3.CO.COOH \rightarrow CH_3.CHO + CO_2$$

(iv) $$\left.\begin{array}{l} CH_3.CO.C\langle^{H}_{O} \\ \qquad\qquad + \begin{array}{c}O\\ \diagup\diagdown \\ H\ \ H\end{array} \\ CH_3.C\langle^{H}_{O} \end{array}\right\} \rightarrow CH_3.CO.COOH + \underset{\text{ethyl alcohol}}{CH_3.CH_2OH}$$

This scheme involves (i) the production by loss of water from the hexose of two molecules of methylglyoxal which may exist in keto and enol forms, (ii) the transformation of methylglyoxal into glycerol and pyruvic acid by addition of water and a Cannizzaro reaction (p. 147), i.e. the oxidation of one aldehyde molecule and the reduction of the other, (iii) the production of acetaldehyde by loss of carbon dioxide from pyruvic acid probably through the agency of the enzyme *carboxylase* present in the fermenting complex and (iv) the transformation of methylglyoxal and acetaldehyde into pyruvic acid and ethyl alcohol respectively by the Cannizzaro transformation; the pyruvic acid produced in this reaction can then undergo reaction (iii). This scheme of reactions indicates the production of a small amount of glycerol, which is actually always found in the fermenting solution. It may be remarked that Cannizzaro transformations in reactions (ii) and (iv) might also be expected to produce $\alpha\beta$-dihydroxypropionic acid, $CH_2OH.CHOH.COOH$,* and hydroxyacetone† or 'pyroracemic alcohol', $CH_3.CO.CH_2OH$ [reaction (ii)], and, also, hydroxyacetone and acetic acid [reaction (iv)]. Another simpler scheme has been put forward; this scheme is:

$$(a)\quad C_6H_{12}O_6 \rightarrow 2CH_3.CO.COOH + 4H$$

$$(b)\quad 2CH_3.CO.COOH \xrightarrow{\text{carboxylase}} 2CH_3.CHO + 2CO_2$$

$$(c)\quad 2CH_3.CHO + 4H \rightarrow 2CH_3.CH_2OH$$

Reaction (*c*) could be catalysed by the enzyme *reductase* known to be present in the fermenting complex. The drawbacks to this scheme are that the mechanism of transformation of the hexose into pyruvic acid is not explained and the presence of glycerol in the fermenting solution is not accounted for.

Fusel oil and succinic acid are in all probability produced from amino acids arising from the hydrolysis of proteins present in yeast. For example, leucine (p. 326) by hydrolysis gives rise to *iso*amyl alcohol (p. 112) or 3-methyl-*n*-butyl alcohol, thus:

$$\underset{\text{leucine}}{(CH_3)_2{>}CH.CH_2.CH(NH_2).COOH} + H_2O \rightarrow \underset{\textit{iso}\text{amyl alcohol}}{(CH_3)_2{>}CH.CH_2.CH_2OH} + NH_3 + CO_2$$

and, by a similar reaction, *iso*leucine gives rise to the optically active (*laevo*rotatory) amyl alcohol or 2-methyl-*n*-butyl alcohol, which always accompanies *iso*amyl alcohol in fusel oil:

$$\frac{CH_3}{C_2H_5}{>}CH.CH(NH_2).COOH + H_2O \rightarrow \underset{\text{amyl alcohol (laevorotatory)}}{\frac{CH_3}{C_2H_5}{>}CH.CH_2OH} + NH_3 + CO_2$$

* Dihydroxypropionic acid is glyceric acid (p. 390), which forms an undistillable syrup. It has been obtained in optically active forms.

† *Hydroxyacetone* is a colourless liquid, b.p. 145–146°.

This method of production of these compounds was demonstrated by Ehrlich in 1907. The lower alcohols present in fusel oil arise from similar reactions with simpler amino acids.

Ehrlich also showed that succinic acid, always present in small amounts in the fermenting solution, arises from the dibasic mono-amino-acid, glutaminic acid, by an oxidation process (induced by an enzyme, *oxidase*, present in the fermenting complex), thus:

$$\underset{\text{glutaminic acid}}{HOOC.CH_2.CH_2.CH(NH_2).COOH} + 2O \rightarrow \underset{\text{succinic acid}}{HOOC.CH_2.CH_2.COOH} + NH_3 + CO_2$$

In normal alcoholic fermentation, the solution is practically neutral or very faintly alkaline in reaction and, in general, enzyme reactions are extremely sensitive to the conditions pertaining. Neuberg and others provided an example of the sensitivity of enzymes in their reaction to the addition of salts by showing that if sodium sulphite be added to an already fermenting solution of hexose by yeast, the yield of ethyl alcohol and carbon dioxide diminishes and considerable quantities of glycerol and acetaldehyde (the latter in the form of its sodium bisulphite compound) are produced, the normal fermentation process being considerably modified. On this discovery, an important industrial process for the production of glycerol is based, the yield of glycerol being about 15 to 20 per cent. of the sugar employed. In these conditions, the reaction taking place may be represented:

$$C_6H_{12}O_6 + Na_2SO_3 + H_2O \rightarrow \underset{\text{glycerol}}{C_3H_8O_3} + \underset{\substack{\text{acetaldehyde-}\\\text{sodium bisulphite}}}{CH_3.CHOH.SO_3Na} + NaHCO_3$$

but the normal reactions leading to the production of carbon dioxide, ethyl alcohol, etc. are not entirely suppressed.

Another industrially important fermentation process is the production of acetone and *n*-butyl alcohol (p. 110), starting with certain types of starchy materials (e.g. maize), the process being induced by enzymes produced by the *Bacillus macerans*. The starch is first converted by hydrolysis into hexose, which then undergoes fermentation. The total reaction has been written as:

$$3C_6H_{12}O_6 \rightarrow \underset{\substack{n\text{-butyl}\\\text{alcohol}}}{2C_4H_9OH} + \underset{\text{acetone}}{CH_3.CO.CH_3} + 7CO_2 + 4H_2 + H_2O$$

but the nature of the intermediate products and, hence, the stages in the reactions have not been fully elucidated.

The production of organic compounds by biochemical or fermentation processes is highly important industrially and, by investigation, is continually increasing. The production of lactic acid, of importance in the dyeing and leather industries, from glucose has already been described (p. 364). In this process, the reaction may proceed via pyruvic acid, of which lactic acid is a reduction product. *n*-Butyric acid is also produced by the fermentation of glucose in the presence of calcium carbonate. The catalytic enzymes for this fermentation are provided in modern processes by pure cultures of *B. subtilis* and *B. boocapricus*. The reaction taking place is generally described, but is not so simple, as:

$$C_6H_{12}O_6 \rightarrow CH_3.CH_2.CH_2.COOH + 2CO_2 + 2H_2$$

Another example is the production of *citric acid*, hydroxytricarballylic acid, 2-hydroxy-1 : 2 : 3-tricarboxypropane, $C_6H_8O_7 + H_2O$, the constitution of which is written:

$$\begin{array}{l} CH_2.COOH \\ | \\ C(OH)COOH \\ | \\ CH_2.COOH \end{array}$$

The molecule of citric acid, like that of a hexose, contains six carbon atoms. The conversion of hexose into citric acid is obviously an oxidation process, which might be represented by an equation such as:

$$2C_6H_{12}O_6 + 3O_2 \rightarrow 2C_6H_8O_7 + 4H_2O$$

but the actual reactions taking place must be highly complex, since the hexose is finally converted into a compound the carbon skeleton of which in its simplest form may be represented:

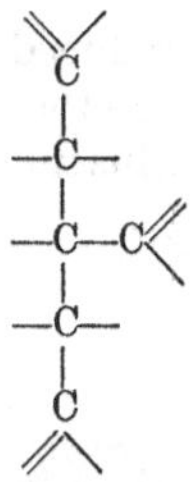

```
   \\ /
    C
    |
   -C-
    |      //
   -C--C<
    |      \
   -C-
    |
    C
   // \
```

The enzymes inducing the transformation of hexose into citric acid are provided by certain species of citromyces, especially *C. pfefferianus*, and certain moulds, e.g. *Aspergillus niger*. The yield of citric acid is as high as 50 per cent., based on the sugar, and almost the whole of the citric acid used industrially and especially in the food industry is prepared by the fermentation process.

As alternative to the production of acetone and *n*-butyl alcohol (butanol) by fermentation, chemical methods for the manufacture of these compounds have been successfully developed in this country and have become highly important.

The starting material for the production of both substances is ethyl alcohol obtained by the fermentation (yeast) of molasses. To obtain acetone, ethyl alcohol vapour and steam are passed over a catalyst (rusted iron impregnated with nickel oxide is mentioned in the patent specification) at a dull red hot (500°). The resulting vapours contain chiefly acetone, hydrogen, excess steam and unchanged alcohol. Acetone of high quality is obtained in a yield as high as 88 per cent. after fractional distillation of the liquid product. Hydrogen is the important by-product in this reaction and is used in the production of *n*-butyl alcohol as described below. The reaction for the conversion of ethyl alcohol into acetone may be summarised:

$$6C_2H_6O \rightarrow 4C_3H_6O + 2H_2O + 4H_2$$

but it is unlikely to be a one-stage process.

The first stage in the production of *n*-butyl alcohol from ethyl alcohol consists in the oxidation of the latter catalytically to acetaldehyde. The alcohol vapour mixed with air is passed over a silver catalyst maintained at a suitable temperature and the acetaldehyde separated finally by fractional distillation. The acetaldehyde is then condensed to aldol (p. 158) by means of alkali

$$CH_3.CHO + CH_3.CHO \rightarrow CH_3.CHOH.CH_2.CHO$$

The aldol on steam distillation is converted into crotonaldehyde

$$CH_3.CHOH.CH_2.CHO \xrightarrow{-H_2O} CH_3.CH{:}CH.CHO$$

and the latter is catalytically reduced (using hydrogen obtained as the by-product in the above acetone process) to *n*-butyl alcohol which only needs fractional distillation for complete purification

$$CH_3.CH{:}CH.CHO + 2H_2 \rightarrow CH_3.CH_2.CH_2.CH_2OH$$

CHAPTER XVI

DERIVATIVES OF CARBONIC ACID—UREA—PURINES

Simple Derivatives of Carbonic Acid

Carbonic acid, which can also be described as hydroxyformic acid, is only known in the form of its derivatives. The constitutions of the hypothetical carbonic acid and of its acid chlorides would be represented:

OH	Cl	Cl		OH	Cl	Cl
|	|	|		|	|	|
C=O	C=O	C=O	compare	SO_2	SO_2	SO_2
|	|	|		|	|	|
OH	OH	Cl		OH	OH	Cl
	(*a*)	(*b*)		sulphuric acid	chloro-sulphonic acid	sulphuryl chloride

The compound (*a*) is unknown, but it would be described as *chloroformic acid*. Compound (*b*) is *carbonyl chloride* or *phosgene*; this is readily prepared by the interaction of chlorine and carbon monoxide in the presence of absorptive charcoal:

$$CO + Cl_2 \rightarrow COCl_2$$

The combination of carbon monoxide and chlorine also takes place in the presence of sunlight (phosgene, from φῶς (light) and γεννάω (I produce)). Phosgene may also be produced by the action of fuming sulphuric acid on carbon tetrachloride:

$$CCl_4 + 2SO_3 \rightarrow COCl_2 + \underset{\text{pyrosulphuryl chloride}}{S_2O_5Cl_2}$$

Phosgene is a colourless gas, b.p. 8·2°. In spite of its highly poisonous properties, it is an important chemical reagent for the manufacture of dye-stuffs belonging to the di- and tri- phenylmethane series. Although the gas does not fume in air it reacts as a typical acid chloride, and when heated with water it is hydrolysed to carbon dioxide and hydrochloric acid.

When it reacts with ethyl alcohol, *ethyl chloroformate* (colourless liquid, b.p. 93°) is obtained:

$$Cl.CO.Cl + HO.C_2H_5 \rightarrow \underset{\text{ethyl chloroformate}}{Cl.CO.OC_2H_5} + HCl$$

The complete reaction of ethyl alcohol with phosgene is to produce the typical ester of carbonic acid, *ethyl carbonate* (colourless liquid, b.p. 126°), ethyl chloroformate being the intermediate product:

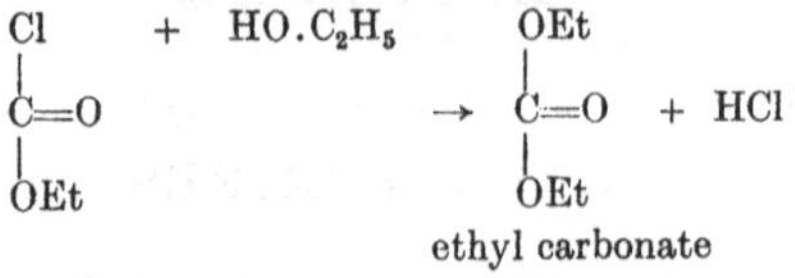

$$\begin{array}{c}Cl\\|\\C{=}O\\|\\OEt\end{array} + HO.C_2H_5 \rightarrow \begin{array}{c}OEt\\|\\C{=}O\\|\\OEt\end{array} + HCl$$

ethyl carbonate

Ethyl carbonate is also obtained when ethyl iodide reacts with silver carbonate:

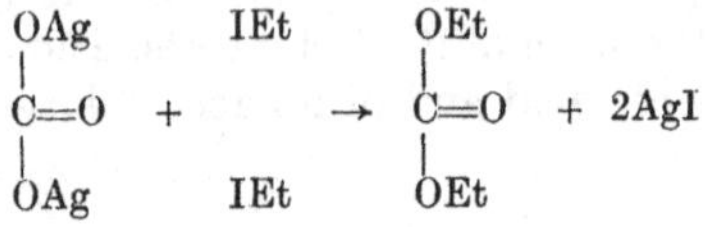

$$\begin{array}{c}OAg\\|\\C{=}O\\|\\OAg\end{array} + \begin{array}{c}IEt\\ \\ \\ \\IEt\end{array} \rightarrow \begin{array}{c}OEt\\|\\C{=}O\\|\\OEt\end{array} + 2AgI$$

Phosgene also reacts with ammonia to produce *urea* (sometimes formerly known as *carbamide*, i.e. the amide of the unknown *carbamic acid* or *aminoformic acid*):

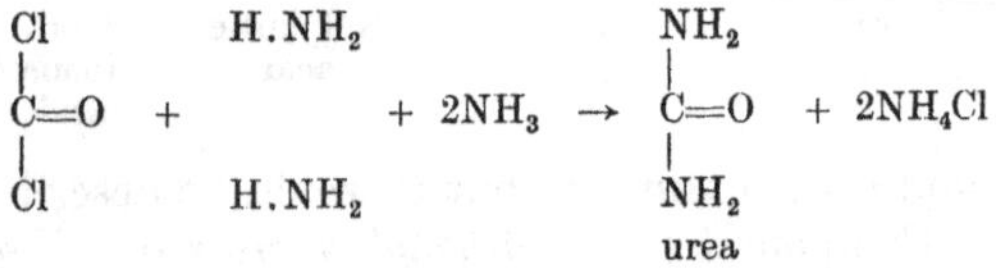

$$\begin{array}{c}Cl\\|\\C{=}O\\|\\Cl\end{array} + \begin{array}{c}H.NH_2\\ \\ \\ \\H.NH_2\end{array} + 2NH_3 \rightarrow \begin{array}{c}NH_2\\|\\C{=}O\\|\\NH_2\end{array} + 2NH_4Cl$$

urea

but other products are also formed at the same time. Ethyl carbonate also reacts with ammonia, giving urea:

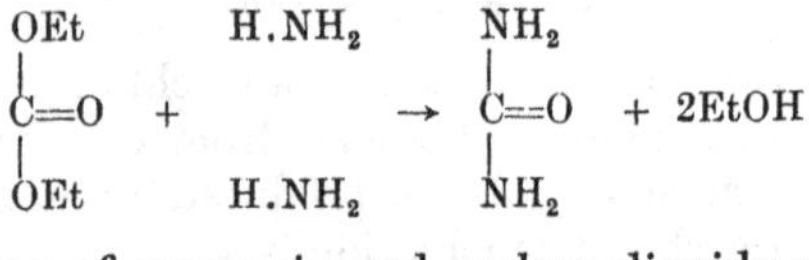

$$\begin{array}{c}OEt\\|\\C{=}O\\|\\OEt\end{array} + \begin{array}{c}H.NH_2\\ \\ \\ \\H.NH_2\end{array} \rightarrow \begin{array}{c}NH_2\\|\\C{=}O\\|\\NH_2\end{array} + 2EtOH$$

The interaction of ammonia and carbon dioxide may lead to the following products, derivatives of carbonic acid, according to the conditions employed:

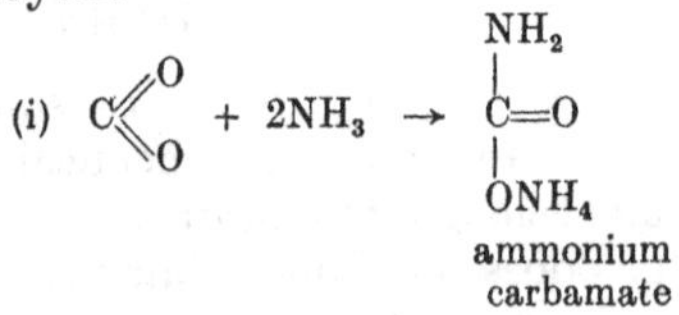

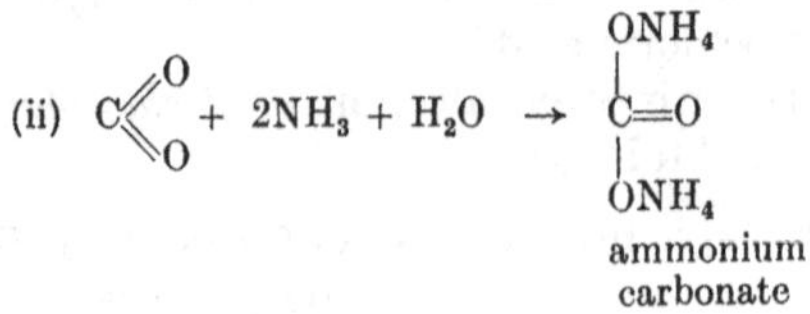

$$\text{(i)}\quad C\begin{array}{l}{\nearrow}O\\{\searrow}O\end{array} + 2NH_3 \rightarrow \begin{array}{c}NH_2\\|\\C{=}O\\|\\ONH_4\end{array}$$

ammonium carbamate

$$\text{(ii)}\quad C\begin{array}{l}{\nearrow}O\\{\searrow}O\end{array} + 2NH_3 + H_2O \rightarrow \begin{array}{c}ONH_4\\|\\C{=}O\\|\\ONH_4\end{array}$$

ammonium carbonate

$$\text{(iii)}\quad C\begin{smallmatrix}\nearrow O\\ \searrow O\end{smallmatrix} + 2NH_3 \rightarrow \begin{matrix}NH_2\\ |\\ C{=}O\\ |\\ NH_2\end{matrix} + H_2O$$

Reaction (iii) is used for the large scale manufacture of urea as an artificial fertiliser.

The three substances—urea, ammonium carbonate and ammonium carbamate—can be converted into each other as indicated:

$$(a)\quad \begin{matrix}ONH_4\\ |\\ C{=}O\\ |\\ ONH_4\end{matrix} \underset{\text{hydrolysis}}{\overset{\text{heat}}{\rightleftharpoons}} \begin{matrix}NH_2\\ |\\ C{=}O\\ |\\ ONH_4\end{matrix} + H_2O$$

$$(b)\quad \begin{matrix}NH_2\\ |\\ C{=}O\\ |\\ ONH_4\end{matrix} \underset{\text{hydrolysis}}{\overset{\text{heat}}{\rightleftharpoons}} \begin{matrix}NH_2\\ |\\ C{=}O\\ |\\ NH_2\end{matrix} + H_2O$$

$$(c)\quad \begin{matrix}ONH_4\\ |\\ C{=}O\\ |\\ ONH_4\end{matrix} \underset{\text{hydrolysis}}{\overset{\text{heat}}{\rightleftharpoons}} \begin{matrix}NH_2\\ |\\ C{=}O\\ |\\ NH_2\end{matrix} + 2H_2O$$

The conversion of ammonium carbamate and of ammonium carbonate into urea by the action of heat was discovered by Basarov in 1869 and the reverse change of urea to ammonium carbonate (in which, presumably, ammonium carbamate is an intermediate stage) takes place during the alkaline fermentation of urea by means of an enzyme produced by the microorganism, *Micrococcus ureae*, isolated by van Tieghem (1864).

Ammonium carbamate is the ammonium salt of the at present unknown carbamic acid (acid amide of carbonic acid):

$$\begin{matrix}NH_2\\ |\\ C\end{matrix}\begin{smallmatrix}\nearrow O\\ \searrow OH\end{smallmatrix}$$

of which many esters have been prepared. These esters are known as *urethanes*. *Ethyl urethane* or, simply, *urethane* is prepared by the action of ammonia on ethyl chloroformate:

$$\begin{matrix}Cl\\ |\\ C\end{matrix}\begin{smallmatrix}\nearrow O\\ \searrow OEt\end{smallmatrix} + \begin{matrix}H.NH_2\\ NH_3\end{matrix} \rightarrow \begin{matrix}NH_2\\ |\\ C\end{matrix}\begin{smallmatrix}\nearrow O\\ \searrow OEt\end{smallmatrix} + NH_4Cl$$

It is a colourless crystalline substance, m.p. 50°, b.p. 184°.

Urea and its Derivatives

Urea was known as an essential constituent of human urine (in which it occurs roughly as a 2 per cent. solution) at the beginning of the eighteenth century, but it was not obtained as a crystalline substance from urine until the end of that century. The fresh urine was carefully concentrated without overheating and, after cooling, concentrated nitric acid was carefully added, when a crystalline mass (urea nitrate) was precipitated. This was separated, dissolved in water and treated with lead or barium oxide. The residue obtained on evaporation to dryness was extracted with hot alcohol, and this solution freed from solid matter deposited colourless crystals of urea on evaporation and cooling. Such material was shown to have the composition CH_4ON_2.*

A number of ways in which urea as a derivative of carbonic acid may be synthesised have been mentioned above. The outstanding fact in connexion with the chemistry of urea is its synthesis by Wöhler in 1828. Wöhler was aiming at the preparation of ammonium cyanate, essentially by allowing potassium cyanate to react with ammonium sulphate. When the aqueous solution of these two salts was evaporated to dryness and the residue extracted with alcohol, potassium sulphate was left undissolved and the alcoholic solution contained a substance which proved to be identical with urea, which had already been isolated from urine. This synthesis of a substance, essentially a product of life processes, was of fundamental importance in the chemistry of organic compounds, which at that time was defined as the chemistry of compounds formed under the influence of life. Shortly afterwards, Liebig and Wöhler (1830) prepared ammonium cyanate and showed it was different from and isomeric with urea. Actually, this was the first definite example of isomerism. This work may be summarised:

$$2KOCN + (NH_4)_2SO_4 \rightarrow K_2SO_4 + \underset{\text{ammonium cyanate}}{2NH_4OCN} \rightarrow \underset{\text{urea}}{2CH_4ON_2}$$

Assuming the constitution of cyanic acid to be H—O—CN, ammonium cyanate is $[NH_4][—O—C{\equiv}N]$, and urea as a derivative of carbonic acid is

* Unless the action of concentrated nitric acid on evaporated urine is carefully controlled, it may lead to various complications, including charring of the mixture. Nitric acid can be effectively replaced by a hot solution of oxalic acid, and when this is added to the concentrated urine the sparingly soluble urea oxalate separates. This is filtered off, boiled in aqueous solution with precipitated calcium carbonate, when insoluble calcium oxalate is formed and urea liberated. The filtered solution may be decolorised with decolorising charcoal, again filtered and evaporated to dryness on the water-bath. The residue, urea, may be conveniently recrystallised from a little alcohol.

$$\begin{array}{c} H \\ H \end{array}\!\!>N-\underset{\underset{O}{\|}}{C}-N\!\!<\!\!\begin{array}{c} H \\ H \end{array}$$

According to this, urea contains the grouping

$$\begin{array}{c} H \\ H \end{array}\!\!>N-CO-$$

and like many other compounds containing this grouping it may react under favourable conditions as if it contained the grouping

$$HN{=}C(OH)-$$

In other words, the substance urea is capable of reacting under suitable conditions according to the alternative constitutions:

$$\begin{array}{c} H-N-H \\ | \\ C{=}O \\ | \\ H-N-H \end{array} \quad \text{and} \quad \begin{array}{c} H-N-H \\ | \\ C-(OH) \\ \| \\ N-H \end{array}$$

which can be derived from each other by the transference of a hydrogen atom from oxygen to nitrogen and *vice versa*. These formulae do not imply the actual existence of two forms of urea but merely that the compound affords an example of tautomerism (p. 373), only one of the tautomeric forms (probably the carbamide form) being capable of isolation.

Calcium cyanamide is an important artificial fertiliser; the technical article contains free carbon and is generally known under the name *nitrolim*. It is manufactured by heating calcium in a current of nitrogen to a high temperature, the reaction being usually represented:

$$Ca\!\!<\!\!\begin{array}{c} C \\ \||| \\ C \end{array} + N_2 \rightarrow \underset{\text{calcium cyanamide}}{Ca{=}N-C{\equiv}N} + C$$

When calcium cyanamide is treated with the calculated quantity of sulphuric acid, *cyanamide* (the amide of cyanic acid, $H-O-C{\equiv}N$) is formed (see below):

$$Ca{=}N-C{\equiv}N + H_2SO_4 \rightarrow CaSO_4 + H_2N-C{\equiv}N$$

and in aqueous solution especially in the presence of dilute mineral acid it is rapidly converted into urea, thus:

$$\begin{array}{c} H-N-H \\ | \\ C \\ \||| \\ N \end{array} + H_2O \rightarrow \begin{array}{c} H-N-H \\ | \\ C-OH \\ \| \\ N-H \end{array} \rightleftarrows \begin{array}{c} H-N-H \\ | \\ C{=}O \\ | \\ H-N-H \end{array}$$

Urea crystallises in long needles, m.p. 132·6°. It is readily soluble in water, in which it has a normal molecular weight, as determined by the lowering of the freezing point of an aqueous solution. It may be recrystallised from hot alcohol and is insoluble in ether.

When heated in a test-tube, urea melts, evolves ammonia, a colourless sublimate being produced; when all the ammonia has been evolved the residue at the bottom of the test-tube solidifies. This residue is *biuret*, which crystallises from water in long needles, m.p. 193°, with decomposition. The sublimate is chiefly *cyanuric acid* or polymerised *iso*cyanic acid. The changes which take place when urea is heated may be expressed:

$$\begin{array}{c} NH_2 \\ | \\ CO \\ | \\ NH_2 \end{array} \rightleftharpoons \boxed{\begin{array}{l} NH_2 \\ | \\ C—O \\ \| \\ N—H \end{array}} H \rightarrow NH_3 + \underset{isocyanic\ acid}{H—N{=}C{=}O} \rightarrow \underset{\substack{cyanuric\ acid \\ (sublimate)}}{(H—N{=}C{=}O)_3}$$

$$\begin{array}{ccccc} H_2N—CO—NH_2 & +NH_2—CO—NH_2 & \rightarrow & NH_3+H_2N—CO—NH—CO—NH_2, \text{ or} \\ \downarrow\uparrow & \downarrow\uparrow & & \downarrow\uparrow \\ HN{=}C(OH)—NH_2 & +NH_2—C(OH){=}NH & \rightarrow & NH_3+HN{=}C(OH)—NH—C(OH){=}NH \\ & & & \text{biuret (residue)} \end{array}$$

The sublimate of cyanuric acid containing some unpolymerised *iso*cyanic acid is distinguished by the intense blue colour produced when drops of an aqueous solution of cobalt nitrate are allowed to flow over it.

The biuret is distinguished by dissolving it in a dilute solution of sodium hydroxide and this solution giving a pink-violet coloration on addition of a few drops of a dilute aqueous solution of copper sulphate. This constitutes the 'biuret test', which, however, is not confined to biuret but is given by the proteins and their partial hydrolytic products (polypeptides) containing at least two —CO—NH— groupings.

Urea, although neutral in reaction in aqueous solution, behaves as a monoacid base and forms well-defined crystalline salts, such as:

$$\underset{\text{hydrochloride}}{CH_4ON_2.HCl} \qquad \underset{\text{nitrate}}{CH_4ON_2.HNO_3} \qquad \underset{\text{oxalate}}{(CH_4ON_2)_2.H_2C_2O_4}$$

In these salts, the urea may have the 'imino' formula, in which case their respective constitutions may be written:

Hydrochloride $[HN{=}C(OH)—NH_3]Cl$

Nitrate $[HN{=}C(OH)—NH_3]NO_3$

Oxalate $[HN{=}C(OH)—NH_3]_2C_2O_4$

Urea nitrate is readily soluble in water and sparingly soluble in the presence of nitric acid. In the dry condition, it detonates in a characteristic manner on being heated. Urea oxalate is sparingly soluble in water. Both these salts may be conveniently employed in the isolation and purification of urea.

The constitutional formula of urea implies that it is a primary amine and also an amide. In the latter connexion its hydrolysis by boiling with alkalies is very slow, and if rough methods are used in investigating the action of hot solutions of sodium hydroxide on urea the ammonia evolved may escape detection. The reaction taking place is usually written:

$$CO(NH_2)_2 + 2NaOH \rightarrow Na_2CO_3 + 2NH_3$$

but this hardly explains what is happening under ordinary conditions.

The hydrolysis of urea to ammonium carbonate

$$CO(NH_2)_2 + 2H_2O \rightarrow (NH_4)_2CO_3$$

is quantitatively effected by the hydrolytic enzyme *urease*, which is present in the soya bean. The action of urease is made the basis of the quantitative estimation of urea in physiological fluids. This estimation is conveniently carried out as described in *Organic and Biochemistry*, by R. H. Plimmer, p. 130 (Longmans, Green and Co., 1933).

With freshly prepared aqueous alkaline solutions of sodium or potassium hypobromite at the ordinary temperature, urea undergoes rapid oxidation and nitrogen is evolved. This forms the most satisfactory test for urea, and the reaction taking place is usually represented:

$$CO(NH_2)_2 + 3NaOBr + 2NaOH \rightarrow 3NaBr + Na_2CO_3 + N_2 + 3H_2O$$

According to this equation, one molecular proportion of urea should give rise to one molecular proportion of nitrogen, and this reaction was (and is still) extensively employed in estimating the urea content of urine. Using all precautions, the volume of nitrogen evolved is always lower than that indicated by the equation; with elaborate apparatus and using a very large excess of alkali in concentrated solution, the volume of nitrogen is generally of the order of 2 per cent. less than the theoretical quantity. More usually, the deficiency of nitrogen is some 8 per cent. In spite of this, the method is so convenient that it will always be employed in clinical examinations. Where however it is important to have a more accurate determination, and especially in determining the urea content of the blood, the urease method of estimation should be employed.

Like all compounds containing the —NH_2 group, urea reacts with nitrous acid in the presence of mineral acids to give nitrogen and, if the urea reacted entirely in the carbamide form, thus:

$$CO(NH_2)_2 + 2HONO \rightarrow CO_2 + 3H_2O + 2N_2$$

two molecular proportions of nitrogen should be obtained from one molecular proportion of urea; from this it would appear that this reaction might be used for the quantitative estimation of urea. This quantity of nitrogen, however, is never obtained and, under the most favourable conditions (large excess of mineral acid present), only some 70 per cent. of the theoretical amount of nitrogen according to the above equation is obtained. This is another example of anomalous reaction of urea which has been discussed by several investigators.* It has been suggested that in the presence of mineral acids urea reacts with nitrous acid according to the 'imino' form, giving rise to *iso*cyanic acid as an intermediate compound, thus:

$$\begin{matrix} NH_2 \\ | \\ CO \\ | \\ NH_2 \end{matrix} \xrightarrow{\text{mineral acids}} \begin{matrix} NH_2.HX \\ | \\ C\text{—}(OH) \\ \| \\ N\text{—}H \end{matrix} + HONO \rightarrow H\text{—}N{=}C{=}O + N_2 + 2H_2O + HX$$

It is further suggested that the *iso*cyanic acid may then react as in (*a*) and (*b*):

$$(a)\quad HNCO + H_2O + HX \rightarrow NH_4X + CO_2$$

$$(b)\quad HNCO + HONO \rightarrow N_2 + CO_2 + H_2O$$

in which case it is implied that the deficiency of nitrogen is due to some of it remaining as ammonium salt of the mineral acid (HX). In view of the fact that ammonium salts in the presence of mineral acid are decomposed by nitrous acid, thus:

$$NH_4X + HONO \rightarrow N_2 + HX + 2H_2O$$

this explanation involves the assumption that this last reaction goes at a negligible rate compared with reactions (*a*) and (*b*) and, further, that a not inconsiderable quantity of the nitrogen remains as ammonium salt.

When urea is heated with an excess of acetic anhydride in the presence of a trace of sulphuric acid, *diacetylurea*, m.p. 214°, is formed and separates when the solution is cooled:

$$\begin{matrix} NH_2 \\ | \\ CO \\ | \\ NH_2 \end{matrix} + 2O(CO.CH_3)_2 \rightarrow \underset{\text{diacetylurea}}{\begin{matrix} H\text{—}N\text{—}CO.CH_3 \\ | \\ CO \\ | \\ H\text{—}N\text{—}CO.CH_3 \end{matrix}} + 2CH_3.COOH$$

Diacetylurea undergoes an interesting isomeric change when dissolved in acetic anhydride at about 60° and after addition of two or three drops of sulphuric acid. Under these conditions, *diacetylisourea*, m.p. 153·5°, is formed and can readily be isolated. This compound would appear to have the constitution:

* See *The Chemistry of Urea*, by Werner (Longmans, Green and Co., 1923).

$$\begin{array}{l} \text{H—N—CO.CH}_3 \\ \quad\;\; | \\ \quad\;\; \text{C—O—CO.CH}_3 \\ \quad\;\; \| \\ \quad\;\; \text{N—H} \end{array}$$

indicating that it is a derivative of the 'imino' form of urea which has been described as *iso*urea in spite of its not having been isolated.

Diphenylurea, a colourless crystalline substance, m.p. 238–239°, is prepared by the interaction of carbonyl chloride (phosgene) and aniline:

$$\begin{array}{l} \text{Cl} \\ | \\ \text{CO} \\ | \\ \text{Cl} \end{array} + \begin{array}{l} \text{HNH(C}_6\text{H}_5) \\ \\ \\ \\ \text{HNH(C}_6\text{H}_5) \end{array} \rightarrow \begin{array}{l} \text{H—N—C}_6\text{H}_5 \\ \quad\;\; | \\ \quad\;\; \text{CO} \\ \quad\;\; | \\ \text{H—N—C}_6\text{H}_5 \end{array} + 2\text{HCl}$$

Alkyl and aryl substituted ureas, having the general formula $R\text{NH.CO.NH}_2$, are easily prepared by the action of monosubstituted ammonium salts, $[\text{N}R\text{H}_3]X$, on potassium cyanate in aqueous solution. The aqueous solution containing equimolecular quantities of the two salts is evaporated to dryness and the residue extracted with hot alcohol, which dissolves the urea derivative and in which the inorganic potassium salt is insoluble (compare Wöhler's synthesis of urea, p. 476, and Fischer's synthesis of uric acid, p. 488). For example, *monoethylurea* is prepared as indicated, omitting the intermediate formation of the substituted ammonium cyanate:

$$\text{KOCN} + [\text{N(C}_2\text{H}_5\text{)H}_3]\text{Cl} \rightarrow \text{KCl} + \underset{\text{monoethylurea, m.p. 92°}}{\text{C}_2\text{H}_5\text{.NH.CO.NH}_2}$$

The urea derivatives of some dibasic acids are of importance; they belong to a class of compounds known as *ureides*. A simple ureide is *malonylurea* or *barbituric acid*, which can be compared constitutionally with malonamide, the amide of malonic acid (p. 358):

$$\underset{\text{malonamide}}{\begin{array}{l} \text{H}_2\text{N.CO} \\ \qquad\;\; | \\ \qquad\;\; \text{CH}_2 \\ \qquad\;\; | \\ \text{H}_2\text{N.CO} \end{array}} \qquad \underset{\text{malonylurea or barbituric acid}}{\begin{array}{l} \text{HN——CO} \\ \;\;|\qquad\;| \\ \text{OC}\quad\;\text{CH}_2 \\ \;\;|\qquad\;| \\ \text{HN——CO} \end{array}}$$

Malonylurea may be prepared by heating together malonic acid, urea and phosphorus oxychloride at 100°:

$$\begin{array}{l} \text{H—N—H} \\ \quad\;\; | \\ \quad\;\; \text{CO} \\ \quad\;\; | \\ \text{H—N—H} \end{array} + \begin{array}{l} \text{HOOC} \\ \qquad | \\ \qquad \text{CH}_2 \\ \qquad | \\ \text{HOOC} \end{array} \xrightarrow{-2\text{H}_2\text{O}} \begin{array}{l} \text{HN——CO} \\ \;\;|\qquad\;| \\ \text{OC}\quad\;\text{CH}_2 \\ \;\;|\qquad\;| \\ \text{HN——CO} \end{array}$$

It crystallises from water in colourless prisms containing two molecules of water of crystallisation, and when it is boiled with alkalies it is hydrolysed, yielding malonic acid (as malonate) and urea. It forms a silver salt, $C_4H_2N_2O_3Ag_2$, which is precipitated on adding a solution of silver nitrate to an aqueous solution of malonylurea in ammonia (i.e. a solution of the ammonium salt).

When the silver salt of malonylurea interacts with ethyl iodide, using two molecular proportions of the latter to one of the former, C-*diethylbarbituric acid* or C-*diethylmalonylurea* is produced. This same compound is also obtained by condensing the ethyl ester of diethylmalonic acid with urea in the presence of sodium ethylate, the sodium salt being formed, thus:

$$\begin{array}{c} HNH \\ | \\ CO \\ | \\ HNH \end{array} + \begin{array}{c} C_2H_5OOC \\ \quad\diagdown \\ C(C_2H_5)_2 \\ \quad\diagup \\ C_2H_5OOC \end{array} \xrightarrow{2NaOC_2H_5} \begin{array}{ccc} NaN & — & CO \\ | & & | \\ OC & & C(C_2H_5)_2 \\ | & & | \\ NaN & — & CO \end{array} + 4C_2H_5OH^*$$

From the sodium salt in aqueous solution the C-diethylbarbituric acid is precipitated by the addition of the equivalent quantity of mineral acid. The compound crystallises in colourless needles, m.p. 212°, from hot water. It is a hypnotic and is employed medicinally under the name of *veronal* or barbitone. Other compounds of the same class, described, generally, as *barbiturates*, are also extensively employed as hypnotics; typical compounds are:

$$\begin{array}{ccc} HN & — & CO \\ | & & | \\ OC & & C(C_3H_7)_2 \\ | & & | \\ HN & — & CO \end{array}$$

C-dipropylbarbituric acid
(Proponal)

$$\begin{array}{ccc} HN & — & CO \\ | & & | \\ OC & & C(C_3H_5)_2 \\ | & & | \\ HN & — & CO \end{array}$$

C-diallylbarbituric acid
(Dial)

$$\begin{array}{ccc} HN & — & CO \\ | & & | \\ OC & & C\langle^{C_2H_5}_{C_6H_5} \\ | & & | \\ HN & — & CO \end{array}$$

C-ethylphenylbarbituric acid
(Gardenal, Luminal, Phenobarbital)

* Bearing in mind that diethylbarbituric acid yields salts with metals, its formation from the silver salt of barbituric acid may be assumed (omitting other possible stages) to take place thus:

$$\begin{array}{ccc} AgN & — & CO \\ | & & | \\ OC & & CH_2 \\ | & & | \\ AgN & — & CO \end{array} \rightarrow \begin{array}{ccc} AgN & — & CO \\ | & & | \\ OC & & C\langle^{H}_{Ag} \\ | & & | \\ HN & — & CO \end{array} \xrightarrow[-AgI]{+C_2H_5I} \begin{array}{ccc} AgN & — & CO \\ | & & | \\ OC & & C\langle^{H}_{C_2H_5} \\ | & & | \\ HN & — & CO \end{array}$$

$$\longrightarrow \begin{array}{ccc} HN & — & CO \\ | & & | \\ OC & & C\langle^{Ag}_{C_2H_5} \\ | & & | \\ HN & — & CO \end{array} \xrightarrow[-AgI]{+C_2H_5I} \begin{array}{ccc} HN & — & CO \\ | & & | \\ OC & & C\langle^{C_2H_5}_{C_2H_5} \\ | & & | \\ HN & — & CO \end{array}$$

diethylbarbituric acid or veronal

which are prepared by analogous methods to those employed for the preparation of veronal, i.e. starting with the ethyl ester of the disubstituted malonic acid. The sodium salts, like that of veronal (*barbitone soluble*), are more soluble than the free acid and their hypnotic action is therefore more rapid.

Malonyl urea or barbituric acid reacts with nitrous acid, yielding *oximinobarbituric acid* or *violuric acid*:

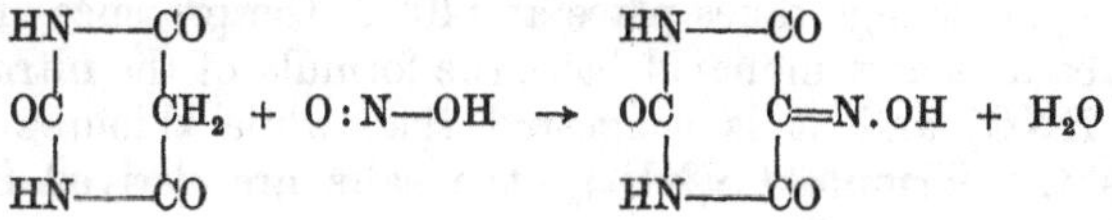

$$\begin{array}{l} HN\text{---}CO \\ OC \quad CH_2 \\ HN\text{---}CO \end{array} + O{:}N\text{—}OH \rightarrow \begin{array}{l} HN\text{---}CO \\ OC \quad C{=}N.OH \\ HN\text{---}CO \end{array} + H_2O$$

violuric acid

and violuric acid on reduction yields C-*aminomalonylurea* or *uramil*:

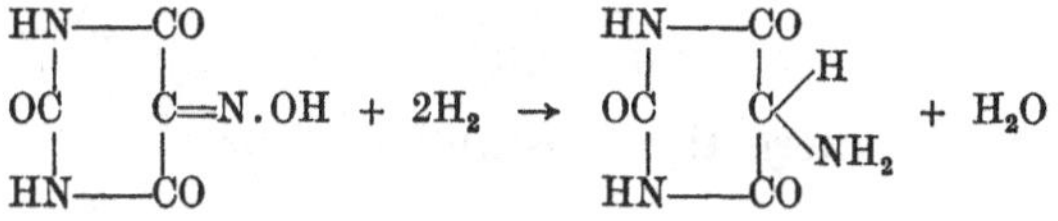

$$\begin{array}{l} HN\text{---}CO \\ OC \quad C{=}N.OH \\ HN\text{---}CO \end{array} + 2H_2 \rightarrow \begin{array}{l} HN\text{---}CO \\ OC \quad C\begin{smallmatrix} H \\ NH_2 \end{smallmatrix} \\ HN\text{---}CO \end{array} + H_2O$$

uramil

Malonylurea or barbituric acid, violuric acid and uramil are substances concerned in the synthesis of uric acid (p. 488).

The ureide of mesoxalic acid (HOOC.CO.HOOC), i.e. *mesoxalylurea* or *alloxan*, is an oxidation product of uric acid. Alloxan on further oxidation yields the ureide of oxalic acid or *oxalylurea*, also known as *parabanic acid*:

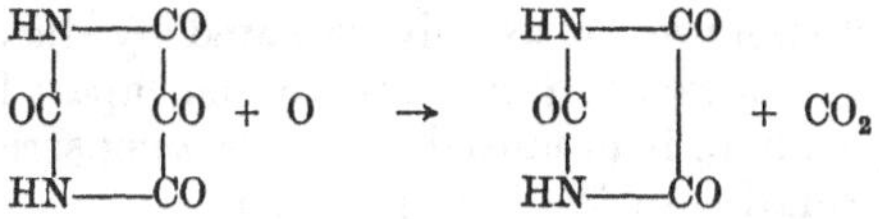

$$\begin{array}{l} HN\text{---}CO \\ OC \quad CO \\ HN\text{---}CO \end{array} + O \rightarrow \begin{array}{l} HN\text{---}CO \\ OC \quad | \\ HN\text{---}CO \end{array} + CO_2$$

alloxan
decomposes at about 175°

parabanic acid
m.p. 242° (decomp.)

These two ureides, like the others, form salts with metals.

Other substances of different types but which are related to urea may be synthesised starting with cyanamide (see above), which is a colourless crystalline substance, easily soluble in water, alcohol and ether and which has m.p. 44°.

By allowing hydrogen sulphide to react with cyanamide, *thiourea* is produced

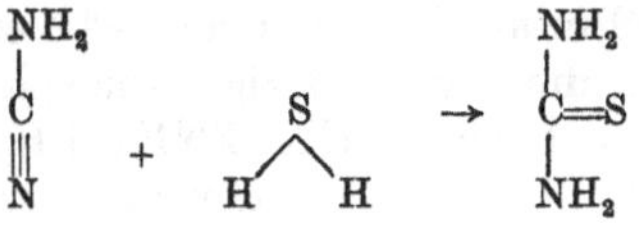

$$\begin{array}{c} NH_2 \\ | \\ C \\ ||| \\ N \end{array} + \begin{array}{c} S \\ H \quad H \end{array} \rightarrow \begin{array}{c} NH_2 \\ | \\ C{=}S \\ | \\ NH_2 \end{array}$$

Thiourea is one of the products of the action of heat (about 180°) on ammonium thiocyanate (compare the action of heat on ammonium cyanate)

$$NH_4.S.C⋮N \rightleftarrows CS(NH_2)_2$$

and it is a colourless crystalline substance, m.p. 182°. The conversion of ammonium thiocyanate into thiourea is never complete, since the latter is reconverted into ammonium thiocyanate at about 160–170° and this same change takes place at 140° in the presence of water.

Thiourea acts as a monacid base, the formula of the nitrate being $CSN_2H_4.HNO_3$ and it is suggested that while thiourea as such possesses the formula $CS(NH_2)_2$, the salts are derived from the alternative or *pseudo*-form

$$\begin{array}{l} NH_2 \\ | \\ C\text{—}SH \\ \| \\ NH \end{array}$$

analogous to the alternative constitution of urea.

When cyanamide is allowed to react with ammonia, direct combination takes place, thus

$$\begin{array}{l} NH_2 \\ | \\ C{\equiv}N \end{array} + NH_3 \rightarrow \begin{array}{l} NH_2 \\ | \\ C{=}NH \\ | \\ NH_2 \end{array}$$

the substance produced being the strong monacid base *guanidine*. This compound is, however, usually prepared by the action of heat on ammonium thiocyanate (see above) when, apart from thiourea, guanidine thiocyanate is produced; the following series of reactions indicate the formation of the final products

$$2NH_4.S.C⋮N \rightarrow CS(NH_2)_2 + \underset{\text{cyanamide}}{H_2N.C⋮N} + H_2S$$

$$\begin{array}{l} NH_2 \\ | \\ C \\ ||| \\ N \end{array} + NH_3 + \underset{\substack{\text{or} \\ NH_4SCN}}{H.S.C⋮N} \rightarrow \begin{array}{l} NH_2 \\ | \\ C{=}NH \\ | \\ NH_2.HSCN \end{array}$$

guanidine thiocyanate
m.p. 118°

Guanidine is a colourless crystalline substance which is highly soluble in water and in alcohol. It reacts as a strong monacid base and absorbs carbon dioxide from the atmosphere when it is converted into the carbonate $[HN{:}C(NH_2)_2]_2H_2CO_3$. The nitrate $[HN{:}C(NH_2)_2]HNO_3$, m.p. 214°, is somewhat sparingly soluble in water and the much less soluble picrate, $[HN{:}C(NH_2)_2]C_6H_2(OH)(NO_2)_3$, is useful in isolating the substance.

Guanidine is closely related constitutionally to urea into which it is converted by boiling with alkalis, ammonia being evolved

$$\underset{NH_2}{\overset{NH_2}{|}}\!C{=}NH + H_2O \xrightarrow{-NH_3} \underset{OH}{\overset{NH_2}{|}}\!C{=}NH \rightarrow \underset{O}{\overset{NH_2}{|}}\!C{-}NH_2$$

Cyanamide is the essential reagent for the synthesis of all guanidine derivatives. For example, the physiologically important substance *arginine* is a guanidine derivative of ornithine and its synthesis is outlined

$$\overset{NH_2}{|}\!C{\equiv}N + \underset{\text{ornithine}\\ \alpha\delta\text{-diaminovaleric acid}}{H_2N.CH_2.CH_2.CH_2.CH(NH_2).COOH}$$

$$\rightarrow \underset{\text{arginine}\\ \alpha\text{-amino-}\delta\text{-guanidovaleric acid}}{HN{=}\overset{NH_2}{\overset{|}{C}}{-}NH.CH_2.CH_2.CH_2.CH(NH_2).COOH}$$

Another example of the use of cyanamide as a synthetic reagent is in the production of *creatine*. This synthesis is indicated

$$\overset{NH_2}{|}\!C{\equiv}N + \underset{\text{sarcosine (p. 319)}}{\underset{COOH}{|}\!CH_2.N\!\left\langle{}^{H}_{CH_3}\right.} \rightarrow \underset{\text{creatine}}{\overset{NH_2}{|}\!C{=}NH \atop CH_3{-}N{-}CH_2.COOH}$$

Creatine, or N-methylguanidoacetic acid, crystallises from water in colourless crystals containing one molecule of water of crystallisation and having m.p. 100°. When anhydrous, the substance has m.p. 295°. It behaves as a weak base forming readily hydrolysable salts with acids. When boiled with alkalies it is finally hydrolysed into urea and sarcosine (N-methylaminoacetic acid)

$$\overset{NH_2}{|}\!C{=}NH \atop CH_3{-}N{-}CH_2.COOH \;+ H_2O \rightarrow CH_3NH.CH_2.COOH + \underset{OH}{\overset{NH_2}{|}}\!C{=}NH \rightarrow \underset{O}{\overset{NH_2}{|}}\!C{-}NH_2$$

Creatine occurs in relatively considerable quantity in voluntary muscle and, to a less extent, in many organs of the body.

By boiling with aqueous solutions of acids, creatine is converted by loss of water into *creatinine*

```
      NH2  HOCO                     NH———CO
       |     |         -H2O          |     |
HN=C         |        ———→   HN=C          |
       |     |                       |     |
      N(CH3)—CH2                    N(CH3)—CH2
      creatine                      creatinine
```

and creatinine may thus be regarded as the *lactam* of creatine.

Creatinine is present in all mammalian urine and in blood. It is present in meat extracts whence it may arise during evaporation from the action of acids present in the muscle extract on creatine. It crystallises from hot aqueous solution in colourless crystals, m.p. 260° (decomp.). It behaves as a strong monacid base and will liberate ammonia from the salts of the latter. When it is gently heated with alkalies, it is converted into creatine and the relationship between these two substances may be illustrated thus

```
      NH2  HOCO                        NH———CO
       |     |        acids             |     |
HN=C         |       ———→      HN=C           |        + H2O
       |     |       ←———               |     |
      N(CH3)—CH2    alkalies           N(CH3)—CH2
      creatine                         creatinine
```

Purines

The constitutional formula of purine has been deduced as a condensed ring system from pyrimidine and glyoxaline (p. 306) to be

```
         H                                       H
         |                                       |
         C                                       C
       //  \                                   //6 \
     N       C———N                         N1    5C———7NH
     |       ||   ||            or         |      ||     |
     HC      C    CH                       HC2  3 4C  9  8CH
       \\  /  \  /                            \\ /  \  //
         N      N                               N     N
                H
```

Since the phenols (monohydroxybenzene, C_6H_6O; dihydroxybenzenes, $C_6H_6O_2$, etc.) are oxidation products of benzene, the following hydroxypurines can be regarded as oxidation products of purine:

6-oxypurine or *hypoxanthine* | 2 : 6-dioxypurine or *xanthine* | 2 : 6 : 8-trioxypurine or *uric acid*

On account of the presence of the grouping,

$$=C(OH)-N= \rightleftarrows -CO-NH-$$

the above alternative formulae are possible and indicate that each of these substances may react in two ways according to the conditions pertaining; in the literature these formulae are more usually written in a slightly modified form, thus:

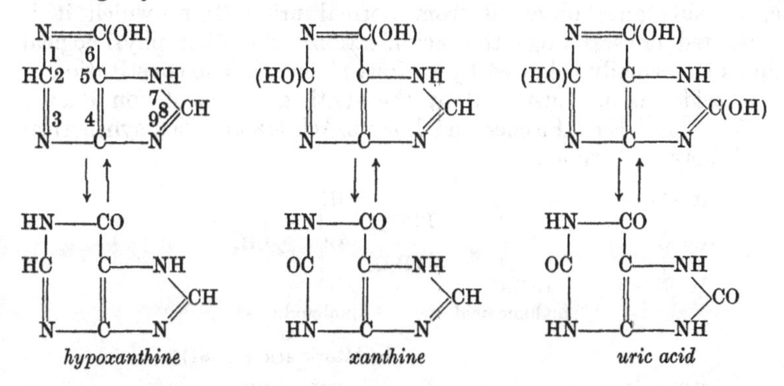

hypoxanthine | *xanthine* | *uric acid*

(N.B. The method of numbering of the atoms in the 'purine' ring is peculiar to the purine compounds. It differs from the modern method of numbering *iso*cyclic and *hetero*cyclic compounds.)

Uric acid may be regarded as the final direct oxidation product of purine; it is excreted in the urine as the end-product of purine metabolism in man, although only its derivatives and not purine itself occur naturally. The direct relationship between purine and uric acid is shown by the action of phosphorus pentachloride on the latter, whereby the three hydroxyl groups are replaced by chlorine forming 2 : 6 : 8-*trichloropurine*; this compound is readily converted into purine by reduction and the reactions concerned are outlined:

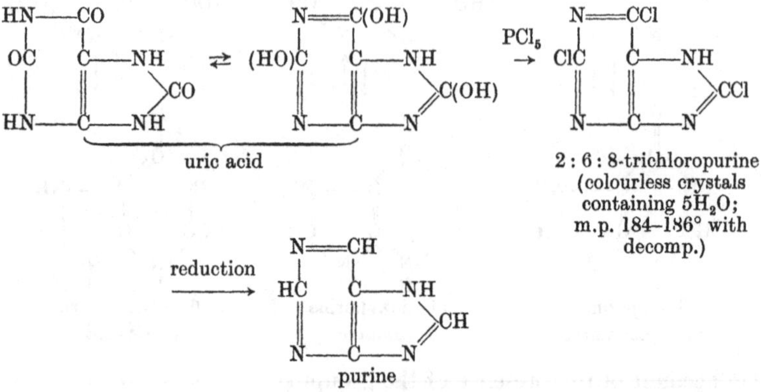

HN—CO | OC C—NH | ‖ >CO | HN—C—NH ⇄ N=C(OH) | (HO)C C—NH ‖ ‖ >C(OH) N—C—N

uric acid

$\xrightarrow{PCl_5}$ N=CCl | ClC C—NH ‖ ‖ >CCl N—C—N

2 : 6 : 8-trichloropurine (colourless crystals containing $5H_2O$; m.p. 184–186° with decomp.)

$\xrightarrow{\text{reduction}}$ N=CH | HC C—NH ‖ ‖ >CH N—C—N

purine

Purine, $C_5H_4N_4$, the fundamental compound of the purine group, is a colourless crystalline substance, m.p. 216°. It is readily soluble in water and reacts both as an acid and as a base.

Uric acid, $C_5H_4O_3N_4$, was discovered by Scheele (1776) in urinary calculi. When pure, it is a colourless microcrystalline substance; the crude substance obtained from normal urine (from which it is deposited on standing after acidification) and other physiological fluids is normally coloured by various pigments. The constitution of uric acid was demonstrated by the synthetic work of von Baeyer (1863–4) and Emil Fischer and Lorenz Ach (1895). This synthetical work may be outlined:

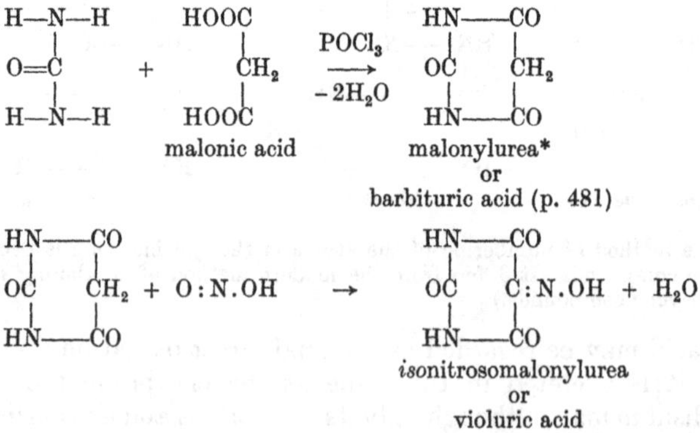

H—N—H | O=C | H—N—H + HOOC | CH_2 | HOOC (malonic acid) $\xrightarrow[-2H_2O]{POCl_3}$ HN—CO | OC CH_2 | HN—CO

malonylurea* or barbituric acid (p. 481)

HN—CO | OC CH_2 | HN—CO + O:N.OH → HN—CO | OC C:N.OH | HN—CO + H_2O

*iso*nitrosomalonylurea or violuric acid

* Malonylurea and its immediate derivatives are described as 1 : 3-diazines.

```
HN——CO                        HN——CO
|    |                        |    |  /H
OC   C:N.OH + 2H2     →       OC   C<        + H2O
|    |                        |    |  \NH2
HN——CO                        HN——CO
```

aminomalonylurea or uramil
(forms salts with bases and acids)

Just as in the conversion of ammonium salts into urea by evaporating their aqueous solutions containing an equivalent amount of potassium cyanate, thus:

$$NH_4Cl + KOCN \rightarrow NH_4(OCN) \rightarrow NH_2.CO.NH_2$$

so uramil is converted into the corresponding urea derivative, viz. *pseudouric acid*, thus:

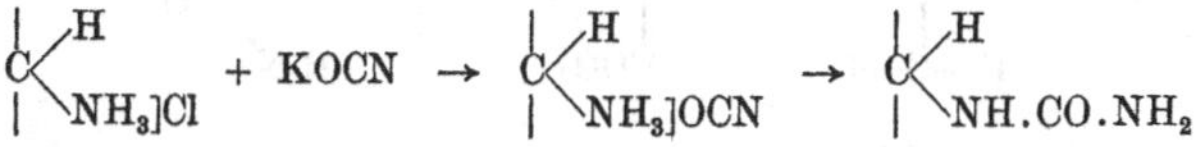

```
|  /H                       |  /H                 |  /H
C<          + KOCN   →      C<            →       C<
|  \NH3]Cl                  |  \NH3]OCN           |  \NH.CO.NH2
```

The final stage in the synthesis is the conversion of *pseudouric acid* into uric acid by heating with oxalic acid or concentrated hydrochloric acid, whereby loss of water takes place as shown:

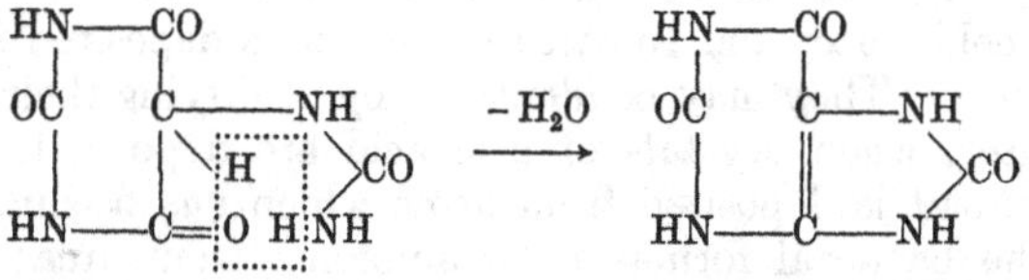

```
HN——CO                          HN——CO
|    |                          |    |
OC   C———NH         - H2O       OC   C——NH
|    | \H    >CO    ————→       |    ||   >CO
HN——C=O H NH                    HN——C——NH
```

The constitutional formula of uric acid is in keeping with its behaviour on oxidation. On moderate oxidation, uric acid is converted into *alloxan* (mesoxalylurea) and urea, thus:

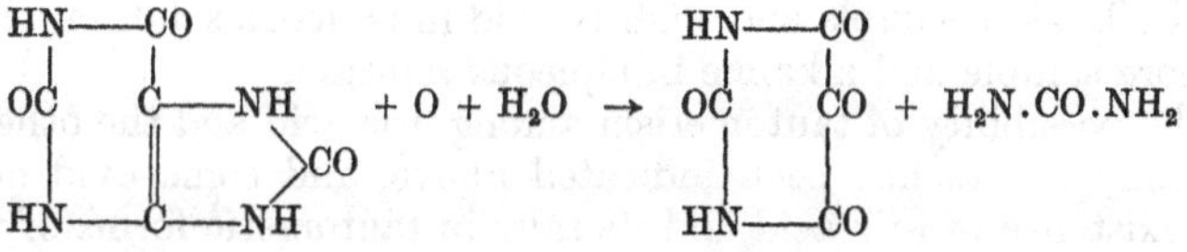

```
HN——CO                          HN——CO
|    |                          |    |
OC   C——NH    + O + H2O   →     OC   CO  + H2N.CO.NH2
|    ||   >CO                   |    |
HN——C——NH                       HN——CO
```

alloxan

On further oxidation, the alloxan is converted into *parabanic acid* (oxalylurea) and carbon dioxide:

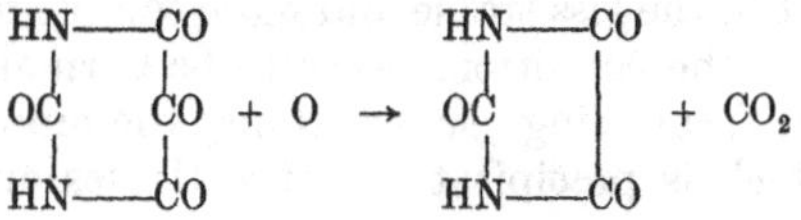

```
HN——CO                  HN——CO
|    |                  |    |
OC   CO  + O   →        OC   |     + CO2
|    |                  |    |
HN——CO                  HN——CO
```

parabanic acid

Parabanic acid can be hydrolysed to urea and oxalic acid:

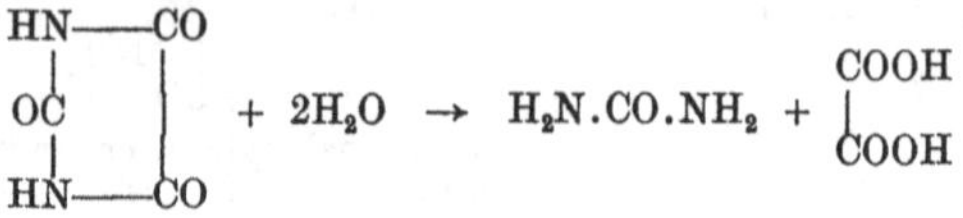

$$\begin{array}{l}HN\text{—}CO\\ \;|\qquad\;\;|\\ OC\qquad|\\ \;|\qquad\;\;|\\ HN\text{—}CO\end{array} + 2H_2O \rightarrow H_2N.CO.NH_2 + \begin{array}{l}COOH\\ |\\ COOH\end{array}$$

This breaking down of urea indicates the possibility of uric acid being formed from two molecules of urea and a three-carbon compound such as lactic acid, thus:

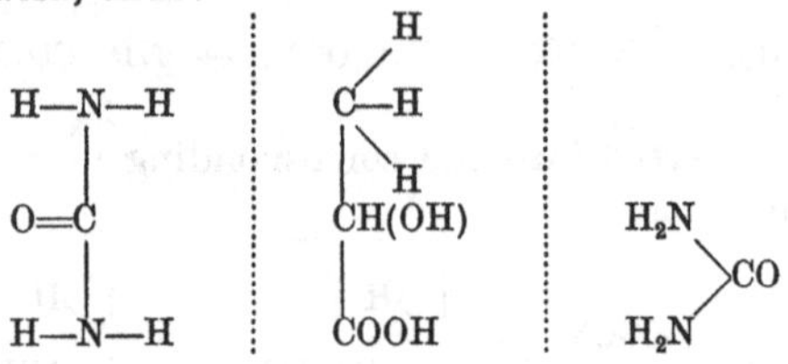

$$\begin{array}{l}H\text{—}N\text{—}H\\ \qquad|\\ O{=}C\\ \qquad|\\ H\text{—}N\text{—}H\end{array} \;\vdots\; \begin{array}{l}\quad H\\ \;/\\ C\text{—}H\\ |\;\diagdown\\ |\quad H\\ CH(OH)\\ |\\ COOH\end{array} \;\vdots\; \begin{array}{l}H_2N\\ \qquad\rangle CO\\ H_2N\end{array}$$

which may be considered to have some physiological significance.

Uric acid is very sparingly soluble in water (at 18°, 1 part of uric acid will dissolve in 40,000 parts of water) and it is insoluble in alcohol and ether. The acid salts, $C_5H_3O_3N_4Na$, $C_5H_3O_3N_4K$ and $C_5H_3O_3N_4(NH_4)$, are the best known. These salts are also sparingly soluble in cold water and constitute the urates deposited from concentrated urine. They may be identified by acidifying their solutions in hot water, when crystals of uric acid are deposited. The acid ammonium salt is deposited from urine which has become alkaline through the bacterial formation of ammonia from urea; this acid ammonium salt is insoluble in solutions of ammonium salts. The normal urates, $C_5H_2O_3N_4X_2$ (X=Na or K), are readily soluble and give alkaline solutions owing to hydrolysis. These acid and normal salts may be compared with the corresponding salts of carbonic acid, $NaHCO_3$—less soluble and slightly acid in aqueous solution, Na_2CO_3—more soluble and alkaline in aqueous solution.

The possibility of tautomerism among uric acid and the other two hydroxypurines has been indicated above, and some evidence for the existence of uric acid and its salts in tautomeric forms of which one is the less stable and therefore the more soluble is provided by the fact that body fluids and, in particular, urine always contain more uric acid (or its salts) than is provided for by the ordinary solubility of the substance. Presumably, the uric acid or urate is normally present in the less stable and more soluble form, and during an attack of gout the conditions cease to be normal, the less stable and more soluble form being converted into the more stable and less soluble form which is precipitated in the tissues and fluids during the attack.

Uric acid is a fairly strong reducing agent; it will reduce solutions of silver salts to the metal and will also reduce Fehling's solution, when cuprous oxide is produced. In testing urine for reducing sugar, the reducing power of uric acid therefore should always be borne in mind. One method (Hopkins) for the estimation of uric acid depends on the reducing power of uric acid in the presence of hot dilute sulphuric acid as indicated by a standard solution of potassium permanganate. The method is an arbitrary one (compare the estimation of sugars by standard solutions of cupric salts, p. 462), but under the established conditions the following relationship has been found to hold:

$$C_5H_4O_3N_4 \equiv 2KMnO_4$$

Uric acid and the urates are recognised by the *murexide test.* This depends on the oxidation of uric acid by nitric acid. On evaporating a small quantity of uric acid or urate with two to three drops of dilute nitric acid *on the water-bath,* a yellow residue is left which is coloured purple on addition of one or two drops of an aqueous solution of ammonia; this colour becomes blue on addition of a little aqueous solution of sodium hydroxide. The name murexide is derived from *Murex,* a mollusc from which the purple colouring matter of the ancients was obtained. Other purine derivatives exhibit the same behaviour, but these are unlikely to be excretory products. The production of murexide from uric acid may be represented thus:

```
HN——CO                            OC——NH
|    |                            |    |
OC   C—:—NH                 HN—:—C     CO
|    ||  :  >CO      +     OC<   :  ||    |
HN——C—:—NH                  HN—:—C——NH

   +3H2O            HN——CO          OC——NH
————————————→       |    | /H       |    |
-2NH2.CO.NH2        OC   C<     +   OC   CO
                    |    | \OH      |    |
                    HN——CO          OC——NH
                  dialuric acid     alloxan

   HN——CO           OC——NH                    HN——CO  H  OC——NH
   |    | /OH   HO\ |    |                    |    |  N   |    |
→  OC   C——————————C    CO    +NH3          OC   C/  \C   CO
   |    |          |    |   ————————→       |    |————|   |
   HN——CO           OC——NH   -2H2O          HN——CO    OC——NH
            alloxantin                              purpuric acid
                                          of which murexide is the
                                               ammonium salt
```

Two other members of the purine group which are of physiological importance are *adenine* and *guanine*:

```
N==C.NH2             N==C(OH)                NH——CO
|  |                 |  |                    |   |
HC C——NH       NH2.C C——NH        ⇄   NH2.C  C——NH
||  ||  \            ||  ||  \               ||  ||  \
||  ||   CH          ||  ||   CH             ||  ||   CH
||  ||  //           ||  ||  //              ||  ||  //
N——C——N              N——C——N                 N——C——N
```

adenine
6-aminopurine ($C_5H_5N_5$)

guanine
2-amino-6-oxypurine ($C_5H_5ON_5$)
2-aminohypoxanthine

Adenine crystallises from water in colourless crystals containing three molecules of water of crystallisation, which when anhydrous have m.p. 360–365° with decomposition. It was synthesised by Emil Fischer (1897). *Guanine*, which is found in guano, forms colourless needles which decompose above 360° and are insoluble in water but soluble in alkalies. These are basic substances from which uric acid is physiologically derived by hydrolysis and oxidation as indicated:

```
N==C.NH2                  N==C(OH)
|  |          +H2O        |  |
HC C——NH    ———————→      HC C——NH
||  ||  \     -NH3        ||  ||  \
||  ||   CH               ||  ||   CH
||  ||  //                ||  ||  //
N——C——N                   N——C——N   \ oxidation
adenine                   hypoxanthine  \
                                          ↘
                                            N==C(OH)
                                            |  |
                                      (HO)C  C——NH
                                            ||  ||  \
                                            ||  ||   C(OH)
                                 oxidation  ||  ||  //
                                         ↗  N——C——N
                                 /            uric acid
N==C(OH)                  N==C(OH)
|  |          +H2O        |  |
NH2.C C——NH  ———————→  (HO)C C——NH
||  ||  \     -NH3        ||  ||  \
||  ||   CH               ||  ||   CH
||  ||  //                ||  ||  //
N——C——N                   N——C——N
guanine                   xanthine
```

Guanine, on oxidation, yields guanidine (p. 484).

Adenine and guanine occur in the nucleoproteins, the most complex of the proteins constituting the essential constituents of cell nuclei.

Three purine derivatives which occur in plants are *theobromine*, *theophylline* and *caffeine*. The constitutions of these compounds are represented:

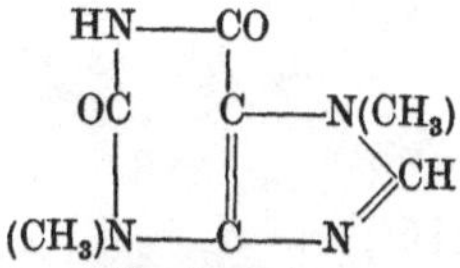

```
HN——CO                          (CH3)N——CO
|    |                                |    |
OC   C——N(CH3)                       OC   C——NH
|    ||    >CH                        |    ||   >CH
(CH3)N——C——N                    (CH3)N——C——N
```

theobromine
3 : 7-dimethyl-2 : 6-dioxypurine
3 : 7-dimethylxanthine

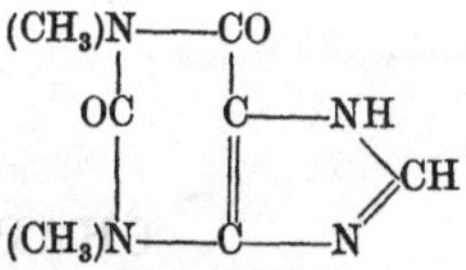

theophylline
1 : 3-dimethyl-2 : 6-dioxypurine
1 : 3-dimethylxanthine

isomeric

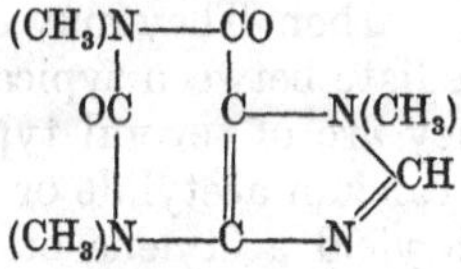

```
(CH3)N——CO
|    |
OC   C——N(CH3)
|    ||    >CH
(CH3)N——C——N
```

caffeine
1 : 3 : 7-trimethyl-2 : 6-dioxypurine
1 : 3 : 7-trimethylxanthine

Theobromine is a sparingly soluble colourless crystalline substance which can be sublimed. It occurs to the extent of about 2 per cent. in cocoa. Its use in medicine depends on its diuretic properties, although it is inferior in this respect to theophylline. Its salts with acids are easily hydrolysed by water. Its silver salt, $C_7H_7O_2N_4Ag$, is converted into caffeine by reaction with methyl iodide.

Theophylline is a colourless crystalline substance, m.p. 268°, occurring along with caffeine in tea. Like theobromine it forms a silver salt which reacts with methyl iodide yielding caffeine.

Caffeine (or theine) occurs in coffee, tea and kola nuts. It is a colourless crystalline substance, m.p. 236·5°, and yields hydrolysable salts with acids. It has considerable therapeutic use as a stimulant.

These three N-methyl purine derivatives are alkaloids, since they occur in plants (N-methyl compounds are rarely of animal origin), possess basic properties and also possess definite pharmacological action. They have been synthesised and their constitution and interrelationships investigated, chiefly by Emil Fischer.

CHAPTER XVII

ORGANO-METALLIC COMPOUNDS

ORGANO-METALLIC compounds include all those substances which contain metals directly united to carbon. They may be considered to be a class of compounds intermediate between typical inorganic and typical organic compounds. They are of several types and include the metallic acetylides, such as calcium acetylide or (as it is usually called) 'calcium carbide', which yield acetylene on treatment with water or mineral acids, aluminium carbide, Al_4C_3, which on similar treatment yields methane, and the unique metallic carbonyls such as nickel carbonyl, $Ni(CO)_4$. The largest class of organo-metallic compounds, however, includes (i) compounds of the type MR_n, where the metal, M, of fundamental valency, n, is united to monovalent alkyl or aryl groups, R, and (ii) compounds of the type $MR_{n-p}X_p$, having the metal, M, united to one or more alkyl or aryl groups and one or more halogen atoms or equivalent radicals, X, again according to the fundamental valency, n, of the metal and where the number of halogen atoms or equivalent groups, p, is at least one and always less than n.

In his investigations on the possible production of hydrocarbon radicals, Sir Edward Frankland showed that when such metals as zinc and tin are heated in sealed tubes with the alkyl iodides, direct combination takes place and colourless solid compounds which are stable in the absence of moisture are produced. Typical solid compounds so obtained are *zinc methyl iodide*, $Zn\begin{cases}CH_3\\I\end{cases}$, *zinc ethyl iodide*, $Zn\begin{cases}C_2H_5\\I\end{cases}$, and *tin diethyl diiodide*, $Sn\begin{cases}(C_2H_5)_2\\I_2\end{cases}$ Frankland's work in this field, which has inspired many other workers, began in 1849 and continued until about 1863.

By heating zinc alkyl iodides in an inert atmosphere, Frankland showed that the zinc alkyls are produced according to the general reaction:

$$2Zn\begin{cases}R\\X\end{cases} \rightarrow Zn\begin{cases}R\\R\end{cases} + ZnX_2 \quad (R = \text{alkyl and } X = \text{halogen, generally iodine})$$

Since the zinc alkyls are volatile, they are readily separated from the zinc halide, generally zinc iodide.

Zinc dimethyl, $Zn(CH_3)_2$, b.p. 46°, and *zinc diethyl*, $Zn(C_2H_5)_2$, b.p. 118°, are colourless liquids with characteristic odours. They are spontaneously inflammable in air or oxygen, burning with a greenish-blue flame, producing dense white fumes of zinc oxide and depositing zinc on a cold surface held in the flame. On account of their spontaneous inflammability, great care has to be exercised in working with these compounds, but, in spite of this, Frankland and co-workers were able to determine their vapour density and other physical properties. Other analogous compounds, such as zinc dipropyl, zinc di*iso*propyl, zinc di*iso*butyl, zinc di*iso*amyl, having similar properties have also been investigated.

In early work on this subject, Frankland and Duppa showed that the zinc alkyls react with metallic halides, yielding alkyl derivatives of other metals. For example, *tin tetraethyl* (colourless liquid, b.p. 181°) is produced by the following reaction:

$$Sn\begin{cases}(C_2H_5)_2\\ I_2\end{cases} + Zn\begin{cases}C_2H_5\\ C_2H_5\end{cases} \rightarrow Sn(C_2H_5)_4 + ZnI_2$$

and mercury dialkyls by the following general reactions:

$$2HgCl_2 + ZnR_2 \rightarrow 2HgRCl + ZnCl_2$$

$$HgCl_2 + ZnR_2 \rightarrow HgR_2 + ZnCl_2$$

The work was also extended to the preparation of boron alkyls and an unsuccessful attempt was made to prepare sodium and potassium alkyls; these latter compounds were obtained in a state of purity much later. Later workers extended the reactions indicated to the preparation of organo derivatives of other metals.

That part of this early work indicating that the zinc alkyls are useful synthetic reagents was of particular importance. Frankland showed that the zinc alkyls react vigorously with water, yielding hydrocarbons, thus:

$$Zn(CH_3)_2 + 2HOH \rightarrow 2CH_4 + Zn(OH)_2$$

and with alkyl iodides, thus:

$$Zn(CH_3)_2 + 2(CH_3)_3CI \rightarrow (CH_3)_4C + ZnI_2$$

In the latter reaction tertiary butyl iodide is converted into *tetramethyl methane* or $\beta\beta$-dimethylpropane, b.p. 9·5°. Further, by carefully regulating its reaction with oxygen, Frankland showed that zinc ethylate is produced and that the latter reacts with water, yielding ethyl alcohol as shown:

$$Zn(C_2H_5)_2 + O_2 \rightarrow \underset{\text{zinc ethylate}}{Zn(OC_2H_5)_2}$$

$$Zn(OC_2H_5)_2 + 2H_2O \rightarrow Zn(OH)_2 + 2C_2H_5OH$$

The use of zinc alkyls for wider synthetic purposes was rapidly extended. Some of the simplest examples of the synthetic uses of zinc alkyls are:

(i) Production of alcohols from aldehydes, e.g.

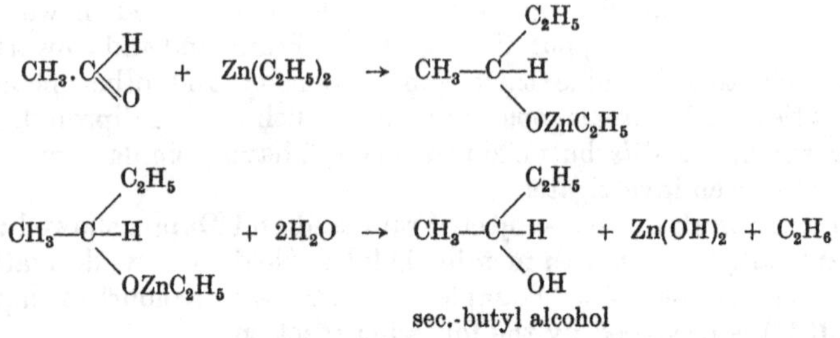

$$CH_3.C\begin{matrix}\diagup H\\ \diagdown\!\!\diagdown O\end{matrix} \quad + \quad Zn(C_2H_5)_2 \quad \rightarrow \quad CH_3-\overset{C_2H_5}{\underset{OZnC_2H_5}{C}}-H$$

$$CH_3-\overset{C_2H_5}{\underset{OZnC_2H_5}{C}}-H \quad + \; 2H_2O \quad \rightarrow \quad CH_3-\overset{C_2H_5}{\underset{OH}{C}}-H \quad + \; Zn(OH)_2 \; + \; C_2H_6$$

sec.-butyl alcohol

If formaldehyde be used in this reaction, a primary alcohol (using zinc diethyl, *n*-propyl alcohol) is obtained.

(ii) Production of tertiary alcohols from ketones, e.g.

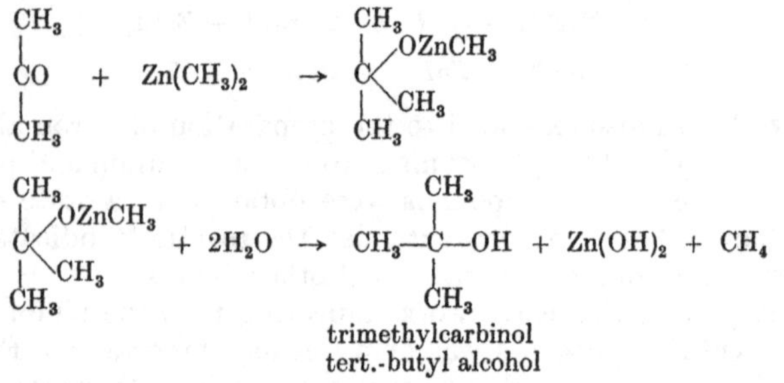

$$\overset{CH_3}{\underset{CH_3}{CO}} \quad + \quad Zn(CH_3)_2 \quad \rightarrow \quad \overset{CH_3}{\underset{CH_3}{C}}\begin{matrix}\diagup OZnCH_3\\ \diagdown CH_3\end{matrix}$$

$$\overset{CH_3}{\underset{CH_3}{C}}\begin{matrix}\diagup OZnCH_3\\ \diagdown CH_3\end{matrix} \quad + \; 2H_2O \quad \rightarrow \quad CH_3-\overset{CH_3}{\underset{CH_3}{C}}-OH \; + \; Zn(OH)_2 \; + \; CH_4$$

trimethylcarbinol
tert.-butyl alcohol

(iii) Production of ketones and tertiary alcohols from acid chlorides, e.g.

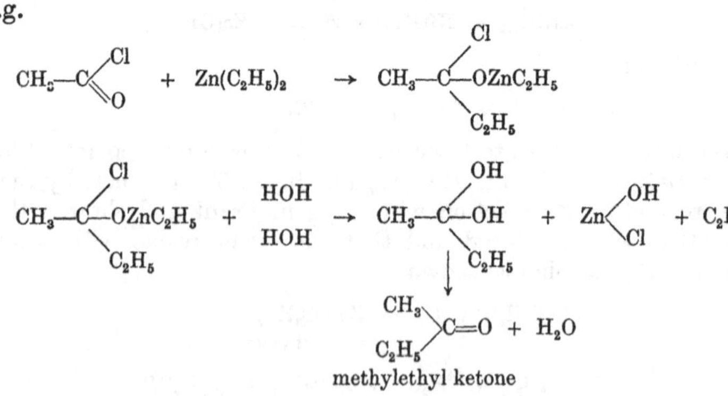

$$CH_3-C\begin{matrix}\diagup Cl\\ \diagdown\!\!\diagdown O\end{matrix} \quad + \quad Zn(C_2H_5)_2 \quad \rightarrow \quad CH_3-\overset{Cl}{\underset{C_2H_5}{C}}-OZnC_2H_5$$

$$CH_3-\overset{Cl}{\underset{C_2H_5}{C}}-OZnC_2H_5 \; + \; \begin{matrix}HOH\\ HOH\end{matrix} \quad \rightarrow \quad CH_3-\overset{OH}{\underset{C_2H_5}{C}}-OH \quad + \quad Zn\begin{matrix}\diagup OH\\ \diagdown Cl\end{matrix} \quad + \; C_2H_6$$

$$\downarrow$$

$$\begin{matrix}CH_3\diagdown\\ C_2H_5\diagup\end{matrix}C{=}O \; + \; H_2O$$

methylethyl ketone

$$CH_3—C(Cl)(C_2H_5)—OZnC_2H_5 + Zn(C_2H_5)_2 \rightarrow CH_3—C(C_2H_5)(C_2H_5)—OZnC_2H_5 + Zn(C_2H_5)(Cl)$$

$$CH_3—C(C_2H_5)(C_2H_5)—OZnC_2H_5 + \begin{matrix}HOH\\HOH\end{matrix} \rightarrow CH_3—C(C_2H_5)(C_2H_5)—OH + C_2H_6 + Zn(OH)_2$$

tert.-hexyl alcohol, methyldiethyl carbinol or 3-methyl-3-hydroxypentane } b.p. 123°

(iv) Production of tertiary alcohols from esters, e.g.

$$CH_3—C(=O)(OC_2H_5) + Zn(CH_3)_2 \rightarrow CH_3—C(OZnCH_3)(OC_2H_5)—CH_3$$

$$CH_3—C(OZnCH_3)(OC_2H_5)—CH_3 + Zn(CH_3)_2 \rightarrow CH_3—C(OZnCH_3)(CH_3)—CH_3 + Zn(CH_3)(OC_2H_5)$$

$$CH_3—C(OZnCH_3)(CH_3)—CH_3 + \begin{matrix}HOH\\HOH\end{matrix} \rightarrow CH_3—C(OH)(CH_3)—CH_3 + Zn(OH)_2 + CH_4$$

tert.-butyl alcohol

(esters of formic acid yield secondary alcohols).

The reactions are usually carried out by gradually mixing the dry reactants dissolved in an inert solvent (usually dry ether) in a dry inert atmosphere under suitable temperature conditions. When the reaction is finished, water is gradually added followed by a suitable acid to dissolve zinc hydroxide or basic zinc salts. If it is soluble, the product is isolated by extraction with a solvent and evaporating the solution. It will be realised that the carrying out of syntheses with the zinc alkyls is attended with the difficulty caused by their inflammability under atmospheric conditions. A further complication is frequently experienced on account of the great vigour of the reaction, which may be difficult to control.

ORGANO-MAGNESIUM COMPOUNDS

In 1899, by using methyl iodide and magnesium, Barbier was able to convert a ketone into a tertiary alcohol; and, soon after, Grignard began his classical work on the use of organo-magnesium compounds for synthetic purposes. Grignard showed that magnesium will dissolve in dry ethereal solutions of alkyl and aryl halides forming magnesium alkyl (aryl) halides which are stable in dry ethereal solution. The essential condition is that the reactants should be dry. He showed that the magnesium alkyl (aryl) halides have the general formula (A), and it is probable that they are present in the formate of 'etherates' having the general formula (B):

$$\mathrm{Mg}\begin{matrix} \diagup R \\ \diagdown X \end{matrix} \quad (A) \qquad \begin{matrix} (C_2H_5)_2O \searrow \\ (C_2H_5)_2O \nearrow \end{matrix} \mathrm{Mg}\begin{matrix} \diagup R \\ \diagdown X \end{matrix} \quad (B) \qquad (R = \text{alkyl or aryl}, X = \text{halogen})$$

in which the magnesium compound is represented as being co-ordinated to the oxygen atoms of two molecules of ether.

The actual magnesium compounds can be isolated and obtained as grey-white solids which are readily decomposed in a moist atmosphere. In actual practice, the magnesium compound is not isolated but used immediately in ethereal solution as prepared.* These organo-magnesium compounds are generally known as *Grignard reagents*, and reactions in which they are used as *Grignard reactions*. Grignard reagents have almost superseded the use of zinc alkyls; they are not only more convenient to use since they are not inflammable, but are of more diverse types since they are not confined to the aliphatic series.

The reagent is prepared by suspending magnesium turnings in dry ether and adding a dry ethereal solution of the alkyl or aryl halide gradually as the magnesium dissolves. It may be necessary to assist the reaction by adding a trace of catalyst (frequently iodine) or by warming. Generally, solution of the magnesium proceeds rapidly and with evolution of heat and cooling may be necessary. The reaction is usually carried out by mixing a dry ethereal solution of the reactant with the ethereal solution of the Grignard reagent, allowing the reaction to proceed, completing it by warming and working up the product as in the case of the use of zinc alkyls. Typical examples of the Grignard reactions are:

* Many workers have shown that, although they are less soluble in ether and react less vigorously as a rule than the Grignard reagents, the zinc alkyl halides may be used for synthetic purposes like the corresponding magnesium compounds. Unlike the latter, the zinc compounds cannot be used for the introduction of aryl groups.

(i) Production of hydrocarbons by decomposition with water, mineral acids, ammonia, alcohols* and primary amines, e.g.

$$\mathrm{Mg}{<}^{R}_{X} + \mathrm{HOH} \rightarrow R\mathrm{H} + \mathrm{Mg}{<}^{\mathrm{OH}}_{X}$$

$$\mathrm{Mg}RX + \mathrm{HCl} \rightarrow R\mathrm{H} + \mathrm{Mg}X\mathrm{Cl}$$

$$\mathrm{Mg}RX + R'\mathrm{OH} \rightarrow \mathrm{Mg}{<}^{\mathrm{O}R'}_{X} + R\mathrm{H}$$

$$\mathrm{Mg}RX + R'\text{—}\mathrm{N}{<}^{\mathrm{H}}_{\mathrm{H}} \rightarrow R'\text{—}\mathrm{N}{<}^{\mathrm{Mg}X}_{\mathrm{H}} + R\mathrm{H}$$

(ii) Production of alcohols from aldehydes, ketones, acid chlorides and esters by reactions similar to those with zinc alkyls:

$$\underset{\text{aldehyde}}{R'\text{—}\mathrm{C}{<}^{\mathrm{H}}_{\mathrm{O}}} + \mathrm{Mg}RX \rightarrow {}^{R'}_{R}{>}\mathrm{C}{<}^{\mathrm{H}}_{\mathrm{OMg}X}$$

$${}^{R'}_{R}{>}\mathrm{C}{<}^{\mathrm{H}}_{\mathrm{OMg}X} + \mathrm{H_2O} \rightarrow \underset{\text{secondary alcohol}}{{}^{R'}_{R}{>}\mathrm{C}{<}^{\mathrm{H}}_{\mathrm{OH}}} + \mathrm{Mg}X(\mathrm{OH})$$

(If $R' = \mathrm{H}$, the aldehyde is formaldehyde and a primary alcohol is obtained.)

$${}^{R'}_{R''}{>}\mathrm{C{=}O} + \mathrm{Mg}RX \rightarrow {}^{R'}_{R''}{>}\mathrm{C}{<}^{R}_{\mathrm{OMg}X} \xrightarrow[-\mathrm{MgOH}]{+\mathrm{H_2O}} \underset{\text{tertiary alcohol}}{{}^{R'}_{R''}{>}\mathrm{C}{<}^{R}_{\mathrm{OH}}}$$

$$\underset{\text{acid chloride}}{R'\text{—}\mathrm{C}{<}^{\mathrm{O}}_{\mathrm{Cl}}} + \mathrm{Mg}RX \rightarrow {}^{R'}_{R}{>}\mathrm{C}{<}^{\mathrm{OMg}X}_{\mathrm{Cl}} \xrightarrow[-\mathrm{MgCl}X]{+\mathrm{Mg}RX} {}^{R'}_{R}{>}\mathrm{C}{<}^{\mathrm{OMg}X}_{R}$$

$$\xrightarrow[-\mathrm{Mg}X(\mathrm{OH})]{+\mathrm{H_2O}} \underset{\text{tertiary alcohol}}{{}^{R'}_{R}{>}\mathrm{C}{<}^{\mathrm{OH}}_{R}}$$

$$\underset{\text{ester}}{R'\text{—}\mathrm{C}{<}^{\mathrm{O}}_{\mathrm{O}R''}} \xrightarrow{+\mathrm{Mg}RX} {}^{R'}_{R}{>}\mathrm{C}{<}^{\mathrm{OMg}X}_{\mathrm{O}R''} \xrightarrow[-\mathrm{Mg}X(\mathrm{O}R'')]{+\mathrm{Mg}RX} {}^{R'}_{R}{>}\mathrm{C}{<}^{\mathrm{OMg}X}_{R}$$

$$\xrightarrow[-\mathrm{Mg}X(\mathrm{OH})]{+\mathrm{H_2O}} \underset{\text{tertiary alcohol}}{{}^{R'}_{R}{>}\mathrm{C}{<}^{\mathrm{OH}}_{R}}$$

(N.B. $\mathrm{Mg}X(\mathrm{O}R'') + \mathrm{H_2O} \rightarrow \mathrm{Mg}X\mathrm{OH} + R''\mathrm{OH}$.)

* Grignard reagents have been used for the estimation of hydroxyl groups in alcohols.

If dibasic esters are used, dihydric alcohols (glycols) are the final product of the reaction.

Tertiary alcohols may also be obtained starting from carbonyl chloride (phosgene), thus:

$$\mathrm{Cl_2C{=}O} \xrightarrow[-2\mathrm{Mg}X\mathrm{Cl}]{+3\mathrm{Mg}RX} R_3\mathrm{C{-}OMg}X \xrightarrow[-\mathrm{Mg}X(\mathrm{OH})]{+\mathrm{H_2O}} \underset{\text{tertiary alcohol}}{R_3\mathrm{C{-}OH}}$$

(iii) Production of carboxylic acids by the action of dry carbon dioxide. For example, *methylethylacetic acid* (α-methyl-*n*-butyric acid), b.p. 174°, is readily obtained in good yield by passing dry carbon dioxide from a cylinder under slight pressure into a cooled ethereal solution of magnesium sec.-butyl iodide:

$$\mathrm{(C_2H_5)(CH_3)CH{-}MgI} + \mathrm{O{=}C{=}O} \rightarrow \mathrm{(C_2H_5)(CH_3)CH{-}C({=}O)OMgI}$$

$$\mathrm{(C_2H_5)(CH_3)CH{-}C({=}O)OMgI} + \mathrm{H_2O} \rightarrow \underset{\text{methylethylacetic acid}}{\mathrm{(C_2H_5)(CH_3)CH{-}COOH}} + \mathrm{MgI(OH)}$$

Analogous reactions take place using carbon disulphide and carbon oxysulphide. In the former case dithio acids, $R\mathrm{{-}C({=}S)SH}$, and in the latter case thiol acids, $R\mathrm{{-}C({=}O)SH}$, are obtained.

These are typical of many reactions of the Grignard reagents and they have been greatly extended and modified.

Grignard reagents will react directly with the halogen derivatives of various non-metals to give organic derivatives. For example, from boron trifluoride in the form of potassium borofluoride (KBF_4), *boron triphenyl* crystallising in long prisms, m.p. 136°, is obtained in good yield by the action of magnesium phenyl bromide; *trimethylarsine* (b.p. 70°), *triethylarsine* (b.p. 140°) and *triphenylarsine* (m.p. 57°) have been prepared by reactions which may be generally represented:

$$\mathrm{AsCl_3} + 3\mathrm{Mg}RX \rightarrow \mathrm{As}R_3 + 3\mathrm{Mg}X\mathrm{Cl} \quad (R = \text{methyl, ethyl, phenyl};\ X = \mathrm{I} \text{ or } \mathrm{Br})$$

and by a similar reaction using silicon tetrachloride and magnesium methyl bromide the interesting compound, *tetramethylsilicane*, $Si(CH_3)_4$, b.p. 26–27° (compare tetramethylmethane), was prepared.

The Grignard reagents have been extensively used for preparing organo compounds of the following metals: silver, copper, zinc, cadmium, mercury, beryllium, aluminium, thallium, tin, lead, antimony, bismuth, chromium, iron, gold and platinum. The case of zinc is interesting, since by acting upon zinc chloride with magnesium phenyl bromide it has been possible to prepare zinc phenyl bromide and *zinc diphenyl* (colourless needles, m.p. 105–106°):

$$2ZnCl_2 + 2Mg(C_6H_5)Br \rightarrow 2Zn(C_6H_5)Br + MgCl_2 + MgBr_2$$

$$ZnCl_2 + 2Mg(C_6H_5)Br \rightarrow Zn(C_6H_5)_2 + MgCl_2 + MgBr_2$$

these compounds were not previously available, since Frankland's original method is not applicable in the aromatic series.

The alkyl compounds of lead and particularly lead tetraethyl are used technically as anti-knock agents in internal combustion engines. *Lead tetraethyl*, for example, is prepared by the following reaction:

$$2PbCl_2 + 4Mg(C_2H_5)Br \rightarrow Pb(C_2H_5)_4 + Pb + 2MgCl_2 + 2MgBr_2$$

it is a colourless liquid, b.p. 83° at 14 mm. It was first prepared by the action of ethyl iodide on lead-sodium alloy and was also obtained by Buckton by the action of zinc ethyl on lead chloride. It has been shown by Paneth and co-workers (1931) that lead tetramethyl and tetraethyl when heated under a low pressure in an inert gas evolve free methyl and ethyl radicals respectively; the evolution of the free radicals was proved by showing that the volatile products of decomposition are able to attack mirrors of lead, tellurium, etc., forming identifiable methyl and ethyl derivatives of these substances. Although Frankland did not succeed in isolating free hydrocarbon radicals, it is significant that their production was first made possible by the decomposition of organo-metallic compounds and he was the first to prepare the latter.

In the case of certain metals, such as beryllium, aluminium, tin, zinc, cadmium, mercury, lead, antimony and bismuth, the organo compounds are of the two types mentioned above. In other cases, such as those of the metals tervalent gold, tervalent thallium and quadrivalent platinum, the organo compounds obtained by the direct reaction of the Grignard reagent are only of the type $MR_{n-p}X_p$. Actually the tervalent thallium compounds are of the type TlR_2X; the tervalent gold compounds are of the type $(AuR_2X)_2$, e.g. *diethylmonobromogold*, colourless needles, m.p. 58°, and the compound obtained by the action of magnesium methyl iodide on anhydrous platinic chloride is *platinum trimethyl iodide*, $Pt(CH_3)_3I$, crystallising in colourless small prisms which explode on being heated.

The organo-metallic compounds are not only interesting in themselves but their investigation has, from the middle of the nineteenth century, contributed considerably to the development of the theory of valency.

APPENDIX

(a) Isolation and Purification of Organic Compounds

For the identification of organic substances already known and for describing newly discovered compounds, the determination of their physical constants is essential. The importance of the accurate determination of these physical constants cannot be too strongly emphasised, and the accumulation of accurate data concerning compounds of all types is essential for the satisfactory development of that part of the subject concerning the relation between physical properties and chemical constitution.

For such purposes the compounds must be pure, and a compound can only be recognised as pure when further attempts to purify it by adequate means do not cause any change in the physical constants.

After a reaction for the preparation of a compound has been completed, the product may be isolated by methods which depend on its physical properties, e.g. solubility in a solvent which is either immiscible or only partly miscible with an aqueous solution. Many organic compounds are soluble in such solvents as ether, benzene and chloroform and, provided that other substances accompanying the desired substance are insoluble, the latter may be isolated from an aqueous reaction mixture by shaking it with a suitable solvent using a *separating funnel* which permits the easy separation of the two liquids. Various types of separating funnels are used; a pear-shape and a cylindrical separating funnel are illustrated in Fig. I. The efficiency of the process depends on the partition coefficient of the substance to be isolated between water and the particular solvent and the number of times the extraction is performed. The solution of the compound may be washed in the separating funnel with water until free from water soluble impurities, the solution freed from most of the water by means of a drying agent which does not react either with the compound or with the solvent and the compound isolated by suitable means, such as evaporation of the solvent. Fused calcium chloride, anhydrous sodium sulphate, phosphorus pentoxide, anhydrous potassium carbonate and fused potassium hydroxide are substances frequently used as 'drying agents'.

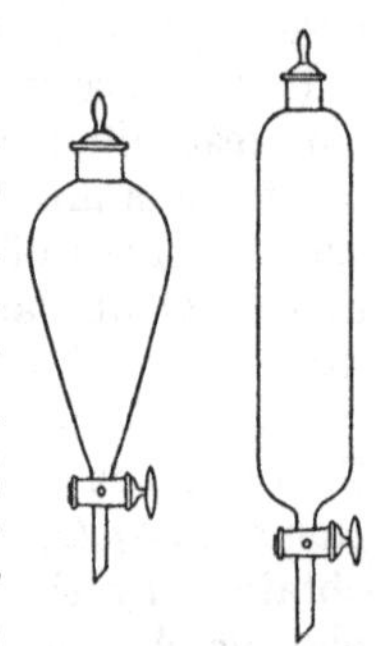

Fig. I. separating funnels

In certain cases, the product of a chemical reaction having an appreciable vapour pressure at 100° may be freed from non-volatile substances by distillation with steam. Steam from a suitable source is passed into the reaction mixture and the vaporised product passes over with excess of steam and is condensed and collects with excess of water in the distillate. If the compound is immiscible with water, then at the boiling point of the mixture the external pressure (P) is equal to the sum of the pressure of water vapour (p) and that (p') of the substance, i.e. $P=p+p'$, and the amounts of water (w) and substance (w') distilling over is given by

$$\frac{w}{w'}=\frac{p\times m}{p'\times m'},$$

where m and m' are the molecular weights of water (18) and the substance respectively. A simple form of apparatus for distillation in steam is shown in Fig. II. The substance, unless it crystallises out,

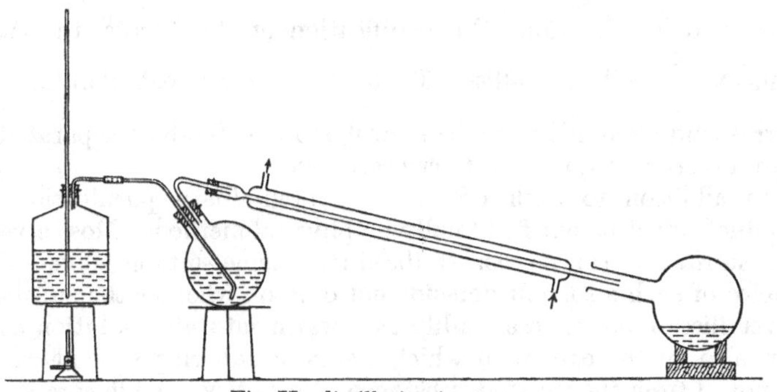

Fig. II. distillation in steam

may be separated from the distillate by extraction with a suitable solvent, as described above.* If the desired product is not volatile in steam, it may be separated from contaminating substances which are volatile in steam by the same process. After the steam distillation is ended, the product can generally be isolated from the contents in the distillation flask by extraction with a solvent.

In many cases, separation and provisional purification of the product of a reaction may have to be carried out by chemical methods. Acids and bases may be separated by taking advantage of their solubility in aqueous solutions of suitable alkalis and mineral acids respectively and then liberating them from these solutions by the

* Steam distillation can be regarded as a special case of distillation of a substance under reduced pressure (p. 510).

action of mineral acids and alkalis respectively. As typical of the kind of procedure that may have to be followed in special cases is the following. The result of alkylation of aniline is to give a mixture of monoalkylaniline, dialkylaniline and unchanged aniline in solution in excess of acid. The mixture of bases separated from this solution with excess of sodium hydroxide cannot be separated by distillation (*v.* below). On treatment with an excess of a suitable acyl (Ac) chloride (benzoyl chloride, benzene sulphonyl chloride, etc.), the aniline is converted into an amide of the type $C_6H_5—N\langle^{H}_{Ac}$ (soluble in alkali), the monoalkylaniline into an amide of the type $C_6H_5—N\langle^{R}_{Ac}$ (insoluble in alkali), while the dialkylaniline is unaffected. On steam distillation of the alkaline reaction mixture the last named is volatile and is separated as indicated above; in the distillation flask the amide $C_6H_5—N\langle^{R}_{Ac}$ will crystallise from the cold mixture and is separated by filtration. On acidification of the filtrate, the amide $C_6H_5—N\langle^{H}_{Ac}$ will crystallise. To obtain the monoalkylaniline, the corresponding amide must be hydrolysed and the base separated by steam distillation, etc. as described above.

In addition to such processes of preliminary purification the products must be purified finally by physical methods. Most crystalline substances can be recrystallised from some suitable solvent. The choice of such a solvent depends not only on whether the substance crystallises more or less readily as a warm saturated solution cools, but also on the ease with which traces of adhering solvent can be removed from the isolated substance. Not a few organic compounds crystallise from aqueous solutions, but frequently the more volatile organic solvents, such as the methyl and ethyl alcohols, diethyl ether, acetone, ethyl acetate, chloroform and benzene, are employed; acetic acid is also frequently employed when its chemical properties do not contra-indicate its use, but any solvent may be used subject to the limitation mentioned above. Solutions of substances in solvents having boiling points below 85° are conveniently made using a water-bath, as shown in Fig. III.

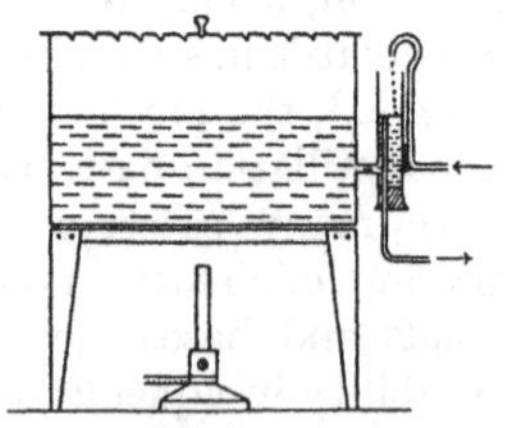
Fig. III. water-bath

Water-baths of various types, which may be heated electrically or by gas, are indispensable in preparing and isolating organic compounds. Having obtained by a small scale experiment an approximate idea of the solubility of the substance which crystallises as the boiling solution cools, an excess of the solid together with the solvent is heated in a round-bottom flask, fitted with a condenser to prevent loss of solvent, until nearly all has dissolved. According to the boiling point of the solvent, the flask and its contents are heated either on the water-bath or over a Bunsen burner, using a wire gauze. A round-bottom flask fitted with a water-cooled 'reflux' condenser is represented diagrammatically in Fig. IV.

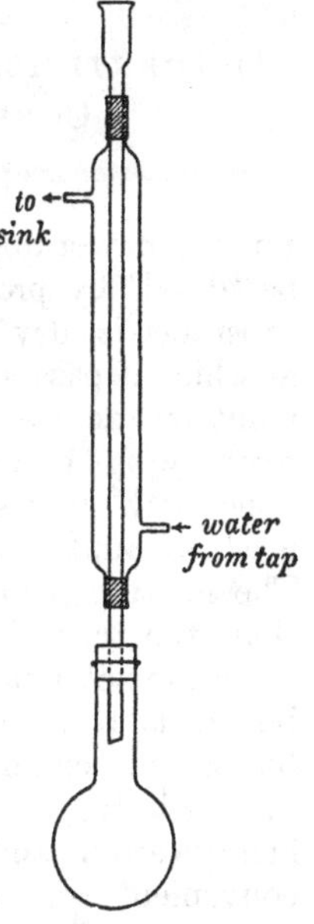

Fig. IV. round-bottom flask and reflux

The hot solution is rapidly filtered from insoluble material and, as the filtrate cools (rarely, if ever, in a basin or a so-called crystallising dish), the substance should crystallise. The substance is filtered under pressure, using a special funnel fitted up as illustrated in Fig. V. The so-called Büchner funnels which may be cylindrical or 'funnel-shape' are made of porcelain with a perforated disc, and the substance is collected on close-fitting filter paper firmly bedded down before beginning the filtration and transference of the solid. Glass funnels having sealed-in sintered glass discs of varying degrees of fineness—rendering the use of filter paper unnecessary—are employed in special cases. Instead of Büchner funnels, etc., ordinary glass funnels supplied with loose filter discs which may be made of glass or porcelain and carefully fitted with filter paper of suitable size may be employed. These have the great advantage of being more easily cleaned than the former.

After transference to the filter, washing with a little pure solvent

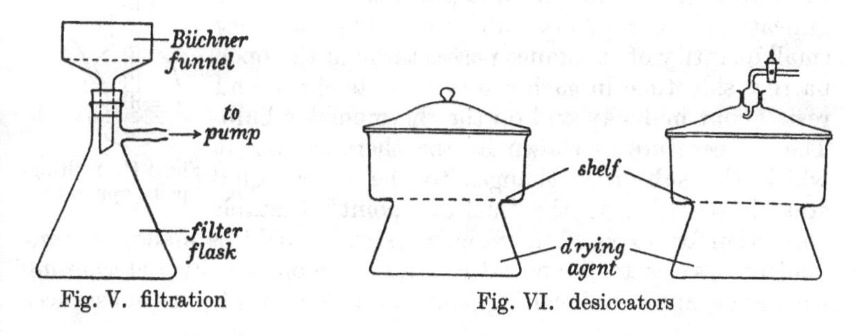

Fig. V. filtration

Fig. VI. desiccators

and pressing down with a spatula, the substance is carefully transferred for 'drying' in a convenient receptacle to a desiccator containing a suitable 'drying agent'.* By 'drying agent' is usually meant a substance capable of absorbing the vapour of adhering solvent. The 'drying' of the substance may be carried out at ordinary pressure or, as is frequently the case, under reduced pressure. Types of desiccators used for 'drying' substances at ordinary temperatures are illustrated in Fig. VI.

(b) Identification and Determination of Physical Constants of Organic Compounds

Melting point

For rapid working with small quantities of substances, these may be 'dried' by pressing out on *clean* unglazed porcelain. When the substance is 'dry', i.e. freed from adhering solvent, the temperature at which it passes from the solid to the liquid condition—its melting point—is determined. By convention, in routine work with organic compounds this is usually done with very small amounts of substance—the least amount easily visible—in capillary tubes.† The melting point so determined is usually higher than the true melting point as determined using a thermometer immersed in the melting substance, but it is recognised as the routine method used for organic compounds, most of the melting points of which are taken thus and employed largely for identification and not quantitative purposes. A convenient type of melting-point apparatus is illustrated in Fig. VII.

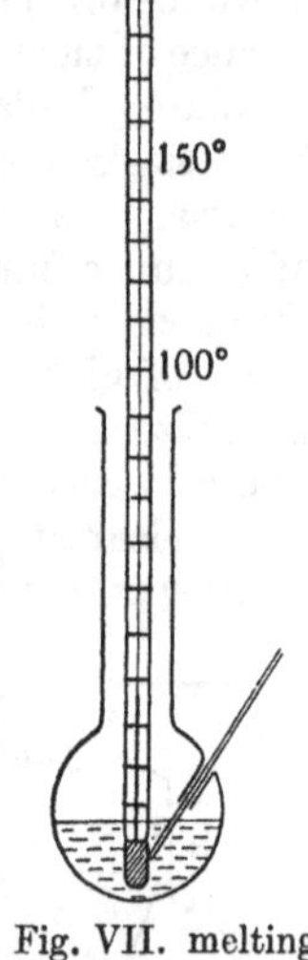

Fig. VII. melting-point apparatus

A suitable colourless heating liquid having a boiling point much higher than the melting point of the substance is directly and gradually heated as indicated. The capillary tube containing the very small quantity of substance passes through the open narrow side tube in such a way that its closed end rests about mid-way and on the thermometer bulb. The temperature as shown by the thermometer at which the substance changes to the clear liquid state is regarded as the 'melting point'. Suitable correction for 'exposed thermometer stem' should be made. A convenient way is to 'standardise' the thermometer by determining a series of 'apparent' melting points of a number of pure substances

* Purification of solid substances by sublimation is rarely resorted to and then only in special cases.

† This is probably the oldest micro method in chemistry.

and plotting them on graph-paper against their *true* melting points as given in the literature. From the curve so constructed for the particular thermometer the *true* melting point of a substance can then be determined from the *observed* melting point.

The melting point of a pure substance must by definition be 'sharp' and well defined. If the conversion from the solid to the liquid condition takes place over (say) more than one degree (Centigrade), the substance in question is generally impure. In such a case, the substance should be submitted to further purification until the melting point is 'sharp' and reaches its highest value.

In most cases the melting point of a substance (A) is depressed and becomes indefinite when admixed with another substance (B).* With the apparatus as ordinarily used for determining the melting point of organic compounds, it is possible to determine with fair accuracy the molecular weight of suitable organic compounds by making use of the quantitative depression of the melting point of particular substances by admixture. The best-known method of molecular weight determination by this method is that of Rast (1922).

By determining the melting points of known homogeneous and dilute mixtures of an organic compound with camphor (i.e. the temperature at which the material in the melting-point tube becomes a clear liquid), the molecular weight of the compound can be calculated, knowing the molecular depression constant (K) of camphor—a hydroaromatic ketone belonging to the dicyclic terpene series†—to be 400, an unusually high value. This method of Rast only requires the ordinary melting-point apparatus and, since the differences in melting point from that of camphor (m.p. 178·7°) are usually large, only a reasonably accurate ordinary thermometer covering the necessary range of temperatures is required. The melting points of quantitatively made up and homogeneous mixtures are taken in the usual capillary tubes. The method has its limitations; for example, salicylic acid gives abnormal results on account of compound formation between it and camphor (cf. p. 10).

The conversion of a substance from the solid to the liquid phase may be attended by decomposition. In such a case the compound is described as having 'm.p. ...° (decomp.)'; this information is

* If substance (A) has a lower melting point than substance (B) and if these two substances form a complete series of solid solutions with each other, then the melting point of (A) is raised and becomes indefinite when mixed with substance (B). On the other hand, the melting point of (B) would be depressed by admixture with substance (A).

The mixing of two substances in the molten condition may also be attended with the formation of eutectic mixtures and the production of compounds. To the elucidation of the problems connected with the equilibria between at least two component systems the Phase Rule has been successfully applied and the equilibrium diagrams of a considerable number of mixtures of two different organic substances have been deduced. A convenient résumé of some of these is given in *Physikalische Tabellen*, by Landolt-Börnstein, under the heading 'Gleichgewichte je zweier organischer Stoffe'.

† The study of the large class of compounds known as the terpenes and their derivatives is beyond the scope of the present work and the student is referred to *The Terpenes* by Professor J. L. Simonsen, F.R.S., Cambridge University Press (1931–1932).

useful in identifying compounds, but the actual temperature quoted must be somewhat indefinite on account of the effects of the product or products of decomposition on the incipient melting point of the undecomposed substance. Not a few solid substances undergo decomposition without melting and, when the decomposition temperature is reasonably definite, this may be quoted for identification purposes.

Boiling point

The purification of compounds which are liquid at the ordinary temperature is carried out by fractional distillation. In the case of a pure substance, the boiling point, i.e. the temperature at which it passes from the liquid to the gaseous condition (without decomposition) at a stated pressure, must be within very narrow limits—1° or less.

The boiling point of a liquid may be determined using different types of apparatus and, as in the case of the determination of melting points, several are available. The simplest type of apparatus used for determining the boiling point of a liquid at ordinary pressures is shown diagrammatically in Fig. VIII.

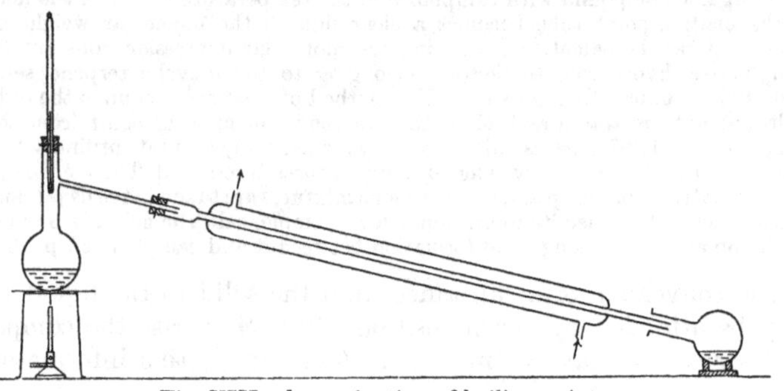

Fig. VIII. determination of boiling point

If the liquid has a boiling point appreciably higher than 100°, the water-cooled condenser is replaced by an air (or unjacketed) condenser. The heating is done by a small flame whose tip just touches the bottom of the distilling flask. In distilling liquids it is generally advisable to place small pieces of *clean* porous porcelain in the liquid, which tend to promote gentle ebullition and prevent superheating and 'bumping'. The rate of heating is carefully controlled, so that drops of the distilled liquid collecting in the receiver fall from the end of the condenser at a definite and timed rate. If superheating be avoided, the boiling point as indicated by the thermometer, so placed that its bulb is just below the side tube of the distilling flask, should remain constant during the distillation of the liquid. For most practical purposes this, after correction for exposed stem, may be taken as the boiling point at the particular pressure (shown by the barometer) to be quoted. If the pressure at which the boiling point is determined is not quoted, it is understood to be at the standard pressure of 760 mm.

If the liquid is pure,* practically the whole should distil at a definite temperature—the boiling point. If the liquid (A) is not pure, the temperature of distillation will be indefinite and, provided the vapour pressures of the impurities (B, C, etc.) at the boiling point of (A) are much less than 760 mm., the compound (A) may be purified by 'fractional distillation'. If the fractional distillation be carried out at ordinary atmospheric pressure, the process may be rendered more efficient by increasing considerably the distance between the surface of the liquid in the distilling flask and the side tube by means of a 'fractionating' column. There are many types of fractionating column, which may be of various lengths; two types of fractionating column are represented in Fig. IX. Such columns tend to prevent the vapours of the less volatile components reaching the condenser and consequently the separation of the components of a mixture of substances having different boiling points may be effected. Supposing, in a 'straight' distillation, the distillation of a liquid mixture takes place chiefly between 90° and 130°. On fractional distillation, the first fraction may be collected up to 90°, the second between 90° and 100°, the third between 100° and 110°, the fourth between 110° and 120° and the fifth between 120° and 130°. Each of these fractions is then redistilled, thus: the first fraction again up to 90° and to the residue remaining in the distillation flask is added the second fraction and this is distilled up to 100°, and to the residue the third fraction is added and distilled up to 110°, and the process continued until the last fraction has been included as indicated. An inspection of the distillate fractions will indicate that one fraction is much the greatest in amount. This probably contains the chief constituent, and this can be submitted to further careful fractionation until a fraction of definite and 'sharp' boiling point is ultimately obtained. On the technical scale, as in the petroleum

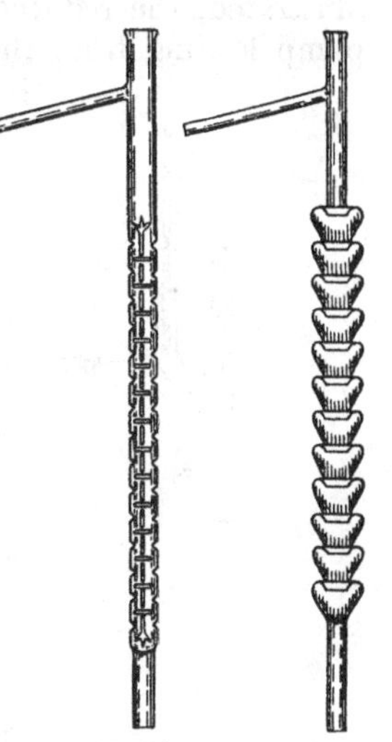

Fig. IX. fractionating columns

* Assuming the possibility of a 'constant boiling mixture' has been excluded.

industry, fractional distillations are carried out with the highest efficiency.

Liquids which undergo decomposition when distilled at the ordinary pressure may be distilled under reduced pressure. A convenient type of apparatus for this purpose is illustrated in Fig. X. The ordinary distillation flask is replaced by a two-neck flask (Claisen) fitted with a thermometer (*t*) and a very fine capillary tube (*c*), as shown. Through *c* a small amount of air or other indifferent gas is admitted through the distilling liquid and this promotes gentle ebullition and tends to avoid 'bumping'. The pressure in the 'circuit' is determined by means of a suitable manometer and, in the type of apparatus illustrated, the reduced pressure is obtained by means of a water pump leading from the 'fractionation triangle'. The apparatus is

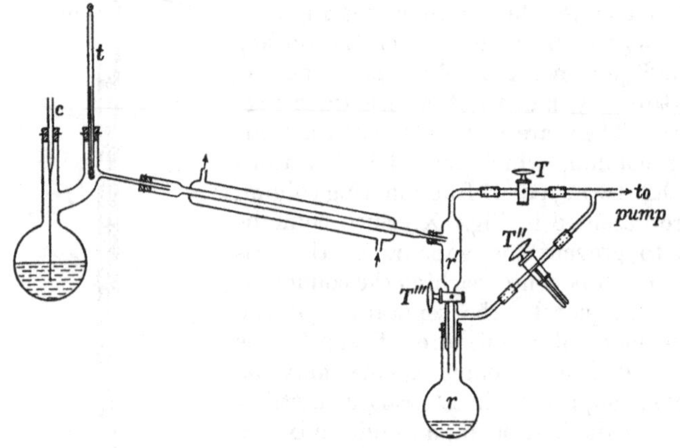

Fig. X. vacuum distillation

carefully set up with well-fitting corks as illustrated, the only inlet of air (or indifferent gas) being through *c*. Taps T'' and T''' being closed, the apparatus is evacuated by means of tap T and the distillation is started. Liquid will collect in the receiver r'. Pressure and temperature of distillation being steady, T is closed and T'' is opened so as to evacuate the receiver r. When the pressure has reached its original value, T is again opened; T''' is also opened, allowing the liquid in r' to run into r; this is the first fraction. Tap T''' is then closed and the two-way tap T'' turned so as to admit air from the atmosphere to r. The receiver r is removed and replaced by a similar receiver and the distillation continued as before, fractions being collected in various receivers at known intervals of temperature, the pressure in the apparatus being kept as constant as possible throughout the distillation. Using the same apparatus each fraction can be re-distilled under similar conditions until a product of constant boiling point (at the particular pressure) is obtained.

This apparatus can be used for the determination of boiling points at suitable reduced pressures. Distillation of liquids under reduced pressure is very frequently employed for purification purposes and, by using one of the modern electrically driven vacuum pumps instead of a water pump, very low pressures can be obtained.

The melting point and boiling point (at stated pressure) are the ordinary constants recorded when possible. In the case of pure liquids, these data may well be supplemented by the results of the determination of other physical constants, such as the density, refractive index (for a particular wave-length of light), etc., all under stated conditions. For pure optically active substances, the specific rotatory power, $[\alpha]$, at a known temperature and for a particular wave-length is an important constant; in the case of optically active substances in solution, the concentration (usually denoted by c, the number of grams in 100 ml. of solution) as well as the name of the solvent must be stated in quoting the specific rotatory power.

The complete identification of a prepared compound with one already described in the literature may require not only the determination of such physical constants as are possible but also analysis of the compound; in this case, the analytical figures obtained and the physical constants must agree within the limits of experimental error with those of the already described substance*. The more detailed the information accumulated about a compound the less likely will an erroneous conclusion be drawn as to the nature of the compound in question. It is obvious that two compounds are identical if they have the same chemical properties; the identity of their physical constants, however, furnishes quantitative as contrasted with qualitative proof of their identity.

Identification by method of mixed melting point

If a particular series of reactions for the production of a compound is always followed, the resulting compound must necessarily always be the same. It not infrequently happens, however, that the same compound may be prepared in various ways and, consequently, it is often necessary to establish the identity of compounds resulting from different reactions. If the compound has a definite melting point, the most rapid and frequently used method for establishing the identity of two compounds is by the *method of mixed melting point*. If the two compounds are identical, the melting point of a specimen of either will not be depressed by admixture with the other. On the

* Compare footnote (p. 416) concerning hexaacetyl-*d*-sorbitol.

other hand, two substances are frequently proved not to be identical by showing that the melting point of one is changed (frequently depressed) by admixture with the other.* Other physical constants may be employed, but generally not so conveniently, for establishing the identity or dissimilarity of organic compounds.

Description of a new chemical compound

The account of any compound described for the first time must include (i) the results of accurately carried out analysis and, when possible, molecular weight determination leading to the establishment of the molecular formula, (ii) such physical constants as may be conveniently determined and (iii) a description of the chief chemical reactions leading to the elucidation of the constitution of the compound. In the case of crystalline substances the importance of their investigation by modern crystallographic methods may be mentioned, although crystallography is a science by itself. The crystallographic investigation of compounds is of the highest importance as contributing to our knowledge of the 'architecture of the molecule' and as indicating the relationship between crystal form and chemical constitution.

(c) Typical Apparatus used in the Preparation of Organic Compounds

The apparatus employed in the preparation of organic compounds is usually comparatively simple and, except in special cases, is generally limited to a few types:

(i) Heating together of two reactants in a suitable solvent. For this a round-bottom flask of convenient size is fitted with a so-called 'reflux' condenser (p. 505). The reflux condenser may be either water cooled or air cooled according to the boiling point of any volatile material which may be originally present in the mixture or produced during the reaction. The flask is heated either on the water-bath or over a wire gauze or sand-bath according to the temperature required. Baths containing oil (such as high boiling paraffin), glycerol, or fusible metal are also frequently employed. If the reaction mixture has to be heated above the boiling point of any constituent present, the reaction has to be carried out by heating 'under pressure'. On the small scale this is done by heating the mixture in sealed glass vessels and, on the large scale, in autoclaves.

(ii) Addition of a reagent as a reaction is proceeding. In such

* This method has certain limitations briefly referred to above (p. 507).

circumstances, the apparatus illustrated in Fig. XI is convenient. The two-neck flask is fitted with a 'dropping funnel' for the addition of the reagent and a suitable reflux condenser. The flask can be heated or cooled as may be found necessary.

(iii) Many reactions are promoted or proceed more smoothly when the reaction mixture is stirred. This stirring is essential when the reaction mixture consists of two immiscible or partly miscible liquids, when one of the constituents is a solid and the other a liquid and when admixture with a gas is necessary. Fig. XII represents, in

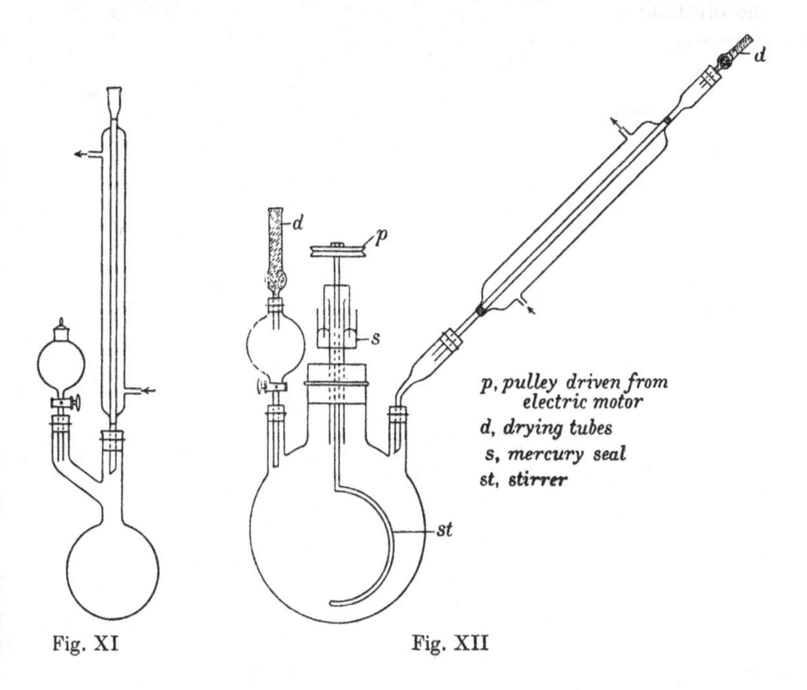

Fig. XI   Fig. XII

outline, the apparatus for carrying out a reaction in a 'dried' atmosphere with stirring and addition of a reagent ás the reaction proceeds.

These three examples of typical apparatus used in preparative work may be sufficient to illustrate general procedure. It will be realised that such apparatus can be considerably modified to suit particular conditions. Reactions may have to be carried out in an indifferent atmosphere, in the absence of moisture, or a certain gas may have to be supplied for the particular reaction. The apparatus necessary to meet such conditions will readily be devised by those familiar with routine chemical manipulation.

Component parts of the apparatus are fitted together with corks and (much less frequently) india-rubber stoppers. Glass tubes may be fixed end to end with rubber tubing. In using corks or rubber stoppers for fixing together apparatus, the action of chemical reagents on them must be taken into account. In many cases what may be called 'organic' joints must be avoided and component parts of apparatus are then fitted together by ground glass joints or actually sealed glass joints. In much modern apparatus for the preparation of organic compounds the use of 'organic' joints is being reduced to the minimum.

Symbols, Atomic Numbers and Atomic Weights of Elements Mentioned in this Volume

		Atomic Number	Atomic Weight			Atomic Number	Atomic Weight
Aluminium	Al	13	26·97	Magnesium	Mg	12	24·32
Antimony	Sb	51	121·76	Manganese	Mn	25	54·93
Arsenic	As	33	74·91	Mercury	Hg	80	200·61
Barium	Ba	56	137·36	Nickel	Ni	28	58·69
Bismuth	Bi	83	209·00	Nitrogen	N	7	14·008
Cadmium	Cd	48	112·41	Oxygen	O	8	16·000
Calcium	Ca	20	40·08	Palladium	Pd	46	106·7
Carbon	C	6	12·00	Phosphorus	P	15	31·02
Chlorine	Cl	17	35·457	Platinum	Pt	78	195·23
Chromium	Cr	24	52·01	Potassium	K	19	39·096
Copper	Cu	29	63·57	Selenium	Se	34	78·96
Fluorine	F	9	19·00	Silicon	Si	14	28·06
Gold	Au	79	197·2	Sodium	Na	11	22·997
Hydrogen	H	1	1·0078	Sulphur	S	16	32·06
Iodine	I	53	126·92	Tellurium	Te	52	127·61
Iron	Fe	26	55·84	Thallium	Tl	81	204·39
Lead	Pb	82	207·22	Tin	Sn	50	118·70
Lithium	Li	3	6·940	Zinc	Zn	30	65·38

1 litre of hydrogen at 0° C. and 760 mm. weighs 0·08982 gram.
1 litre of nitrogen at 0° C. and 760 mm. weighs 1·2507 grams.

REFERENCES

In addition to those authoritative works mentioned in the text and 'Index', the following are also suggested for consultation regarding particular aspects or portions of organic chemistry:

GENERAL

Electronic Theory of Valency, by N. V. Sidgwick. (Oxford University Press.)
Physical Aspects of Organic Chemistry, by W. A. Waters. (Routledge.)
Dictionary of Applied Chemistry, edited by J. F. Thorpe and (Miss) M. A. Whiteley. (Longmans.)

DYE-STUFFS

Synthetic Dye-Stuffs, by J. F. Thorpe and R. P. Linstead. (Griffin.)

FATS

The Fats, by J. B. Leathes and H. S. Raper. (Longmans.)

NITROGEN

The Organic Chemistry of Nitrogen, by N. V. Sidgwick, revised and rewritten by T. W. J. Taylor and Wilson Baker. (Oxford University Press.)
Plant Alkaloids, by T. A. Henry. (Churchill.)
Numerous original papers on alkaloids by R. Robinson and (the late) W. H. Perkin in the *Journal of the Chemical Society*.

NUCLEIC ACIDS

Nucleic Acids, by Walter Jones. (Longmans.)

PHENANTHRENE, STEROLS, BILE ACIDS, HORMONES, ETC.

The Chemistry of Natural Products related to Phenanthrene, by L. F. Fieser. (Reinhold Publishing Co., New York.)

STEREOCHEMISTRY

Stereochemie, by Georg Wittig (in German).
Numerous original papers by (Sir) W. J. Pope, W. H. Mills and others in the *Journal of the Chemical Society*. Progress in Stereochemistry is well summarised in the *Annual Reports on the Progress of Chemistry*. (The Chemical Society.)

PREPARATION OF ORGANIC COMPOUNDS

Organic Syntheses, by various editors. (Chapman and Hall.)

ANALYSIS, ETC., OF ORGANIC COMPOUNDS

Students' Manual of Organic Chemical Analysis, Qualitative and Quantitative, by J. F. Thorpe and (Miss) M. A. Whiteley. (Longmans.)
Quantitative Organic Microanalysis, by F. Pregl, translated by Fyleman. (Churchill.)
Laboratory Methods of Organic Chemistry, by L. Gattermann, revised by H. Wieland, translated by McCarthy. (Macmillan.)

INDEX

(The more important of several references are indicated in heavy type.)

* The terminal '-ine' signifies a basic substance, e.g. amine, pyridine, etc.

Printed in the United States
By Bookmasters